Research and Applications of Bio-based Degradable Materials

Liming Zhao
Editor

Research and Applications of Bio-based Degradable Materials

Springer

Editor
Liming Zhao
State Key Laboratory of Bioreactor Engineering
East China University of Science and Technology
Shanghai, China

ISBN 978-981-95-1187-7 ISBN 978-981-95-1188-4 (eBook)
https://doi.org/10.1007/978-981-95-1188-4

Translation from the Chinese language edition: "生物基可降解材料的研究与应用" by Liming Zhao, published by Shanghai Scientific & Technical Publishers, 2022.

The original submitted manuscript has been translated into English. The translation was done using artificial intelligence. A subsequent revision was performed by the author(s) to further refine the work and to ensure that the translation is appropriate concerning content and scientific correctness. It may, however, read stylistically different from a conventional translation.

Jointly published with Shanghai Scientific and Technical Publishers

This Springer imprint is published by the registered company Springer Nature Singapore Pte Ltd.
The registered company address is: 152 Beach Road, #21-01/04 Gateway East, Singapore 189721, Singapore

Preface

The increasing global concern over environmental pollution and the depletion of nonrenewable resources has driven an urgent shift toward sustainable materials. Among these, bio-based materials have the characteristics of being environmentally friendly and renewable, and they have great advantages over traditional fossil-based chemical materials in terms of carbon reduction, sustainability, and promoting development. They play a significant role in achieving the "carbon peak, carbon neutrality" goal and are gradually becoming a new leading industry driving world scientific and technological innovation and economic development. The academic, industrial, and financial sectors attach great importance to and actively invest in the field of bio-based materials. Among them, bio-based degradable materials, which are obtained through enzymes or cell factories to obtain monomers or polymers, and can naturally degrade in a certain environment and time, are the hotspots of development in the field of bio-based materials and the bio-manufacturing industry.

Several academic monographs on bio-based materials have been published, which have played a positive role in promoting scientific and technological progress and industrial development. However, there has not yet been a monograph specifically focusing on the preparation of bio-based degradable materials through green production methods based on biological systems as basic units, and systematically focusing on the intersection of bioengineering and materials engineering. In 2019, as the chairman of the conference, I hosted the "High-Performance Bio-based Materials Manufacturing and Applications" Oriental Science and Technology Forum co-sponsored by the Shanghai Municipal People's Government, the Chinese Academy of Sciences, and the Chinese Academy of Engineering. The government and industry leaders, authoritative experts in the field of bio-based materials, and entrepreneurs who attended the meeting reached a forum consensus on the development positioning and innovation direction of China's bio-based materials. This also sparked the idea of summarizing and sorting out new concepts, insights, technologies, and products in this field, and began to prepare this monograph focused on the research and applications of bio-based degradable materials. The content of this book emphasizes green biological manufacturing of materials or material monomers, synthesis and modification of materials and their applications, material intelligence, and material lifecycle management. The main materials partly come from the results accumulated by the author's research team and the co-editors in related research practices, and summarized the latest research progress globally.

The book is divided into three parts and ten chapters. It first introduces and prospects the concepts, categories, research, and application progress, as well as development prospects of bio-based degradable materials, and then focuses on the biological manufacturing, synthesis and application, and lifecycle evaluation of bio-based materials. The book comprehensively and systematically introduces the entire cycle process of bio-based degradable materials from manufacturing to recycling, from the concept of source monomer biological manufacturing, metabolic pathways, chassis construction to fermentation process, separation and purification, and intelligence, to the chemical synthesis and modification methods of materials, applications, and material intelligence, and finally introduces the degradation, recycling, and comprehensive utilization of materials in the lifecycle. This book also specifically includes engineered living materials in the field of vision and introduces several recently developed bio-based degradable materials, which have significant advantages in future applications. It is our hope that this monograph will serve as a valuable resource for graduate students entering the field, researchers seeking to expand their knowledge, and industry professionals aiming to implement sustainable solutions.

The authors' research in this field has been funded by the National High Technology Research and Development Program (863 Program) (2014AA021202) and the National Key R&D Program (2017YFB0309300, 2019YFD0901800, and 2022YFC2104500). The contributors to this book all come from the State Key Laboratory of Bioreactor Engineering at East China University of Science and Technology and the Key Laboratory of Biobased Material Engineering of China Light Industry. More than a dozen of the authors' postdocs and graduate students participated in the writing of this book. Dr. Wang Yibing and Dr. Wang Mingda assisted the authors in finalizing and revising the manuscript. We would like to express our gratitude to the numerous scientists and engineers whose contributions have shaped this field. During the compilation of this book, we received guidance and help from relevant authoritative experts, industry leaders, and corporate experts in the field, for which we express our gratitude. Due to the limited level of the editors, there are inevitably omissions and deficiencies in the book; we welcome feedback and insights from readers, as continuous dialogue is essential for driving innovation in bio-based biodegradable materials. Together, we strive to advance the adoption of these materials and contribute to a more sustainable future.

As this work goes to press, it coincides with the 70th anniversary of the School of Biotechnology, East China University of Science and Technology, and this book is dedicated to the occasion!

Shanghai, China Liming Zhao
July 2025

Contents

The original version of the book has been revised. A correction to this book can be found at https://doi.org/10.1007/978-981-95-1188-4_11

Chapter 1
Introduction

Yongjun Qiu, Jia Liu, Yibing Wang, and Liming Zhao

Biobased materials, mainly referring to biobased polymer materials here, are polymer materials obtained from renewable biomass. They include polymers directly extracted from raw materials (such as starch, cellulose, and other polysaccharides, as well as silk, collagen, and other proteins), polymers synthesized from bio-derived chemicals (such as polylactic acid), and polymers synthesized by microbial organisms (such as polyhydroxyalkanoates). These are used in biobased fibers, biobased plastics, biobased rubbers, biobased coatings, etc., and are beneficial supplements to traditional petroleum-based polymer materials. Correspondingly, biobased material engineering utilizes the theories and methods of biology, engineering, and materials science to research, develop, and apply materials based on renewable biomass according to the needs of human society. The research and development of biobased materials involves many different disciplines, including chemistry, engineering, physics, microbiology, materials science, physiology, botany, and structural biology. Any progress in basic research in these disciplines may promote the research and development of biobased materials.

According to the definition in GB/T 20197—2006, degradability refers to the process where the structure of a material undergoes significant changes in a specific external environment over a certain period of time and through one or more steps,

Y. Qiu · L. Zhao (✉)
State Key Laboratory of Bioreactor Engineering, East China University of Science and Technology, Shanghai, China

Key Laboratory of Biobased Material Engineering of China Light Industry, East China University of Science and Technology, Shanghai, China
e-mail: qiuyongjun@ecust.edu.cn; zhaoliming@ecust.edu.cn

J. Liu · Y. Wang
State Key Laboratory of Bioreactor Engineering, East China University of Science and Technology, Shanghai, China
e-mail: ybwang@ecust.edu.cn

L. Zhao (ed.), *Research and Applications of Bio-based Degradable Materials*,
https://doi.org/10.1007/978-981-95-1188-4_1

and its properties (such as integrity, relative molecular mass, and structure or mechanical strength) gradually disappear. This usually includes photodegradation, chemical degradation, thermal degradation, hydrolysis, and biodegradation; biodegradation is the degradation of materials caused by biological activity, especially the action of enzymes, which allows them to be gradually dissolved by microorganisms or certain organisms as a nutrient source, leading to a decrease in their relative molecular mass, loss of mass, and decline in physical properties.

Biobased degradable materials possess both biobased and degradable characteristics. The biobased degradable materials mainly discussed in this book refer to biodegradable polymer materials obtained from biomass raw materials through cell factories or biocatalytic conversion, or biodegradable polymer materials obtained by chemical polymerization after obtaining monomer substances through the aforementioned biological processes. It must be emphasized that not all biobased materials are biodegradable, and not all materials with biodegradability are biobased materials. Therefore, this book focuses on materials that are both biobased and biodegradable. These types of biobased materials can be completely converted into small molecular compounds such as CO_2, H_2O, and CH_4 under the action of microorganisms, animals, and plants, respectively, and the biodegradation process of the materials is conducive to the establishment of a benign cycle of land and environmental systems. Therefore, biobased degradable materials, which are renewable and environmentally friendly, are truly green, clean, and renewable materials. They have outstanding advantages in terms of resources and environment, and have been used in agriculture, food, medical, and other fields for the manufacture of agricultural films, food packaging, medical devices, sanitary materials, etc., showing great application prospects. The development of biobased degradable materials also provides a reasonable solution to the global environmental problem of "white pollution." As the whole society strives to achieve the goal of "carbon peak, carbon neutrality," the field of biobased degradable materials will usher in great development opportunities and is gradually becoming a pillar industry of technological innovation and economic development in the field of materials.

This book mainly introduces the concept, scope, types, research, and application progress, and development prospects of biobased degradable materials, emphasizing the green biological manufacturing and intelligent manufacturing of these materials or material monomers, the synthesis and modification of materials and their applications, the intelligence of materials, and the life cycle of materials. At the same time, it also provides a special introduction to bioengineering living materials for the readers' enjoyment.

1.1 Classification and Characteristics of Biobased Polymer Materials

Depending on the synthesis process, biobased polymer materials are mainly divided into natural polymers, chemically synthesized polymers, and microbially synthesized polymers (Fig. 1.1). Among them, natural polymers refer to high molecular

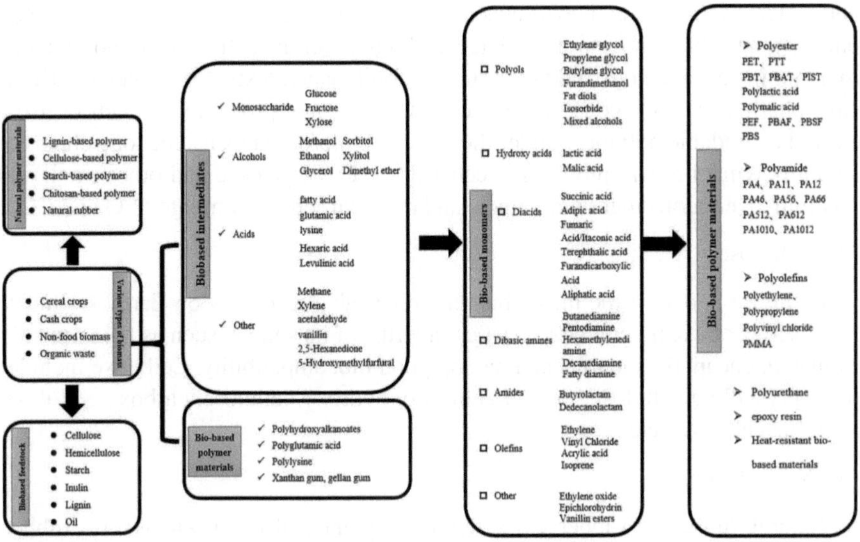

Fig. 1.1 From biomass raw materials to biobased polymer materials
PET polyethylene terephthalate, *PTT* polytrimethylene terephthalate, *PBT* polybutylene terephthalate, *PBAT* poly(butylene adipate-co-terephthalate), *PIST* polyisoprene terephthalate, *PEF* polyethylene furanoate, *PBAF* poly(butylene adipate-co-furanoate), *PBSF* poly(butylene succinate-co-furanoate), *PBS* polybutylene succinate, *PA* polyamide, *PMMA* polymethyl methacrylate

materials that can be directly obtained from organisms (animals, plants, microorganisms, etc.) in the natural environment, such as starch, cellulose, chitin, lignin, natural rubber, proteins, polypeptides, polysaccharides, and nucleic acids. Chemically synthesized polymers refer to the use of renewable biomass as raw materials (such as grains, legumes, straw, bamboo, and wood powder), through biosynthesis, bioprocessing,

Basic biobased chemicals (such as bio-alcohols, organic acids, alkanes, and alkenes) obtained through the bio-refining process, and biobased polymers prepared by chemical synthesis methods, including biobased polylactic acid (PLA), biobased polyamide 4 (PA4), biobased furan polyester, etc. Microbially synthesized polymers refer to high-molecular materials directly biosynthesized from biomass substrates through biosynthesis and biofermentation, such as polyhydroxyalkanoates (PHA).

1.1.1 *Classification of Biobased Polymers*

1. *Natural Biobased Polymers*

(1) Starch

Starch is a natural polysaccharide high-molecular compound, widely present in the fruits, roots, stems, and leaves of plants, serving as a nutrient storage in plants and a major source of carbohydrates in human food. Starch resources are abundant, cheap, renewable, and closely related to human survival. It is an important raw material in various industrial fields such as food, paper, textiles, petroleum, chemicals, pharmaceuticals, construction, and environmental protection. Starch is also a natural degradable polymer. Under the action of microorganisms, the long chains of starch polymers break into monosaccharides such as glucose and other small and medium-sized molecular compounds, and eventually transform into H_2O and CO_2.

(2) Cellulose

Cellulose is one of the most abundant renewable resources on Earth. Cellulose has a series of excellent characteristics in terms of its source, such as being readily available, cheap, degradable, and having good biocompatibility. Cellulose includes microcrystalline cellulose, methyl cellulose, hydroxy cellulose, carboxy cellulose, bacterial cellulose, etc.

(3) Natural Rubber

Natural rubber is rubber extracted from plants, also known as raw rubber. According to statistics, there are about 9 families, more than 300 genera, and more than 2500 species of plants worldwide that can produce natural rubber. Natural rubber can be divided into cis-natural rubber, i.e., cis-1,4-polyisoprene; trans-natural rubber, i.e., trans-1,4-polyisoprene; some plants synthesize both cis and trans natural rubber, and the mixed rubber produced can be called cis-trans-natural rubber. Utilizing renewable biological resources to develop biobased elastomers will be an important way for the sustainable development of the rubber industry in our country and even the world.

2. *Chemically Synthesized Biobased Polymers*

Chemically synthesized biobased polymers refer to those whose monomers come from biomass. Most of them are biobased monomers obtained from biomass through biofermentation, separation, and purification, and then prepared into biobased polymers through chemical synthesis. Currently, the more mature biobased polymers are mostly prepared in this way, such as polylactic acid, biobased polyamides, and polyfuran resin series.

(1) Polylactic Acid (PLA)

Polylactic acid is currently one of the most cost-effective biodegradable polymer materials, with the largest production scale and widest application in the emerging bioplastic market. The raw material of PLA, lactic acid, is prepared by microbial fermentation of glucose and other raw materials. Its process is mature and cost-effective, and it is widely used in the field of food additives. PLA has good biocompatibility and bioabsorbability and has been widely used in drug release, surgical sutures, tissue scaffolds, orthopedic repair, sports medicine fixation materials, and other fields.

(2) Biobased Polyester

Biobased polyester refers to the polyester obtained by reacting biobased diols and biobased dicarboxylic acids as raw materials. It can be used in fibers, fabrics, thermoplastic engineering plastics, electronic connectors, and wiring harnesses. The core issue of biobased polyester is the preparation of its monomers, among which biobased diols can be prepared by the fermentation of glucose, sucrose, and other biomasses.

(3) Biobased Polyamide (PA)

Polyamide, commercially known as nylon, is a type of polymer whose molecular main chain contains amide bonds (—NH—CO—). Since the industrialization of nylon in 1939, polyamide has been widely used in textiles, automobiles, electronics, packaging, sports products, and other fields. PA11 is recognized as the first industrialized biobased polyamide. The production of PA11 uses castor oil as a raw material and is made into ω-undecanoic amino acid through cracking, alcoholysis, high-temperature cracking, hydrolysis, bromination, ammonolysis, and other steps, and then polymerized into PA11. Biobased polybutyramide (PA4) is usually based on glutamic acid, and butyramide (2-pyrolidone) is obtained through biocatalytic transformation, separation and purification, dehydration, refining, and other steps, and then PA4 is obtained through ring-opening polymerization under the action of a catalyst.

3. *Microbial Synthesis of Biobased Polymers*

The microbial synthesis of biobased polymers refers to the use of biomass to directly synthesize biobased polymer materials with high molecular characteristics through microbial cells. Polyhydroxyalkanoates (PHA) and others are typical biologically synthesized degradable polymers. Polyamino acids refer to polymers formed by connecting a type of amino acid through amide bonds, with a structure similar to but different from proteins. Polyamino acids have excellent water solubility, biocompatibility, biodegradability, and easy structural modification, showing very broad application prospects in the fields of medicine, chemical industry, environmental protection, and agriculture.

1.1.2 Characteristics of Biobased Polymers

The strong modifiability of the repeating structural units is an important feature of biobased polymer materials. Starting from the structural units or derivatives of these materials, high-molecular materials with special functions and uses can be obtained through biological or chemical methods. This especially refers to materials that can respond to external physical, chemical, and biological stimuli (such as temperature, pressure, pH, light, electromagnetic fields, and chemical substances, enzymes), including drug loading or targeted delivery, slow release or controlled release, pesticide, fertilizer, and herbicide loading, slow release or controlled release, adsorption and flocculation of pollutants and toxins, medical polymer materials used for human tissue organ repair or functional regeneration, artificial plasma or artificial blood, and blood dialysis materials. The response to external stimuli includes changes in volume and shape, degree of cross-linking, chemical composition, water absorption state, dissolution or aggregation state, etc.

Biobased materials can be applied in the fields of environmental protection, medical treatment, agriculture, packaging materials, etc. Among the many characteristics of biobased materials, solving environmental problems will become the ultimate goal of biobased material research and development.

1.2 Measurement and Evaluation of Biobased Content

The biobased content can be used to evaluate the proportion of biobased in polymer materials, or to distinguish between biobased polymer materials and fossil-based polymer materials. The percentage of modern carbon in polymer materials to the total carbon of the polymer is defined as the biobased content. Based on the ^{14}C in the standard substance of modern carbon-containing materials, and assuming that the intensity of cosmic rays has not changed for a long time (i.e., the production rate of ^{14}C is constant), as long as the ratio or reduction degree of ^{14}C in the carbon-containing material to ^{14}C in the modern carbon-containing standard material is measured, the content of modern carbon in the carbon elements of the measured material can be calculated; that is, the biobased content can be obtained. The biobased content will become a technical requirement for some developed countries to purchase biobased products and may even become a technical barrier. Therefore, while promoting the development of biobased polymer materials, it is extremely necessary to formulate methods for the biobased content in polymer materials and establish standards. The Technical Committee of the American Society for Testing and Materials (ASTM) established the test method for the biobased content of materials (ASTM D6866) as early as 2004 and has been continuously updated. The current effective version of this standard is ASTM D6866-21, which took effect in January 2021. China established the national standard: Determination of Bio-based Content in Bio-based Materials by Liquid Scintillation Counter Method (GB/T 29649—2013) in 2013, using the liquid scintillation counter (LSC) to measure the ^{14}C radioactive element technology, by counting the ^{14}C. The method of measuring the β particles emitted by ^{14}C decay is used to determine the ^{14}C content. In 2021, the new national standard "Plastic Biobased Content Part 2: Determination of Biobased Carbon Content (GBT 39715.2—2021)" was issued and implemented in March 2022.

1.3 Life cycle and Chemical Recycling of Biobased Materials

Materials have a life cycle. When the act of "use" occurs, the life cycle of the polymer material as a consumer product begins. After use, the material has realized its use value; it is discarded as waste, and its life ends. However, during this process, the physical properties and performance of the polymer material do not change. Therefore, how to reasonably recycle discarded materials, simplify the life cycle as much as possible, and achieve efficient recycling and reuse of materials is worth in-depth study and discussion. Studying the life cycle of biobased degradable materials is an important task to ensure the resource utilization, low carbon, and environmental protection of biobased materials, and it is also key to reflecting the performance advantages of biobased degradable materials.

Biobased material recycling technology can be divided into two major categories: physical recycling (material recycling and reuse, and thermal recycling and reuse) and chemical recycling (chemical recycling and reuse). Among them, material recycling and reuse are low-cost and convenient to operate. However, this method cannot avoid the quality loss of regenerated products, which greatly limits the reuse. Chemical recycling and reuse can be divided into gasification, oilification, methanol recovery, and monomer recovery of waste plastics, as well as the recycling and treatment of biodegradable biobased materials by biodegrading plastics.

This book will focus on the biorecycling and reuse of post-use biodegradable materials and biobased polymers.

1.4 Current Status of Biobased Material Development

1.4.1 Industrialization Status of Biobased Materials

From 1990 to 2020, the global production of polymer materials increased from millions of tons to billions of tons, and petroleum resources were consumed in large quantities. Since the 1990s, countries around the world have successively developed renewable chemicals to partially replace petroleum-based materials. After entering the twenty-first century, countries around the world have increased their efforts to develop renewable resources and further accelerated the research and development of non-grain renewable resources. For example, since 1996, the US Department of Energy has cooperated with the Corn Wet Milling Association, the Corn Growers Association, and American companies, including GENENCOR, to formulate a sustainable development plan for renewable resources from crops by 2020. China has also included the development of new renewable resources to replace oil in the national key scientific and technological project.

Many multinational companies have developed biobased polymer products, the most representative of which is polylactic acid (PLA) from NatureWorks in the

United States, which has built a production line with a capacity of 160,000 tons in the United States; ADM in the United States and Metabolix jointly developed a large-scale production line for polyhydroxyalkanoates (PHA) with a capacity of 50,000 tons; Procter & Gamble (P&G) in the United States and Kaneka in Japan have also developed a series of Nodax^{TM} polyhydroxyalkanoates products.

China has now become one of the countries with the most varieties and the largest production of PHA in the world. Some domestic companies, such as Tianjin Guoyun Biotechnology Company, Ningbo Tianan Biomaterials Company, and Guangdong Lianyi Bioengineering Company, all use microbial fermentation technology to produce PHA. In addition, the polybutyramide (PA4) developed by our team has also been trial-produced in Hengtian Fiber Group. At present, with the implementation of the "plastic ban" at home and abroad, domestic capital has begun to actively invest in the industrialization of biobased degradable material products. According to incomplete statistics, the planned production capacity of polylactic acid products is close to 3 million tons. However, most of the current research on biobased materials in China is mainly supported by government funding, and large companies' R&D investment in the development and application of biobased materials is still insufficient. Therefore, the government should take the lead, with enterprises as the main body, to vigorously promote the R&D and application of biobased materials. Overall, the commercial demand for biobased polymers is enormous, with great room for development. In the future, biobased polymers will undoubtedly become a huge pillar industry.

1.4.2 Application and Development of Biobased Materials

The harm of various discarded plastics to the ecology is becoming more and more serious, and "white pollution" has become a prominent harmful issue. The development of renewable, environmentally friendly plastics to replace nondegradable traditional plastic products and alleviate environmental pollution is a green development approach strongly advocated by countries around the world. Biobased degradable polymers use abundant, renewable biomass as raw materials, and under the action of microorganisms or animals and plants, can completely convert the materials into CO_2, H_2O, CH_4, and other small molecule compounds. They have a wide range of applications in medicine, agriculture, and packaging.

1. *Medical Field*

In the drug-controlled release system, polymer materials can be used in different controlled release systems, such as gel-controlled release, microsphere- and microcapsule-controlled release, in vivo implant-controlled release, and targeted controlled release. Because these polymers have the characteristic of being degradable in the body, compared with traditional nondegradable drug carriers, they have advantages such as less dependence of the release rate on drug properties, better adaptation to the release requirements of unstable drugs, and more stable release rates.

2. *Agricultural Field*

The second largest application field of biobased degradable polymers is agriculture. China is a large agricultural country, and the consumption of agricultural films, fresh-keeping films for agricultural by-products, seedling bowls, and fertilizer packaging bags is considerable every year. Traditional, ordinary agricultural films are difficult to recycle and do not easily degrade in the natural environment. Discarded agricultural films not only pollute the environment, but the gradual accumulation of plastic films in the soil can reduce soil permeability and hinder the development of crop root systems and the absorption of water and nutrients, leading to crop yield reduction. After the agricultural film made of biobased degradable polymers is discarded, it can be organically degraded under appropriate conditions, and the degradation products can be used as mixed fertilizers or composted with organic waste.

3. *Packaging Materials*

In the field of packaging, biodegradable biobased polymers have completely or partially replaced traditional nondegradable plastic products. Various molded products or films made by specific processing techniques using degradable polymer materials can be used for the packaging of food, cosmetics, detergents, and daily necessities. For example, with starch-based materials, by controlling specific molecular chain structures and modification schemes such as blending and copolymerization, materials with both water resistance and mechanical strength can be obtained, which can be used to produce garbage bags. In food packaging, degradable materials can be directly used to process packaging products, or a small amount of additives can be added to make nonbiodegradable plastic products decompose into small molecules under the action of light, and further be biodegraded.

4. *Other Aspects*

Biobased degradable polymer materials have also been widely used in other fields, such as tree planting bags in forestry, greening protection rolls, nursery films, fishing nets in fisheries, fishing lines, and films for construction and civil engineering. In addition, in disposable daily necessities, diapers, sanitary napkins, cosmetics, gloves, shoe covers, head covers, tablecloths, and gardening, there is also a good development prospect.

1.4.3 The Significance of Developing Biobased Plastics

At present, the development of the polymer material industry is diversified, sustainable, and environmentally friendly. The goal of "Carbon Peaking, Carbon Neutrality" has focused people's attention on renewable biomass energy. Therefore, biobased polymers have received much attention and have become the focus and highlight of the development of a green, circular, and sustainable economy. The fundamental starting point for developing biobased plastics is based on the finiteness of the large

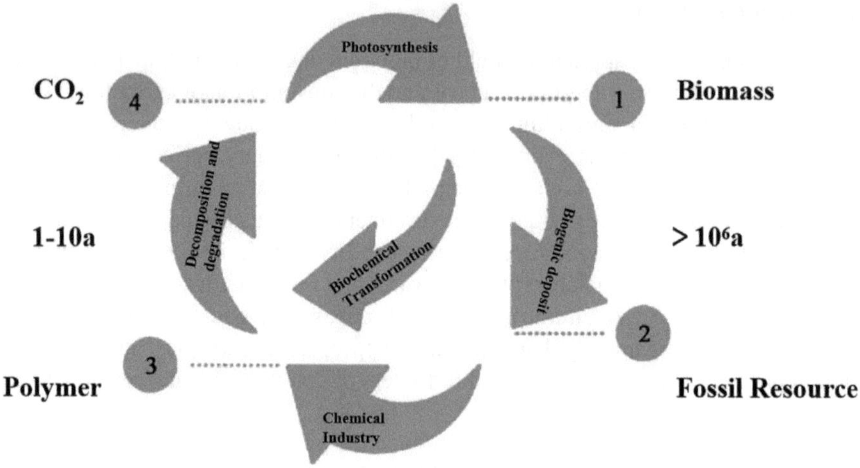

Fig. 1.2 Earth's carbon balance

amount of fossil resources currently used (Fig. 1.2). CO_2 is converted into biomass by biological processors, such as plant photosynthesis, and biomass is converted into fossil resources under certain conditions, a process that generally takes more than 1 million years; fossil resources are turned into polymers, chemicals, and fuels through the chemical industry and are used, and their transformation into CO_2 only takes 1–10 years. Obviously, the speed at which the products of the chemical industry form CO_2 far exceeds the speed at which CO_2 is converted back into fossil resources through biological processing, which will eventually lead to the depletion of oil resources. Biobased plastics are directly converted from renewable biomass into polymers, shortening the conversion process from CO_2 to polymers, and this process is sustainable. Therefore, research on biobased plastics is of great significance for maintaining the long-term security of national energy.

Currently, the raw materials for biobased plastics mainly come from biomass that uses solar energy as energy and CO_2 as a carbon source. Compared with ordinary plastics, biobased plastics can reduce the consumption of petroleum resources by 30–50%, reducing our dependence on petroleum resources for the synthesis of plastics that require petroleum-based refining to obtain monomers.

1.5 Outlook

Biobased biodegradable materials have unique features of being biologically sourced and biodegradable, and many categories also have biocompatibility and adjustable physical and chemical properties. They have broad application prospects in various fields such as textiles and clothing, agriculture, medical care, packaging, and energy. The global demand for biobased biodegradable materials is strong, and research and application development are in full swing. Especially in recent years, the field of biobased biodegradable materials has developed rapidly, with breakthroughs in key technologies, a rapid increase in product types, enhanced product economy, and has become a hot spot for industrial investment, showing a strong development momentum. Dozens of production lines with a capacity of over 10,000 tons are already under construction or in the planning stages. In the short term, due to technical reasons, the production cost of biobased biodegradable materials is still relatively high, but because biobased biodegradable materials have biodegradability, they meet the requirements of plastic bans in various countries, and comply with the policy requirements of ecological civilization construction and harmonious development of humans and the environment, so even if the cost is high, there is a large market space. In the long run, with the rapid development of synthetic biology, artificial intelligence and other scientific and technological advancements, the level of synthesis chassis or biocatalysts for biobased materials or monomers will be rapidly improved, and the biomanufacturing system will achieve scale, precision, and intelligence, thereby greatly improving the conversion rate and yield of biobased materials, significantly reducing production costs and energy consumption, and the large-scale application of biobased materials will become a reality; at the same time, more and more new products will be developed to enrich the variety of biobased materials and meet the application needs of different fields. Although the biobased biodegradable materials industry is currently in the stage of large-scale research and development and transitioning to industrial production and scale application, and will gradually become a bulk material in industrialization, there is still a need to increase research and development efforts in the construction of microbial synthetic strains, the design of biocatalysts, optimization of metabolic pathways and fermentation process optimization and amplification, separation and purification, product molding and processing technology and equipment, scale application demonstration, bio-recycling and reuse, and continue to conduct in-depth research.

In the future, the development path of the biobased materials industry will continue to be optimized and clarified, with more emphasis on specific measures and results to solve resource and environmental problems, and continuously introduce demonstrable, low-cost, high-benefit scale application products. The life cycle assessment of biobased materials will receive attention and emphasis, clarifying the advantages of biobased materials from both resource and environmental aspects, and promoting the greening and sustainable development of the entire process from source to waste. Promoting the non-grainization of biobased material raw materials is crucial to industrial development, and it is necessary to develop functional biological monomers, combine biological fermentation with chemical synthesis technology, achieve scale and cost reduction of existing biobased monomers, and functionalize and add high value to biobased materials. As a category of material, biobased materials should develop synergistically with materials from other sources, complement each other's advantages, explore their own application characteristics and prospects, and leverage the advantages of biobased materials in environmental protection and resource recycling.

Chapter 2
Biomanufacturing of Biobased Material Monomers

Xiwei Tian, Yuanchan Luo, Juefeng Lu, Jiaping Jiang, Yijia Chen, Liming Zhao, and Hui Wu

Biobased materials are new materials made from renewable biomass resources such as grains, straws, and legumes. Their monomers include bio-alcohols, organic acids, and olefins. In recent years, with the increasing scarcity of global petroleum resources, bioenergy, biobased chemicals, and biobased materials industries have become hotspots for global development. Biobased materials, due to their green, environmentally friendly, and resource-friendly characteristics, have attracted great interest from the scientific and industrial communities, as well as the financial investment field. The tools for biomanufacturing of biobased materials are microorganisms, cell lines, and enzymes, with the pathways being large-scale cell culture and metabolism, biocatalytic transformation, and separation and purification.

X. Tian · Y. Chen
State Key Laboratory of Bioreactor Engineering, East China University of Science and Technology, Shanghai, China
e-mail: xiweitian@ecust.edu.cn

Y. Luo · J. Lu · J. Jiang
School of Biotechnology, East China University of Science and Technology, Shanghai, China
e-mail: luoyuanc@ecust.edu.cn

L. Zhao
State Key Laboratory of Bioreactor Engineering, East China University of Science and Technology, Shanghai, China

Key Laboratory of Biobased Material Engineering of China Light Industry, East China University of Science and Technology, Shanghai, China
e-mail: zhaoliming@ecust.edu.cn

H. Wu (✉)
State Key Laboratory of Bioreactor Engineering, East China University of Science and Technology, Shanghai, China

MOE Key Laboratory of Bio-Intelligent Manufacturing, School of Bioengineering, Dalian University of Technology, Dalian, China
e-mail: hwu@dlut.edu.cn; hwu@ecust.edu.cn

L. Zhao (ed.), *Research and Applications of Bio-based Degradable Materials*,
https://doi.org/10.1007/978-981-95-1188-4_2

2.1 Biomanufacturing and Bioprocess Optimization

The rise of the biomanufacturing industry will have significant changes to the overall competitiveness of countries around the world and the global economic landscape. Seizing the commanding heights of biomanufacturing technology and industry has become the focus of global competition. According to the Organization for Economic Co-operation and Development (OECD), by 2030, approximately 35% of chemicals and other industrial products will come from biomanufacturing. In recent years, China's ability to independently construct high-performance industrial strains has been continuously improved, but there is still a significant gap in the efficient optimization and rational scale-up of industrial bioprocesses compared to foreign countries, with relatively low industrial production efficiency and high pollution and energy consumption (Weichen et al. 2014). As a core link of biomanufacturing, the greening of biological processes is an essential path to achieve green biomanufacturing. The National Research Council of the US National Academy of Sciences emphasizes that research funding agencies should support "major fundamental scientific and technological research such as new biological process models and experimental methods." Since the production of biological process products based on microbial cells is actually a life process centered on cell metabolism, it has high complexity and typical nonlinear characteristics. The optimization of the bioprocesses requires starting from the physiological metabolic characteristics of microbial cells, fully understanding the cell's response to environmental disturbances, and developing suitable processes and matching equipment to achieve optimal cell metabolism status. In addition, when the strains or processes optimized at a small scale are transferred to large-scale bioreactors for actual industrial production, an insufficient understanding of the complex relationship between cell metabolism and flow field environment may lead to unsatisfactory scale-up effects or even failure. Therefore, it is necessary to integrate the physiological metabolic characteristics of cells and the flow field characteristics of bioreactors to achieve rational scale-up of bioprocesses.

In the process of industrial biomanufacturing, in order to obtain higher yield of the target product and better understand the intelligent regulation system of cells, it is necessary to carry out research in five aspects, including mining and development of chassis cell components (Mine), modeling and design (Model), component assembly and pathway construction (Manipulate), system testing (Measure), and efficient industrial manufacturing (Manufacture), namely the "5M" strategy (Le Shi et al. 2019). First, according to the target molecule, the biochemical reaction database is used to intelligently search and output potential metabolic pathways. Second, among the massive data, the optimal metabolic pathway search algorithm is adopted to explore the optimal metabolic pathway and its internal mechanism from substrate to target products. Finally, through the integration of multi-omics data and utilization of the basic principles of metabolic regulation mechanisms, the metabolic phenotype prediction and experimental verification can be achieved with the whole genome metabolic network model. The intelligence of cells can be divided into three parts, namely molecular intelligence, pathway intelligence, and chassis cell intelligence.

2.1.1 Molecular Intelligence

The process of molecular intelligence involves the mining, modification, and utilization of industrial enzymes in the era of big data, which is the core of mastering the design and modification of industrial enzymes (Lei et al. 2016). With the rapid development of artificial intelligence technology, protein design technology has been greatly improved, revolutionarily promoting the rapid development of green biomanufacturing. Traditional protein engineering mainly relies on laboratory evolution. Currently, computational virtual design is more frequently used to assist in predicting protein structures and designing new functional enzymes, thus utilizing the advantages of interdisciplinary intersections of biology, chemistry, physics, mathematics, etc., to achieve rational and efficient design. The team led by Wu Bian at the Institute of Microbiology, Li et al. (2018) used artificial intelligence technology to construct a series of new enzyme proteins and realized the discovery of catalytic reactions that have not been reported before. On this basis, they obtained industrial microbial strains completely under the guidance of computational assistance, thus achieving the first breakthrough in the industrial application of artificial intelligence-driven biomanufacturing. This research not only reduces the stringent requirements for reaction conditions in traditional chemical synthesis, but more importantly, it solves the pollution problems brought by chemical synthesis, opening up a new situation for artificial intelligence to empower traditional biotechnology.

2.1.2 Pathway Intelligence

Pathway intelligence refers to the design, prediction, and construction of biological components or modules with expected functions through the adaptive research of synthetic biology, and organic combination of them into a system network to assemble, test, and optimize in chassis cells, achieving efficient intelligent biomanufacturing of artificially constructed cell factories. The rapid development of synthetic biotechnology greatly aids the excavation of key components, enriches the basic module library necessary for intelligent biomanufacturing, and also provides an efficient platform for the performance testing and optimization of chassis cells. “Design-Build-Test-Learn” is the core research content of creating an intelligent cell factory. System metabolic engineering and synthetic biology of natural hosts are the basis for efficient biomanufacturing.

2.1.3 *Chassis Cell Intelligence*

Chassis cell intelligence mainly involves the development of biosensors and the intelligent switching of cellular metabolic pathways (Wei et al. 2019). Developing new sensors that can perceive the physiological metabolic characteristics of cells can accelerate the understanding and recognition of the physiological metabolic characteristics of cells in the process of dynamically responding to environmental disturbances in the bioreactor, thereby helping to analyze the response mechanism of cells to disturbances from seconds to hours. In addition, highly sensitive metabolic probes can also be well used for high-throughput screening of high-performance strains. Single-cell level detection and screening of enzyme reactions or target metabolites can be carried out in uniformly sized droplets of only pL level, achieving a screening speed of up to 10^8 clones/day. This greatly enhances the ability to obtain efficient catalytic elements and high-performance strains (Qiang and Yu 2017).

In addition to the microbial cells themselves, the microbial intelligent biomanufacturing process also needs to consider the complex system composed of microbial cells and bioreactors. At the industrial scale, besides paying attention to the physiological characteristics of microbial cells, it is also necessary to pay attention to the complex relationship with the external environment.

2.2 Biological Processes and Their Intelligent Approaches

The biological process is a systematic life process involving complex cellular metabolism. For microbial cells, the reaction process involves multiple inputs and outputs under multi-scale conditions, such as gene scale, cell scale, and bioreactor scale. Among them, the flows of matter, energy, and information determine the characteristics of the biological process. Therefore, it is necessary to observe and regulate the flows of matter, energy, and information across scales for the bioprocess optimization and scale-up (Ping et al. 2018). The development and application of new sensors can provide real-time and online detection of key macro-physiological metabolic parameters, thereby understanding the metabolic characteristics of cells. And with the combination of the analyses of micro-physiological metabolic characteristics in cells, as well as the flow field characteristics of the bioreactor, a multi-scale biological process optimization and scale-up method combining macroscopic and microscopic metabolism, cell physiological and flow field characteristics can be formed (Zhang et al. 2004).

2.2.1 Macro-physiological Metabolic Characteristics of Microbial Cells

For the biological process, in addition to the commonly monitored parameters such as temperature, agitation, aeration, pH, and dissolved oxygen (DO), in order to gain a deeper understanding of the changes in the macro-physiological metabolic characteristics of cells in the bioreactor, a series of advanced online sensors needs to be introduced. For example, a process exhaust gas mass spectrometer can be used to detect the concentration of O_2 and CO_2 in the exhaust gas, thereby calculating the oxygen uptake rate (OUR), CO_2 evolution rate (CER), and respiratory quotient (RQ). Electronic nose can online detect volatile substances produced during the fermentation process, such as methanol, ethanol, and propanol, thereby achieving online monitoring of key substrates and products. Viable cell sensors can accurately characterize the content of living cells by online detection of the capacitance value of the fermentation broth, which is of great significance for a deep understanding of the cell growth status. In addition, low-field nuclear magnetic resonance, near-infrared spectroscopy, mid-infrared spectroscopy, Raman spectroscopy, and other online sensors have been applied in the online detection of the fermentation process, thereby forming a comprehensive detection system for substrates, products, and intermediate metabolites. On this basis, using a computer-automated data collection and visualization system, a complete control system is formed with the bioreactor. Otherwise, sensitive parameters of the fermentation process can be mined from massive process data, thereby improving the efficiency of bioprocess optimization and scale-up.

2.2.2 Micro-physiological Metabolic Characteristics of Microbial Cells

All biochemical reaction processes are completed within cells, so various metabolites in the cell metabolism process will form a dynamic balance inside and outside the cell. The changes in the concentration of these metabolites characterize the changes in cellular metabolic flux, which is the core of the fermentation process regulation. With the development of metabolomics technology, more intracellular metabolite concentrations can be accurately detected; however, concentration is only a state variable and cannot actually represent the change in metabolic flux. From the perspective of multi-scale system theory of biological processes, process parameter correlation may be caused by simple changes at a certain scale or structural changes in the multi-dimensional scale, so analysis from a single scale may not explain many phenomena in the fermentation process. It is necessary to integrate the physiological metabolic characteristics inside and outside the cell to possibly obtain clues for potential fermentation process regulation.

Cross-scale observation and operation refer to the regulation at the molecular level, cellular level, and bioreactor engineering level through the adjustment of the environment or operating parameters, enabling cells to achieve optimal growth or metabolic states. The key is to use a multi-parameter correlation analysis method to systematically analyze various online and offline parameters, direct and indirect parameters, state and process parameters, etc., and to identify certain correlation characteristics between the parameters. Parameter correlation is the result of balance or imbalance between the flows of material, energy, and information in the bioreactor. Regardless of changes at the microscopic scale of cells and the macroscopic scale of bioreactors, they will eventually be reflected in the changes of macroscopic parameters, which also provides a theoretical basis for intelligent correlation analysis of big data in biological processes.

2.2.3 Combination of Bioreactor Flow Field Characteristics with Cell Physiological Characteristics

Although conventional stirred bioreactors have a simple structure, their internal flow fields are closely related to bioreactor size, impeller type, and operating conditions, showing significant differences under different conditions. When optimized biological processes are obtained at the laboratory scale, the key to reproducing the results in large-scale bioreactors lies in how to reproduce the physiological state of cells. Even though the same process operating conditions are adopted in large and small bioreactors, the results can sometimes vary greatly due to differences in the flow field characteristics of bioreactors. This requires advanced sensing technology to characterize the physiological metabolic characteristics of cells during the fermentation process, so that key sensitive parameters in large bioreactors exhibit a trend similar to that in small bioreactors, achieving the scale-up of optimized processes. Conversely, it is necessary to optimize the flow field characteristics of a large-scale bioreactor and achieve efficient production by adjusting the bioreactor structure or developing the optimal process adaptive to the existing bioreactor structure.

Although traditional methods based on multi-scale parameter correlation analysis can solve optimization and scale-up problems in biomanufacturing processes, the mining of sensitive parameters still mainly relies on manual experience analysis, which poses a great challenge to efficiency and accuracy. There are two key fundamental issues in traditional data processing. First, in terms of microbial production processes, it undergoes a process from complex intracellular metabolism to extracellular operation, generating data of different dimensions, types, and structures, including genomes, transcriptomes, proteomes, metabolomes, metabolic flux, environmental parameters, operating parameters, etc. Some of these parameters can be interrelated, but others cannot be directly analyzed at a single level, and their complexity has exceeded the boundaries of knowledge. Therefore, data processing based on manual analysis has become an insurmountable limitation for massive data. Second, there are fluctuations in the industrial production process, including strains, seed quality, raw material sources, and operating conditions, so manual intervention and processing are often required, which will introduce a lot of limitations due to human factors, resulting in a large amount of unrelated experimental or production data, forming a "data island," which is difficult to support efficient decision-making for the production process (Siliang 2016).

With the advent of the big data era, the intelligence of the production process can connect the laboratory to the factory throughout the entire chain, guiding R&D and production at all stages with data flow, thereby achieving the optimal allocation of production resources. Knowledge graphs and deep learning are the basic methods to achieve artificial intelligence. The former is a white box, capable of reasoning and tracing back, and excels in scenarios. The latter is a black box that cannot be traced or attributed, but is superior to massive data analysis (Guilin et al. 2017). A knowledge graph is a knowledge base required for intelligent search engines that is composed of a large amount of computer-readable, structured, and transferable knowledge (Sowa 1991; Berners-Lee et al. 2001). The construction of a knowledge graph consists of three parts. The first is knowledge acquisition, which primarily involves obtaining the knowledge required for the knowledge base from unstructured, semi-structured, and structured data. Next is data fusion, which mainly involves merging knowledge from different data sources and establishing relationships between the data. Finally, there is computation and application, which mostly involves developing the computational capabilities of the knowledge graph and its application scenarios. During the knowledge acquisition process, technologies such as natural language recognition are used to identify entity recognition and relationship recognition between entities in unstructured data. Then, appropriate tools are used to reorganize, clean, and check the data to obtain the desired data that meets the requirements. Data fusion is based on a unified terminology structure, using data mapping technology to establish mapping relationships between terms in the ontology and vocabulary from different data sources. At the same time, entity matching is also used to fuse data from different sources with the same object, and finally, a knowledge base with management solutions is formed (Otero-Cerdeira et al. 2015; Dong et al. 2015). Knowledge computation mainly involves obtaining more implicit knowledge based on the information provided by the graph. Deep learning is reflected in the recognition of relationships between entities during knowledge acquisition. The data mapping technology space, which uses ontology as a unified term, forms the virtual space of information technology, known as "Cyberspace," which dynamically reflects the state of entities driven through real-time data-driven mirror space.

The process-level intelligence of biomanufacturing is based on existing large-scale biological process data, and a knowledge base can be formed through case study learning and knowledge generation. At the same time, real-time fuzzy inference is used to detect abnormal states in the current biological process, and then hidden parameter correlations between process big data are automatically found, inferring the sensitive parameters for biological process optimization.

Molecular-level intelligence refers to the study of possible molecular mechanisms of cell metabolism on the basis of process-level intelligence, fundamentally elucidating the correlation of multiple parameters in the biological process. In addition, molecular-level intelligent systems combine synthetic biology with artificial intelligence, utilizing knowledge reasoning techniques in the design, construction, and testing stages of synthetic biology to predict and validate various hypotheses, thereby improving the availability of high-performance producing strains (Guojun et al. 2020). In addition, from the perspective of enterprises, intelligent biomanufacturing is based on intelligent bioreactors as the key core, with the Internet of Things (IoT) data flow as the basis, targeting enterprise resources, production processes, and production management, and forming an adaptation of manufacturing execution system (MES) and enterprise resource planning (ERP), in order to achieve optimal production and intelligent management (Siliang 2016).

2.3 Biomanufacturing of Organic Acid-Based Biomaterial Monomers

2.3.1 Biomanufacturing of Lactic Acid

Lactic acid (LA), also known as 2-hydroxypropanoic acid, is a naturally occurring hydroxycarboxylic acid. It is a colorless and odorless liquid with hygroscopic properties. As one of the common metabolic products in nature, it exists in many biological metabolic reactions, such as the production of lactic acid in humans during intense exercise. In 1780, the Swedish chemist Scheele first discovered lactic acid in sour milk, but he believed at the time that lactic acid was just a component of milk. In 1789, Lavoisier named it "acide lactique," which may be the origin of the current name "lactic acid." In 1857, Pasteur discovered that lactic acid was not a component of milk but was produced by the fermentation of specific microorganisms (Konings et al. 2000). The molecular structure of lactic acid contains a chiral carbon atom, so lactic acid exhibits two optical configurations, namely L(+)-lactic acid and D(−)-lactic acid (Fig. 2.1). Since the human body lacks the dehydrogenase to metabolize D-lactic acid, it can cause acidosis and decalcification reactions, so the US FDA considers L-lactic acid to be safe and harmless when used as a food additive. Depending on the demand and use in different fields, lactic acid is often divided into pharmaceutical grade, food grade, and industrial grade.

Fig. 2.1 Structure of L-lactic acid (left) and D-lactic acid (right)

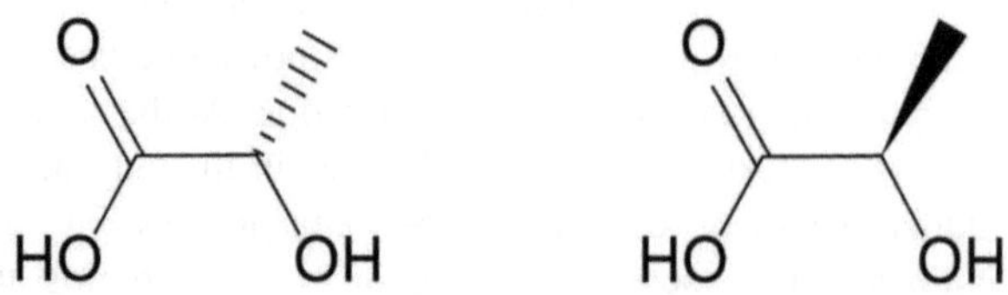

Lactic acid is mainly used in food, cosmetics, and other chemical industries. As a flavoring agent, pH regulator, acidifier, and preservative, etc., lactic acid is most widely used in the food industry, accounting for more than 80% of its usage. In addition, as one of the natural ingredients in cosmetics, lactic acid not only has pH regulating functions but also possesses antimicrobial activity and skin glossing and moisturizing effects. Since lactic acid naturally exists in the human body, it and its salts as cosmetic ingredients increasingly align with the current pursuit of natural and safe ingredients. In the pharmaceutical industry, lactic acid is typically used as a solvent to enter the human vein to replenish body fluids. When lactic acid is used as a platform compound, it can be used to synthesize many other compounds, such as esterification to form lactates, dehydrogenation to form acrylic acid, hydrogenation to form 2,3-propanediol, polymerization to form polylactic acid (PLA), and oxidation to form pyruvic acid (Gao et al. 2011; Kumar et al. 2013).

(PLA) is a typical biobased plastic with good biocompatibility and biodegradability. With the strengthening of global environmental protection awareness and the increasing depletion of fossil resources, the idea of using biodegradable plastics to replace traditional petrochemical plastics is gaining more and more attention; thus, the demand for PLA has greatly increased. NatureWorks LLC is a global leader in the PLA field, and its PLA resins and PLA fibers are best-selling worldwide. Many other famous lactic acid production companies in the world, including Purac in the Netherlands, Galactic in Belgium, Cargill in the United States, and many Chinese companies, are accelerating the construction of a new industry based on the lactic acid polymerization platform, promoting environmentally friendly PLA products to replace traditional plastics (Datta et al. 1995).

1. *Production Methods of Lactic Acid*

Lactic acid can be obtained through chemical synthesis and microbial fermentation. The chemical synthesis method is mainly based on the strong acid hydrolysis of lactonitrile to produce lactic acid; other approaches include the catalysis of sugar under alkaline conditions, the oxidation of propylene glycol, the reaction of acetaldehyde, carbon monoxide, and water under high temperature and pressure, the hydrolysis of cyanic acid, and the nitric acid oxidation of propylene. However, the lactic acid produced by these processes is racemic DL-lactic acid, which greatly limits the application of the chemical synthesis method.

The fermentation method of lactic acid production not only solves the problem of lactic acid optical purity, and the obtained lactic acid can be optically pure L-lactic acid or optically pure D-lactic acid, but also can use sustainable raw materials as substrates (Abdel-Rahman et al. 2013); therefore, it has an economic advantage compared to the chemical method, and can effectively solve the problems of environmental pollution during production and reduce the use of fossil raw materials. Compared with the chemical method, the fermentation method of lactic acid production has the following advantages: ① low substrate cost, ② mild production temperature, ③ low energy consumption, and ④ single optical specificity (Abdel-Rahman et al. 2013; John et al. 2007; Wee et al. 2006a). Currently, 90% of the world's lactic acid production comes from microbial fermentation, with the yield of lactic acid in batch fermentation reaching 90–95% (w/w).

2. *Lactic Acid-Producing Strains*

Lactic acid-producing strains are an important component of the lactic acid fermentation process, and generally have the following characteristics: (1) belong to gram-positive bacteria, (2) are aerotolerant, (3) do not produce spores, (4) do not produce catalase, (5) are picky about the environment and acid-resistant, and (6) can use carbohydrates as substrates to produce lactic acid (Abdel-Rahman et al. 2011; Okano et al. 2009; Ou et al. 2011; Maas et al. 2008; Chang et al. 1999; Zhou et al. 2003; Yukawa et al. 2007; Biddy et al. 2016). High-performance strains (high optical purity and high production capacity) are often the key to efficient production of lactic acid. Currently, there are many microorganisms that can produce lactic acid, mainly including bacteria, fungi, yeasts, microalgae, and photosynthetic microorganisms. Bacteria, compared to other microorganisms, have advantages such as high productivity, high yield, and high optical purity, therefore commonly being used for commercial lactic acid production. The bacteria used for lactic acid production mainly include four types: lactic acid bacteria (Abdel-Rahman et al. 2011; Okano et al. 2009), *Bacillus* (Ou et al. 2011; Maas et al. 2008), *Escherichia coli* (Chang et al. 1999; Zhou et al. 2003), and *Corynebacterium glutamicum* (Yukawa et al. 2007; Biddy et al. 2016).

When lactic acid bacteria undergo lactic acid fermentation, the optimal conditions are related to strain specificity. According to the different fermentation products, lactic acid fermentation can be divided into homolactic acid fermentation and heterolactic acid fermentation. Commercially applied lactic acid bacteria are mostly homolactic fermenters, and their theoretical yield of lactic acid from glucose can reach 1.0 g/g, while the theoretical yield of heterolactic fermenters can only reach 0.5 g/g. Due to its simple culture medium composition and high temperature resistance (>50 °C), *Bacillus* is attracting more and more attention from researchers and companies. The types of lactic acid-producing strains can be divided according to different metabolic modes into ① homolactic fermenters that can convert more than 85% of sugars into lactic acid, ② heterolactic fermenters that can only convert about half of the sugars into lactic acid while producing CO_2, acetic acid, ethanol, and other by-products, and ③ a few heterolactic fermenters that can produce D, L-lactic acid and acetic acid as well as CO_2. Recent reports related to lactic acid biosynthesis, show that the biosynthesis of lactic acid generally uses homolactic fermenters of the genus *Lactobacillus*, *Streptococcus,* and *Pediococcus* (Biddy et al. 2016), because homolactic fermenters almost only produce lactic acid and do not produce other by-products during the fermentation process, while heterolactic fermenters produce a mixture of various by-products. Researchers believe that the reason for this product difference is that homolactic fermenters use the Embden–Meyerhof–Parnas (EMP) pathway, which can convert six-carbon sugar molecules into two- to three-carbon molecules and then convert them into lactic acid, almost without going through other metabolic pathways, while heterolactic fermenters use the phosphopentose pathway (PP pathway) for carbon metabolism, six-carbon sugar is converted into five-carbon sugar and CO_2, and then five-carbon sugar is converted into two-carbon and three-carbon molecules, where the three-carbon molecules are further converted into lactic acid, and the two-carbon molecules are converted into by-products such as ethanol and acetic acid (Vinderola et al. 2004). However, some researchers have found that some facultative microorganisms have two types of carbon metabolism at the same time, they can metabolize glucose to produce lactic acid through the EMP pathway, and also can produce lactic acid and some two-carbon by-products through the PP pathway. Therefore, these microorganisms will use the first pathway (EMP) for homolactic acid fermentation when hexose is present and use the second pathway (PP pathway) for heterolactic acid fermentation when pentose is present.

In the industrial production of lactic acid, *Lactobacillus* and *Bacillus* are the most widely used lactic acid-producing bacteria. As for *E. coli* and *C. glutamicum*, due to their clear genetic background and easy gene manipulation characteristics, they are still used in laboratory research for lactic acid production.

3. *Lactic Acid Fermentation Process*

The common lactic acid fermentation modes include batch culture, fed-batch culture, continuous culture, cell recycle culture, etc.

Batch culture is the simplest and most widely used fermentation mode. During the fermentation process, apart from adding neutralizers to regulate pH, no other carbon and nitrogen sources are added. Therefore, this closed system can effectively reduce the risk of contamination and achieve a higher concentration of lactic acid. Factors affecting the efficiency of lactic acid batch fermentation mainly include nitrogen source, pH, neutralizer, aeration, etc. The selection of different nitrogen sources and their concentrations is an effective approach to increase the lactic acid titer, and replacing expensive nitrogen sources with appropriate, cheap ones is also an important way to achieve economical production of lactic acid. Food waste, corn steep liquor, soybean flour, etc. are currently widely used as cheap nitrogen sources (Rivas et al. 2004). The addition of neutralizers can effectively control the pH within an appropriate range, which is conducive to cell growth and metabolism, thereby improving fermentation efficiency. Calcium hydroxide, sodium hydroxide, potassium hydroxide, ammonium hydroxide, calcium carbonate, etc. are commonly used neutralizers in the lactic acid fermentation process (Tashiro et al. 2011; Romero-Garcia et al. 2009; Moon et al. 2012). The choice of different neutralizers has a significant impact on the efficiency of lactic acid fermentation (Karp et al. 2011; Nakano et al. 2012). The control of aeration rate is crucial in the lactic acid fermentation by some strains (such as *Bacillus* and *E. coli*). A high aeration rate will lead to the production of by-products, while a low aeration rate affects the cell growth; therefore, an appropriate aeration rate during the fermentation process is an important guarantee for efficient production (Qin et al. 2010; Zhu et al. 2007).

Fed-batch culture can effectively alleviate the inhibitory effect of high substrate concentration on cell growth and metabolism, but it does not affect the inhibition problem of high lactic acid concentration. In order to obtain a higher product concentration during the fermentation process, the feeding timing and type of substrate, the concentration of substrate in the fermentation broth, and the feeding mode are all issues that should be considered. In the lactic acid fermentation process, common feeding modes include intermittent feeding, constant speed feeding, and exponential feeding (Bai et al. 2003; Ding and Tan 2006; Li et al. 2010). Research has shown that the final concentration of lactic acid can reach up to 226 g/L in the fed-batch culture (Wang et al. 2011). Although this culture mode can result in high lactic acid concentration, the problem of low lactic acid productivity has always been a key factor limiting its widespread application.

Cell recycle culture can effectively solve the problem of low productivity by separating cells through modules such as centrifugation and membrane filtration, and then reusing them for production. This mode can not only achieve high lactic acid concentration, high lactic acid yield, and shorten the overall fermentation cycle, but also improve the productivity compared to batch culture and fed-batch culture (Kim et al. 2006; Wee et al. 2006b; Zhao et al. 2010).

Due to the close relationship between cell growth and lactic acid production, the inhibition problem of high product concentration in batch culture and fed-batch culture can be solved through continuous culture. At the same time, the yield of lactic acid can be optimized by changing the dilution rate. In traditional continuous culture, unused substrates and cells are wasted as they flow out with the fermentation broth. Therefore, cell immobilization and cell recycling techniques can be introduced during continuous culture to greatly increase the dilution rate compared to traditional continuous culture, thereby increasing the yield and reducing the outflow of unused substrates (Chang et al. 2011; John and Nampoothiri 2011).

4. *Cell Metabolism in Lactic Acid Fermentation Process*

In the actual industrial lactic acid production process, the presence of residual sugar in the fermentation broth has a significant impact on the quality of lactic acid products. Therefore, batch culture and fed-batch culture are more suitable fermentation modes. An efficient and economical lactic acid fermentation process aims for high concentration, high productivity, high yield, and high optical purity. Thus, optimizing the lactic acid fermentation process based on understanding the metabolic characteristics of cells is possible to achieve the above goals.

The research on cell metabolism in the lactic acid fermentation process has a long history. Researchers often use the model bacterium *Lactococcus lactis* (*L. lactis*) as an example to investigate the metabolic regulation of lactic acid bacteria. In *L. lactis*, glucose is linearly converted into pyruvate through the glycolytic pathway, and ATP is generated by substrate-level phosphorylation. The NADH produced at the glyceraldehyde-3-phosphate dehydrogenase (GAPDH) site is reoxidized to NAD^+ in the process of LDH-catalyzed pyruvate to lactic acid, thus maintaining the redox balance. Despite the relatively simple metabolism, a comprehensive understanding of the metabolic regulation of lactic acid bacteria has not yet been formed (Neves et al. 2005; Ramos et al. 2002).

In the absence of external electron acceptors (such as O_2), the metabolism of lactic acid bacteria is limited by the balance between NADH production and consumption in cells, and the metabolism of carbon flow is closely coupled with

NADH. Under aerobic conditions, the metabolism of carbon flow is not closely coupled with redox metabolism due to the presence of NADH oxidase for regenerating NAD^+ (Garrigues et al. 1997). Simultaneously, due to the lack of a functional electron transfer chain, *L. lactis* mainly relies on substrate-level phosphorylation in the fermentation process to produce energy under low oxygen levels, so ATP/ADP (the energy charge state inside the cell) may also have a significant impact on cellular metabolism (Koebmann et al. 2002).

Although many reports in the past have claimed that lactic acid bacteria produce lactic acid under anaerobic conditions, some studies have found that a moderate oxygen environment is conducive to cell growth. Therefore, it is very important to study the oxygen metabolism of lactic acid bacteria.

5. *Oxygen Metabolism Research During Lactic Acid Fermentation Process*

The strains used in lactic acid fermentation are mostly facultative anaerobic lactic acid bacteria, and the entire fermentation is mostly a microaerobic process. Due to trace amounts of O_2 being needed, microaerobic fermentation is often achieved by providing a very small aeration rate in a stirred bioreactor. During the entire fermentation process, the DO quickly drops to a very low level, even to zero. Therefore, the role of DO electrode in lactic acid fermentation is not as obvious as it is in aerobic fermentation.

The role of O_2 in lactic acid fermentation has been studied for a long time, and it has made significant progress. Although many researchers have proposed that lactic acid bacteria can grow well under anaerobic conditions, and an aerobic environment may inhibit the cell growth to some extent, there are also reports that under aerobic conditions, the cell growth rate of lactic acid bacteria is significantly faster than that under anaerobic conditions. This may be attributed to the fact that under aerobic conditions, the sugar metabolism shifts from lactic acid synthesis to acetic acid synthesis, which doubles the efficiency of ATP production, thereby promoting cell growth (Condon 1987). Qin et al. (2010) compared the cell growth and lactic acid production of *Bacillus* sp. Na-2 strain under different aeration conditions and proposed a two-stage aeration control strategy, thereby increasing the yield and improving fermentation efficiency.

In past studies on aerobic lactic acid fermentation, more attention has been paid to the comparison of anaerobic, aerobic, and respiratory conditions (adding heme in an aerobic environment). The analyses from the gene level, transcription level, protein level, and metabolite level have been carried out to explain the oxygen metabolism mechanism on cell growth and product formation. Duwat et al. (2001) found that for facultative anaerobic *L. lactis*, the presence of O_2 is not conducive to cell growth and survival, but the addition of heme in an aerobic environment can greatly improve the growth and survival periods. The Gruss research group (Pedersen et al. 2008) compared the transcriptional information of *L. lactis* under anaerobic, aerobic, and respiratory conditions using gene chip technology. The results showed: ① stress response genes are mainly affected by aerobic fermentation, and the respiratory environment helps to alleviate oxidative stress, ② the gene expression required for respiratory metabolism is similar under the three conditions, ③ only 11 genes are expressed differently under respiratory condition compared to aerobic and anaerobic conditions, among which the *ygf*CBA operon is only induced by heme, so this is a gene related to heme homeostasis. Vido et al. (2004) compared the proteome maps of *L. lactis* under anaerobic, aerobic, and respiratory conditions, confirming that under these conditions, the levels of 21 proteins differed, and two main groups of proteins involved in carbon metabolism and nitrogen metabolism were identified. Murphy and Condon (1984) examined the cell growth and glucose metabolism of *L. plantarum* under aerobic and anaerobic conditions. The results showed that aerobic culture would cause a decrease in cell growth rate, and the starting time of the decrease in growth rate would be earlier with the increase in oxygen consumption rate, which was related to the accumulation of H_2O_2. In addition, aerobic culture would also cause a shift in glucose metabolism from lactic acid accumulation to acetic acid accumulation.

At the same time, the oxidative stress produced during the aerobic lactic acid fermentation has also received widespread attention. These studies cover the mechanisms underlying oxidative stress, its impact on cell growth and metabolism, and the responses and protective mechanisms of cells to oxidative stress. The types of oxygen that cause oxidative stress mainly include superoxide, hydrogen peroxide, and hydroxyl radicals. Under aerobic conditions, the production of superoxide is due to the reduction of an O_2 molecule by an electron, usually as a result of the oxidation of flavin adenine dinucleotide (FADH2), flavin mononucleotide (FMNH2), or phenol (demethylated naphthoquinone, methyl naphthoquinone, pantothenic acid) (Huycke et al. 2001). Hydrogen peroxide comes from the reduction of two electrons from the O_2 molecule, caused by some flavoproteins, phenols, and enzymes in the metabolic process (NADH peroxidase, pyruvate oxidase, and lactate oxidase) (Goffin et al. 2006; Barre et al. 2007). Superoxide dismutase can convert superoxide into hydrogen peroxide in the presence of manganese. Hydroxyl radicals are mainly obtained from hydrogen peroxide through the Fenton reaction (Temple et al. 2005). The oxidative stress effects of these reactive oxygen species (ROS) are much stronger than the O_2 molecule and can cause damage to proteins and DNA in cells, thereby affecting cell growth and metabolism. Abbott et al. (2009) found that in the lactic acid fermentation process by *S. cerevisiae*, the presence of lactic acid inhibits cell growth under aerobic conditions, which is due to the increase in the level of ROS in cells, and overexpression of cytoplasmic catalase can effectively reduce the ROS level, thereby increasing cell growth rate. Like other microorganisms, the response of lactic acid bacteria to oxidative stress intersects with responses to other environmental stresses. The Porro research group (Branduardi et al. 2007) pointed out that expressing ascorbic acid in yeast can effectively protect cells from environmental oxidative and acid stresses. Temple et al. (Abbott et al. 2009) also pointed out that for yeast, the adaptive response to oxidative stress also helps the response to other stresses (heat stress, acid stress, etc.) and vice versa.

Although great progress has been made in the study of aerobic effects in the lactic acid fermentation process, there are few reports on the changes in oxygen metabolism and its characterization parameters. As for an industrial lactic acid fermentation process, as lactic acid is a growth-related product, a micro-aerobic environment often improves cell growth, thereby promoting the efficient production of lactic acid. However, an oxygen-excessive environment will cause changes in metabolic flux, thereby affecting the lactic acid conversion efficiency. Appropriate parameters to quantitatively characterize the micro-aerobic environment are the key to an efficient lactic acid fermentation process. Tachon et al. (2010) reported the trend of oxidation–reduction potential (ORP) during the fermentation process by *L. lactis*, and revealed that NADH oxidase and electron transfer chain are the main causes of ORP changes through genetic engineering. However, since ORP is not a direct characterization parameter of O_2, it is also affected by many other substances in the fermentation broth, so ORP is not the best choice for oxygen metabolism research in the lactic acid fermentation process.

2.3.2 Biomanufacturing of Long-Chain Dicarboxylic Acid

Long-chain dicarboxylic acid (LCDCA) refers to straight-chain aliphatic saturated dicarboxylic acid [HOOC-(CH_2)n-COOH] containing 10 or more carbon atoms. Long-chain dicarboxylic acids are important fine chemical intermediates, which can synthesize a series of high-value-added products such as fragrances, special nylons, and polyamide hot-melt adhesives. They are widely used in the chemical, light industry, national defense, automotive industry, agriculture, medicine, and other fields (Gang and Yuantong 2000; Zhongmao et al. 2009) .

The production methods of long-chain dicarboxylic acid are mainly chemical synthesis and microbial fermentation. The chemical synthesis method, such as dodecanedioic acid, which can be chemically synthesized using butadiene as a raw material, is complex, and the conditions are harsh. It requires high temperature and high pressure, as well as explosion-proof and anti-poison equipment, with low production, high cost, and serious environmental pollution. Another production method of chemically synthesized long-chain dicarboxylic acid is to oxidize fatty acids with oxidants such as nitric acid and potassium permanganate, but the product is a mixture that needs to be separated and has a large amount of wastewater discharge, so its application is greatly limited.

The microbial fermentation method for producing long-chain dicarboxylic acid is simple in process, mild in reaction, and low in cost. In addition, the chemical method cannot synthesize all long-chain dicarboxylic acids from C11 to C18, while the microbial fermentation method can obtain them. Microbial fermentation utilizes enzymes in cells to catalyze n-alkanes or fatty acids into long-chain dicarboxylic acids. Using mixed alkanes/fatty acids of varying chain lengths as substrates, the products are mixed long-chain dicarboxylic acids, that is, they contain several long-chain dicarboxylic acids with different carbon numbers. Using a single alkane/fatty acid as a substrate, the product is a single long-chain dicarboxylic acid. At present, the production of long-chain dicarboxylic acid abroad mainly uses chemical synthesis, while China is one of the few countries in the world that uses microbial fermentation for industrial production of long-chain dicarboxylic acid, with a leading position in the research and production. Shanghai Kasei Biotechnology Co., Ltd. is one of the largest suppliers of biologically produced long-chain dicarboxylic acid products in the world.

The research on the industrial application of long-chain dicarboxylic acid fermentation technology mainly focuses on three aspects: screening of high-yield strains, optimizating of the fermentation process, and optimizating of the purification process.

1. *Biosynthetic Pathway of Long-Chain Dicarboxylic Acids*

The process of *Candida tropicalis* transforming alkanes into long-chain dicarboxylic acids is mainly completed through ω-oxidation, and cytochrome P450 monooxygenase (CYP450), fatty alcohol oxidase (FAO), fatty alcohol dehydrogenase (ADH), and fatty aldehyde dehydrogenase (FALDH) are involved in the related reactions (Hettema and Tabak 2000; Mauersberger et al. 1992; Gurvitz et al. 2001; Kawamoto et al. 1978). After alkanes (or fatty acids) are uptaken by cells, the terminal methyl is oxidized under the catalysis of CYP450 to form fatty alcohol, which then forms monocarboxylic acid (or dicarboxylic acid) under the successive action of FAO (or ADH) and FALDH. It is finally secreted outside the cell, or stored in the cell in the form of lipids (Yamada et al. 1980), or decomposed and metabolized through the β-oxidation pathway, providing materials and energy for cell growth.

CYP450 is usually the key enzyme and rate-limiting enzyme in the ω-oxidation process, so the production capacity of cells can be improved by increasing the enzyme activity of CYP450, such as adding iron, barbiturates, and penicillin (Chen et al. 2005; Yunzhi and Peng 2001). The β-oxidation system of *C. tropicalis* degrading long-chain dicarboxylic acids includes acyl coenzyme A oxidase, enoyl coenzyme A hydratase, 3-hydroxyacyl coenzyme A dehydrogenase, 3-ketoacyl coenzyme A thiolase, and acetyl coenzyme A thiolase (Kawamoto et al. 1978). The β-oxidation of fatty acids is a major factor in reducing the conversion rate of alkanes, so blocking the β-oxidation pathway can effectively accumulate long-chain dicarboxylic acids (Picataggio et al. 1992). Acyl coenzyme A oxidase is usually the rate-limiting enzyme in the degradation of long-chain dicarboxylic acids, so the β-oxidation process can also be regulated by inhibiting acyl coenzyme A oxidase, such as adding acyl coenzyme A oxidase inhibitors, Ag^+, Pb^+, and nonionic surfactants or by blocking the β-oxidation pathway through genetic engineering.

2. *Selection of High-Yield Strains*

There are many types of bacteria, fungi, and actinomycetes that can utilize petroleum hydrocarbons. The reaction of alkane oxidation to fatty acids has been observed in many microorganisms such as *Pseudomonas* and *Bacillus*, but their production of dicarboxylic acids is very low, and most of them are short-chain dicarboxylic acids (Kester and Foster 1963). Yeast genera reported to utilize aliphatic alkanes for cell growth include *Candida*, *Cryptococcus*, *Endomyces*, *Hansenula*, *Pichia*, *Rhodotorula*, *Saccharomycopsis*, *Sporobolomyces*, *Saccharomyces*, and *Zygosaccharomyces*. Among them, *Candida* is a high-yield microorganism for the production of long-chain dicarboxylic acids from alkanes.

Up to now, the *Candida* used for the production of long-chain dicarboxylic acids are all wild strains screened from nature, which are obtained after multiple rounds of mutation and selection, and can oxidize alkanes from decane to octadecane to generate corresponding chain length dicarboxylic acids (Chengyong and Jian 2002). Due to the strong β-oxidation ability of the wild strains, it is difficult to accumulate a large amount of long-chain dicarboxylic acids. In order to accumulate more long-chain dicarboxylic acids and prevent them from being degraded by β-oxidation, it is necessary to reduce the β-oxidation ability of the microorganisms and block the β-oxidation pathway.

Mutation breeding is a routine method for the selection of high-yield strains, mainly including mutation and selection. By treating the starting strain with physical or chemical mutagens, the mutation frequency of the strain is increased, thereby obtaining a large number of mutants, and then using rapid and simple screening methods to select the desired target mutants. Mutagens mainly include physical mutagens and chemical mutagens.

In terms of physical mutation, UV mutation technology was applied as early as the 1970s. The Japanese Mining Company used UV mutation and other methods to screen the *C. tropicalis* mutant M2030, which can produce 130 g/L of acid after 120 h of fermentation (Xiaozeng and Mingyou 1996). Zhao (Xingqing 2003) used sucrose (only) carbon source medium, alkane (only) carbon source medium, and phenol red colorimetric medium for a composite screening of *Candida* strains producing DCA13, and further improved the acid production performance through UV mutation methods, achieving relatively ideal acid production effects in shake flasks. Chen et al. (2000) used low-energy particle beam injection technology to treat *C. tropicalis* SCB412 and screened a high-yield strain that can use dodecane to produce DCA12. The acid production increased from 43.6 to 73.3 g/L. In terms of chemical mutation, Chen et al. (Yuantong and Yuechuan 1997) used $NaNO_2$ mutation in the process of screening octadecenedioic acid-producing strains and also achieved a relatively good result on acid production. In addition, a patent pointed out that using nitrosoguanidine as a chemical mutagen, the screened high-yield *Candida* strain CH-14-204 can reach an acid production of 192.3 g/L after 139 h of fermentation (Yuantong et al. 2007).

Genetic engineering technology is also one of the important means to obtain high-yield strains. *C. tropicalis* is mostly used for the industrial production of long-chain dicarboxylic acids. The construction usually begins with the deletion of genes related to the metabolic pathways of fatty acid and DCA degradation, namely the knockout of genes involved in the β-oxidation pathway and the overexpression of key enzyme genes encoding the ω-oxidation pathway. Gao et al. (Hong et al. 2003) constructed single-copy and double-copy deletion strains of the botulinum acyltransferase gene and found that the double-copy deletion strain of the botulinum acyltransferase gene could not metabolize alkanes. Long-chain fatty acids and DCA depend on the activation of acyl-CoA synthetase, which is then transported to the peroxisome. Therefore, mutant strains lacking acyl-CoA synthetase have defects in fatty acid uptake. In various yeasts, the knockout of the gene for acyl-CoA oxidase (*pox* gene) is the first step in blocking the β-oxidation pathway. Okazaki et al. (1986)

pointed out that there are three *pox* genes (*pox2*, *pox4*, and *pox5*) in the *C. tropicalis* pK233. After the isozymes encoded by *pox4* and *pox5* were knocked out, DCA enrichment could be found in this double deletion strain. *Yarrowia lipolytica* is another yeast that can use alkanes to produce DCA. Since 1998, six *pox* homologous genes have been characterized in *Y. lipolytica*, and different mutants with *pox* gene deletions have been constructed by genetic engineering (double, triple, and quadruple deletion strains of *pox* 2, *pox* 3, *pox* 4, and *pox* 5genes). All gene deletion strains can accumulate DCA12 (7–20 g/L), and the higher the number of *pox* gene deletions, the higher the yield of DCA compared to the wild-type strain (Fickers et al. 2010).

3. *Optimization of the Fermentation Process*

The fermentation process of long-chain dicarboxylic acids is nongrowth coupled; that is, there is no obvious correlation between cell growth and the accumulation of dicarboxylic acids. The cell growth and the synthesis of products occur in two different stages. At present, the optimization of the fermentation process for long-chain dicarboxylic acids mainly focuses on medium, fermentation conditions, and feed control.

(1) Medium Optimization

The dicarboxylic acid fermentation process is divided into two stages: the cell growth period and the acid production period. The carbon source for the cell growth period is usually sucrose, and the carbon source for the acid production period is alkanes. The fermentation medium for long-chain dicarboxylic acids requires carbon source, nitrogen source, essential inorganic salts, and trace growth factors. Liu et al. (Zutong and Zhongxiang 1989) have pointed out that when using *C. tropicalis* to produce long-chain dicarboxylic acids, urea is the best nitrogen source. However, research by Shen et al. (Yongqiang et al. 1979) found that when the urea content is too high, long-chain dicarboxylic acids do not accumulate. The presence of high concentrations of urea may enhance the pathways of phosphoric acid oxidation, tricarboxylic acid cycle, and glyoxylate cycle, thereby promoting the assimilation of alkanes, but at the same time, it may inhibit the synthesis of cytochrome enzymes, thereby reducing the accumulation of long-chain dicarboxylic acids.

In addition to carbon and nitrogen sources, the seed medium should contain essential inorganic substances and trace growth factors, such as phosphates, sodium chloride, magnesium sulfate, vitamins, corn syrup, and yeast extract. Mobley et al. (1999) pointed out that yeast extract is essential for increasing acid production, and the optimal amount is 0.05–0.1%. When this limit is exceeded, the acid production significantly decreases. The minimum addition of magnesium ions is 0.05%; if magnesium ions are lacking, the acid production decreases by about 75%. Shen et al. (Yongqiang et al. 1979) and Xu et al. (Xiaozeng and Mingyou 1996) also found that when the content of sodium dihydrogen phosphate is 0.1–0.2%, and the content of dipotassium hydrogen phosphate is 0.4–0.6%, it is conducive to the accumulation of long-chain dicarboxylic acids.

To improve the production efficiency of long-chain dicarboxylic acids, normal alkanes and sugars, such as glucose and sucrose, can be used to form a dual-substrate carbon source, which can increase the cell growth rate, so that alkanes can be converted into dicarboxylic acids in a shorter time (Borowitzka 1988). In addition, adding an appropriate amount of surfactants and emulsifiers to the fermentation medium can enhance the mixing of alkanes and H_2O during the fermentation process, which is also beneficial to the production of long-chain dicarboxylic acids to a certain extent.

(2) Fermentation Technology Optimization

In the oxygen-consuming fermentation process, oxygen is the key substrate to maintain microbial cell growth and metabolism. Liu et al. (Shuchen et al. 2000, 2002) found that enhancing the stirring and mixing effect and controlling low DO is an effective method to stably increase the yield of long-chain dicarboxylic acids. Controlling the DO level at 20–30% is most suitable for acid production. The effect of high DO levels on increasing acid production is not significant, but it increases energy consumption.

(3) Feed Control Optimization

In terms of feed control, sugar supplementation can effectively inhibit the β-oxidation process and provide energy for cells instead of fatty acids. Gao et al (Hong et al. 2006) found that the yields of DCA13 by the original strain and recombinant bacteria increased by 25.0% and 23.4%, respectively, and the alkane conversion rates increased by 35.5% and 28.6%, respectively, with the sugar supplementation strategy. In addition, the initial alkane concentration, alkane supplementation amount, and timing are closely related to the yield of long-chain dicarboxylic acids. When alkanes are added at the beginning of fermentation, high concentrations of alkanes have a certain inhibitory effect on cell growth, so it is best not to add alkanes or add a small amount of alkanes (<2%) at the beginning of fermentation, allowing for rapid cell growth. After the cell concentration reaches a certain level, a sufficient amount of alkane is added to induce the expression of the cytochrome P450 enzyme system, thereby increasing the yield of long-chain dicarboxylic acids (Xiaoshu et al. 2017).

Moreover, due to the hydrophilic property of carbohydrates and the hydrophobic property of alkanes, the uptake of alkanes and fatty acids in fermentation broth largely depends on the degree of emulsification of the medium.

Therefore, controlling the stirring speed and aeration rate during the fermentation process is crucial, which not only affects the oxygen supply level but also has a significant effect on the mass transfer, thereby affecting the cell growth and product synthesis.

4. *Extraction Method of Long-Chain Dicarboxylic Acids*

The extraction methods of long-chain dicarboxylic acids are mainly divided into two categories: aqueous phase extraction and solvent extraction (Shanghai Kasei Biotcchnology R&D Center Co., Ltd 2014).

(1) Aqueous Phase Extraction

The aqueous phase extraction and purification method uses activated carbon or ultrafiltration membranes to remove pigments, proteins, and organic impurities from the fermentation clear liquid, and then uses concentrated H_2SO_4 for acidification, following crystallization, cooling, and filtration to obtain long-chain dicarboxylic acid crystals, but their purity is usually less than 97.5%, and the pigment content is still relatively high. The advantages of this method include low investment, simple operation, safety, and environmental protection, but the disadvantages are also obvious, mainly low product purity, poor color, uneven crystal grain size, high protein, and other impurity content. Therefore, to obtain high-quality products, further improvements in the methods of impurity removal from fermentation clear liquid and crystallization control are needed.

(2) Solvent Extraction

The alcohol solution crystallization method is to remove the cells from the fermentation broth by membrane filtration, decolorize with activated carbon, acidify, crystallize, filter, and the wet filter cake obtained is dissolved in heated C_2–C_5 alcohols (such as ethanol, octanol, and C_2–C_5 diols), then decolorize and filter with activated carbon, cool, and crystallize to obtain pure long-chain dicarboxylic acids. As alcohols have a relatively high solubility for long-chain dicarboxylic acids at room temperature, this method has problems such as high solution loss rate (>15%), high product loss rate (~10%), a small amount of ester impurities remaining in the product, and small crystalline particles.

Shanghai Kasei Biotechnology R&D Center Co., Ltd. uses alkanes as the solvent for recrystallization to refine crude dicarboxylic acids (Shanghai Kasei Biotechnology R&D Center Co., Ltd 2014). The crude dicarboxylic acid is mixed with alkanes and heated to above 110 °C to dissolve the dicarboxylic acid. Most of the impurities in the crude product are insoluble in alkanes, forming a precipitate and separating out. Then, the dicarboxylic acid alkane solution is cooled and crystallized. The product obtained has a high total acid and monocarboxylic acid purity, and the total nitrogen content can be reduced to below 70 mg/kg. The alkane solvent used in this process has low corrosiveness and high safety. Otherwise, the impurity content in the alkane crystalline mother liquor is very low, and there is no need to distill and purify the solvent, simplifying the process procedure. In addition, inorganic acid is added to the dicarboxylic acid fermentation broth to form a mixed solution with cells and dicarboxylic acid as the main insoluble substances. Then, solid–liquid separation is carried out, and most of the water in the solid mixture is evaporated. Acetic acid or butyl acetate is added as a solvent for dissolution and extraction. The solid impurities are filtered out, the crystals are precipitated by cooling, and the dicarboxylic acid product is obtained. Zhanzhao et al. (2009) used acetic acid recrystallization to refine dodecanedioic acid. When the solid–liquid ratio of dodecanedioic acid crude product to acetic acid is 1:12.5, the dissolution temperature is 90 °C, the amount of decolorizer is 1% (w/v), and the decolorization time is 60 min, the monocarboxylic acid purity of the product dodecanedioic acid can reach more than 99.6%.

At present, the main Chinese dicarboxylic acid-producing enterprises all use acetic acid as a solvent to refine dicarboxylic acid crude products, but acetic acid is extremely corrosive, leading to high equipment costs, and the reuse of acetic acid solvent requires distillation dehydration, which has high energy consumption. The Fushun Petrochemical Research Institute of Sinopec uses ethers such as ethyl ether, propyl ether, butyl ether, pentyl ether, or hexyl ether as solvents. After acidifying the fermentation broth and filtering to obtain the dicarboxylic acid filter cake, it is mixed and extracted with the solvent, and then the solvent phase and water phase are separated. The solvent phase is added with an adsorbent for decolorization treatment, and the purity of monocarboxylic acid obtained after filtering out solid matter and crystallization is >98.5% (China Petroleum, and Chemical Corporation, Fushun Petrochemical Research Institute of China Petroleum, and Chemical Corporation 2013). In addition, Sinopec Yangzi Petrochemical Co., Ltd. also uses ether solvents such as diethyl ether, methyl tert-butyl ether, and ethyl tert-butyl ether to extract the dicarboxylic acid acidified liquid, which can obtain a total acid purity of >99% with low total nitrogen content and good appearance in refined dicarboxylic acid (China Petroleum, and Chemical Corporation, Sinopec Yangzi Petrochemical Co., Ltd 2013). The corrosiveness of ether solvents to equipment is significantly lower than that of acetic acid solvents, and they do not react with the product dicarboxylic acid, resulting in good product quality. However, ether solvents are usually flammable and volatile, and their vapors have a special irritating smell. Therefore, in actual production process, fire, anti-explosion, and anti-poison devices are needed. Moreover, there are also solvent recovery issues, resulting in higher production cost.

2.3.3 Biosynthesis of Succinic Acid

1. *Basic Properties and Applications of Succinic Acid*

Succinic acid, scientifically known as butanedioic acid, is a four-carbon dicarboxylic acid. As an intermediate product of the tricarboxylic acid cycle (TCA cycle), it is widely present in humans, animals, plants, and microorganisms. The molecular formula of succinic acid is $C_4H_6O_4$, with a molecular weight of 118.09, a melting point of 184 °C, and a boiling point of 235 °C. Succinic acid is a colorless crystal, sour in taste, and flammable. Its relative density is 1.572, easily soluble in H_2O, slightly soluble in ethanol, ether, acetone, glycerin, etc., and almost insoluble in benzene, carbon disulfide, carbon tetrachloride, and petroleum ether.

Succinic acid is an important chemical raw material, widely used in chemical, food, pharmaceutical and other industries (Choi et al. 2015). Currently, succinic acid is mainly used as a surfactant, cleaner, additive, and foaming agent. In addition, succinic acid can also be used as an ion chelating agent, used in the electroplating industry to prevent metal corrosion and pitting. In the food industry, succinic acid can be used as an acidifier, flavor enhancer, pH modifier, antibacterial agent, etc. In the pharmaceutical industry, succinic acid can be used to produce health products, such as antibiotics, amino acids, and vitamins. (Zeikus et al. 1999). Succinic acid has a linear saturated dicarboxylic acid structure, so it can be used as an intermediate for the synthesis of many compounds, such as further transformation into 1,4-butanediol, tetrahydrofuran, γ-butyrolactone, and adipic acid (Zeikus et al. 1999; Xiaobei and Yihui 2003)., which are compounds with important industrial uses.

Another potential market for succinic acid is as a monomer for biobased polymer materials. Succinic acid is an important monomer of polybutylene succinate (PBS), which is synthesized mainly by esterification of 1,4-butanediol and succinic acid under certain conditions, followed by polycondensation under the action of a catalyst, high temperature, and pressure. PBS has physical properties similar to polyethylene terephthalate (PET), is biodegradable, and can be used to make packaging materials and biomedical materials. Compared with other biodegradable materials such as polylactic acid, PBS has better mechanical properties and heat resistance, and is currently the best-performing material in degradable materials (Choi et al. 2015). Currently, the price of PBS is high and the production scale is limited, but due to the many advantages of PBS, it can still compete with PET in the future.

2. *Production of Succinic Acid by Biofermentation*

The traditional production method of succinic acid is chemical synthesis, which mainly relies on petrochemical processes, and various substances can be transformed into succinic acid under certain conditions. Chemical synthesis methods mainly rely on petroleum raw materials, and the main production methods include electrolysis (Pinazo et al. 2015), catalytic hydrogenation (Xingquan 1984), acetylene carbonylation (Rui et al. 2015), and oxidation (Yingjie et al. 2013), all of which have the disadvantages of high energy consumption and large pollution. The production of succinic acid by microbial fermentation can overcome the disadvantages of chemical synthesis to a certain extent; therefore, the chemical synthesis process of succinic acid is gradually being replaced by biofermentation.

(1) Production Pathway of Succinic Acid

Succinic acid is an intermediate metabolite of the TCA cycle and is also the end product of many microbial anaerobic fermentation pathways, playing an important role in microbial cell metabolism. Through metabolic engineering, a variety of microorganisms can synthesize succinic acid under aerobic or anaerobic conditions (Fig. 2.2). The main production pathways used in this method to synthesize succinic acid include the TCA cycle reduction arm, the glyoxylate cycle, and the TCA cycle oxidation arm (Zhu and Tang 2017).

Under anaerobic conditions, the reduction arm of the TCA cycle is the main pathway for succinic acid production. In this pathway, phosphoenolpyruvate carboxylase (PEP carboxylase) and phosphoenolpyruvate carboxykinase (PEP carboxykinase) can convert phosphoenolpyruvate (PEP) into oxaloacetate (OAA), which is then successively converted into succinic acid through reactions catalyzed

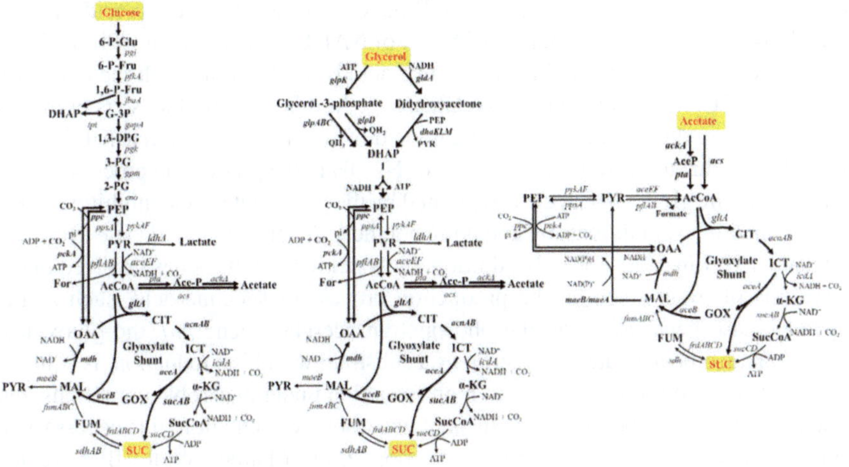

Fig. 2.2 Aerobic and anaerobic pathways associated with succinate synthesis. (**a**) Succinate synthesis pathway with glucose as carbon source; (**b**) Succinate synthesis pathway with glycerol as carbon source; (**c**) Succinate synthesis pathway with acetate as carbon source; 6-P-Glu: Glucose-6-phosphate; 6-P-Fru: Fructose-6-phosphate; 1,6-P-Fru: Fructose-1, 6-diphosphate; DHAP: dihydroxyacetone phosphate; G-3P: Glyceraldehyde-3-Phosphate; 1,3-DPG: 1, 3-diphosphoglycerate; 3-PG: 3-phosphoglycerate; 2-PG: 2-phosphoglycerate; PEP: Phosphoenolpyruvate; PYR: Pyruvate; AcCoA: Acetyl-CoA; For: Formate; Acc-P/AceP: Acetyl phosphate; CIT: Citrate; ICT: Isocitrate; GOX: Glyoxylate; α-KG: α-ketoglutarate; SucCoA: Succinyl-CoA; SUC: Succinate; FUM: Fumarate; MAL: Malate; OAA: Oxaloacetate; ADP: Adenosine 5′-diphosphate; ATP: Adenosine 5′-triphosphate; NAD^+/NADH: Oxidized/Reduced nicotinamide adenine dinucleotide; $NADP^+$/NADPH: Oxidized/Reduced nicotinamide adenine dinucleotide phosphate; *pgi*: glucose-6-phosphate isomerase gene; *pfkA*: 6-phosphofructokinase gene; *fbaA*: fructose-bisphosphate aldolase gene; *gapA*:glyceraldehyde-3-phosphate dehydrogenasegene; *gpsA*: glycerol-3-phosphate dehydrogenase gene; *pgk*: phosphoglycerate kinase gene; *gpm*: 2,3-bisphosphoglycerate-dependent phosphoglycerate mutase gene; *eno*: enolase gene; *ppc*: phosphoenolpyruvate carboxylase gene; *pckA*: phosphoenolpyruvate carboxykinase (ATP) gene; *ppsA*: phosphoenolpyruvate synthetase gene; *pykAF*: pyruvate kinase genes; *IdhA*: D-lactate dehydrogenase gene; *pfLAB*: pyruvate formate-lyase activating enzyme genes; *aceEF*: pyruvate dehydrogenase genes; *pta:* phosphate acetyltransferase gene; *ackA*: acetate kinase gene; *gltA*: citrate synthase gene; *acnAB*: aconitate hydratase genes; *aceA*: isocitrate lyase gene; *aceB*: malate synthase gene; *icdA*: isocitrate dehydrogenase gene; *sucAB*: 2-oxoglutarate decarboxylase genes; *sucCD*: succinyl-CoA synthetase genes: *frdABCD*: fumarate reductase genes; *maeB*: malate dehydrogenase (oxaloacetate-decarboxylating) (NADP+); *mdh*: malate dehydrogenase; *glpK*: glycerol kinase gene; *glpABC*: anaerobic glycerol 3-phosphate dehydrogenase genes; *gldA*: L-1,2-propanediol dehydrogenase / glycerol dehydrogenase; *dhaKLM*: dihydroxyacetone kinase genes; *acs*: acetyl-CoA synthetase gene

by malate dehydrogenase, fumarase, and fumarate reductase. Through this pathway, the synthesis of 1 mol of succinic acid can fix 1 mol of CO_2, while consuming 2 mol of NADH. However, when glucose is used as a carbon source, 1 mol of glucose can only provide 2 mol of NADH through the glycolysis pathway. Therefore, NADH is often a limiting factor in succinic acid production. The glyoxylate cycle converts acetyl coenzyme A (acetyl CoA) and OAA into succinic acid and malic acid through a series of enzyme-catalyzed reactions. Producing 1 mol of succinic acid through the glyoxylate cycle only requires 1.25 mol of NADH (Sánchez et al. 2005). Under anaerobic fermentation conditions, succinic acid is produced through the reduction arm of the TCA cycle and the glyoxylate cycle. At this time, if glucose is used as a carbon source, the maximum theoretical yield of succinic acid can reach 1.71 mol/mol glucose due to the reduced demand for NADH in the production process of succinic acid. The glyoxylate cycle is regulated by the glyoxylate cycle inhibitor (IclR). Under anaerobic conditions, by knocking out the *iclR* gene, the glyoxylate cycle can be activated (Lin et al. 2005). In addition, by knocking out the genes involved in the synthesis of by-products in the production process of succinic acid, such as the acetyl kinase gene *ackA* and the phosphotransacetylase gene *pta*, the glyoxylate cycle can also be activated. Glycerol, as a substrate with high reduction, is suitable for the production of products with high reduction under anaerobic conditions, but due to the large number of by-products of anaerobic fermentation, it is necessary to further metabolically engineer the production strain. Matthew et al. (Blankschien et al. 2010) constructed strains lacking the ethanol dehydrogenase gene (*adhE*), phosphotransacetylase gene (*pta*), pyruvate oxidase gene (*poxB*), and lactate dehydrogenase gene (*ldhA*) and the PEP carboxykinase gene (*ppc*) of the *Escherichia coli* defective strain, while overexpressing the pyruvate carboxylase gene (*pyc*) from *Lactococcus lactis*; when the strain uses glycerol as a substrate, the yield of succinic acid reaches 14 g/L, with a yield rate of 54%.

With the continuous improvement of biorefining technology requirements, the aerobic fermentation substrates of succinic acid have evolved from grain-based substrates such as glucose to non-grain-based fermentation substrates such as glycerol and acetic acid. Kang Zhen et al. (2008) started with the *E. coli* MG1655 strain, knocked out the encoding genes of succinic acid dehydrogenase (*sdh*), phosphotransacetylase (*pta*), pyruvate oxidase (*poxB*), isocitrate lyase repressor (*iclR*) and glucose-phosphotransferase (*ptsG*), then constructed an aerobic succinic acid-producing strain—*E. coli* QZ111. When using glucose as a carbon source in shake flask fermentation, this strain yielded 26.4 g/L of succinic acid , and acetic acid was the only detected by-product (2.3 g/L). Li et al. (2017) also used MG1655 as the starting strain, knocked out encoding genes of the pyruvate formate lyase (*plB*) and lactate dehydrogenase (*ldhA*), then solved the problem of low growth and yield under anaerobic conditions through a two-stage fermentation strategy. At the same time, further overexpressing the endogenous PEP carboxykinase gene (*pckA*), when using glycerol as a substrate, the yield of succinic acid in shake flask fermentation reached 118.1 mM, and the yield in a 1.5 L fermenter could reach 360.2 mM, with a high yield rate of 0.93 mol/mol. In addition, Li et al. (2016a) studied the synthesis of succinic acid using acetic acid as a carbon source. Since acetic acid activation requires energy, and energy is limited under anaerobic conditions. Therefore, succinic acid can be produced from acetic acid under aerobic conditions. Under aerobic conditions, both the glyoxylate pathway and the oxidative arm of the TCA cycle can produce succinic acid, but the oxidative arm of the TCA cycle will release CO_2, causing carbon loss, so the *iclR* gene can be knocked out to increase the flux of the glyoxylate pathway, thereby increasing the yield of succinic acid. The final constructed MG03(pTrc99a- *gltA*) strain can produce 16.45 mM of succinic acid after 72 h of shake flask fermentation, with a yield rate of 0.46 mol/mol, reaching 92% of the theoretical yield, and no other by-products were found during the fermentation process. Huang et al. (2018), starting further from the MG03 mutant strain, by knocking out the PEP carboxykinase gene (*pckA*) and overexpressing formate dehydrogenase, found that the succinic acid yield reached 30.9 mM. Since the absence of the isocitrate dehydrogenase gene (*icdA*) would lead to severe growth defects, Huang et al. further used the resting cell transformation method to increase the succinic acid yield to 194 mM, with a yield increase to 0.44 mol/mol.

(2) Succinic Acid-Producing Strains

Succinic acid, as an intermediate product of the TCA cycle and one of the end products of anaerobic metabolism, can be synthesized by almost all microorganisms, animals, and plant cells. Microorganisms that can be used for efficient production of succinic acid mainly include fungi and bacteria (Table 2.1). *Aspergillus niger*, *Pichia kudriavzevii*, *Saccharomyces cerevisiae*, *Yarrowia lipolytica*, and other fungi can synthesize succinic acid as a metabolic by-product under both anaerobic and aerobic conditions (Ahn et al. 2016; Gao et al. 2016). Among them, the research on yeast in the production process of succinic acid is more thorough. A series of yeast mutant strains have been constructed by strategies such as inactivating the key enzymes of by-product synthesis pathways, and some mutant strains can show better succinic acid fermentation production performance (Arikawa et al. 1999; Raab et al. 2010). Gao et al. (2016) use crude glycerol as a carbon source to produce succinic acid with a genetically engineered *Y. lipolytica* strain (inactivating SDH5). After optimizing fermentation conditions, the highest yield of succinic acid

Table 2.1 Biosynthesis of succinic acid

Host	Carbon source	Yield (g/L)	Space–time yield (g/L/h)	Yield (g/g)	References
M. succiniciproducens	Glycerol, glucose	97.10	3.01	0.83	Ahn et al. (2016)
M. succiniciproducens	Sucrose, glycerol	69.2	2.50	1.02	Lee et al. (2002)
Y. lipopytica	Crude glycerol	160.2	0.4	0.4	Gao et al. (2016)
Y. lipopytica	Glycerol	110.7	0.8	0.53	Cui et al. (2017)
A. succiniciproducens	Glucose	32.2	1.2	0.99	Nghiem et al. (1997)
A. succinogenes 130Z	Algae hydrolysate	36.8	3.9	0.92	Marinho et al. (2016)
Basfia succiniciproducens	Corn starch hydrolysate	30.0	0.69	0.43	Salvachúa et al. (2016)
E. coli	Algae hydrolysate	17.4	–	0.87	Bai et al. (2015)
E. coli	Acetic acid	3.65	–	0.99	Hang et al. (2018)

with this strain can reach 160 g/L, and the production rate is 0.4 g/(L·h), which is currently the highest yield of succinic acid produced by biological fermentation. However, due to many problems with using fungi as host strains for succinic acid's fermentative production, such as difficulties in fermentation operation, separation, purification, and low production efficiency, methods using fungi to produce succinic acid are mostly limited to the food and beverage manufacturing industry.

Many bacteria can also serve as host strains for succinic acid production, which can be divided into natural hosts and metabolically engineered strains. Natural hosts, such as *Actinobacillus succinogenes*, *Mannheimia succiniciproducens*, and *Anaerobiospirillum succiniproducens*, can accumulate high concentrations of succinic acid under certain conditions. These host bacteria are mostly isolated from the rumen of ruminants, and the anaerobic environment formed by CO_2, CH_4, H_2 gases creates favorable conditions for the production of succinic acid (Kamra 2005; Park and Zeikus 1999). Since the genetic information of some natural host bacteria is clear, their succinic acid production capacity can be further improved through genetic engineering.

(3) Metabolic Engineering Strategies for Succinic Acid Production

Under aerobic fermentation conditions, succinic acid is a product of the microbial TCA cycle and the glyoxylate cycle. The consumption pathway of succinic acid needs to be blocked to accumulate succinic acid. Under anaerobic conditions, succinic acid is the end product of microbial redox, and its generation pathway can be considered as part of the reductive TCA cycle. In addition, the precursor of succinic acid, succinyl-CoA, is an intermediate product in the 3-hydroxypropionic acid cycle and the dicarboxylic acid/4-hydroxybutyric acid double cycle in the natural carbon fixation pathway (Liu et al. 2020). Therefore, the biological synthesis yield of succinic acid can be improved through the following metabolic engineering strategies:

① Block the Synthesis Pathway of By-products

When *E. coli* undergoes by-product fermentation, the main products are acetic acid, lactic acid, formic acid, and ethanol, while only a small amount of succinic acid accumulates. The synthesis of by-products not only consumes carbon sources but also requires the consumption of NADH, which is not conducive to the production of succinic acid. Wang et al. (Wei et al. 2009) in the lactate dehydrogenase gene (*ldhA*) and pyruvate formate lyase gene (*pflB*) double mutant strain NZN111 of *E. coli*, on the basis of reducing the by-products lactic acid and formic acid, the two-stage succinic acid fermentation and overexpression of malate dehydrogenase (MDH) were used to solve the problem of the strain's inability to grow and produce under anaerobic conditions. In addition, ethanol is one of the main by-products of anaerobic fermentation, and knocking out the ethanol dehydrogenase-encoding gene *adhE* can effectively improve the yield of succinic acid (Donnelly et al. 2001). Acetic acid is the main by-product of aerobic fermentation of succinic acid. In order to improve the yield of succinic acid by reducing the accumulation of acetic acid, researchers usually choose to knock out *pta*, *ackA*, and *poxB* genes (De Mey et al. 2007).

② Enhancing the Expression of Key Enzymes in the Succinic Acid Synthesis Pathway and Reducing Carbon Loss in the Synthesis Pathway

A common strategy in metabolic engineering is to increase the expression of key enzymes in the synthesis pathway to enhance the metabolic flow toward succinic acid. The main pathways for the production of succinic acid include the TCA cycle and the glyoxylate cycle. The main input nodes of the TCA cycle are acetyl-CoA and OAA. Overexpression of the citrate synthase encoding gene can significantly increase the carbon flux into the TCA cycle, increase the yield of succinic acid (Li et al. 2016a), and effectively reduce the accumulation of acetic acid (Tovilla-Coutiño et al. 2020). Currently, there is a lot of research on enhancing the supply of precursor OAA. In *Escherichia coli*, OAA mainly comes from two sources: one is the synthesis of OAA by PEP carboxylase (PPC) and PEP carboxykinase (PCK) catalyzing PEP to fix CO_2; the other is the synthesis of PEP by overexpressing exogenous pyruvate carboxylase (PYC) catalyzing pyruvate, thus forming OAA to enter the TCA cycle, increasing the yield of succinic acid (Vemuri et al. 2002). Although both PPC and PCK can catalyze the production of OAA from PEP, there are still significant differences in their enzymatic properties. In *Escherichia coli*, PPC is the main enzyme for fixing CO_2 to synthesize OAA, and its Km value for CO_2 is relatively low (about 0.1 μM) (Kai et al. 1999), while PCK has a lower affinity for CO_2, with a Km value of 13 μM, so PCK mainly works in the absence of PPC or under high concentration of CO_2. PYC is an enzyme that does not exist in *Escherichia coli* itself, but is widely present in fungal cells, so the pathway to synthesize OAA in *E. coli* consumes ATP. Li et al. (2016b) compared the overexpression of *pck*, *ppc*, and *pyc*. The impact of time on the production of succinic acid was studied, and it was found that overexpression of *pyc* and *pck* can significantly promote the production of succinic acid. This may be because PYC increases the production rate of succinic acid by increasing the consumption rate of the substrate glycerol, while PCK generates more ATP in the process of catalyzing CO_2, effectively slowing down the phenomenon of inhibited cell growth in mutant strains, maintaining a larger cell density, and thus producing more succinic acid.

The loss of carbon metabolic flow in the TCA cycle is mainly caused by the glycolytic pathway of malic acid and the release of CO_2 from the TCA oxidation arm, both of which reduce the utilization rate of carbon sources. Huang et al. (Hang et al. 2018) When producing succinic acid with acetic acid as the carbon source, the related genes of the glycolytic pathway *maeB* and *pckA* were knocked out, which can effectively improve the yield and efficiency of succinic acid. In addition, by knocking out the genes of the TCA oxidation arm *sucC* or *icdA*, the utilization rate of carbon atoms can be improved. However, knocking out *icdA* will severely affect cell growth, and static regulation can only be applied to the production method of succinic acid resting cells, so dynamic regulation of *icdA* growth is needed (Bing 2019).

In addition, NADH is an important coenzyme in the synthesis of succinic acid. San et al. (Balzer et al. 2013) overexpressed the formate dehydrogenase gene (*fdh*) from Candida boidinii in the production of succinic acid fermentation, making 1 mol of formic acid react to produce 1 mol of NADH and 1 mol of CO_2, not only reducing the accumulation of formic acid during fermentation, but also providing more reducing power to increase its yield.

③ Enhance Substrate Uptake and Product Transport Pathways

By improving the ability of microorganisms to take up substrates and speeding up their utilization of substrates, the rate of succinic acid production in microorganisms can be effectively increased. Researchers have used different strategies to enhance the utilization of substrates by microorganisms when using different substrates, such as using genome editing technology to construct a strong constitutive acetic acid utilization pathway in *E. coli*, which can effectively accelerate the utilization of acetic acid and the transformation of succinic acid in *E. coli* (Hang et al. 2018). In addition, replacing the endogenous *glpK* glycerol utilization pathway in *E. coli* with the energy-saving *glpA* pathway from *Klebsiella pneumoniae* accelerates the synthesis of succinic acid (Yu et al. 2019). Chen et al. (2014) studied the impact of the four-carbon carboxylic acid efflux protein on the production of succinic acid; it was found that after knocking out the endogenous four-carbon carboxylic acid efflux protein gene, overexpression of *ducB* and *ducC* can significantly promote the production of succinic acid. After optimizing expression through RBS combination, the yield of succinic acid can be increased by 34%.

④ Laboratory Evolution

Metabolic engineering and evolutionary engineering are commonly used methods and strategies for strain construction and optimization, but they each have their advantages and disadvantages. In recent years, more and more researchers have combined these two methods to promote the biosynthesis of succinic acid, which not only fully utilizes the rational design of metabolic engineering, but also includes the irrational design of evolutionary engineering. For example, the Ingram research group (Jantama et al. 2008) knocked out the pathways of by-products such as ethanol, acetic acid, and lactic acid in succinic acid-producing strains, making the succinic acid production pathway the only pathway in the cells that absorbs NADH. In this way, cell growth and ATP production are coupled with the production of succinic acid through the consumption of NADH, so that the strains with accelerated growth selected through evolutionary engineering strategies are also the strains with increased succinic acid yield. After multiple rounds of gene knockout and adaptive evolution, two high-yield succinic acid strains were selected, and their succinic acid yield and efficiency could reach 73.4–86.5 g/L and 1.2–1.6 mol/mol glucose.

⑤ Microbial Electrosynthesis

Microbial electrosynthesis refers to the process where microorganisms absorb external electric energy as reducing power and convert it into intracellular usable reducing power forms NAD(P) or transfer it to corresponding electron transfer media, electron acceptors to store as chemical energy (Chen et al. 2020). With the surplus of electric energy and the development needs of the new generation of bio-refining, more and more scholars have begun to study the metabolic engineering applications of microbial electrosynthesis. Under anaerobic conditions, succinate dehydrogenase, as the main pathway for succinic acid synthesis, can absorb and utilize external electrons to reduce fumaric acid to succinic acid. Wu et al. (2019a) heterologously expressed cytochrome proteins with electron transfer activity, transforming *E. coli* into electroactive microorganisms with certain electron absorption capabilities, and further improved the succinic acid production capacity of the engineered strains under anaerobic fermentation conditions. In addition, by modifying the endogenous nicotinamide mononucleotide synthesis pathway in *E. coli*, *E. coli* can produce succinic acid through electrosynthesis fermentation using its own secreted electron mediators (Wu et al. 2019b).

2.3.4 Biosynthesis of Furan-2,5-Dicarboxylic Acid

1. *Basic Properties and Applications*

2,5-Furandicarboxylic acid (FDCA) is an important dicarboxylic acid, which was listed as one of the 12 most potential high-value bio-refinery products by the US Department of Energy in the early twentieth century. Its molecular formula is $C_6H_4O_5$, with a melting point of 342 °C and a boiling point of 419.2 °C, it is stable in nature. In recent years, FDCA has attracted much attention due to its wide application in various fields, especially as a substitute for terephthalic acid, a petrochemical derivative. Although FDCA has certain chemical stability, it can easily undergo typical carboxylic acid reactions, producing carboxyl dihalides, esters, and amides. The current global market value of FDCA materials, including plastics, plasticizers, thermosetting materials, and coatings, can reach several billion euros. Polyethylene furandicarboxylate (PEF), synthesized from FDCA as a monomer, can be used to replace polyethylene terephthalate (PET), which is widely used in industry. Compared with PET, PEF has better heat resistance, higher mechanical strength, stronger antioxidant capacity, and about an order of magnitude higher gas barrier properties (Rodriguez and Jing 2018; Mckenna et al. 2015).

2. *Production Method*

Currently, the synthesis routes of FDCA can be divided into glycolic acid route, 5-hydroxymethylfurfural (HMF) route, hexonic acid route, furan route, etc., based on the type of reaction starting materials. These starting materials can be synthesized into FDCA through oxidation–reduction reactions, disproportionation reactions, etc. (Jinggang et al. 2017). At present, there is relatively more research on the route starting with HMF as the reaction material. The reaction route starting with glycolic acid has a simple production process, but the synthesis path of glycolic acid needs to start from epoxethane, which makes the price of glycolic acid much higher than other starting materials, hindering its application in industry. The HMF route is considered one of the most potential routes for industrial production of FDCA. At the same time, since HMF can be produced from renewable resources (Motagamwala et al. 2018), the production costs of HMF and even FDCA will not fluctuate dramatically due to changes in raw material costs and supply, making the FDCA synthesis route starting with HMF more suitable for large-scale industrial production.

The existing methods for catalyzing HMF to synthesize FDCA mainly include chemical catalysis, biological catalysis, and electrochemical catalysis. The intermediate products involved in the synthesis route include 5-hydroxymethyl-2-furan carboxylic acid (HMF acid), 2,5-diformylfuran (DFF), and 5-formyl-2-furoic acid (FFA) (Yan et al. 2018; Zhou et al. 2021). Among the three synthesis methods, the research on chemical catalysis is the most in-depth, while the research related to electrocatalysis and biocatalysis is less. In chemical catalysis, precious metal catalysts need to be added to catalyze the reaction system, and high temperature and pressure need to be maintained in the reaction system. In contrast, the advantage of biological catalysis is that it does not require precious metal catalysts, and the reaction can proceed at normal temperature and pressure (Fig. 2.3) (Boyue 2020).

3. *Biocatalysis*

(1) Microbial Catalysis of HMF to FDCA

HMF was initially discovered as a compound that inhibits microbial growth and metabolism during the pretreatment process of lignocellulose; its presence is detrimental to the fermentation process of lignocellulose. In the process of studying the removal of fermentation inhibitors in lignocellulose, researchers found a large number of strains that can oxidize and degrade HMF, and then conducted research on the HMF metabolic pathway in these strains, and applied these researches to the biosynthesis of FDCA. Frank et al. (Koopman et al. 2010) studied the metabolic pathways of HMF and furfural in the strain HMF14 of *Cupriavidus basilensis*; they identified two gene clusters related to the above metabolic pathways, *hmfABCDE* and *hmfFGH*, through transposon mutagenesis. These two gene clusters were expressed in *Pseudomonas putida* S12, clarifying the metabolic pathways of furfural and HMF under aerobic conditions in *C. basilensis* HMF14 and *P. putida* S12. It was later found that the mutant of *C. basilensis* HMF14 strain lacking the *hmfFG* gene could accumulate FDCA, indicating that the product of this gene can react with HMF and FDCA as substrates. Gazi Sakir Hossain et al. (2016) screened a strain of *Raoultella ornithinolytica* BF60 that can use HMF to produce FDCA from soil with high HMF content, and optimized its fermentation pH, temperature, HMF substrate concentration, and biomass. Further knockout of the HMF, FDCA degradation genes *aldR* and *dcaD* in the *R. ornithinolytica* BF60 strain, while

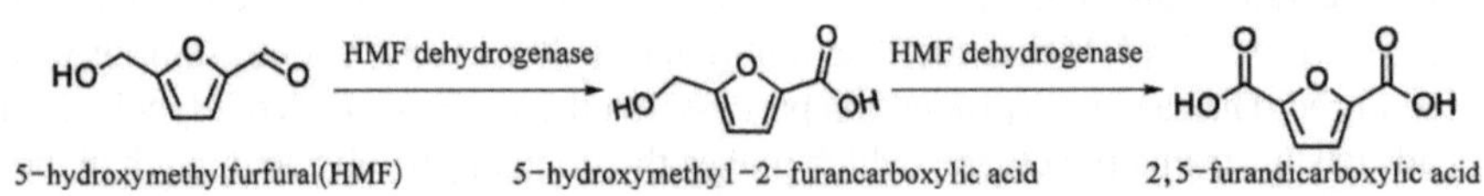

Fig. 2.3 Bioconversion pathway of furandicarboxylic acid. (Koopman et al. 2010) HMFdehydrogenase, 5-hydroxymethylfurfural dehydrogenase

overexpressing the endogenous FDCA synthesis gene aldh1 in the BF60 strain, the HMF oxidoreductase gene hmfH from *C. basilensis* HMF1 strain, and the FAD-dependent oxidase gene *hmfO* from *Methylovorus sp.* MP688 strain, thereby effectively improving the yield of FDCA (Yuan et al. 2018).

(2) Enzymatic Catalysis of HMF to FDCA

The existing enzymatic catalysis methods often involve multiple enzymes in the reaction process of producing FDCA, and there is less research on the single-enzyme catalysis of HMF to FDCA. The enzymes used in the single-enzyme catalysis reaction are mainly based on DFF as the substrate, including lipase B (CALB) from Antarctic yeast (*Candida antarctica*), wild-type aryl alcohol oxidase (AAO), and FAD-dependent flavonoid protein oxidase HMFO. Wild-type HMFO can directly oxidize HMF to FDCA with the participation of coenzyme flavin mononucleotide (FAD), and some mutants obtained through directed evolution can even get rid of the dependence on FAD and directly oxidize HMF to FDCA (Wu et al. 2020).

The two-step multi-enzyme catalysis production of FDCA is a more researched strategy at present. It achieves a higher FDCA yield by combining enzymes that can efficiently catalyze reactions in a single step. Karich et al. (2018) used wild-type AAO, unspecific peroxygenase (AaeUPO) extracted from Agrobacterium, and recombinant galactose oxidase (GAO) for enzyme tandem reactions, and the FDCA yield reached 80% after 24 h. Wu et al. (2020) combined HMFO with lipase Novozym 435, resulting in an FDCA yield of up to 94.0%. Although in vitro enzymatic synthesis of HMF to FDCA can achieve a higher yield under relatively mild reaction conditions, it often requires multiple enzymes for synergistic catalysis or the combination of chemical catalysts and enzymes to achieve ideal results. Moreover, the enzymes need to be purified first, which increases the difficulty of production operations and poses a significant obstacle to the industrial production of FDCA. Whole-cell catalysis does not require additional purification of enzymes, and for enzyme-catalyzed reactions that need to add cofactors (such as FAD and NADH), they can be supplemented in situ in the cell. In addition, the various enzymes in the whole-cell catalysis process are less affected by toxic substrates, can maintain enzyme activity well, and can improve catalytic efficiency and reduce production operation difficulty by modifying the production strain to express enzymes that catalyze different reactions heterogeneously or integrate multiple enzymes into one strain. Therefore, compared to in vitro enzymatic catalysis, whole-cell catalysis is undoubtedly more suitable for the industrial production of FDCA.

2.3.5 Biosynthesis of Adipic Acid

1. *Properties and Applications*

Adipic acid (or adipate) is a common saturated dicarboxylic acid, also known as fatty acid, with the molecular formula $C_6H_{10}O_4$. It is easily soluble in ethanol and acetone, slightly soluble in ether, soluble in H_2O, insoluble in benzene and petroleum ether, and is considered by the International Energy Agency as one of the most industrially valuable dicarboxylic acids. The global annual production of adipic acid is about 3 million tons, which can bring about 600 million US dollars in market revenue, and its production capacity is still increasing at a rate of 3–5% per year. Adipic acid is mainly used in the nylon industry and also occupies a certain share in the fields of plasticizers, polyurethane, food, and medicine (Xichun and Yuan 2021).

2. *Production Method*

The synthesis of adipic acid was originally derived from fat oxidation, and the current production mainly relies on the synthesis of petrochemical precursors (such as the most commonly used benzene). The production of adipic acid based on benzene will produce a large amount of N_2O, exacerbating the greenhouse effect. Therefore, there is an urgent need for a new efficient production process for environmentally friendly adipic acid to replace traditional chemical synthesis processes.

In recent years, due to the convenience, reliability, and environmental friendliness of biosynthesis, more and more researchers have begun to explore the biosynthesis methods of adipic acid. There is currently no known biological synthesis pathway for adipic acid in nature, although it has been reported that thermophilic actinomycetes (*Thermobifida fusca*) have a natural adipic acid synthesis pathway, but research has found that its cell lysate can only produce adipic acid after the addition of endogenous 5-carboxy-2-pentenoyl-CoA reductase (Xichun and Yuan 2021). Although natural production strains have certain advantages in adipic acid production, they are still difficult to apply to the industrial production of adipic acid due to the lack of a clear genetic background and acid tolerance. *Corynebacterium glutamicum*, *E. coli,* and *S. cerevisiae*, as common model strains, have a clear genetic background and have been used by researchers as chassis cells for the production of adipic acid. The main production pathways of adipic acid include the reverse reaction of the adipic acid degradation pathway (Xinxing et al. 2009), the production pathway combined with the β-oxidation pathway (or β-reverse oxidation pathway) and the ω-oxidation pathway, the 2-oxo-heptanedioic acid pathway, the 2-oxo-adipic acid pathway, and the lysine pathway, etc. (Skoog et al. 2018).

The reverse reaction of the adipic acid degradation pathway naturally exists in *Penicillium chrysogenum*, and this pathway has a theoretical yield of up to 0.92 mol adipic acid/mol glucose (Mingsong 2018). This pathway synthesizes adipoyl CoA from metabolic intermediate succinyl CoA, and its reaction process includes ① succinic acid CoA and acetyl CoA produce 3-oxoadipic acid CoA (Yaqin and Lijuan 2019); ② 3-oxoadipic acid CoA is reduced to 3-hydroxyadipic acid; ③ 3-hydroxyadipic acid is dehydrated to produce 5-carboxy-2-pentenoyl CoA; ④ 5-carboxy-2-pentenoyl CoA is further reduced to produce adipoyl CoA (this step is the rate-limiting step of this pathway); and ⑤ Finally, adipoyl-CoA is enzymatically catalyzed to form adipic acid. These enzymatic reactions can be catalyzed by succinyl-CoA synthetase or acetyl-CoA esterase, or they can be a two-step catalytic reaction composed of butyryl phosphate transferase and butyrate kinase (Skoog et al. 2018). Several reverse degradation pathways of adipic acid have been constructed in *E. coli* using enzymes from different organisms, and in vivo/in vitro tests have been conducted. It was found that some enzyme combinations, after optimization, promote the production of adipic acid. When glycerol is used as the carbon source, the yield of adipic acid can reach 68 g/L in metabolically engineered strains optimized by the pathway, with a productivity of 0.81 g/L· h^{-1}, and a yield of 0.378 g/g (Xinxing et al. 2009). It is worth mentioning that the NADH/NAD^+ ratio of glycerol under microaerobic conditions is higher than that of glucose, which can explain why the yield of adipic acid is lower when glucose is used as the carbon source. The reason may be due to the relatively less supply of coenzymes during the reaction. In addition, the supply of acetyl coenzyme CoA and succinyl CoA is another key factor affecting the yield of adipic acid. When the ratio of the two is 1:1, it is most ideal for the production of adipic acid. Therefore, it is also crucial to metabolically regulate the content and ratio of acetyl CoA and succinyl CoA in the chassis cells (Xinxing et al. 2009).

When *E. coli* and certain strains of Antarctic yeast (*C. antarctica*) are used as chassis cells, the β-oxidation pathway or reverse β-oxidation pathway is often combined with the ω-oxidation pathway to produce adipic acid. In this method of adipic acid production, the ω-oxidation pathway can oxidize the ω carbon of fatty acids with different chain lengths to carboxyl to form dicarboxylic acids, and the dicarboxylic acids can be degraded by the β-oxidation pathway, forming an acetyl coenzyme A in each cycle. Therefore, in the production of dicarboxylic acids, it is necessary to produce dicarboxylic acids of a specific length to recognize acyl-CoA oxidase with different carbon chain lengths. The POX5 acyl-CoA oxidase from *C. antarctica* strain can recognize dicarboxylic acids other than adipic acid, allowing these dicarboxylic acids to be degraded by β-oxidation. In addition, the yeast strain (*Candida*). After knocking out the POX4 acyl oxidase that can recognize C4–C20 substrates in the body, the degradation pathway of adipic acid is blocked, allowing adipic acid to accumulate. However, the β-oxidation pathway for the production of adipic acid is a carbon loss reaction, which is not ideal from an economic and environmental perspective. If *E. coli* is used as the host cell for production, acetyl CoA is first extended to six-carbon fatty acids through the β-reverse oxidation pathway, and then adipic acid is formed through the ω-oxidation pathway. Compared with the β-oxidation pathway, this pathway has a higher yield, but it also requires the participation of more specific enzymes. For example, thiolase and thioesterase play a role in controlling the carbon chain length in this adipic acid production pathway. The 3-ketoacyl CoA thiolase from Ralstonia eutropha can specifically generate C6–C10 carboxylic acids, and the endogenous thioesterase YdiI in *E. coli* can specifically recognize C6, C8, and C10 acyl CoA and generate corresponding dicarboxylic acids. Previous studies have reported that when overexpressing the 2-oxoheptanedioic acid pathway in *E. coli* and producing aminopropionic acid with glucose as the carbon source, adipic acid is a by-product of this pathway. This pathway starts from 2-oxopentanedioic acid, extends to form 2-oxoadipic acid through the Aks series of enzymes, and 2-oxoadipic acid can further extend to form 2-oxoheptanedioic acid and α-ketooctanedioic acid in methanogenic archaea. 2-oxoheptanedioic acid forms adipic acid semialdehyde by regulating the carbon chain specificity of AksA in the Aks series of enzymes and the activity of downstream 2-oxoheptanedioic acid decarboxylase, and then forms adipic acid catalyzed by an unknown enzyme. In addition to using constructed metabolic engineering production strains, adipic acid can also be synthesized using its in vitro cell-free system. This system is based on the Streptomyces-derived loose helix PKS system (composed of ketone synthesis enzyme, acyl transferase, ketone reductase, and acyl carrier protein), starting with succinic acid CoA, with propionyl CoA as the carbon extension unit, and the final adipic acid ACP is generated by thioesterase to form adipic acid. Although the yield of the in vitro adipic acid synthesis system is low, only 0.3 mg/L, this system provides us with a new idea for in vitro synthesis, different from traditional metabolic engineering and a completely new pathway for adipic acid synthesis (Ruiqi and Yifan 2021). In addition, besides the aforementioned realized adipic acid production pathways, metabolic engineering direct synthesis of adipic acid also has many theoretical adipic acid production pathways, such as the 3-oxoadipic acid pathway, 2-oxoadipic acid pathway, ω-oxidation pathway of hexanoic acid, methyl ester carbon extension pathway, and the synthesis pathway starting from lysine (Skoog et al. 2018).

2.3.6 *Biobased Terephthalic Acid*

1. *Properties and Applications of Terephthalic Acid*

Terephthalic acid (also known as 1,4-benzenedicarboxylic acid, PTA), with the molecular formula $C_8H_6O_4$, and the structure shown in Fig. 2.4, is currently the largest produced dicarboxylic acid and one of the important bulk organic materials. It is a solid at room temperature, does not melt when heated, and sublimates above 300 °C. Terephthalic acid is used in the production of polyester, especially polyethylene terephthalate (PET), and is widely used in the fields of chemical fiber, light industry, electronics, construction, etc. Over 90% of PTA is used to produce PET, while the rest is used as raw material for polytrimethylene terephthalate (PTT) and polybutylene terephthalate (PBT) and their products. PTA is mainly a petrochemical product, both an end product of petroleum and a front-end product of polyester. From the perspective of the industrial chain, PTA is connected to paraxylene (PX) and crude oil upstream, and polyester, polyester staple fiber, and filament

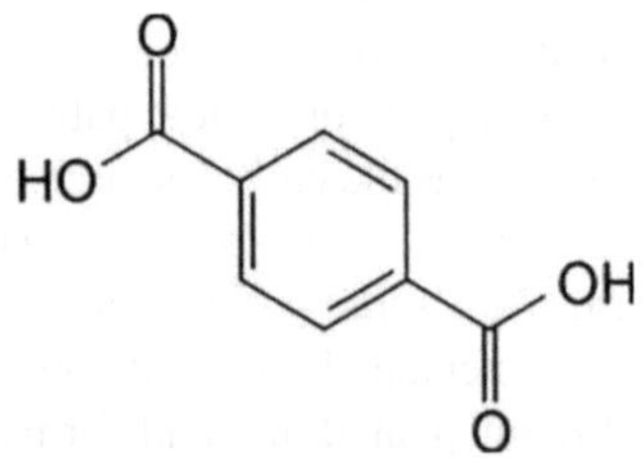

Fig. 2.4 Terephthalic acid structure

downstream. It is the watershed of the petrochemical and polyester industrial chain, playing a bridging role (Xichun and Yuan 2021).

After 2000, the localization of PTA production technology has received attention from enterprises and relevant state departments, and the process of localization has begun to accelerate. Several major achievements have been made: Jinan Zhenghao Chemical Fiber New Material Co., Ltd., as the first domestic PTA project, has completed studies on oxidation reaction kinetics, crystallization rules, and heat and mass transfer, among other topics. At the same time, relying on the existing PTA device of Jinan Zhenghao, through transformation, a number of industrial trials of domestic technology have been carried out. Fifteen sets of key equipment will be manufactured and applied domestically for the first time, achieving the expected results. The complete technology of terephthalic acid from Yangzi Petrochemical, a key project of the "Ten Dragons" of Sinopec undertaken by Yangzi Petrochemical, Zhejiang University, and East China University of Science and Technology, has developed a new type of oxidation reactor and more advanced oxidation process with independent intellectual property rights. This can save 20–30% of the total investment in new PTA projects compared to imported devices, and the energy consumption of its process is more than 20% lower than the current imported PTA devices. So far, 12 patents have been applied for (6 have been authorized). Shanghai Petrochemical uses the oxygen-rich oxidation PTA production technology that it has digested and absorbed to design and build on its own, expanding the original annual production of 225,000 tons of PTA devices to 400,000 tons, and has made breakthroughs in parts such as pumps and heat exchangers, realizing localization. The PTA production center of Yihua Company has used a new process for the catalyst, independently developed a process technology to extend the life of the palladium-carbon catalyst, treated it in the early stage of possible catalyst poisoning, used its own technology to restore the activity of the catalyst, and extended its service life, setting a record for the longest use of MPB5 type palladium-carbon catalyst in the country, reaching the advanced level of Japan and other countries (Xinxing et al. 2009). With the construction and commissioning of large-scale integrated refining and chemical projects in China,

The shortage of raw material PX in China will be improved, which will balance the industrial benefits of aromatics-polyester (PX-PTA-PET). With the gradual recovery of the world economy, the PTA industry will gradually enter a benign operating state. Under the requirement of strengthening environmental protection, the overall advantages of upstream and downstream facilities will be brought into play, and the development of high-quality, low-cost products will be persisted (Mingsong 2018).

2. *Production Methods of Terephthalic Acid*

(1) Petroleum-based terephthalic acid production method

Using air as the source of O_2, and xylene as the raw material, terephthalic acid is produced. Under the joint action of catalysts, a liquid phase oxidation reaction is carried out at a certain temperature, first generating crude terephthalic acid (CTA), and then under the action of the catalyst, hydrogenation is carried out to obtain high-purity terephthalic acid (PTA). The first step is the oxidation reaction, which uses acetic acid as a solvent and air as an oxidant to oxidize PX into CTA in the liquid phase; then refining, dissolving the CTA reactant in deionized H_2O under high temperature and high pressure, under the action of the catalyst, adding H_2 can remove impurities; the reaction level of the oxidation stage will determine the material consumption level and energy consumption level in the PTA production process. The oxidation process is divided into high temperature, medium temperature, and low temperature conditions according to the different temperatures of the oxidation reaction (Yaqin and Lijuan 2019).

Currently, the world's technology for the high-temperature oxidation and hydrogenation refining of xylene to prepare PTA, as well as the liquid-phase oxidation process of PX, all use cobalt acetate and manganese acetate as catalysts, and bromides as promoters, allowing PX to oxidize in acetic acid solvent under certain pressure and temperature to produce crude terephthalic acid. The Mitsui-Amoco technology has its uniqueness in the oxidation reaction conditions (temperature, pressure, catalyst ratio, etc.) and the structure of the oxidation reactor; the oxidation reactor of Lur-gi-Eastman technology is a typical bubble column, with the lowest oxidation reaction pressure and temperature; other technologies represented by Amoco and ICI, the oxidation reactors are all stirred reactors, with higher oxidation reaction pressure and temperature. The PTA process of CTA hydrogenation refining is basically the same, all using the Amoco process, only differing in the heating method and PTA solid separation method (Lin 2019).

(2) Biobased Terephthalic Acid Production Method

The research on biobased PTA is an effective supplement to the petroleum-based PTA preparation method. There are mainly two routes, one is to transform biomass raw materials into biobased PX by chemical method, and then refine and catalyze oxidation and hydrogenation to prepare biobased PTA; the other is to use biomass as raw material to prepare PTA through microbial fermentation.

The production of biobased PX from lignocellulosic raw materials has the advantage of being green and environmentally friendly, and its efficient catalytic synthesis technology is a focus of attention in academia and industry. The process of synthesizing PX from renewable biomass was first highly valued by the United States and Japan and has made significant progress. Internationally famous companies such as Coca-Cola and DuPont have also invested a lot of resources in research and development, forming a number of PX production processes with different raw materials: Anellotech has developed a biomass pyrolysis process to prepare aromatics (benzene, toluene and PX); its core technology is the Bio-TCat™ rapid pyrolysis technology for biomass and has been applied for ten PCT patents. With the funding of strategic partners such as Suntory and Toyota Tsusho, Anellotech built a pilot verification device Tcat-8 in 2016 and announced in March 2018 that it had successfully run continuously for 2 weeks. Virent uses sugars to prepare hydrocarbon compounds by aqueous phase reforming and then further aromatizes to prepare biobased PX. It has developed a process for converting sugars into PX by catalytic conversion and has applied for five patents. In 2016, Virent formed a strategic alliance with

Tesoro, Toray, Cofco, and Coca-Cola to upgrade and scale up its biobased PX preparation technology. Gevo converts biobased alcohols into hydrocarbon compounds including PX, has applied for five patents for biobased PX, and is also cooperating with Coca-Cola and Toray to develop and expand biobased PX and biobased PET. Its biobased isobutanol aromatic process is currently in the pilot stage.

Domestic research has also been carried out on the development of a fully biobased PX preparation process using 2,5-dimethylfuran (DMF, which can be prepared from lignocellulose) or 2,5-hexanedione (which can be prepared from lignocellulose) and ethylene (which can be prepared from straw ethanol) as raw materials. The research group of Professor Tan Tianwei from Beijing University of Chemical Technology has developed a series of phosphorus-containing catalysts with mesoporous structure and matched active center composition, such as niobium phosphate and tin phosphate. The PX selectivity is 93%, the carbon balance is as high as 97%, and the prepared niobium phosphate and tin phosphate have excellent stability. After ten continuous cycles (60 h), the yield of the target product does not decrease, showing good industrial application prospects (Feng et al. 2017). Professor Zhang Weiping of Dalian University of Technology has conducted research on beta molecular sieves and SO_3 H-based catalysts (Zhao et al. 2021). The use of biological methods to produce chemical substances has advantages such as mild reaction conditions, less waste, less harmful impact on the environment, simple technical operations, and simple separation and purification. Therefore, as early as 1974, scholars invented a method to produce PTA from p-xylene through microbial transformation (SU 419509). The microorganisms (soil filamentous fungi) screened in this invention can synthesize oxidase to oxidize p-xylene to produce PTA under the induction of hexadecane. In 1995, Japanese scholars invented a method to produce aromatic carboxylic acids by oxidizing various aromatic compounds through microbial methods (JP 9023891). The microbial strains screened with various aromatic compounds as the only carbon source can oxidize various aromatic compounds to produce aromatic carboxylic acids. In 2002, Mogen and others reported a method for producing terephthalic acid by microbial enzyme oxidation of p-benzenedimethanol: using chloroperoxidase and xanthine oxidase to catalyze p-benzenedimethanol, the highest yield of PTA reached 65% (John et al. 2002). From the current research progress, the biological method of preparing PTA has problems such as low yield, low conversion rate, and many by-products, and it is still in the laboratory stage, and industrial implementation still requires more investment.

2.4 Other Biobased Polyester Monomers

2.4.1 Biobased Ethylene Glycol

1. *Physical and Chemical Properties and Applications of Ethylene Glycol*

Ethylene Glycol (EG), also known as glycol or 1,2-ethanediol, abbreviated as EG. Its chemical formula is $(CH_2 OH)_2$, is a type of diol with two carbon atoms, and is the simplest diol. Ethylene glycol is a colorless, odorless, sweet-tasting liquid with low toxicity to animals. Ethylene glycol can dissolve in H_2 O and acetone, but has a lower solubility in ethers. Due to its low molecular weight, EG is active and can undergo esterification, etherification, alcoholization, oxidation, aldehyde condensation, and dehydration reactions. EG is an important basic organic raw material in the chemical industry, mainly used for the production of polyester, antifreeze, lubricants, plasticizers, etc., with over 90% of ethylene glycol used for polyester production. In recent years, driven by the demand for polyester, China's consumption of ethylene glycol has grown rapidly, with an average annual growth rate of 8.04% (Ruiqi and Yifan 2021) from 2010 to 2020.

2. *Preparation Methods of Ethylene Glycol*

(1) Non-biobased Ethylene Glycol Preparation Methods

Depending on the raw materials used, the production process of ethylene glycol is mainly divided into oil-based (petroleum), gas-based (natural gas), and coal-based (coal). It can also be divided into ethylene method and oxalate ester method. The ethylene method can be further divided into petroleum ethylene method, ethane ethylene method, and methanol-to-ethylene method (MTO method); the oxalate ester method involves producing dimethyl oxalate from coal or natural gas, which is then used to produce ethylene glycol (Liangquan 2008; Yupei et al. 2006).

The advantage of the oil-based ethylene glycol production process is its maturity, but it has the disadvantages of high water consumption, high energy consumption, and high cost. Currently, its key technologies are mainly controlled by Royal Dutch Shell, Scientific Design Company of the United States, and Union Carbide Corporation. In the Middle East and North America, the ethane ethylene method, which uses ethane as a raw material to produce ethylene glycol, has the strongest competitiveness due to abundant resources and low prices (Xiaoming 2012). Coal-based ethylene glycol is a unique production process in China, usually using coal as a raw material to produce CO and H_2; CO is synthesized and refined to produce oxalate, which is then hydrogenated and refined to obtain polyester-grade ethylene glycol. However, this process produces impurities such as 1,2-butanediol and vinyl carbonate, and its acceptance needs to be improved. The natural gas route first converts CH_4 into synthesis gas mainly composed of CO and H_2 through steam reforming, and then synthesizes ethylene glycol using the oxalate ester route. In addition to these main process routes, there is also a unique route in China for producing ethylene glycol from coke oven gas. Using coke oven gas as a raw material, through gas cabinet storage, pressurization, pure oxygen conversion, desulfurization and decarbonization, precision desulfurization, synthesis gas separation, ethylene glycol synthesis; and distillation, ethylene glycol can also be produced (Qun and Decai 2017).

(2) Biobased Ethylene Glycol Preparation Methods

In recent years, with increasingly strict environmental regulations and the rapid pace of energy transition, the bio-economy has developed rapidly, and biobased ethylene glycol has attracted attention worldwide. Biobased ethylene glycol is derived from sugar or other non-grain biomass resources. Compared to petroleum-based ethylene glycol, it has advantages such as shorter process flow, higher environmental friendliness, and greater flexibility in raw material selection. There are mainly two routes for biobased ethylene glycol: one is to produce ethylene glycol through chemical transformation of biomass raw materials, which is further divided into a one-step method and an ethylene route; the other is to prepare ethylene glycol through fermentation of biomass raw materials.

Internationally, the production capacity of biobased ethylene glycol is mainly concentrated in North America, Europe, and other regions, with related production companies including Braskem, Avantium, UPM Finn O'Huichuan, etc. Under government support, China's biobased ethylene glycol industry has also achieved certain accomplishments in recent years. In 2021, biobased materials and biodegradable materials were included in the "Fourteenth Five-Year" national key research and development plan, which will further promote the development of the biobased ethylene glycol industry.

The one-step method uses lignocellulosic material as the raw material, and ethylene glycol is prepared by catalytic hydrogenolysis under hydrothermal conditions. This method first breaks the C–C bond in the hydrolysis product molecules of cellulose under the action of a catalyst to generate acetaldehyde, which is then hydrogenated to produce ethylene glycol under the action of a catalyst. This route does not require a large amount of O_2 consumption, nor does it produce a large amount of wastewater and exhaust gas emissions, making it an environmentally friendly and sustainable green technology (Na et al. 2008).

The ethylene route uses ethanol prepared by fermentation of biomass as the raw material, dehydrates to prepare ethylene, and then catalyzes ethylene to prepare ethylene glycol. Anhui Fengyuan Group adopts the process route of corn → starch → fuel ethanol → ethylene → ethylene oxide → ethylene glycol, and constructs a production device with an annual output of 180,000 tons of ethylene glycol using corn, cassava, and other starch raw materials. Jilin Boda Biochemical Co., Ltd. uses corn as the raw material and constructs a production device with an annual output of 100,000 tons of ethylene glycol. Changchun Dacheng Group has developed a process to prepare a mixture of polyols mainly composed of ethylene glycol using starch as the raw material and has built a polyol production line with an annual output of 200,000 tons.

Biofermentation is another route for the preparation of biobased ethylene glycol. By using *Escherichia coli*, *Clostridium glutamicum*, *Saccharomyces cerevisiae*, or *Kluyveromyces lactis* as hosts, several synthetic pathways for microbial production of EG from pentose, hexose, or ethanol have been established (Tong et al. 2018). Xylonic acid is an intermediate product of this process, and many microorganisms, including *Acetobacterium*, *Pseudomonas aeruginosa*, *Pseudomonas fragi*, and Gram-positive bacteria, have the ability to convert xylose into xylonic acid. Xylonic acid can be used as a carbon source for wild-type *E. coli* to produce ethylene glycol and acetic acid. Using this technology, high substrate conversion rates and high production rates have been achieved. *E. coli* is a commonly used host for constructing ethylene glycol-producing bacteria (Zivkovic et al. 2009).

2.4.2 Biomanufacturing of 1,4-Butanediol

1. *Properties and Applications of 1,4-Butanediol*

1,4-Butanediol (BDO) has a molecular formula of $C_4 H_{10} O_2$ and a molecular weight of 90.12. It appears as a colorless or pale yellow oily liquid, is flammable, and is soluble in methanol, ethanol, and acetone, slightly soluble in ether. BDO is an important organic chemical raw material. Chemical products produced from it, such as tetrahydrofuran (THF), γ-butyrolactone (GBL), 1,3-butadiene (BDE), polybutylene terephthalate (PBT), and polyurethane elastomer (PU), can be widely used in pharmaceutical intermediates, agricultural pesticides, industrial solvents, daily chemicals, fine chemicals, and other fields (Yongli et al. 2020). With the completion and production of several new or expanded facilities in China and South Korea, the global production capacity of 1,4-butanediol has increased dramatically. It is estimated that by 2025, the global 1,4-butanediol market will reach 12.6 billion US dollars, and China's 1,4-butanediol manufacturing capacity will exceed 55% of the world's capacity (Ruqing et al. 2018).

2. *Production Methods of 1,4-Butanediol*

1,4-Butanediol can be obtained through chemical synthesis and microbial fermentation. The chemical production process of 1,4-butanediol mainly includes the Reppe method (Xinming 2017), Davy-Nokee method, Mitsubishi Chemical method, Lyondell method, etc., among which the Reppe method can start from coal-based raw materials, generate butynediol through the reaction of coal-based formaldehyde and acetylene, and then produce 1,4-butanediol after hydrogenation. This method, with its wide range of raw material sources and good economic benefits, has become the most widely used 1,4-butanediol production process in China (Xiuli 2006).

(1) Propylene Oxide Production Method

The catalyst used in the propylene oxide process can be recycled, the production process is simple, the cost input is less, the impact on the environment is small, it can reduce steam consumption, and ensure the production rate of 1,4-butanediol, and can effectively adjust the production load, with strong market change adaptability. However, the use of phosphoric acid lithium, triphenylphosphine, hexarhodium hexacarbonyl, and other raw materials in the use of the propylene oxide production process faces certain difficulties in transportation and storage, and the intermediate products produced during the production process have greater toxicity, such as hydroxybutyraldehyde itself is toxic, and it will decompose upon heating to produce toxic castor aldehyde gas; for this reason, more reasonable transportation and storage measures need to be taken, and the toxicity of intermediate products needs to be reduced.

(2) Butadiene Production Method

The butadiene production method includes chlorination and acetic acid processes (Li 2003). The production of 1,4-butanediol using chlorination technology is carried out at a temperature of 260–300 °C, through the gas-phase chlorination of butadiene to produce 1,4-dichloro-2-butene, which is then hydrolyzed under alkaline conditions to produce butenediol and hydrogenated to produce 1,4-butanediol. In the production of 1,4-butanediol via the acetic acid process, butadiene and acetic acid are first reacted with oxygen in the presence of a palladium-tellurium catalyst to form 1,4-diacetoxy-2-butene through hydroacetoxylation. This intermediate is then catalytically hydrogenated over a palladium catalyst to yield 1,4-diacetoxybutane. Finally, hydrolysis catalyzed by a sulfuric acid-type cation exchange resin affords 1,4-butanediol.

(3) Maleic Anhydride Production Method

The production process includes esterification, hydrogenation, and direct hydrogenation methods. In the esterification hydrogenation preparation process, maleic anhydride and excess ethanol are mixed, and through a monoesterification reaction, the intermediate product reacts with ethanol to produce a substance that is distilled off water and then hydrogenated in the gas phase to produce 1,4-butanediol; directly through the hydrogenation preparation process, nickel–rhenium is used as a catalyst, and maleic anhydride is treated with liquid phase hydrogenation to produce γ-butyrolactone after hydrogenation, and finally 1,4-butanediol is produced (Li 2003).

(4) Acetylene Aldehyde Production Method

Methanol is oxidized to produce formaldehyde, which reacts with acetylene to produce acetylene aldehyde, which is then hydrogenated to produce 1,4-butanediol. This technology has a certain possibility of explosion during the application process, and there may be phenomena such as pipeline blockage caused by polyacetylene materials, affecting production safety, production efficiency, and production schedule. Therefore, the application of this technology is gradually being replaced by other technologies with higher efficiency and better safety (Fangfang and Xinmin 2018).

The production of BDO by biological fermentation has a very important prospect. 1,4-Butanediol is a new natural compound, and there are currently no known organisms that have the ability to naturally synthesize 1,4-butanediol. In addition, 1,4-butanediol is a highly reduced compound containing two hydroxyl groups, and its biosynthesis process requires a large amount of intracellular reduction power and energy supply. Therefore, by introducing artificial metabolic pathways that can convert endogenous precursor libraries or exogenous cheap renewable raw materials into the desired products into the chassis microorganisms, a high-efficiency and sustainable 1,4-butanediol biosynthesis process can be achieved (Wu et al. 2017).

So far, some studies have proven that by introducing different synthetic pathways into *E. coli*, the cells can use glucose or xylose as the main carbon source for 1,4-BDO production, and 1,4-BDO biosynthesis can be achieved. These studies use plasmid expression of key genes in the artificial pathway that produces 1,4-BDO. At the same time, in order to avoid the use of expensive and unnecessary antibiotics and chemical inducers, it is necessary to integrate pathways or auxiliary genes in this strain and express them under constitutive promoters. In addition, the final production strain usually needs to undergo tedious and repeated genome engineering through mutation, deletion, and insertion of multiple genes (Burgard et al. 2016).

Simultaneously, when constructing engineering strains, it is necessary to reduce the production of metabolic by-products, thereby improving carbon yield and reducing downstream separation costs. The work of strain construction first focuses on reducing the production of acetate, ethanol, glutamate, and 4-aminobutyrate. Acetate is directly produced from the CoA transferase step of the BDO pathway. By enhancing the expression of *ackA-pta*, which encodes acetate kinase and phosphotransacetylase, respectively, through constitutive promoters, the accumulation of acetate is alleviated. In cells producing BDO, the oxidative tricarboxylic acid cycle provides a strong irreversible pull, keeping the concentration of acetyl coenzyme A at a sufficiently low level, thereby promoting the ackA-pta balance toward the production of acetyl coenzyme A. Ethanol is another by-product of acetyl coenzyme A, produced by Ald and Adh enzymes that catalyze 4-hydroxybutyryl coenzyme A to produce BDO. When *ppc* expression increases, ethanol production decreases by 50%, which increases the flux from pyruvate to oxaloacetate and into the oxidative TCA cycle, and reduces the concentration of acetyl coenzyme A to the lowest. By constitutively overexpressing AKGDH subunit genes, such as *Cab-lpdA*, glutamate overflow can be overcome. Finally, by deleting the *gabT* and *puuE* genes encoding aminotransferase, the formation of succinic semialdehyde is catalyzed, thereby essentially eliminating 4-aminobutyrate (Butler 2005).

2.4.3 Biomanufacturing of 1,3-Propylene Glycol

1. *Properties and Applications of 1,3-Propylene Glycol*

1,3-Propylene glycol (1,3-propanediol, 1,3-PDO) is a colorless, transparent viscous liquid, odorless, hygroscopic, and miscible with H_2O, alcohol, and various organic solvents. It can be used directly as an antifreeze and is a synthetic raw material for various plasticizers, detergents, preservatives, and emulsifiers. It is also used in the food, cosmetics, and pharmaceutical industries. In addition, 1,3-PDO can also serve as a monomer for polyesters, polyethers, and polyurethanes. Currently, the most important use of 1,3-PDO is as a raw material for the production of polytrimethylene terephthalate (PTT). Compared with polyethylene terephthalate (PET) and polybutylene terephthalate (PBT), PTT fiber has excellent resilience, dyeability, fluffiness, toughness comparable to nylon, stain resistance, antistatic properties, and UV resistance due to the suitable carbon chain length and special angular configuration of 1,3-PDO. It is widely used in engineering plastics, clothing fabrics, and carpets, and has a very broad development prospect.

2. *Production Method of 1,3-Propanediol*

1,3-Propanediol can be obtained through chemical synthesis and microbial fermentation. The chemical synthesis method includes the production of 1,3-propanediol from epoxide ethanal under the action of a formaldehyde hydrogenation catalyst, the hydration-hydrogenation synthesis of 1,3-propanediol from acrolein, and the synthesis of 1,3-propanediol from ethylene via the Prins reaction. Among them, the catalyst system for the epoxide ethanal hydrogenation route is complex, the production process is harsh and unstable, the ligand is highly toxic, the synthesis process of 1,3-propanediol requires high pressure, and the equipment requirements are high. This technology route requires a high level of comprehensive technical expertise and industrial group quality. The yield of 1,3-propanediol in this route is about 44%. The total yield of the acrolein route is nearly 70%, the hydration process conditions are mild, the hydrogenation process is mature, the equipment requirements are not high, and it is in line with the national conditions of our country. The ethylene route uses cheap and readily available raw materials, and the yield is also considerable. The by-product acetate can be converted into 1,3-propanediol (Limei et al. 2000).

Among the microorganisms used in the fermentation production of 1,3-PDO, *Klebsiella pneumoniae*, *Butyric acid Clostridium*, and *Fusarium citric acid bacteria* have received more attention due to their high production efficiency and 1,3-PDO yield. Wild strains produce a variety of by-products in the process of preparing 1,3-PDO due to their complex enzyme system, which poisons the strains and brings difficulties to separation and purification. Therefore, it is necessary to modify the strains through genetic engineering (Shaobo et al. 2021).

The raw materials for the fermentation production of 1,3-PDO are mainly glycerol and glucose. DuPont in the United States has constructed an engineered strain that uses glucose as a raw material to produce 1,3-PDO. It introduces the glycerol synthesis pathway from brewing yeast and the 1,3-PDO synthesis pathway from *Klebsiella* into *E. coli*, while reducing the carbon metabolic flow into the TCA cycle, promoting glucose metabolism to glycerol, significantly improving the accumulation of 1,3-PDO, with a yield of 135 g/L, a production rate of 3.5 g/L/h, a molar conversion rate of 121%, and a 40% reduction in energy consumption and CO_2 emissions compared to the traditional petrochemical route (Nakamura and Whited 2003). Currently, DuPont's biobased 1,3-PDO production capacity has reached 75,000 tons/year, and sales account for one third of the global 1,3-PDO market.

China has taken a different approach, using glycerol, a by-product of biodiesel production (100 kg of glycerol is produced for every 1 ton of biodiesel produced), as a raw material to carry out technical research on the fermentation production of biobased 1,3-PDO. Glycerol fermentation is usually carried out under anaerobic conditions and is mainly divided into oxidation and reduction pathways. In the oxidation pathway, under anaerobic conditions, glycerol, with NAD^+ as a coenzyme, is catalyzed by glycerol dehydrogenase (GDH) to form dihydroxyacetone (DHA). Adenosine triphosphate (ATP) and 2-hydroxyacetone kinase (DHAK) work together to convert DHA into dihydroxyacetone phosphate (DHAP). DHAP is further oxidized to produce metabolites such as ethanol, 2,3-butanediol, and acetic acid, as well as ATP and $NADH_2$. In the reduction pathway, under the action of glycerol dehydratase (GDHt), glycerol produces 3- HPA is then reduced to 1,3-PDO under the action of 1,3-PDO dehydrogenase (PDOR), consuming the $NADH_2$ produced in the oxidation pathway, maintaining the balance of redox substances in the cell. The enzymes GDH, GDHt, and PDOR play a key role in the formation of 1,3-PDO. Moreover, the concentration of glycerol affects the activity of the enzyme, among which GDHt is the rate-limiting enzyme for glycerol consumption and 1,3-PDO formation.

Institutions such as Tsinghua University, East China University of Science and Technology, Dalian University of Technology, Shandong University, Jiangnan University, Nanjing University of Technology, Southeast University, Shenyang Agricultural University, Anhui Science Park (Group) Co., Ltd. have carried out research on strain selection, genetic modification, fermentation process optimization, efficient reactor development, refining and purification, and process integration around the route of fermentative production of 1,3-PDO using glycerol as a raw material. During the "Tenth Five-Year Plan" period, Professor Liu Dehua of Tsinghua University cooperated with Heilongjiang Chen Energy Bioengineering Co., Ltd. to build a production line for the fermentation of glycerol to produce bio-based 1,3-PDO with a capacity of 2500 tons/year, which has now been expanded to 20,000 tons/year; Professor Liu Dehua's newly developed process for the co-production of biodiesel and 1,3-PDO by the all-biological method has been industrialized at Suzhou Suzhen Bioengineering Co., Ltd. of Shenghong Group, with a biobased 1,3-PDO production line of 20,000 tons/year. Professor Gong Heng of East China University of Science and Technology cooperated with Huamei Bioengineering Co., Ltd. to build a biobased 1,3-PDO production line with a capacity of 20,000 tons/year. Professor Xiu Zhilong of Dalian University of Technology has completed a thousand-ton production trial using probiotic butyric acid Clostridium to ferment glycerol to produce 1,3-propanediol, and is cooperating with enterprises to build a production line with a capacity of 15,000 tons/year.

2.5 Preparation of Biobased Polybutyramide (PA4) Monomer

With the increasing scarcity of petroleum resources and the gradual increase in environmental pollution, the demand for green degradable materials is increasing, and polybutyramide (PA4) as a degradable material has attracted attention. The polymerization monomer of polybutyramide is butyramide. At present, butyramide is mainly based on petroleum-based raw materials, prepared by chemical methods through high-temperature and high-pressure reactions, and the cost is very high, so the production cost of butyramide is also high. Due to the limitation of monomer sources, PA4, an excellent synthetic fiber, has not yet been industrialized worldwide. The production of biobased polybutyramide using biobased γ-aminobutyric acid as a raw material has attracted attention in recent years. Through biological transformation, biobased raw materials (starch, Jerusalem artichoke, straw hydrolyzed sugar, etc.) are fermented into glutamic acid, which is then converted into γ-aminobutyric acid by glutamic acid decarboxylase, and γ-aminobutyric acid is cyclized to form the monomer butyramide of polybutyramide, and finally biobased polybutyramide is obtained through anionic ring-opening polymerization (De et al. 2019). The biobased route of polybutyramide provides a feasible path for its large-scale low-cost manufacturing, and its monomer biobased butyramide and precursor γ-aminobutyric acid are the keys to affecting the industry of polybutyramide.

2.5.1 Polybutyramide and Its Properties

Polybutyrolactam, also known as polyamide 4 (PA4), is a semi-transparent or milky white thermoplastic resin obtained by anionic ring-opening polymerization of butyrolactam. It is a new type of polyamide (also known as nylon) polymer compound; its molecular structure is shown in Fig. 2.5. The relative density of PA4 is 1.22–1.24, and the melting point is 260–265 °C. It is soluble in zinc chloride or other inorganic salt solutions at room temperature, and can also be dissolved in superheated water. It undergoes hydrolysis at 100 °C in 0.1 mol/L sodium hydroxide or hydrochloric acid.

The structure of PA4 is mainly composed of amide bonds and secondary methyl groups. The secondary methyl group is a hydrophobic group with excellent hydrophilicity. Its hygroscopicity is close to natural cotton and silk, and it has anti-static characteristics. PA4 has better thermal stability than other polyamides (nylons) and also has excellent wear resistance and mechanical properties. Artificial leather made from PA4 is elastic, porous, and does not generate static electricity. It can be used to prepare synthetic fibers, artificial leather, synthetic paper, etc. Compared with PA4, nylon 610, 1010, 12 and other polymers with more than seven secondary methyl

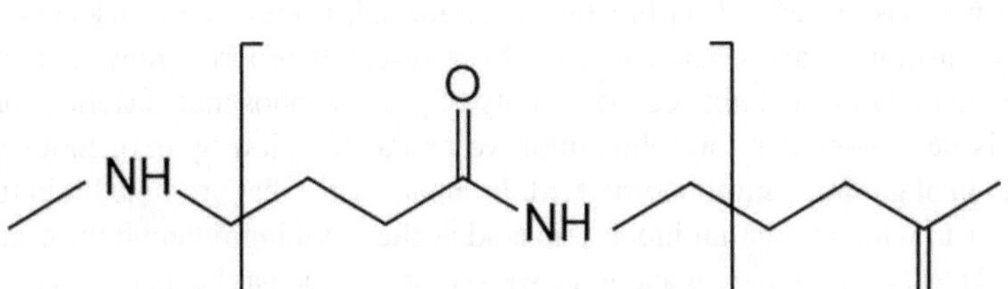

Fig. 2.5 Structure of polybutyrolactam

groups have strong hydrophobicity and extremely poor water absorption; nylon 2, 3 and other polymers, although they have excellent hygroscopicity, their processability is poor. Considering the hygroscopicity and processability of the materials, among all nylon products, PA4 fiber is closer to natural fiber than other synthetic fibers, so it can replace cotton and silk fibers to meet human needs (De et al. 2019).

General polyamide products, such as nylon 6, nylon 66, cannot naturally degrade in the natural environment. However, PA4 has excellent biodegradability. It can naturally degrade in marine environments and soil, and is a typical biodegradable polymer material, so it has great potential application prospects. Although PA4 has many attractive material characteristics, there are also some limitations that need to be overcome. For example, the thermal decomposition temperature of PA4 is very close to the melting point, which makes it difficult for PA4 to be processed and applied through thermoplastic molding processes. It is expected to improve its thermal processing performance through copolymerization, blending, or end-group derivatization or modification. However, since PA4 is insoluble in common organic solvents suitable for anionic polymerization (such as toluene and tetrahydrofuran), the above-mentioned synthetic modification is somewhat difficult, and many properties of polybutyrolactam have not yet been studied and explored.

2.5.2 γ-Aminobutyric Acid and Its Biosynthesis

γ-Aminobutyric acid (GABA), also known as 4-aminobutyric acid, is a naturally occurring, four-carbon nonprotein amino acid with the chemical formula $C_4H_9NO_2$, and a molecular weight of 103.1 Da, with a melting point of 203 °C. As an active amino acid, GABA mainly exists in organs such as the brain, spinal cord, and liver of animals, and has important functions in the brain and spinal cord of mammals. It is an important inhibitory neurotransmitter in the central nervous system.

Gamma-aminobutyric acid plays a crucial role in information transmission in the central nervous system of mammals, with approximately 30% of central nervous synapse neurotransmission mediated by gamma-aminobutyric acid and its synaptic target GABA-A receptor. The GABA-A receptor can regulate the effects of various medicinal drugs and is the target of barbiturates, benzodiazepines, and alprazolam. The GABA-A receptor also has a significant impact on the mental activity of the central nervous system. If its function is abnormal, it may cause diseases such as anxiety, insomnia, and even epilepsy. New research reports show that gamma-aminobutyric acid can enhance the activity of glucose phosphate esterase, promote brain tissue metabolism, and thus improve neural function by exogenous supplementation of gamma-aminobutyric acid. In Japan, drugs that promote brain function metabolism with gamma-aminobutyric acid as the main ingredient have been introduced. They are used to treat and improve symptoms such as headaches and tinnitus caused by cerebral infarction and can also be used to improve brain cell function disorders (Chae et al. 2017).

Gamma-aminobutyric acid has physiological activities such as lowering blood pressure, improving cerebral blood circulation, and calming nerves. It is an excellent raw material for medical drugs and health products. Japan was the first country to start researching the application of gamma-aminobutyric acid in food, and the most representative product is the tea beverage Gabaron. China also approved gamma-aminobutyric acid as a new resource food (now renamed as new food raw material) in 2009, and it is now widely used in flavor drinks, jam pastries, biscuits, and seasoning ingredients. In addition, high-concentration gamma-aminobutyric acid beverages produced by probiotics such as lactic acid bacteria and food ingredients rich in gamma-aminobutyric acid can be directly used for a daily diet to lower blood pressure, enhance brain function, and improve liver function.

1. *Microbial Fermentation Method*

The biological preparation of gamma-aminobutyric acid is mainly obtained through plant enrichment and microbial fermentation. Microbial fermentation and immobilized enzyme technology are currently the mainstream methods for enriching gamma-aminobutyric acid through microbial methods (Qinlu et al. 2008). It was first produced using *E. coli* as the production strain, and the crude product of gamma-aminobutyric acid obtained was purified for chemical production or material preparation. However, if it is to be used in food, the use of *E. coli* poses safety issues. So far, microorganisms such as lactic acid bacteria, *Aspergillus*, and yeast have been reported for the production of gamma-aminobutyric acid and have become a research hotspot. The industrialized method is to obtain glutamate decarboxylase (GAD, EC4.1.15) through microbial fermentation, using glutamate (GA) or its sodium salt or glutamate-rich substances as substrates, and then GAD catalyzes the decarboxylation of glutamate and other substrates to produce GABA. Many microorganisms can produce gamma-aminobutyric acid, such as brewing yeast, lactic acid bacteria, *E. coli*, and *Monascus*. Through the modification and enhancement of the strain and the optimization of the culture conditions, the ability of microorganisms to produce gamma-aminobutyric acid can be greatly improved (Jun 2007).

Another method is to prepare γ-aminobutyric acid through immobilized enzyme technology. Immobilized enzymes not only have the specificity and efficiency of enzymes, but also have the advantages of being recyclable and reusable. The specific process of this method is to use microorganisms such as *Escherichia coli* and brewing yeast as expression hosts, construct genetically engineered bacteria with high expression of glutamate decarboxylase, extract and purify glutamate decarboxylase after fermentation, obtain immobilized enzymes with appropriate immobilization methods, and use them for the transformation and preparation of γ-aminobutyric acid. This method has high conversion efficiency, few by-products, no endotoxins, less wastewater, and high purity of γ-aminobutyric acid, making it an excellent choice for the preparation of high-purity γ-aminobutyric acid (Shucong 2010).

Although the research on γ-aminobutyric acid in China started relatively late, it has developed very rapidly, especially in the fields of biofermentation and enzyme transformation, and the current technical level has reached the international leading level. Jiang Bo, Xu Jianjun, and others from Jiangnan University in China have

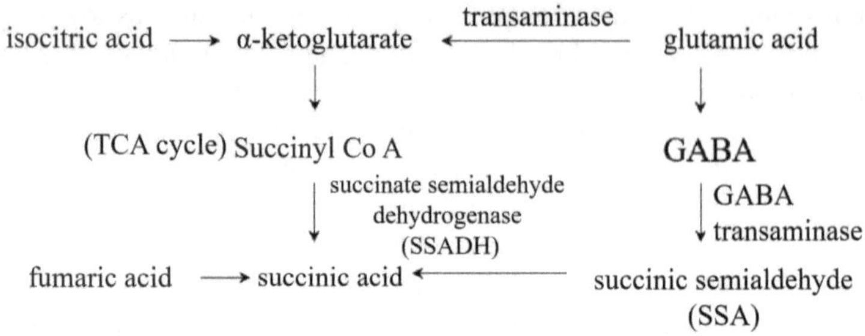

Fig. 2.6 GABA shunt. (Reprinted with permission from Zhejiang Agricultural Science, 2010, 2: 225–229 (Hongmiao et al. 2010))

conducted research on biofermentation production, using lactic acid bacteria or a combination of lactic acid bacteria and yeast as the bacterial strain, using L-glutamic acid sodium as the substrate, adding carbon source, nitrogen source, and inorganic salts to form the fermentation medium, and using the fermentation method to biotransform and prepare γ-aminobutyric acid. After fermentation, the mass concentration of γ-aminobutyric acid in the fermentation liquid is as high as 300–500 mg/100 mL. In addition, Liu Qing from Jiangnan University (Qing et al. 2004) and Xu Dongyun from Zhejiang University of Technology (Dongyun et al. 2006), through the screening of food safety grade (GRAS) lactic acid bacteria, obtained a strain that can produce a high yield of glutamate decarboxylase, and optimized the fermentation medium and fermentation conditions of this strain; the content of γ-aminobutyric acid in the fermentation liquid can reach more than 3.10 g/L.

The main pathway for the synthesis of γ-aminobutyric acid is the decarboxylation of L-glutamic acid (L-Glu), which is catalyzed by GAD. In some cases, γ-aminobutyric acid can be converted from ornithine and putrescine, but these substances are all produced from glutamic acid, so it can be said that glutamic acid is the only source of γ-aminobutyric acid. In bacteria and the brains of mammals, γ-aminobutyric acid first undergoes transamination with α-ketoglutarate to form succinic semialdehyde (SSA) and glutamic acid under the catalysis of γ-aminobutyric acid transaminase (7-aminobutyrate transaminase), then succinic semialdehyde is oxidized to form succinic acid entering the tricarboxylic acid cycle under the action of succinate semialdehyde dehydrogenase (SSADH). These reactions, together with the decarboxylation reaction of glutamic acid catalyzed by GAD, constitute another pathway for α-ketoglutarate to be oxidized to succinic acid, known as the GABA shunt (Zhang et al. 2016), as shown in Fig. 2.6.

2. *Plant Enrichment Method*

As a nonprotein natural amino acid, γ-aminobutyric acid is widely distributed in various plants. It is found in the seeds, rhizomes, and tissue fluids of plants such as ginseng, beans, and traditional Chinese herbs. It is also present in the root nodules of leguminous plants and the phloem of plant xylem. Moreover, in many plants such as large-leaf tea trees, γ-aminobutyric acid exists at high concentrations (0.03–32.5 μmol/g), exceeding many protein amino acids. The plant enrichment method is to enrich γ-aminobutyric acid through the stress metabolism of plant tissues. Under the stress of adversity, the content of γ-aminobutyric acid in the plant body will significantly increase. With this characteristic, a certain amount of γ-aminobutyric acid can be enriched in plants. There are two ways for plants to enrich γ-aminobutyric acid (Zhang et al. 2016): one is to use glutamate decarboxylase to catalyze the decarboxylation of glutamate to produce γ-aminobutyric acid; the other is the degradation of chemical polyamines—putrescine is converted to GABA by the γ-aminobutyraldehyde intermediate in the degradation reaction of diamine oxidase (DAO) Pas. The first method is the main one, which is produced by the stress metabolism of plants in response to external stressors such as temperature, pressure, oxygen concentration, and mechanical damage during growth. It is currently mainly used for the development of plants with high γ-aminobutyric acid content, such as brown rice, tea, and soybeans. The plant enrichment method is safe and environmentally friendly, and can also improve the nutritional value of raw materials. After grains such as rice, beans, wheat, and other economic crops are enriched with γ-aminobutyric acid, they can be directly processed into food, functional beverages, and livestock feed. This production method can better meet the consumer's demand for γ-aminobutyric acid and is one of the future directions for the deep processing of food. However, the current production efficiency is very low and cannot meet market demand.

3. *Chemical Synthesis Method*

The chemical synthesis method is the earliest method used to produce γ-aminobutyric acid. There are mainly two ways: one is to react o-benzoyl isocyanate with γ-chlorobutyronitrile, then reflux the product with concentrated sulfuric acid, and then crystallize and purify to obtain the finished product; the second is to use calcium hydroxide and ammonium bicarbonate to hydrolyze pyrrolidone to obtain GABA (Segal 1984). The chemical synthesis method has severe reaction conditions, high energy consumption, and many by-products. The reagents used are often toxic and corrosive, which does not conform to the concept of green and sustainable development. It is mainly used in the chemical and pharmaceutical fields and is not suitable for application in the food field.

2.5.3 Preparation and Synthesis of PA4 Monomer Butyrolactam

Currently, the monomer used for ring-opening production of PA4 is butyrolactam (BL). Butyrolactam, also known as 2-pyrrolidone or pyrrolidone, is an organic compound with the molecular formula C_4H_7NO. It is a colorless crystal that can be used as a solvent and an intermediate in organic synthesis. It has weak alkalinity and can form hydrochloride with hydrochloric acid; it also has weak acidity and can form sodium salt with sodium hydroxide (Bo 1990). Butyrolactam can be used to prepare PA4 and vinyl pyrrolidone, and it is also an important industrial solvent and raw material for pharmaceuticals and chemicals. It is commonly used as a solvent for animal injections, a component of active pharmaceutical ingredients, an optical auxiliary solvent for water-based ink formulations, a process solvent copolymer for membrane filters, and a floor polish. The annual consumption of butyrolactam worldwide is 150,000 tons, while the domestic annual production is less than 500 tons. With the development of various applications, butyrolactam has become a product with huge commercial potential.

1. *Chemical Synthesis Method*

Most lactams are synthesized from fossil raw materials. For the synthesis of various lactam compounds, a series of effective methods have been developed, among which atom transfer radical cyclization (ATRC) is a powerful means to construct lactams (Segal 1984). The industrial production of butyrolactam mainly uses chemical methods to produce it from fossil-based raw materials. There are several production processes: ① Tafel first synthesized 4-butyrolactam by electrolytic reduction of succinimide in the laboratory in 1907. Due to high electricity consumption, low product yield, and difficult-to-obtain raw materials, this method has not been industrialized; ② The Reppe method (Bo 1990) uses acetylene and formaldehyde as raw materials, hydrogenates under high temperature and pressure to produce 1,4-butanediol, dehydrogenates to form 4-butyrolactone, and then ammonolysis to produce butyrolactam. The Reppe method is the earliest industrialized production method. Both the American General Aniline and Film Company and Germany's BASF Company have used this route to produce 4-butyrolactam (Bo 1990); ③ Maleic anhydride method (Hui and Rui 2011; Jean 1952) , this method is divided into one-step and two-step methods. The American Petrochemical Company uses the one-step method, using maleic anhydride (hereinafter referred to as maleic anhydride) with hydrogen and chlorine, heating and pressurizing to obtain butyrolactam in one step. Mitsubishi Chemical Company of Japan uses the two-step method, catalyzing maleic anhydride to hydrogenate to produce 4-butyrolactone, and then ammoniating to produce 4-butyrolactam; ④ Methyl or ethyl acrylate reacts with hydrogen cyanide to produce cyanoacrylic acid methyl or ethyl ester, and then hydrogenates to produce butyrolactam; ⑤ Butyrolactam can also be prepared from acrylonitrile or butadiene as raw materials. These methods all require the use of nonrenewable fossil resources as raw materials, and butyrolactam is produced through high-temperature and high-pressure reactions, making the production cost of butyrolactam high.

2. *Microbial Synthesis Method*

With the development of synthetic biology technology, the production of chemicals and materials from renewable resources using microorganisms is becoming increasingly important for sustainable chemical industry. Exploring the preparation of butyrolactam by biological fermentation is an effective alternative to overcome the various disadvantages of the current production by chemical methods from fossil raw materials.

The Bioengineering Research Institute of the University of California, Berkeley, reported a two-step biosynthetic pathway for glutamic acid in 2016. In this scheme, by fusing the *E. coli mutant GadB_ΔHT* (expressing the variant of glutamate decarboxylase) and ORF27-MBP (expressing the variant of 2-pyrrolidone synthase), through ORF27, an auxiliary enzyme in the biosynthetic cluster of Streptomyces aizunensis, the γ-aminobutyric acid activation step is performed under mild fermentation conditions to form butyrolactam, achieving the purpose of direct synthesis of butyrolactam using *E. coli*. This is also the first time that ORF27 has been coupled with glutamate decarboxylase in *E. coli,* using recombinant *E. coli* strains to directly synthesize butyrolactam. When cultured in ZYM-5052 medium containing 9 g/L L-glutamic acid, about 7.7 g/L L-glutamic acid was converted into 1.1 g/L butyrolactam within 31 h, achieving a molar yield of about 25% (Youn 2014). Figure 2.7 shows the two-step enzymatic process of butyrolactam biosynthesis from glutamic acid.

In 2017, Sang Yup Lee and others from the National Laboratory of Biomolecular Engineering in Korea reported a solution for the de novo synthesis of butyrolactam using glucose as a substrate (Chae et al. 2017), constructing a new efficient platform metabolic pathway for the production of butyrolactam. This pathway uses γ-aminobutyric acid as a precursor, including two steps, namely the activation of γ-aminobutyric acid catalyzed by the β-alanine CoA transferase (Act) of Clostridium propionicum, followed by the spontaneous cyclization of the activated γ-aminobutyric acid to form butyrolactam. Specifically, the natural glutamate decarboxylase GadB gene was mutated into the mutant GadB E89Q Δ452–466 to enhance the pH range of the enzyme activity and was overexpressed together with the Act gene (encoding the ω-amino acid cyclase β-alanine CoA transferase) to create a complete metabolic pathway. The final supplement batch fermentation experiment of this experiment produced 54.14 g/L of butyrolactam in glucose basic medium, with a productivity of 0.58 g/L/h and a yield of 0.12 g/g glucose.

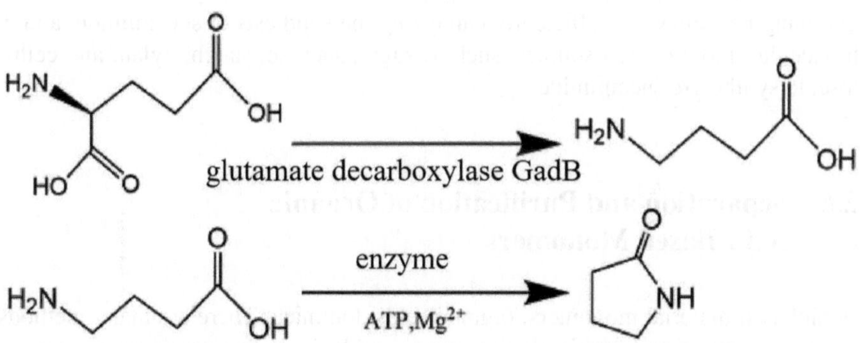

Fig. 2.7 Glutamic acid decarboxylation to form γ-aminobutyric acid (GABA) and then enzymatically cyclized to butyrolactam

In 2020, Jiangnan University published a research result in "Food and Fermentation Industries" for the first time, reporting a butyrolactam synthesis pathway established in glutamic acid bacillus. This pathway first blocks the L-arginine synthesis pathway by knocking out the N-acetylglutamate kinase gene (*argB*), prompting more glucose to flow to L-glutamate; then by expressing, the glutamate decarboxylase (Gad) mutant that converts L-glutamate into γ-aminobutyric acid is obtained to synthesize γ-aminobutyric acid recombinant bacteria. At the same time, heterologous expression of the optimized N-terminal RBS of *Anaerotignum propionicum*. The coenzyme A transferase from the source achieved the conversion of γ-aminobutyric acid to succinimide. The final constructed *C. glutamicum EAGAN2* recombinant strain accumulated (8 ± 0.3) g/L of succinimide after 72 h of fed-batch fermentation in a 5 L fermenter. This study also realized the one-step synthesis of succinimide using Corynebacterium glutamicum with cheap raw material glucose as the substrate, providing a new idea for the production of succinimide using renewable resources (Zhenfeng et al. 2020).

3. *Prospect*

At present, compared with the traditional petroleum-based synthesis of succinimide, the yield of microbial synthesis is low, and the production efficiency is still low, but its advantage lies in the direct synthesis of succinimide through cheap substrate glucose, ensuring the greenness of the substrate and production process. One of the main challenges faced by amide synthesis is the lack of efficient (specific) enzymes that can cyclize ω-amino acids. An efficient enzyme must have high activity, good thermal stability, required substrate selectivity, and high enantioselectivity. Identifying new enzymes by batch testing enzyme activity in various organisms, calculating detection enzymes by using algorithms to analyze biological databases, and enzyme engineering calculation methods have always been the main means of discovering amide cyclase. In the future, the research on the microbial synthesis of succinimide will focus on the following aspects: ① By rationally modifying the CoA transferase to broaden the substrate and enzyme binding pocket to improve the catalytic efficiency of the enzyme; ② adopting irrational strategies to randomly mutate the CoA transferase, high-throughput screening of highly mutated enzymes; ③ mining new enzymes, efficiently catalyzing the synthesis of succinimide; and ④ broadening more carbon sources, such as sugar, sucrose, starch, xylan, and cellulose, to synthesize succinimide.

2.6 Separation and Purification of Organic Acid-Based Monomers

In biobased material monomers, organic acids dominate. There are many methods for the separation and purification of organic acids, including precipitation, extraction, membrane separation, and ion exchange.

2.6.1 Precipitation Method

The precipitation method is a method of separating compounds or ions in a solution by changing the conditions of the solution based on different solubilities. The precipitation method is a classic method for recovering organic acids from fermentation broth. Because this method has the advantages of high selectivity and a low ratio of toxic by-products, it has been used and improved for many years. Choosing the right precipitant for specific precipitation of the target product is the core challenge in the development of this method. Organic acids must form stable solutions in water under certain conditions, and changing the physicochemical parameters of the solution will destroy the stability of its aqueous solution (Liming and Xiulai 2020). Common precipitation methods include salting out, crystallization, organic solvent precipitation, isoelectric point precipitation, polymer precipitation, and affinity precipitation, etc.

Common precipitants for the industrial-scale recovery of organic acids include calcium hydroxide, calcium oxide, and ammonia. In the fermentation process, to precipitate and recover 1 M of organic acid, an equivalent amount of $Ca(OH)_2$/$CaCO_3$ and H_2SO_4 is consumed, but this method produces a low-value by-product—calcium sulfate (Sadare et al. 2021). The specific steps for separating and purifying succinic acid using the calcium precipitation method are: ① The target product fermentation broth is centrifuged or membrane filtered to remove bacterial cells; ② The resulting supernatant is mixed with 3% activated carbon for 1 h for decolorization; ③ The suspension containing activated carbon is filtered, and about 20% of calcium hydroxide solution is added to the filtered clear liquid, the pH of the clear liquid is adjusted to about 13.5, and then incubated at 200 rpm, 39 °C for 20 h; ④ The precipitated calcium succinate is filtered out and washed with water to remove impurities; ⑤ Excess H_2SO_4 is added to the calcium succinate precipitate to form calcium sulfate, and when the pH reaches 2.5, the calcium sulfate precipitate is filtered out; ⑥ The calcium sulfate precipitate is washed twice with distilled water, the residual succinic acid is washed out, and the wash liquid is combined with the filtrate containing succinic acid; ⑦ The filtrate containing succinic acid is vacuum evaporated to remove residual volatile carboxylic acids, and the solution is concentrated to 20% of the initial solution volume; ⑧ The concentrated succinic acid solution is crystallized at 4 °C for 24 h, the crystals are filtered out, then washed with cold saturated succinic acid solution, and then dried at 70 °C for 12 h, finally obtaining solid succinic acid product (Maria et al. 2019).

Crystallization is a method in which substances crystallize out based on their different solubilities in the same solvent at different temperatures. The steps to obtain succinic acid by crystallization include the target product fermentation broth is centrifuged, filtered, decolorized, and acidified; then vacuum evaporated at 60 °C to remove volatile carboxylic acids (such as acetic acid and formic acid); continuous vacuum evaporation until the solution is concentrated to about 20% of its original volume; the concentrated solution is crystallized at 4 °C for 24 h; the resulting slurry is filtered, dried, and the crystals obtained are carefully washed with saturated succinic acid synthesis solution to remove impurities, thus obtaining the succinic acid product. Compared with the calcium precipitation method, the yield and purity of succinic acid obtained by the direct crystallization method are higher (Sadare et al. 2021).

2.6.2 Extraction Method

Extraction is a method that utilizes the differences in solubility or distribution ratio of a substance in two immiscible (or slightly soluble) solvents to achieve separation, extraction, or purification. Depending on the extraction mechanism, extraction methods can be divided into physical extraction and chemical extraction; based on the different phases involved in solute distribution, extraction methods can be divided into solid–liquid extraction and liquid–liquid extraction (Liming and Xiulai 2020). The extraction of target products from fermentation broth is based on the immiscibility and density difference between the solvent used in the fermentation broth and the target compound. The choice of an appropriate solvent depends on the distribution coefficient, toxicity to the microbial production strain, degree of immiscibility between the two solvents, selectivity for the target product, and solvent stability. In addition, the combination of extraction and counter-extraction can improve the recovery rate of organic acids in the fermentation broth (Kurzrock and Weuster-Botz 2010). Hu Manman and others (Manman et al. 2018) used a combination of extraction and counter-extraction to obtain 2,5-furandicarboxylic acid after filtering, activated carbon decolorization, organic solvent extraction, alkali water counter-extraction, and combined organic solvent crystallization from the fermentation broth containing 2,5-furandicarboxylic acid. This method is flexible and has a high recovery rate. In addition to the above extraction methods, researchers have proposed an electrochemically induced extraction method that can recover succinic acid from aqueous solutions. This method mainly changes the pH value through electrochemical operations to complete the separation of succinic acid. Under the condition of a total pH of 5–7, electrochemical extraction can recover succinic acid from an aqueous solution with a concentration of about 50 g/L. Replacing the addition of alkali and acid operations to adjust the pH with an electrochemical operation pH change device can provide a more environmentally friendly separation strategy for biobased succinic acid production. Moreover, the particle size distribution of succinic acid obtained by electrochemical crystallization is uniform, which is more in line with the commercial standards of the product. The electrochemical-driven separation process of succinic acid is based on three principles: creating a pH gradient spatially separated by water electrolysis and ion exchange; performing extraction/counter-extraction reactions; and realizing pH shift crystallization due to the different solubilities of succinic acid at different pHs. In the long run, the application of electrochemical devices provides a method for purifying carboxylic acids using electrical energy. Carboxylic acids like succinic acid are important fine chemical raw materials and are basic units for the production of high-value-added chemicals based on biology. Various studies have shown that the operating pH value chosen in the extraction is very important for the extraction efficiency. Only by acidifying the aqueous solution to a lower pH condition can a higher extraction rate be obtained. However, the pH suitable for product extraction may not match the pH required for optimal fermentation conditions (López-Garzón and Straathof 2014), especially when inorganic salts are formed by acids and bases, and pH adjustment is needed between fermentation and separation, which limits the use of in situ separation methods. Most carboxylic acid fermentations are carried out at temperatures of 35–43 °C, with a required pH of 5.5–7.0. Now, in situ or online separation can be achieved by in situ precipitation, or by adjusting the pH online before extraction, and then further adjusting the pH in the cycle of the fermentation process (Gorden et al. 2017; Gausmann et al. 2019).

2.6.3 Ion Exchange Method

The ion exchange method is a technical means of separating or purifying substances by utilizing the difference in binding ability between the active groups in the ion exchanger and the charged particles in the solution. Ion exchange resin is a solid polymer compound that is insoluble in acids, alkalis, and organic solvents. It is composed of a polymer skeleton, ion exchange groups, and pores. It has good chemical stability and ion exchange capability. According to the classification of active groups, ion exchange resins can be divided into cation exchange resins (containing acidic groups) and anion exchange resins (containing basic groups). They can be further subdivided into strong acid cation exchange resins, weak acid cation exchange resins, strong base anion exchange resins, and weak base anion exchange resins. Different ion exchange resins can be used to separate and purify different organic acids from fermentation broth.

Deng Yu et al. (2017) developed a method to extract adipic acid from fermentation broth. This method involves solid–liquid separation and ultrafiltration concentration of the fermentation broth, followed by ion exchange treatment, to finally obtain adipic acid products. The specific steps include: after solid-liquid separation and ultrafiltration concentration of the fermentation broth containing adipic acid, it is added to a cation exchange resin that has been alkali washed, deionized water washed, acid washed, and deionized water washed to neutrality; after stirring evenly to allow the resin to fully adsorb the fermentation broth, the filtrate is collected by filtration; the collected filtrate is acidified to a pH of acidity with acid; then it is extracted with an organic solvent, and the extract is collected; finally, the remaining liquid from rotary evaporation is crystallized at low temperature (4–8 °C), and after crystallization, it is left to dry at 20–40 °C to obtain the adipic acid product. Succinic acid can be separated using both cation exchange resins and anion exchange resins.

Lin et al. (2010) used a cation exchange resin (sulfonic acid type cation resin Amberlite I) for the separation of succinic acid. The pH of the effluent after the fermentation broth passed through the Amberlite I resin was about 2.0, and then the acidified medium was vacuum evaporated, crystallized, and dried to obtain succinic acid. Song et al. (2007) chose a basic anion exchange resin for the adsorption separation of succinic acid. They found that under pH 4–5 conditions, the strong base 201 resin has good adsorption performance for succinic acid and determined that the most suitable operating conditions for the 201 resin for succinic acid are a solid-liquid ratio of 1(g):20(g), pH 4.14.

2.6.4 Membrane Filtration Method

As early as the eighteenth century, people discovered the existence of semi-permeable membranes in nature. Later, it was further recognized that semi-permeable membranes are widely present in various animals and plants. The membrane in the so-called membrane filtration method is a thin condensed phase in the flowing phase, which can separate the fluid. According to the different driving forces, membrane separation technology can be divided into dialysis, electrodialysis, microfiltration, ultrafiltration, nanofiltration, and reverse osmosis, etc.

The membrane recovery technology of succinic acid is considered to be the most promising technology because the separation membrane has better permeability and elasticity, and has better selectivity for the target product. In addition, the membrane separation technology is environmentally friendly and sustainable, giving it great potential to transition from the laboratory to industrial production. Currently, different researchers have begun to use membrane technology to separate solids and liquids at the molecular level through physical and biochemical processes.

The bipolar membrane electrodialysis (EDBM) technology is a separation technique that combines the advantages of bipolar membranes and electrodialysis, and has been proven to be one of the technologies for recovering and concentrating succinic acid from fermentation broth. This method has been successfully used on a laboratory scale to convert succinate into succinic acid. It uses the latest membrane separation technology, which is environmentally friendly. EDBM technology can recover succinic acid from non-ionized compounds obtained from electrodialysis using suitable ion exchange membranes (Fu et al. 2014). The bipolar membrane consists of two layers; one layer carries cations (cationic membrane), and the other layer carries anions (anionic membrane). The gap between the anion and cation composite layers is about 2 nm. When the electrolyte between the anion and cation composite layers is H_2O, it has the important function of dissociating H_2O molecules into ions (H^+, OH^-), H^+, OH^- can pass through the anionic and cationic membranes, respectively, serving as sources of H^+ and OH^- ions. The desalination membrane used to separate succinic acid is a type of bipolar membrane. When desalination electrodialysis is carried out, the potential separates the ionic compounds from the nonionic compounds in the fermentation broth containing succinate. The positively charged side of the desalination membrane only adsorbs anions (such as succinate) in the fermentation broth, while the negatively charged side only adsorbs cations. The H^+ obtained from the electrolysis of H_2O on the cationic membrane combines with the adsorbed succinate to form succinic acid. The yield of succinic acid obtained by the bipolar membrane electrodialysis method can reach about 77% (Thuy and Boontawan 2017). Although this technology makes the separation process of succinic acid more environmentally friendly, it also has major drawbacks, such as high energy consumption and high cost of membrane materials. In addition, since electrodialysis cannot dissolve some divalent ionic compounds, some fermentation broths that need to use Mg^{2+} and Ca^{2+} to adjust the pH cannot use this method for acidification of succinate and separation of succinic acid. At the same time, so far, the fouling accumulated on the membrane is still another unsolved problem in EDBM technology, but research shows that pre-treating the fermentation broth with ultrafiltration or nanofiltration before using EDBM technology can alleviate the impact of fouling on EDBM.

The ultrafiltration technology for the recovery of succinic acid refers to the technology that uses pressure or concentration gradient as the driving force to separate succinic acid from different components of the mixture through a semi-permeable membrane (Arola et al. 2019). Apart from the difference in the size of the molecules retained, ultrafiltration has no essential difference from reverse osmosis, microfiltration, or nanofiltration. The pressure used in the ultrafiltration process is generally 0.1–1 MPa. Compared with the permeate obtained by centrifugation, ultrafiltration can obtain a clearer and more transparent permeate (Sadare et al. 2021). Ultrafiltration has been proven to be used for clarifying the fermentation broth of succinic acid, and due to its advantages such as environmental protection, low energy consumption, normal temperature operation, and high separation efficiency, it has great potential for application in industrial production.

Reverse osmosis is also a pressure-driven membrane separation process, commonly used for seawater desalination and water purification. It is highly energy-efficient, usually operable at ambient temperature, and can overcome osmotic pressure by using additional pressure. Reverse osmosis membrane is one of the most popular and widely commercialized water treatment technologies (Pandey et al. 2012); however, there are few reports on the recovery of organic acids from fermentation broth by reverse osmosis. Li et al. (2008) used a combination of nanofiltration and reverse osmosis membranes to separate lactic acid from whey fermentation broth. The operation is divided into two steps: the first step uses a nanofiltration membrane to separate the lactic acid aqueous solution from the whey fermentation broth from lactose and cells, and the second step uses a reverse osmosis membrane to concentrate the obtained lactic acid permeate to obtain the finished lactic acid.

Nanofiltration is also one of the methods suitable for the recovery of succinic acid (Choi et al. 2008). Nanofiltration uses a membrane with a pore size of nanometers (1×10^{-9} m) level for filtration, but the removal or separation mechanism of nanofiltration membrane for substances is not pure filtration, but osmosis. Therefore, the nanofiltration membrane is a pressure-driven membrane between ultrafiltration and reverse osmosis. Nanofiltration is widely used in the steps of partial desalination and removal of some non-dissociative compounds (such as glycerol and ethanol) in the fermentation broth. In the nanofiltration process, the permeate flux and the retention rate of each component of the feed liquid depend on the nature of the membrane used and the initial composition and pH of the feed liquid. On the one hand, due to the sieve effect (based on size exclusion), the nanofiltration membrane will retain molecules with a molar mass greater than the retention value of the membrane used, molecules larger than the pore size of the membrane are retained by the membrane, and molecules smaller than the pore size of the membrane can pass through. On the other hand, due to the electrostatic interaction between ions and the charged surface of the membrane, the retention rate of multivalent ions by the nanofiltration membrane is higher than that of monovalent ions. However, the main problem with using pressure-driven membranes is that as the membrane usage time increases, the permeate flux will decrease under constant pressure, and the decrease in permeate volume will increase the time and cost of the separation process, and further reduce its separation effect (Antczak et al. 2019). Polyimide (PI) nanofiltration membranes are used for the recovery of succinic acid in fermentation broth due to their excellent mechanical properties, high chemical corrosion resistance, good thermal stability, and ease of processing. Zaman et al. (2016) The separation of succinic acid from artificially configured simulated fermentation broth was carried out using a polyamide nanofiltration membrane. The effect of different polyamide polymer compositions on the performance of the nanofiltration membrane was investigated. The results showed that when the concentration of organic salts was 10–50 g/L, the nanofiltration membrane with a polyamide polymer content of 15 wt% had the highest selective retention rate for succinate salts, reaching up to 73%. Choi et al. (2008) also used the NF270 nanofiltration membrane to effectively separate succinic acid from wastewater. Regardless of changes in operating pressure and feed concentration, due to the molecular weight of succinic acid being very close to the relative molecular weight of the NF270 membrane used, its retention rate can reach over 90%.

2.6.5 *Molecular Distillation Method*

As early as the eighteenth century, a liquid–liquid separation technology that can be carried out under high vacuum conditions, namely the molecular distillation method, also known as short-range distillation, was discovered. This method has the characteristics of low temperature, high vacuum, short material heating time, and high degree of separation, and is suitable for the separation of high-boiling-point, heat-sensitive, and easily oxidized substances (Park and Zeikus 1999).

Succinic acid recovery distillation technology is a process that separates mixtures based on the different volatilization temperatures of components. The carbonyl groups in most organic acid structures have a strong electron adsorption effect, making their boiling points usually higher than H_2O, thus making the extraction distillation of organic acids more advantageous (Sadare et al. 2021). Fermentation broth containing high concentrations of organic acids will promote the formation of azeotropes during the distillation process. Therefore, when the concentration of organic acids in the fermentation broth is relatively low, distillation can usually effectively separate the organic acids. In this case, continuous reaction distillation can be used to reduce the formation of azeotropes (Maria et al. 2019). Nylon acid (containing a large amount of succinic acid and glutaric acid) will be transformed into a system that is easier to separate after esterification or anhydride formation, and then the components can be separated by vacuum distillation; succinic acid and glutaric acid can be obtained by esterification of nylon acid, anhydride, and then separated by vacuum distillation (Tian and Jianhua 2017).

The extraction and purification process of lactic acid is the key to obtaining high-purity lactic acid, which is of great importance to the entire polylactic acid (PLA) production process. The crude lactic acid product produced by microbial fermentation contains a variety of impurities, usually including acids (formic acid, acetic acid), alcohols (methanol, ethanol), esters, metal elements, and a small amount of sugars or other nutrients. These residual impurities have a great impact on the subsequent preparation of PLA and product performance. For example, the presence of acidic substances (such as acetic acid and propionic acid), alcoholic substances (such as methanol and ethanol), and ester substances in lactic acid will cause lactic acid products to have an off-flavor, while the presence of sugar impurities makes PLA products more prone to discoloration. The traditional extraction process uses a neutralization method to extract lactic acid, that is, first adding an excess of calcium carbonate to the fermentation broth to neutralize lactic acid to form calcium lactate, then filtering and separating, and using strong acids such as sulfuric acid to dissolve calcium lactate to obtain dilute lactic acid. The dilute lactic acid is then filtered through activated carbon, evaporated and crystallized, and other concentration and purification processes to obtain lactic acid products (Demao et al. 2019).. In addition, there are many purification techniques that can be used to purify lactic acid, and the specific needs of the production line can be selected in actual production (Lee et al. 1998).

2.7 Intelligentization of Fermentation Process

The biotechnological fermentation process involves two types of reactor systems, the cellular bioreactor system and the large-scale cellular bioreactor system. The genes, proteins, and metabolic products in the cellular metabolic process are massive, and it can be said that the cell itself is a complex reactor system. The key to achieving industrial fermentation optimization and amplification is to study the engineering of these two reactor systems. The life activities of microorganisms are the main body of this system, and understanding the life activities of living cells in the industrial biological fermentation process is the basis for achieving efficient biological manufacturing. The cellular reaction process is a complex system with multiple scales, multiple inputs, and multiple outputs, such as gene scale, cell scale, and reactor scale. To achieve the optimization and amplification of the biological process, it is necessary to carry out cross-scale detection and regulation of the biological reaction process. Because the regulation network of living cells is very

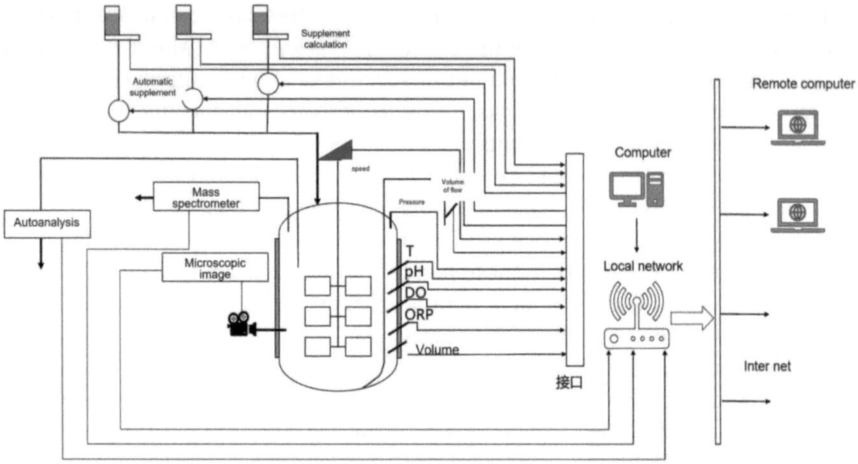

Fig. 2.8 Bioreactor detection parameters with computer system. (Reprinted with permission from Zhang Siliang, Multi scale microbial process optimization, 2003 (Siliang 2003)). *T* temperature, *pH* acidity and alkalinity, *DO* dissolved oxygen, *ORP* oxidation–reduction potential; *P* tank pressure, *LAN* local area network

complex, macro process control technology is difficult to accurately control the micro life activities of living cells.

Intelligent manufacturing is a new type of production method based on the deep integration of a new generation of information and communication technology and advanced manufacturing technology, running through all aspects of design, production, management, service and other manufacturing activities, with self-sensing, self-learning, self-decision-making, self-execution, and self-adaptive functions. The integration and development of intelligent manufacturing and the biotechnological fermentation industry can promote the intelligent and informatization of the fermentation process. In the industrial biological manufacturing process, a database of multi-scale cell physiological metabolism detection data and parameter change information is established, and big data analysis is carried out; modeling, deep learning, and data mining of the fermentation process are carried out, real-time intelligent analysis, diagnosis and precise control of the biological process are realized, thus achieving intelligent manufacturing of the fermentation process (Fig. 2.8) (Siliang 2003).

2.7.1 Fermentation Process Parameters

The detection of fermentation process parameters can understand the diversity, time variation, correlation coupling, and uncertainty between various parameters of the fermentation process, understand the complex fermentation process, obtain the reaction characteristics of the cell scale, molecular scale, and engineering scale, and distinguish different scale problems through analysis. Through the detection and

Table 2.2 Fermentation process parameters

Direct parameters			*Indirect parameters*
Temperature Pressure Power input Ventilation flow rate Foam level Feeding rate Substrate Culture medium weight Culture medium volume Bioheat Appearance of the culture medium	pH Redox potential Dissolved oxygen concentration Dissolved CO_2 concentration Exhaust O_2 partial pressure Exhaust CO_2 partial pressure Other exhaust components	Component concentration Sugar Nitrogen Precursor Inducer Product Intermediate metabolites Metal ions Dehydrogenase activity Various enzyme activities Intracellular components	Oxygen uptake rate (OUR) CO_2 release rate (CER) Respiratory quotient (RQ) Total oxygen utilization Volumetric oxygen transfer coefficient (K_{La}) Cell concentration (X) Cell growth rate Specific growth rate (μ) Cell yield ($Y_{a/b}$) Sugar utilization rate Oxygen utilization rate
Viscosity Accumulated consumption Substrate Acid Alkali Defoaming agent Cell count Bubble content Bubble surface area Surface tension		Proteome Transcriptome Metabolome	Specific substrate consumption rate Precursor utilization rate Product quantity (P) Specific productivity

Reprinted with permission from Zhang Siliang, Multi scale microbial process optimization, 2003 (Siliang 2003)

related analysis of parameters, the fermentation process can be further optimized, ultimately improving product yield.

The fermentation process data acquisition system mainly consists of sensors, computer hardware, and other peripheral systems. According to the data acquisition method, the fermentation process parameters can be divided into direct parameters and indirect parameters. Direct parameters refer to parameters that directly convert nonelectrical changes into electrical changes through sensors and deliver them to the computer for data acquisition in real time. Indirect parameters are parameters of various reaction process characteristics calculated from some direct parameters (Table 2.2) (Siliang 2003).

2.7.2 Detection Technology

1. *Process Exhaust Gas Mass Spectrometer*

The control of industrial-scale fermentation processes is mainly based on the chemical and physical analysis of process control samples. In order to effectively control the process, it is necessary to know exactly when and where changes occur in the process, which creates a demand for fast, accurate, and high-throughput analysis methods. Offline methods are time-consuming and prone to errors, and usually cannot meet these requirements. Therefore, various online analysis methods, such as liquid chromatography (HPLC), infrared spectroscopy (IR), gas chromatography (GC), and mass spectrometry (MS), are increasingly used in fermentation monitoring (Tarkiainen et al. 2005).

The process exhaust gas mass spectrometer can realize the real-time online detection of the components of the fermentation process exhaust gas, such as N_2, O_2, and CO_2, which is of great significance for understanding the metabolic activity of cells during the fermentation process. A mass spectrometer mainly consists of four parts: a sampling system, an ion source, a preset magnetic field, and a detector. Wang Ranming and others (2013) successfully applied the process exhaust gas analysis mass spectrometer and corresponding processing software in the lactic acid fermentation process, used the exhaust gas analysis mass spectrometer to control the oxygen supply level of the oxygen-consuming fermentation process, and successfully realized the online control of the yield and quality of L-lactic acid fermentation.

2. *Cell Size Detection Technology*

The distribution of cell size can provide valuable information for the development, optimization, and control of processes, which are closely related to the cell cycle and specific production rate. By observing changes in cell size, it is possible to monitor whether cells have been infected by viruses during the cultivation process. Typically, the distribution of particle size is determined by offline methods, which include a series of manual steps such as sample dilution and breaking up of aggregates. These steps may greatly affect the accuracy of the measurement data. However, performing online particle size distribution analysis directly in the bioreactor is more conducive to obtaining real-time cell growth information and avoiding interference caused by external sample processing operations. There is also a certain relationship between the frequency of online capacitance and the changes in cell size distribution. The partial least squares (PLS) model can be used to predict the distribution of cell diameter and size. After applying an electric field to the biological cell suspension, the cells become polarized due to the insulating nature of the cytoplasmic membrane. The ions contained in the suspension medium and the cytoplasm will move toward the electrode of the opposite charge, and when they are blocked by the intact membrane, a dielectric double layer will form on the cell surface. Therefore, at this time, the biological cells are like miniature capacitors in the electric field. Measuring the capacitance on alternating currents of different frequencies will produce a spectrum known as "β dispersion."

Complete polarization of the cell membrane will produce a low-frequency platform, within the range of β dispersion, starting from the low-frequency platform, the capacitance is decreasing. When the excitation frequency is low, the cells have enough time to fully polarize, and the cell suspension solution has a high capacitance. However, as the excitation frequency increases, the incomplete polarization

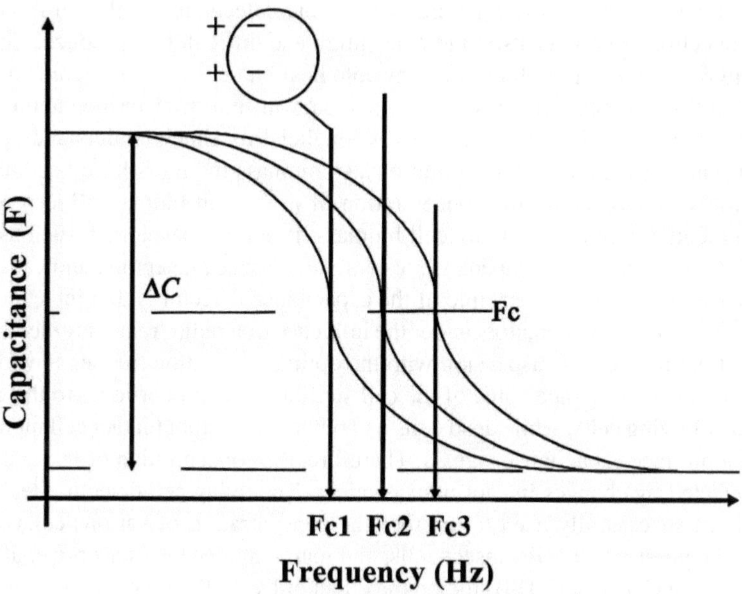

Fig. 2.9 Principle of capacitance measurement at several frequencies. (Reprinted with permission from IFAC Proceedings Volumes, 2007, 40(4): 277–282. © 2007 International Federation of Automatic Control (Henry et al. 2007))
ΔC function of biomass concentration, *Fc* critical frequency

of the cells leads to a decrease in the capacitance of the cell suspension solution. The high-frequency platform is produced when the excitation frequency is too high for the cells to polarize, and the suspension medium or liquid becomes polarized. As shown in Fig. 2.9, ΔC increases proportionally with biomass, while the critical frequency (Fc, the frequency at which half of the polarization rate is completed, the capacitance is half between the high and low platform values) mainly depends on the cell diameter and conductivity (Henry et al. 2007). As the cell size increases, the low frequency of β dispersion will increase, eventually leading to a decrease in Fc. In addition, a decrease in cell size also reduces the low-frequency capacitance of β dispersion, thereby increasing Fc (Li et al. 2014). Ansorge et al. (2010) and others have applied this concept to the production of adenovirus vectors, because in adenovirus vectors, changes in cell size are very important for measuring the success rate of infection. This technology can not only be used to predict online whether the virus has successfully infected, but also can be used to optimize the process of the virus infecting cells.

3. *Live Cell Detection Technology*

In biological processes, the biomass of living cells is an important physiological parameter, closely related to cell growth, metabolism, and productivity. Dry cell weight (DCW), optical density (OD), and mycelial growth amount (PMV) are routine offline detection indicators of biomass. However, these detections are time-consuming and cannot accurately reflect the viability of microorganisms. Although the number of colony-forming units (CFU) only detects the number of living cells, this detection method is also time-consuming and difficult to reproduce. Currently, monitoring and controlling the physiological state of microorganisms using advanced instruments has become mainstream in industrial fermentation. Online detection of physiological parameters is beneficial for timely understanding of cell status and changes in culture conditions, so biomass monitors have become effective tools for evaluating the concentration of living cell biomass. Recently, some instruments for detecting living cell biomass have been developed, such as in situ near-infrared analyzers, impedance testers, fluorescence detectors, and capacitance detectors. The working principle of the capacitance detector is that intact cells will polarize into micro-capacitors under the influence of a radio frequency electric field. After treating the cell suspension with the optimal excitation frequency with β dispersion, the capacitance value of the cell suspension is proportional to the concentration of living cells, while dead cells and other cells without intact cell membranes do not produce capacitance signals. Therefore, the concentration of living cells can be detected by changes in capacitance values. The living cell detection technology has been successfully used to monitor the concentration of various cells, such as bacteria, yeast, plant cells, insect cells, and mammalian cells (Ansorge et al. 2010). Mengmeng Cai et al. (2018) studied the effect of live cell online monitoring feeding technology on L-hydroxyproline fermentation. They installed a live cell online detector on a 30 L fermenter, adjusted the feeding strategy according to the number of live cells, analyzed the effect of live cell online monitoring feeding on the sugar addition rate, yield, sugar acid conversion rate, and metabolic flow distribution of L-hydroxyproline fermentation, and finally effectively reduced the accumulation of by-products and improved the yield of L-hydroxyproline using the new feeding strategy.

4. *Near-Infrared Spectrometer*

Near-infrared (NIR) spectroscopic analysis is a nondestructive analysis technique. Compared with other analysis techniques, it requires less sample preparation and/or pretreatment, fast spectral acquisition (one second or shorter), can simultaneously detect and/or quantify multiple components, and can be coupled with fiber optic probes for on-site measurements; thus it is increasingly attracting the attention of the biotechnology industry. Near-infrared spectroscopic analysis is usually carried out in three modes, namely transmittance, reflectance, and transflectance. Transmittance is suitable for low cell density fermentation broth, transparent liquids, and suspensions; reflectance is suitable for high cell density fermentation broth, opaque liquids, solids, and powders; while transflectance combines transmission and reflection modes, suitable for measuring transparent and turbid liquids and semi-solid samples. The spectral information obtained during the fermentation process includes information generated by all near-infrared active components of the fermentation broth. The complexity of near-infrared spectra usually requires multivariate statistical analysis, such as principal component analysis (PCA) and partial least squares (PLS) analysis (PLS can be used for spectral interpretation and modeling). Near-infrared spectroscopy has been used for modeling and monitoring of various microbial biological processes, such as *Staphylococcus xylosus*. Monitoring of biomass, glucose, lactic acid, and acetic acid content during the fermentation process of ES13 strain; *Vibrio cholerae* monitoring of biomass, glucose, and acetic acid content during the production of cholera toxin under the batch feeding culture mode; *Pichia pastoris* monitoring of biomass, glycerol, and methanol content during the production of monoclonal antibodies, etc. (Cruz et al. 2015).

5. *Membrane Inlet Mass Spectrometer*

Membrane inlet mass spectrometry (MIMS) is a specific, sensitive, and convenient method for analyzing volatile organic compounds in water and gaseous samples. It separates volatile organic analytes directly from water or gas samples through a sampling needle equipped with a gas filtering membrane. The organic compounds dissolve in the membrane, evaporate into the mass spectrometer after permeating the membrane, and are then accurately analyzed by the mass spectrometer. The most commonly used membrane type in MIMS is the organosilicon membrane, which has higher selectivity for organic compounds compared to the main components of water or air. The enrichment effect of the organosilicon membrane on organic compounds can reach 10 to 100 times that of water or air. The greatest advantage of MIMS is that it can monitor multiple different fermentation products online at the same time, and all operations of the MIMS online monitoring system can be automated. It can also connect automatic feedback control to the monitoring instrument. However, MIMS also has its disadvantages, such as the lack of chromatographic resolution of components and the inability to obtain multi-component spectra of all analytes passing through the membrane. When analyzing biological samples containing a large number of different compounds, it can also bring certain difficulties to the analysis. Beer fermentation liquid is a very complex mixture, containing sugars, yeast cells, dissolved gases, and a small amount of salts. Despite the complex composition of the beer fermentation liquid, which increases the difficulty of using MIMS to analyze its volatile components, there have been successful reports of using MIMS instruments to monitor the continuous fermentation process of beer (Tarkiainen et al. 2005).

6. *Low-Field Nuclear Magnetic Resonance Instrument*

Low-field nuclear magnetic resonance (LF-NMR), as a convenient method for quantifying fats, has been widely used in industries such as oil exploration, agriculture, and food. The states of hydrogen protons in different samples are different. When the radio frequency pulse emitted by LF-NMR is absorbed by hydrogen protons, it will cause resonance. Different states of hydrogen protons in the sample will produce different relaxation times (in the phenomenon of resonance, after the termination of the radio frequency pulse, the process of hydrogen protons returning to the original equilibrium state), and LF-NMR is based on the above principle to quantify fats. In various samples, carbohydrates and proteins have the shortest relaxation times, at the microsecond level, while the relaxation times of bound water, free water, and fats are relatively longer, at several hundred microseconds, several seconds, and several hundred milliseconds, respectively. The recent development of

NMR permanent magnets has greatly improved the performance of LF-NMR equipment, so LF-NMR instruments can be smaller than traditional NMR instruments and also cheaper. At the same time, it uses a smaller frequency (<60 MHz); these advantages make it possible to connect it to a bioreactor for simple and economical real-time fermentation monitoring in large-scale biological fermentation (Wang et al. 2016). Wang et al. (2016) Utilizing LF-NMR technology, a rapid and accurate method for quantifying the intracellular lipid content of *Chlorella protothecoides* has been established.

7. *Online Microscopic Imaging Technology*

The fermentation production process of secondary metabolites is closely related to the morphological changes of microbial production strains, especially filamentous bacteria such as actinomycetes and molds. Therefore, statistical analysis and real-time recording of cell morphology during fermentation are of great significance for the control of its fermentation process.

Micro-spectrophotometry, a technology that combines microscopy and spectrophotometry, is a technique that uses the light absorption, fluorescence emission, and light reflection characteristics of substances as the basis for measurement, and quantitatively detects important biomolecules in cells. During the fermentation process, this detection technology can provide a basis for observation at the cellular level (Siliang 2003).

2.7.3 Prospects

The biotech industry is still significantly behind other industries in adopting advanced process control (APC), especially in using innovative process analytical technologies (PAT) solutions. In other highly complex industries such as oil and natural gas, semiconductors, and automobiles, automation and lean manufacturing have been better integrated into company practices and culture. Over the past 20 years, the development of first-principle mathematical models simulating complex industrial processes has aided the development and deployment of APC solutions, allowing them to test and validate new control strategies in simulated processes, but not implemented in real processes. The future era of Industry 4.0 envisions a highly intelligent data-driven manufacturing environment that integrates various advanced online process analyses, so for the modern biomanufacturing industry, simulation is a crucial need (Goldrick et al. 2019).

1. *Artificial Neural Networks*

The artificial neural network (ANN) model is an information processing black box model that simulates the structure of brain neural networks. It is established by the combination of input and output data, and reflects the nonlinear relationship between input and output through the connection matrix of neurons in the network.

It can be used for process control and process optimization during biological fermentation (Lei et al. 2021).

ANNs have achieved satisfactory results in many different research fields because they can model highly nonlinear complex processes. ANNs have the advantage of effectively handling noise and strong nonlinear data in data processing and modeling, and ANNs have the ability to generalize, estimate, and learn processes without prior knowledge (PK), which makes this technology very useful in industrial bioproduction processes. In many biological process modeling studies, ANNs are often combined with particle swarm optimization (PSO) to optimize the target output variables of the model. This algorithm uses a swarm of particles scattered on the multi-dimensional domain of the model, allowing particles to find a set of values of input variables to obtain the optimal solution of output variables. The movement of particles in search of the optimal solution in the search space is influenced by their better-known positions and the better positions reached by other particles in each interaction. Wong et al. (2013) The performance of a biodiesel engine was optimized using the particle swarm algorithm, considering fuel economy, emission standards, and engine operating range in the selection of the objective function. Compared to another optimization method, the particle swarm algorithm has a lower standard deviation and uses less computational processing time. Therefore, a framework based on artificial intelligence technology can be constructed, combining ANN and particle swarm algorithm, to obtain better operating conditions for improving bio-yield. This method can be used to improve different bio-production processes, especially those with highly nonlinear behavior (Pereira et al. 2020).

2. *Process Information Sheet*

Fermentation biotechnology research is generating an unprecedented flow of unstructured information, thus creating new process intelligence. This situation makes it difficult for researchers to quickly obtain the latest information, so some researchers hope to develop a Web-based structured fermentation information repository to solve this problem. Fermentation information is stored in the form of a process information sheet (PIS), which is an abstraction of process intelligence behavior in the upstream, production, and downstream stages, rather than a large amount of unstructured information. The repository uses the Transmission Control/Internet Protocol (TCP/IP), and researchers can globally communicate all fermentation processes through the central repository. The new fermentation intelligence submits the PIS standard format to the server, and the information is retrieved from the server in the form of PIS or process intelligence trend (PIT) to clarify possible trends in the fermentation process. A complete PIS includes general information related to fermentation (such as substrates used, expected yield, and microorganisms used), culture medium components, inoculation strains, and preparation of culture medium, fermentation process, product recovery strategy, main observed results, and a small amount of cross-referenced literature about the product or production strain. PIS is not intended to replace traditional publications of fermentation research results, but is a simple and fast format for scientists to exchange fermentation information through a global repository (Kana 2011). PIS allows researchers to quickly obtain more accurate fermentation process information about their target strains or products of interest.

References

Abbott DA, Suir E, Duong GH et al (2009) Catalase overexpression reduces lactic acid-induced oxidative stress in *Saccharomyces cerevisiae*. Appl Environ Microbiol 75(8):2320–2325

Abdel-Rahman MA, Tashiro Y, Zendo T et al (2011) Efficient homofermentative L-(+)-lactic acid production from xylose by a novel lactic acid bacterium, Enterococcus mundtii QU 25. Appl Environ Microbiol 77(5):1892–1895

Abdel-Rahman MA, Tashiro Y, Sonomoto K (2013) Recent advances in lactic acid production by microbial fermentation processes. Biotechnol Adv 31(6):877–902

Ahn JH, Jang YS, Lee SY (2016) Production of succinic acid by metabolically engineered microorganisms. Curr Opin Biotechnol 42:54–66

Ansorge S, Henry O, Aucoin M et al (2010) Monitoring the cell size distribution of mammalian cell cultures using on-line capacitance measurements. Springer, Netherlands

Antczak J, Szczygiełda M, Prochaska K (2019) Nanofiltration separation of succinic acid from post-fermentation broth: impact of process conditions and fouling analysis. J Ind Eng Chem 77:253–261

Arikawa Y, Kuroyanagi T, Shimosaka M et al (1999) Effect of gene disruptions of the TCA cycle on production of succinic acid in *Saccharomyces cerevisiae*. J Biosci Bioeng 87(1):28–36

Arola K, Bruggen B, Manttari M et al (2019) Treatment options for nanofiltration and reverse osmosis concentrates from municipal wastewater treatment: a review. Crit Rev Environ Sci Technol 49(19/24):2049–2116

Bai D, Wei Q, Yan Z et al (2003) Fed-batch fermentation of *Lactobacillus lactis* for hyper-production of L-lactic acid. Biotechnol Lett 25(21):1833–1835

Bai B, Zhou JM, Yang MH et al (2015) Efficient production of succinic acid from macroalgae hydrolysate by metabolically engineered *Escherichia coli*. Bioresour Technol 185:56–61

Balzer GJ, Thakker C, Bennett GN et al (2013) Metabolic engineering of *Escherichia coli* to minimize byproduct formate and improving succinate productivity through increasing NADH availability by heterologous expression of NAD(+)-dependent formate dehydrogenase. Metab Eng 20:1–8

Barre O, Mourlane F, Solioz M (2007) Copper induction of lactate oxidase of *Lactococcus lactis*: a novel metal stress response. J Bacteriol 189(16):5947–5954

Berners-Lee T, Hendler J, Lassila O (2001) The semantic web: a new form of web content that is meaningful to computers will unleash a revolution of new possibilities. Sci Am 284(5):34–43

Biddy MJ, Scarlata C, Kinchin C (2016) Chemicals from biomass: a market assessment of bio-products with near-term potential. National Renewable Energy Lab.(NREL), Technical Report, Golden

Bing H (2019) System metabolic engineering research on succinic acid production from acetic acid by *Escherichia coli*. East China University of Science and Technology, Shanghai

Blankschien MD, Clomburg JM, Gonzalez R (2010) Metabolic engineering of *Escherichia coli* for the production of succinate from glycerol. Metab Eng 12(5):409–419

Bo S (1990) Preparation and application of 4-butyrolactam. Chem Eng 2:17–19

Borowitzka MA (1988) Fats, oils and hydrocarbons. In: Micro algal biotechnology. Cambridge University Press

Boyue Z (2020) Research on whole cell catalysis of 5-hydroxymethylfurfural to synthesize 2,5-furandicarboxylic acid. Dalian University of Technology, Dalian

Branduardi P, Fossati T, Sauer M et al (2007) Biosynthesis of vitamin C by yeast leads to increased stress resistance. PLoS One 2(10):e1092

Burgard A, Burk MJ, Osterhout R et al (2016) Development of a commercial scale process for production of 1,4-butanediol from sugar. Curr Opin Biotechnol 42:118–125

Butler M (2005) Animal cell cultures: recent achievements and perspectives in the production of biopharmaceuticals. Appl Microbiol Biotechnol 68(3):283–291

Chae TU, Ko YS, Hwang KS et al (2017) Metabolic engineering of *Escherichia coli* for the production of four-, five- and six-çarbon lactams. Metab Eng 41:82–91

Chang DE, Jung HC, Rhee JS et al (1999) Homofermentative production of D-or L-lactate in metabolically engineered *Escherichia coli* RR1. Appl Environ Microbiol 65(4):1384–1389

Chang HN, Kim NJ, Kang J et al (2011) Multi-stage high cell continuous fermentation for high productivity and titer. Bioprocess Biosyst Eng 34(4):419–431

Chen Q, Sanglard D, Vanhanen S et al (2005) Candida yeast long chain fatty alcohol oxidase is a type haemoprotein and plays an important role in long chain fatty acid metabolism. Biochim Biophys Acta 2735(3):192–203

Chen J, Zhu X, Tan Z et al (2014) Activating C4-dicarboxylate transporters DcuB and DcuC for improving succinate production. Appl Microbiol Biotechnol 98(5):2197–2205

Chen H, Dong F, Minteer S (2020) The progress and outlook of bioelectrocatalysis for the production of chemicals, fuels and materials. Nature Catalysis 3:225–240

Chengyong X, Jian Z (2002) Research progress in the production of long-chain dicarboxylic acids by fermentation. Adv Bioeng 22:66–69

China Petroleum & Chemical Corporation, Fushun Petrochemical Research Institute of China Petroleum & Chemical Corporation (2013) A method for obtaining high-purity dicarboxylic acid. CN102911036

China Petroleum & Chemical Corporation, Sinopec Yangzi Petrochemical Co., Ltd (2013) Method for extracting and refining long-chain dicarboxylic acid. CN103121935

Choi J, Fukushi K, Yamamoto K (2008) A study on the removal of organic acids from wastewaters using nanofiltration membranes. Sep Purif Technol 59(1):17–25

Choi S, Song CW, Shin JH et al (2015) Biorefineries for the production of top building block chemicals and their derivatives. Metab Eng 28:223–239

Condon S (1987) Responses of lactic acid bacteria to oxygen. FEMS Microbiol Lett 46(3):269–280

Cruz MV, Sarraguça MC, Freitas F et al (2015) Online monitoring of P(3HB) produced from used cooking oil with near-infrared spectroscopy. J Biotechnol 194:1–9

Cui Z, Gao C, Li J et al (2017) Engineering of unconventional yeast *Yarrowia lipolytica* for efficient succinic acid production from glycerol at low pH. Metab Eng 42:126–133

Datta R, Tsai SP, Bonsignore P et al (1995) Technological and economic potential of poly(lactic acid) and lactic acid derivatives. FEMS Microbiol Rev 16:221–231

De Mey M, De Maeseneire S, Soetaert W et al (2007) Minimizing acetate formation in *E. coli* fermentations. Ind Microbiol Biotechnol 34(11):689–700

De W, Liangchen T, Songchao T et al (2019) Synthesis and properties of bio-based adipamide and polyadipamide. J Funct Polym 32(1):110–116

Demao L, Wuxi C, Wei G et al (2019) Industrial bio-fermentation process simulation: progress and development trend. J Bioeng 35(10):1974–1985

Deng Y, Zhao M, Mao Y et al (2017) Extracting adipic acid from fermentation liquor comprises e.g. carrying out solid-liquid separating from the fermentation broth thalli and impurity, collecting the clear solution of the zymotic fluid and carrying out ultra-filtering. CN107382706-A

Ding S, Tan T (2006) L-lactic acid production by *Lactobacillus casei* fermentation using different fed-batch feeding strategies. Process Biochem 41(6):1451–1454

Dong XL, Gabrilovich E, Heitz G et al (2015) From data fusion to knowledge fusion. Proc Vldb Endow 7(10):881–892

Dongyun X, Liping Z, Zhenyu T et al (2006) Isolation and screening of lactic acid bacteria producing γ-aminobutyric acid. Modern Food Sci Technol 3(59-61):64

Donnelly M, Millard C, Stols L (2001) Mutant *E. coli* strain with increased succinic acid production. US Patent, RE37393
Duwat P, Sourice S, Cesselin B et al (2001) Respiration capacity of the fermenting bacterium *Lactococcus lactis* and its positive effects on growth and survival. J Bacteriol 183(15):4509–4516
Fangfang L, Xinmin Z (2018) Discussion on the production process and technological progress of 1,4-butanediol. Chem Manag 27:125–126
Feng X, Shen C, Ji K et al (2017) Production of p-xylene from bio-based 2,5-dimethylfuran over high performance catalyst WO_3/SBA-15. Catal Sci Technol 7(23):5540–5549
Fickers P, Benetti PH, Waché Y et al (2010) Hydrophobic substrate utilisation by the yeast *Yarrowia lipolytica*, and its potential applications. FEMS Yeast Res 5(6-7):527–543
Fu L, Gao X, Yang Y et al (2014) Preparation of succinic acid using bipolar membrane electrodialysis. Sep Purif Technol 127:212–218
Gang R, Yuantong C (2000) Research on the fermentation of dodecanedioic acid. J Biol Eng 16(2):82–86
Gao C, Ma C, Xu P (2011) Biotechnological routes based on lactic acid production from biomass. Biotechnol Adv 29(6):930–939
Gao C, Yang X, Wang H et al (2016) Robust succinic acid production from crude glycerol using engineered *Yarrowia lipolytica*. Biotechnol Biofuels 9(1):179
Garrigues C, Loubiere P, Lindley ND et al (1997) Control of the shift from homolactic acid to mixed-acid fermentation in *Lactococcus lactis*: predominant role of the NADH/NAD+ ratio. J Bacteriol 179(17):5282–5287
Gausmann M, Kocks C, Doeker M et al (2019) Recovery of succinic acid by integrated multi-phase electrochemical pH-shift extraction and crystallization. Sep Purif Technol 240(4):116489
Goffin P, Muscariello L, Lorquet F et al (2006) Involvement of pyruvate oxidase activity and acetate production in the survival of *Lactobacillus plantarum* during the stationary phase of aerobic growth. Appl Environ Microbiol 72(12):7933–7940
Goldrick S, Duran-Villalobos C, Jankauskas K et al (2019) Modern day monitoring and control challenges outlined on an industrial-scale benchmark fermentation process. Comput Chem Eng 130
Gorden J, Geiser E, Wierckx N et al (2017) Integrated process development of a reactive extraction concept for itaconic acid and application to a real fermentation broth. Eng Life Sci 17(7):809–816
Guilin Q, Huan G, Tianxing W (2017) Progress in knowledge graph research. Inf Eng 3(1):4–25
Guojun Y, Zhonghua Y, Li L (2020) Preliminary exploration of the management system and operation mechanism of China's fuel ethanol industry. Brew Sci Technol 307:114–116
Gurvitz A, Hamilton B, Ruis H et al (2001) Peroxisomal degradation of trans-unsaturated fatty acids in the yeast *Saccharomyces cerevisiae*. J Biol Chem 276(2):895–903
Hang B, Yang H, Fang G et al (2018) Central pathway engineering for enhanced succinate biosynthesis from acetate in *Escherichia coli*. Biotechnol Bioeng 115(4):943–954
Henry O, Ansorge S, Aucoin M et al (2007) On-line monitoring of cell size distribution in mammalian cell culture processes. IFAC Proc Vol 40(4):277–282
Hettema EH, Tabak HF (2000) Transport of fatty acids and metabolites across the peroxisomal membrane. Biochim Biophys Acta 1486(1):18–27
Hong G, Chun L, Yutao H et al (2003) The role of botulinum toxin acyltransferase in long-chain dicarboxylic acid metabolism. Chin J Biotechnol 23(6):14–16
Hong G, Ming L, Yingming H et al (2006) Sugar supplementation fermentation experiment of tropical Candida and its recombinant strain producing tridecane dicarboxylic acid. J Process Eng 6(5):814–817
Hongmiao S, Yue T, Huizhong W et al (2010) Synthesis, metabolism and biological function of GABA in plants. Zhejiang Agric Sci 2:225–229
Hossain GS, Yuan H, Li J et al (2016) Metabolic engineering of *Raoultella ornithinolytica* BF60 for production of 2,5-furandicarboxylic acid from 5-hydroxymethylfurfural. Appl Environ Microbiol 83(1):e02312–e02316

Hui Y, Rui G (2011) Research on the synthesis of γ-butyrolactam by Ugi/atom transfer radical cyclization reaction. Org Chem 31(10):1683–1686

Huycke MM, Moore D, Joyce W et al (2001) Extracellular superoxide production by *Enterococcus faecalis* requires demethylmenaquinone and is attenuated by functional terminal quinol oxidases. Mol Microbiol 42(3):729–740

Jantama K, Haupt MJ, Svoronos SA, Ingram LO et al (2008) Combining metabolic engineering and metabolic evolution to develop nonrecombinant strains of *Escherichia coli* that produce succinate and malate. Biotechnol Bioeng 99(5):1140–1153

Jean M (1952) Chemie and technik der acetylen-druck-reaktionen. Anal Chim Acta 6:98–99

Jinggang W, Xiaoqing L, Jin Z (2017) Research progress on the synthesis of bio-based aromatic platform compound 2,5-furandicarboxylic acid. Chem Industry Progr 36(2):672–682

John RP, Nampoothiri KM (2011) Co-culturing of *Lactobacillus paracasei* subsp. *Paracasei* with a *Lactobacillus delbrueckii* subsp. *delbrueckii* mutant to make high cell density for increased lactate productivity from cassava bagasse hydrolysate. Curr Microbiol 62(3):790–794

John AM, Lu Z, Douglas SC (2002) Toward the development of a biocatalytic system for oxidation of p-xylene to terephthalic acid: oxidation of 1,4-benzenedimethanol. J Mol Catal B Enzym 18:147–154

John RP, Nampoothiri KM, Pandey A (2007) Fermentative production of lactic acid from biomass: an overview on process developments and future perspectives. Appl Microbiol Biotechnol 74(3):524–534

Jun H (2007) Biosynthesis of γ-aminobutyric acid (GABA) using immobilized whole cells of *Lactobacillus brevis*. World J Microbiol Biotechnol 23(6):865–871

Kai Y, Matsumura H, Inoue T et al (1999) Three-dimensional structure of phosphoenolpyruvate carboxylase: a proposed mechanism for allosteric inhibition. Proc Natl Acad Sci USA 96(3):823–828

Kamra DN (2005) Rumen microbial ecosystem. Curr Sci 89(1):124–135

Kana E (2011) A repository of intelligence of industrial fermentation (Riif) using web enabled technology. Biotechnol Biotec Eq 25(1):2290–2294

Karich A, Kleeberg SB, Ullrich R et al (2018) Enzymatic preparation of 2,5-furandicarboxylic acid (FDCA)-a substitute of terephthalic acid-by the joined action of three fungal enzymes. Microorganisms 6(1):5

Karp SG, Igashiyama AH, Siqueira PF et al (2011) Application of the biorefinery concept to produce L-lactic acid from the soybean vinasse at laboratory and pilot scale. Bioresour Technol 102(2):1765–1772

Kawamoto S, Nozaki C, Tanaka A et al (1978) Fatty acid beta-oxidation system in microbodies of n-alkane-grown Candida tropicalis. Eur J Biochem 83(2):609–613

Kester AS, Foster JW (1963) Ditermical oxidation of long-chain alkanes by bacteria. J Bacteriol 85:859–869

Kim HO, Wee YJ, Kim JN et al (2006) Production of lactic acid from cheese whey by batch and repeated batch cultures of *Lactobacillus* sp. RKY2. Appl Biochem Biotechnol 132:694–704

Koebmann BJ, Solem C, Pedersen MB et al (2002) Expression of genes encoding F-1-ATPase results in uncoupling of glycolysis from biomass production in *Lactococcus lactis*. Appl Environ Microbiol 68(9):4274–4282

Konings WN, Kok J, Kuipers OP et al (2000) Lactic acid bacteria: the bugs of the new millennium. Curr Opin Microbiol 3(3):276–282

Koopman F, Wierckx N, de Winde JH et al (2010) Identification and characterization of the furfural and 5-(hydroxymethyl) furfural degradation pathways of *Cupriavidus basilensis* HMF14. Proc Natl Acad Sci USA 107(11):4919–4924

Kumar V, Ashok S, Park S (2013) Recent advances in biological production of 3-hydroxypropionic acid. Biotechnol Adv 31(6):945–961

Kurzrock T, Weuster-Botz D (2010) Recovery of succinic acid from fermentation broth. Biotechnol Lett 32(3):331–339

Le Shi L, Yi TG, Shan WW et al (2019) Microorganisms, high IQ, big industry—synthetic biology aids in the efficient intelligent manufacturing of avermectin. Life Sci 31(5):516–525

Lee EG, Moon SH, Chang YK et al (1998) Lactic acid recovery using two-stage electrodialysis and its modelling. J Membr Sci 145(1):53–66

Lee PC, Lee SY, Hong SH et al (2002) Isolation and characterization of a new succinic acid-producing bacterium, Mannheimia succiniciproducens MBEL55E, from bovine rumen. Appl Microbiol Biotechnol 58(5):663–668

Lei YH, Jun ZZ, Xiu LC et al (2016) The discovery, modification and utilization of industrial enzymes in the era of big data. Bio-industry Technol 9(2):48–55

Lei C, Jingyu W, Junchao H et al (2021) Optimization of high-density fermentation process of recombinant *E. coli* pullulanase based on artificial neural network and genetic algorithm. Food. Science 42(10):73–78

Li K (2003) Capacity of various production processes of 1,4-butanediol (BDO) in the world. Petrol Natural Gas Chem Industry 2:106

Li Y, Shahbazi A, Williams K et al (2008) Separate and concentrate lactic acid using combination of nanofiltration and reverse osmosis membranes. Appl Biochem Biotechnol 147(1-3):1–9

Li Z, Lu J, Zhao L et al (2010) Improvement of L-lactic acid production under glucose feedback controlled culture by *Lactobacillus. rhamnosus*. Appl Biochem Biotechnol 162(6):1762–1767

Li L, Wang ZJ, Chen XJ et al (2014) Optimization of polyhydroxyalkanoates fermentations with on-line capacitance measurement. Bioresour Technol 156:216–221

Li Y, Huang B, Wu H et al (2016a) Production of succinate from acetate by metabolically engineered *Escherichia coli*. ACS Synth Biol 5(11):1299–1307

Li Q, Wu H, Li Z et al (2016b) Enhanced succinate production from glycerol by engineered *Escherichia coli* strains. Bioresour Technol 218:217–223

Li Q, Huang B, Wu H et al (2017) Efficient anaerobic production of succinate from glycerol in engineered *Escherichia coli* by using dual carbon sources and limiting oxygen supply in preceding aerobic culture. Bioresour Technol 231:75–84

Li Y, Zhong Z, Hou P et al (2018) Resistance to nonribosomal peptide antibiotics mediated by d-stereospecific peptidases. Nat Chem Biol 14:381–387

Liangquan W (2008) Current status and discussion of research and development of ethylene glycol production technology from non-petroleum route. Shanghai Chem Industry 5:18–22

Limei X, Hai W, Dawei H et al (2000) Chemical synthesis of 1,3-propanediol. Chem Adhesion 1:38–40

Liming L, Xiulai C (2020) Organic acid technology. China Light Industry Press, Beijing, pp 138–157

Lin S (2019) Current status and development trend of domestic terephthalic acid (PTA) industry. Chem Industry Manag 20:1–2

Lin H, Bennett GN, San KY (2005) Metabolic engineering of aerobic succinate production systems in *Escherichia coli* to improve process productivity and achieve the maximum theoretical succinate yield. Metab Eng 7(2):116–127

Lin SKC, Du C, Blaga AC et al (2010) Novel resin-based vacuum distillation-crystallisation method for recovery of succinic acid crystals from fermentation broths. Green Chem 12(4):666–671

Liu X, Feng X, Ding Y et al (2020) Characterization and directed evolution of propionyl-CoA carboxylase and its application in succinate biosynthetic pathway with two CO_2 fixation reactions. Metab Eng 62:42–50

López-Garzón CS, Straathof AJ (2014) Recovery of carboxylic acids produced by fermentation. Biotechnol Adv 32(5):873–904

Maas RH, Springer J, Eggink G et al (2008) Xylose metabolism in the fungus *Rhizopus oryzae*: effect of growth and respiration on L+-lactic acid production. J Ind Microbiol Biotechnol 35(6):569–578

Manman H, Xiao H, Lixiang W et al (2018) Method for extracting 2,5-furandicarboxylic acid from fermentation broth. CN108558799A

Maria A, Anestis V, Harris P et al (2019) Downstream separation and purification of succinic acid from fermentation broths using spent sulphite liquor as feedstock. Sep Purif Technol 209:666–675

Marinho GAS, Alvarado-Morales, Merlin Angelidaki I (2016) Valorization of macroalga *Saccharina latissimaas* novel feedstock for fermentation-based succinic acid production in a biorefinery approach and economic aspects. Algal Res 16:102–109

Mauersberger S, Drechsler H, Oehme G et al (1992) Substrate specificity and stereoselectivity of fatty alcohol oxidase from the yeast *Candida maltosa*. Appl Microbiol Biotechnol 37(1):66–73

Mckenna SM, Leimkühler S, Herter S et al (2015) Enzyme cascade reactions: synthesis of furandicarboxylic acid (FDCA) and carboxylic acids using oxidases in tandem. Green Chem 17(6):3271–3275

Mengmeng C, Hongtong H, Ziqiang L et al (2018) Study on the fermentation process of L-hydroxyproline supplement with live cell online monitoring. Food Ferment Industries 44(5):10–15

Mingsong W (2018) Application and development of terephthalic acid industry and new technology. Fine Petrochem 35(3):71–75

Mobley DP, Shank GK (1999) Fermentation medium and method for producing α, ω-alkanedicarboxylic acids. US6004784 A

Moon SK, Wee YJ, Choi GW (2012) A novel lactic acid bacterium for the production of high purity L-lactic acid, *Lactobacillus paracasei subsp paracasei* CHB2121. J Biosci Bioeng 114(2):155–159

Motagamwala AH, Won W, Sener C et al (2018) Toward biomass-derived renewable plastics: production of 2,5-furandicarboxylic acid from fructose. Sci Adv 4(1):9722

Murphy MG, Condon S (1984) Comparison of aerobic and anaerobic growth of *Lactobacillus plantarum* in a glucose medium. Arch Microbiol 138:49–53

Na, Ji, Tao (2008) Cover picture: direct catalytic conversion of cellulose into ethylene glycol using nickel-promoted tungsten carbide catalysts. Angew Chem Int Edit 47(44):8321–8321

Nakamura CE, Whited GM (2003) Metabolic engineering for the microbial production of 1,3-propanediol. Curr Opin Biotechnol 14(5):454–459

Nakano S, Ugwu CU, Tokiwa Y (2012) Efficient production of D-(-)-lactic acid from broken rice by *Lactobacillus delbrueckii* using $Ca(OH)_2$ as a neutralizing agent. Bioresour Technol 104:791–794

Neves AR, Pool WA, Kok J et al (2005) Overview on sugar metabolism and its control in *Lactococcus lactis*—the input from in vivo NMR. FEMS Microbiol Rev 29(3):531–554

Nghiem NP, Davison BH, Suttle BE et al (1997) Production of succinic acid by Anaerobiospirillum *succiniciproducens*. Appl Biochem Biotechnol 63–65:565–576

Okano K, Yoshida S, Tanaka T et al (2009) Homo-D-lactic acid fermentation from arabinose by redirection of the phosphoketolase pathway to the pentose phosphate pathway in L-lactate dehydrogenase gene-deficient *Lactobacillus plantarum*. Appl Environ Microbiol 75(15):5175–5178

Okazaki K, Takechi T, Kambara N et al (1986) Two acyl-coenzyme A oxidases in peroxisomes of the yeast Candida tropicalis: primary structures deduced from genomic DNA sequence. Proc Natl Acad Sci 83(5):1232–1236

Otero-Cerdeira L, Rodriguez-Martinez FJ, Gomezrodriguez A (2015) Ontology matching: a literature review. Expert Syst Appl 42(2):949–971

Ou MS, Ingram LO, Shanmugam KT (2011) L(+)-Lactic acid production from non-food carbohydrates by thermotolerant *Bacillus coagulans*. J Ind Microbiol Biotechnol 38(5):599–605

Pandey SR, Jegatheesan V, Baskaran K et al (2012) Fouling in reverse osmosis (RO) membrane in water recovery from secondary effluent: a review. Rev Environ Sci Biotechnol 11(2):125–145

Park DH, Zeikus JG (1999) Utilization of electrically reduced neutral red by *Actinobacillus succinogenes* : physiological function of neutral red in membrane-driven fumarate reduction and energy conservation. J Bacteriol 181(8):2403–2410

Pedersen MB, Garrigues C, Tuphile K et al (2008) Impact of aeration and heme-activated respiration on *Lactococcus lactis* gene expression: identification of a heme-responsive operon. J Bacteriol 190(14):4903–4911

Pereira R, Badino A, Cruz A (2020) Framework based on artificial intelligence to increase industrial bioethanol production. Energy Fuel 34(4):4670–4677

Picataggio S, Rohrer T, Deanda K et al (1992) Metabolic engineering of Candida tropicalis for the production of long-chain dicarboxylic acids. Nat Biotechnol 10(8):894–898

Pinazo JM, Domine ME, Parvulescu V (2015) Sustainability metrics for succinic acid production: a comparison between biomass-based and petrochemical routes. Catal Today 239:17–24

Ping ZY, Wei TX, Liang ZS (2018) Optimization and amplification of cell culture process based on multi-scale parameter correlation analysis. Biotechnol Industry 11(1):49–55

Qiang MF, Yu YG (2017) Application of ultra-high throughput screening system based on droplet microfluidic technology in synthetic biology. Biotechnol Bull 33(1):83–92

Qin J, Wang X, Zheng Z et al (2010) Production of L-lactic acid by a thermophilic Bacillus mutant using sodium hydroxide as neutralizing agent. Bioresour Technol 101(19):7570–7576

Qing L, Huiyuan Y, Hui Z (2004) Selection and fermentation condition optimization of lactic acid bacteria for γ-aminobutyric acid production. Amino Acids Biotic Resour 1:40–43

Qinlu L, Jing W, Haijun C (2008) Research progress on γ-aminobutyric acid. Modern Food Sci Technol 5:496–500

Qun L, Decai N (2017) Process and cost analysis of ethylene glycol production from coke oven gas. Petrochem Technol 24(5):46

Raab AM, Gebhardt G, Bolotina N et al (2010) Metabolic engineering of *Saccharomyces cerevisiae* for the biotechnological production of succinic acid. Metab Eng 12(6):518–525

Ramos A, Neves AR, Santos H (2002) Metabolism of lactic acid bacteria studied by nuclear magnetic resonance. Anton Leeuw 82(1–4):249–261

Ranming W, Zejian W, Xiwei T et al (2013) Application of biological process exhaust gas mass spectrometer in lactic acid fermentation process control. J East China Univ Sci Technol (Nat Sci Ed) 39(3):289–295

Rivas B, Moldes AB, Dominguez JM et al (2004) Development of culture media containing spent yeast cells of *Debaryomyces hansenii* and corn steep liquor for lactic acid production with *Lactobacillus rhamnosus*. Int J Food Microbiol 97(1):93–98

Rodriguez DM, Jing L (2018) 2,5-Furandicarboxylic acid (FDCA)—a very promising basic raw material. Int Textile Bull 10:4–6

Romero-Garcia S, Hernandez-Bustos C, Merino E et al (2009) Homolactic fermentation from glucose and cellobiose using *Bacillus subtilis*. Microb Cell Factories 8:23–30

Rui L, Xinyuan M, Xumao X et al (2015) Progress in the study of catalysts for acetylene hydroxylation reaction. Natural Gas Chem Industry 40(5):76–80

Ruiqi L, Yifan S (2021) Forecast and development suggestions for China's ethylene glycol market. Synthetic Fiber Industry 44(5):77–82

Ruqing S, Shifang M, Gang N et al (2018) Market analysis of coal-based 1,4-butanediol and its derived fine chemicals. Modern Chem Industry 38(2):11–13

Sadare OO, Ejekwu O, Moshokoa MF et al (2021) Membrane purification techniques for recovery of succinic acid obtained from fermentation broth during bioconversion of lignocellulosic biomass: current advances and future perspectives. Sustainability 13(12):6794

Salvachúa D, Mohagheghi A, Smith H et al (2016) Succinic acid production on xylose-enriched biorefinery streams by *Actinobacillus succinogenes* in batch fermentation. Biotechnol Biofuels 9:28

Sánchez AM, Bennett GN, San KY (2005) Novel pathway engineering design of the anaerobic central metabolic pathway in *Escherichia coli* to increase succinate yield and productivity. Metab Eng 7(3):229–239

Segal SA (1984) Blockade of central nervous system GABAergic tone causes sympathetic mediated increases in coronary vascular resistance in cats. Circ Res 55:404–408

Shanghai Kasei Biotechnology R&D Center Co., Ltd (2014) Refining method of long-chain dicarboxylic acid. CN103965035

Shaobo W, Yang X, Xin H et al (2021) Research progress on the preparation technology of bio-based polypropylene terephthalate fiber. J Textile Res 42(4):16–25

Shuchen L, Shulan L, Xiangchen F (2000) Research on the fermentation technology of tridecane dicarboxylic acid. Microbiology 40:318–322

Shuchen L, Shulan L, Ping J et al (2002) Research on the fermentation of dodecane dicarboxylic acid. Microbiology 42(6):748–750

Shucong Z (2010) Preliminary study on the synthesis and purification of γ-aminobutyric acid by immobilized *Lactobacillus plantarum*. Jiangnan University, Wuxi

Siliang Z (2003) Multi scale microbial process optimization. Chemical Industry Press

Siliang Z (2016) Research on biological processes in the era of big data. Biotechnol Industry 9(3):34–39

Skoog E, Shin JH, Saez-Jimenez V et al (2018) Biobased adipic acid—the challenge of developing the production host. Biotechnol Adv 36(8):2248–2263

Song L, Zhong Y, Hui L et al (2007) Study on the properties of basic resin separation of succinic acid. Food Ferment Industries 7:54–57

Sowa JF (1991) Principles of semantic networks: exploration in the representation of knowledge. Frame Problem Artif Intell 5:135–157

Tachon S, Brandsma JB, Yvon M (2010) NoxE NADH oxidase and the electron transport chain are responsible for the ability of *Lactococcus lactis* to decrease the redox potential of milk. Appl Environ Microbiol 76(5):1311–1319

Tarkiainen V, Kotiaho T, Mattila I et al (2005) On-line monitoring of continuous beer fermentation process using automatic membrane inlet mass spectrometric system. Talanta 65(5):1254–1263

Tashiro Y, Kaneko W, Sun Y et al (2011) Continuous D-lactic acid production by a novel thermotolerant *Lactobacillus delbrueckii subsp. Lactis* QU 41. Appl Microbiol Biotechnol 89(6):1741–1750

Temple MD, Perrone GG, Dawes IW (2005) Complex cellular responses to reactive oxygen species. Trends Cell Biol 15(6):319–326

Thuy N, Boontawan A (2017) Production of very-high purity succinic acid from fermentation broth using microfiltration and nanofiltration-assisted crystallization. J Membr Sci 524:470–481

Tian L, Jianhua Z (2017) Application and separation and purification technology progress of adipic acid by-products. Chem Technol Dev 46(10):34–38

Tong UC, Choi SY, Ryu JY et al (2018) Production of ethylene glycol from xylose by metabolically engineered *Escherichia coli*. AICHE J

Tovilla-Coutiño DB, Momany C, Eiteman MA (2020) Engineered citrate synthase alters Acetate Accumulation in *Escherichia coli*. Metab Eng 61:171–180

Vemuri GN, Eiteman MA, Altman E (2002) Effects of growth mode and pyruvate carboxylase on succinic acid production by metabolically engineered strains of *Escherichia coli*. Appl Environ Microbiol 68(4):1715–1727

Vido K, Le Bars D, Mistou MY et al (2004) Proteome analyses of heme-dependent respiration in *Lactococcus lactis*: involvement of the proteolytic system. J Bacteriol 186(6):1648–1657

Vinderola G, Ouwehand A, Salminen S et al (2004) Lactic acid bacteria: microbiological and functional aspects. CRC Press

Wang L, Zhao B, Li F et al (2011) Highly efficient production of D-lactate by *Sporolactobacillus* sp. CASD with simultaneous enzymatic hydrolysis of peanut meal. Appl Microbiol Biotechnol 89(4):1009–1017

Wang T, Liu T, Wang Z et al (2016) A rapid and accurate quantification method for real-time dynamic analysis of cellular lipids during microalgal fermentation processes in *Chlorella protothecoides* with low field nuclear magnetic resonance. J Microbiol Methods 124:13–20

Wee Y, Kim J, Ryu H (2006a) Biotechnological production of lactic acid and its recent applications. Food Technol Biotechnol 44:163–172

Wee YJ, Yun JS, Kim D et al (2006b) Batch and repeated batch production of L(+)-lactic acid by *Enterococcus faecalis* RKY1 using wood hydrolyzate and corn steep liquor. J Ind Microbiol Biotechnol 33(6):431–435

Wei W, Li Z, Xie J et al (2009) Production of succinate by a *pflB ldhA* double mutant of *Escherichia coli* overexpressing malate dehydrogenase. Bioprocess Biosyst Eng 32(6):737

Wei TX, Guan W, Liang ZS et al (2019) Progress in the principles and methods of intelligent control of industrial biological processes. J Biotechnol 35(10):2014–2024

Weichen S, Dan G, Tao L (2014) Current status and development suggestions of China's bio-fermentation industry. Fine Specialty Chem 22:7–11

Wong K, Wong P, Cheung C et al (2013) Modeling and optimization of biodiesel engine performance using advanced machine learning methods. Energy 55(15):519–528

Wu MY, Sung LY, Li H et al (2017) Combining CRISPR and CRISPRi systems for metabolic engineering of E.coli and 1,4-BDO biosynthesis. ACS Synth Biol:2350

Wu Z, Wang J, Liu J et al (2019a) Engineering an electroactive *Escherichia coli* for the microbial electrosynthesis of succinate from glucose and CO_2. Microb Cell Factories 18(1):15

Wu Z, Wang J, Zhang X et al (2019b) Engineering an electroactive *Escherichia coli* for the microbial electrosynthesis of succinate by increasing the intracellular FAD pool. Biochem Eng J 146:132–142

Wu S, Liu Q, Tan H et al (2020) A novel 2,5-furandicarboxylic acid biosynthesis route from biomass-derived 5-hydroxymethylfurfural based on the consecutive enzyme reactions. Appl Biochem Biotechnol 191(4):1470–1482

Xiaobei Z, Yihui Z (2003) Fermentation process of succinic acid and its product market. Food Sci Technol 2:44–49

Xiaoming C (2012) Progress of ethylene glycol production technology and domestic market analysis. Polyester Industry 25(1):1–6

Xiaoshu L, Dacheng G, Kai Q et al (2017) Research progress on the separation and extraction of long-chain dicarboxylic acids by fermentation. Petrochemicals 46(9):1214–1218

Xiaozeng X, Mingyou T (1996) Fermentation technology of long-chain dicarboxylic acids. Modern Chem Industry 16(1):22–24

Xichun L, Yuan L (2021) Brief analysis of domestic terephthalic acid technology and production status. Synthetic Fiber 50(9):1–4

Xingqing Z (2003) Experimental research on the production of tridecane dicarboxylic acid by fermentation. Northeast University

Xingquan M (1984) A new process for the preparation of succinic anhydride by one-step catalytic hydrogenation of maleic anhydride. Appl Chem Industry 1:4–8

Xinming Z (2017) Analysis of the current status of 1,4-butanediol preparation process. Chem Manag 26:49

Xinxing Y, Shifang Y, Xuepu Z (2009) Introduction to PTA production process. Guangdong Chem Industry 36(12):76–77

Xiuli S (2006) Evaluation of 1,4-butanediol production technology. Chem Eng 7:67–70

Yamada T, Nawa H, Kawamoto S et al (1980) Subcellular localization of long-chain alcohol dehydrogenase and aldehyde dehydrogenase in n-alkane-grown Candida tropicalis. Arch Microbiol 128(2):145–151

Yan D, Wang G, Gao K et al (2018) One-pot synthesis of 2,5-furandicarboxylic acid from fructose in ionic liquids. Ind Eng Chem Res:1851–1858

Yaqin G, Lijuan X (2019) Research on the development of terephthalic acid production process. Chem Industry Manag 33:194

Yingjie L, Lei A, Tianyu T et al (2013) Discussion on the preparation technology of succinic acid. Shandong Chem Industry 42(9):54–57

Yongli M, Zijian F, Chunhai Y et al (2020) Dynamic process and optimization of intermittent distillation separation of 1,4-butanediol dehydration products. Chem Industry Progr 39(8):2972–2979

Yongqiang S, Chunju L, Keren X et al (1979) Study on the fermentation of long-chain dicarboxylic acids from petroleum—fermentation conditions for the production of tridecane 1:3 dicarboxylic acid by the polyploid variant of tropical Candida (Candida tropicalis). Plant Physiol J 4:70–78

Youn DJW (2014) Metabolic engineering of *Escherichia coli* for the biosynthesis of mandelic acid. Chemie Ingenieur Technik 86(9):1421

Yu Y, Zhu X, Xu H et al (2019) Construction of an energy-conserving glycerol utilization pathways for improving anaerobic succinate production in *Escherichia coli*. Metab Eng 56:181–189

Yuan H, Li J, Shin HD et al (2018) Improved production of 2,5-furandicarboxylic acid by overexpression of 5-hydroxymethylfurfural oxidase and 5-hydroxymethylfurfural/furfural oxidoreductase in *Raoultella ornithinolytica* BF60. Bioresour Technol 247:1184–1188

Yuantong C, Yuechuan P (1997) Screening and mutation of Δ9-1,18-octadecadienoic acid producing strains. Microbiology 37(1):65–67

Yuantong C, Xiuzhen H, Jun X (2007) A new method for the biosynthesis of dodecane dicarboxylic acid. CN1928100

Yukawa H, Omumasaba CA, Nonaka H et al (2007) Comparative analysis of the *Corynebacterium glutamicum* group and complete genome sequence of strain R. Microbiology 153(4):1042–1058

Yunzhi X, Peng J (2001) Induction of P450 enzyme by alkanes and improvement of dicarboxylic acid fermentation process. J Biol Eng 17:218–220

Yupei Z, Dinghua L, Xiaoqin L et al (2006) Progress and outlook of syngas to ethylene glycol process. Natural Gas Chem Industry 3:56–60

Zaman N, Rohani R, Mohammad A (2016) Polyimide membranes for organic salts recovery from model biomass fermentation. Malays J Anal Sci 20:1481–1490

Zeikus JG, Jain MK, Elankovan P (1999) Biotechnology of succinic acid production and markets for derived industrial products. Appl Microbiol Biotechnol 51(5):545–552

Zhang S, Chu J, Zhuang Y (2004) A multi-scale study on industrial fermentation processes and their optimization. Adv Biochem Eng Biotechnol 87:97–150

Zhang J, Kao E, Wang G et al (2016) Metabolic engineering of Escherichia coli for the biosynthesis of 2-pyrrolidone. Metab Eng Commun 3:1–7

Zhanzhao L, Guang C, Changyuan Y (2009) Study on the recrystallization refining process of dodecanedioic acid. Synthetic Fiber Industry 32:31–34

Zhao B, Wang L, Li F et al (2010) Kinetics of D-lactic acid production by *Sporolactobacillus* sp. strain CASD using repeated batch fermentation. Bioresour Technol 101(16):6499–6505

Zhao R, Zhang L, Xu L et al (2021) One-pot selective synthesis of renewable p-xylene by completely biomass-based ethanol and dimethylfuran with functionalized mesoporous MCM-41. ChemistrySelect 6

Zhen K, Yanping G, Yuanyuan Z et al (2008) Construction of an aerobic fermentation strain for the production of succinic acid. J Biotechnol 12:78–82

Zhenfeng M, Meijuan X, Taowei Y et al (2020) Construction and fermentation study of recombinant glutamate rod bacteria synthesizing 2-pyrrolidone from glucose. Food Ferment Industries 46(11):1–8

Zhongmao Z, Jianqi L, Dizhang C et al (2009) Application and market prospects of long-chain dicarboxylic acids. Shandong Food Ferment 13(3):14–17

Zhou S, Causey TB, Hasona A et al (2003) Production of optically pure D-lactic acid in mineral salts medium by metabolically engineered *Escherichia coli* W3110. Appl Environ Microbiol 69(1):399–407

Zhou W, Wu Z, Kong Z et al (2021) Efficient oxidation of biomass derived 5-Hydroxymethylfurfural into 2,5-Diformylfuran catalyzed by NiMn layered double hydroxide. Catal Commun 151(10):106279

Zhu LW, Tang YJ (2017) Current advances of succinate biosynthesis in metabolically engineered *Escherichia coli*. Biotechnol Adv 35(8):1040–1048

Zhu Y, Eiteman MA, DeWitt K et al (2007) Homolactate fermentation by metabolically engineered *Escherichia coli* strains. Appl Environ Microbiol 73(2):456–464

Zivkovic J, Zekovic Z, Mujic I et al (2009) EPR spin-trapping and spin-probing spectroscopy in assessing antioxidant properties: example on extracts of catkin, leaves, and spiny burs of castanea sativa. Food Biophys 4(2):126–133

Zuhua C, Qing Y, Guanglin Y et al (2000) The effect of N^+ ion implantation on the yield of long-chain dicarboxylic acids in tropical Candida. Microbiol Bull 27(3):174–177

Zutong L, Zhongxiang G (1989) Microbial fermentation production of dicarboxylic acids. Pet Sci 5(2):84–90

[illegible] ... (2017) ... [illegible] ... engineering ... [illegible]

[illegible] ... (2018) ... [illegible] ... CRISPR ... [illegible]

[illegible] ... (2020) ... [illegible]

[illegible] ... (2008) ... [illegible]

[illegible] ... (2018) ... [illegible]

[illegible] ... (2012) ... [illegible]

[illegible]

Chapter 3
Biomanufacturing of Biobased Polymers

Yuanchan Luo, Liqiang Fan, Yibing Wang, Ju Wu, Yijia Chen, Jinqi Shi, Peng Fei, Jianing Li, Mingcheng Xu, Pengye Guo, Fei Zhao, Zhitian Qin, Liming Zhao, and Hui Wu

Biobased degradable materials focus on the bio-derived renewable raw materials, avoid the dependence on traditional polymer materials based on fossil raw materials, and have the dual effects of environmental protection and resource conservation. The production methods of biobased degradable materials are mainly divided into three types: one is to use chemical methods alone to transform biobased raw materials for polymer production; the second is to use biological methods alone to transform biobased raw materials for polymer production, and the third is to use biological and chemical methods together to transform biobased raw materials for polymer production. At present, most biobased polymers are produced by the third method. The second method, that is, directly transforming biobased raw materials into a polymer through biological methods, reduces the steps of intermediate chemical synthesis, is the best option for polymer production in terms of energy saving, safety, and economy, and is an important direction for future research and development of biobased materials. This chapter focuses on the more common biobased degradable polymer materials that can be directly prepared by biological transformation, including

Y. Luo · J. Wu · P. Fei · J. Li · M. Xu · P. Guo · F. Zhao · Z. Qin
East China University of Science and Technology, Shanghai, China
e-mail: luoyuanc@ecust.edu.cn

L. Fan · Y. Wang · Y. Chen · J. Shi · L. Zhao
State Key Laboratory of Bioreactor Engineering, R & D Center of Separation and Extraction Technology in Fermentation Industry, East China University of Science and Technology, Shanghai, China
e-mail: fanglq@ecust.edu.cn; ybwang@ecust.edu.cn; zhaoliming@ecust.edu.cn

H. Wu (✉)
East China University of Science and Technology, Shanghai, China

MOE Key Laboratory of Bio-Intelligent Manufacturing, School of Bioengineering, Dalian University of Technology, Dalian, China
e-mail: hwu@dlut.edu.cn; hwu@ecust.edu.cn

L. Zhao (ed.), *Research and Applications of Bio-based Degradable Materials*,
https://doi.org/10.1007/978-981-95-1188-4_3

polyhydroxyalkanoates/poly-β-hydroxybutyrate (PHA/PHB), polylactic acid (PLA), chitin, polybutyramide (PA4), and polylactic acid. The background, advantages, and synthesis methods of these polymers are introduced.

3.1 Synthetic Biology Pathways and Construction Methods of Chassis Cells for Biobased Macromolecules

Compared to using the chemical production method alone, the production of biobased macromolecular materials using the biological method alone or in combination with the chemical method has advantages such as being environmentally friendly, safe to operate, and renewable. When using the biological method alone or in combination with the chemical method for the production of biobased macromolecules, choosing the right chassis cell can double the production of the macromolecules. The development of many industrial microbial chassis cells provides excellent cell factories for the synthesis of biobased macromolecular materials, which is conducive to the production of renewable and environmentally friendly biobased macromolecular materials. This section will introduce some chassis cells that have been applied to or have great development potential in the production process of biobased macromolecular materials, and briefly introduce common molecular tools used to modify different types of chassis cells, so that readers can systematically understand these chassis cells and their construction methods. Using genetic recombination technology to purposefully modify the cell metabolic pathway can improve the metabolic characteristics of the cell and achieve the purpose of efficient production of specific target products. Understanding the metabolic pathways related to the synthesis of biobased macromolecules or their monomers in cells and common modification strategies can improve the production efficiency of biobased macromolecular materials, reduce their production costs, and further promote their industrial development. Because *Escherichia coli* has a clear genetic background, simple culture and molecular operation, high transformation and transduction efficiency, fast growth and reproduction, and low cost, it can be used as the most common chassis cell in the production of biobased macromolecules. This section will take *Escherichia coli* and other chassis cells as examples to introduce common metabolic pathways and common metabolic modification strategies related to the synthesis of biobased macromolecules and their monomers in cells.

3.1.1 Chassis Cells

1. *Escherichia coli*

Escherichia coli (*E. coli*) is a Gram-negative bacterium, belonging to the phylum Proteobacteria, the family Enterobacteriaceae, and the genus *Escherichia*. It is one of the most commonly used prokaryotic hosts in biological fermentation, with a clear genetic background, fast reproduction, simple culture and molecular operation, and low cost. It can efficiently express exogenous genes (Terpe 2006). Because of its clear genetic metabolic background, *E. coli* is often used as a chassis cell in the process of biobased macromolecule generation and can use different gene editing tools, suitable for the production of different substances. Therefore, *E. coli* is widely used in the synthesis process of biobased macromolecular materials (Solaiman and Foglia 2002).

At present, there are many studies on the production of biobased macromolecular material monomers or polymers using *E. coli* as a chassis cell, some of which have achieved industrial-scale production. For example, genetically engineered *E. coli* can be used as a chassis cell for PHA synthesis, degrading biomass raw materials to directly synthesize PHA (Terpe 2006); using glucose to produce 1,3-propanediol (PDO) (Przystałowska et al. 2015), providing monomers for biobased polyethylene terephthalate. In addition, *E. coli* can also produce ethylene glycol as a raw material for bio-fibers (Yuhui 2018), and produce 1,4-butanediol, the polymer monomer of polybutylene terephthalate (PBT) (Liu and Lu 2015).

Homologous recombination is the most commonly used gene modification method in *E. coli*. RecA and RecBCD proteins encoded by the RecA homologous recombination system of *E. coli* itself can mediate exogenous genes to undergo homologous recombination. Among the Rec proteins, RecA is the core of this system and can bind to single-stranded DNA to recognize homologous sequences in DNA (Datta et al. 2006). Due to the disadvantages of the RecA homologous recombination system, such as the need for long homologous arms, low recombination efficiency, and difficult operation, it is difficult to be widely used. However, the Red recombination technology using the λ phage gene (Fig. 3.1) has been widely used due to its advantages of high efficiency, short cycle, and short necessary homologous arms (Murphy 1998). Red recombination technology uses PKD46 as a functional plasmid, composed of three genes (*exo*, *bet*, and *gam*), where *exo* encodes the double-stranded nucleic acid exonuclease Exo, *bet* encodes the single-strand annealing protein Beta, and *gam* encodes the auxiliary protein Gam. Exo can degrade single-stranded DNA from the 5′ end to the 3′ end, produce sticky ends at the 3′ end, and then Beta mediates single-strand annealing repair to complete homologous recombination. Gam can inhibit the activity of the endogenous RecBCD nucleic acid exonuclease in *E. coli*, preventing the exogenous DNA from being degraded in the cell (Datsenko and Wanner 2000).

CRISPR/Cas9 is a gene editing system that can be used in *E. coli* (Fig. 3.2). This system can be used to knock out endogenous genes and insert exogenous genes in *E. coli* and is often used to construct engineered *E. coli* (Jiang et al. 2013; Lin et al. 2015). This system is an immune system used by archaea and bacteria to deal with viruses and other exogenous genes. The most common dual-plasmid CRISPR/Cas9 system consists of Cas9 nuclease and guide RNA (guideRNA, i.e., gRNA, including tracrRNA that activates Cas protein and crRNA that recognizes target DNA sequences). The guide RNA can specifically recognize specific DNA sequences, and under the action of Cas9 nuclease, it causes DNA double-strand breaks at

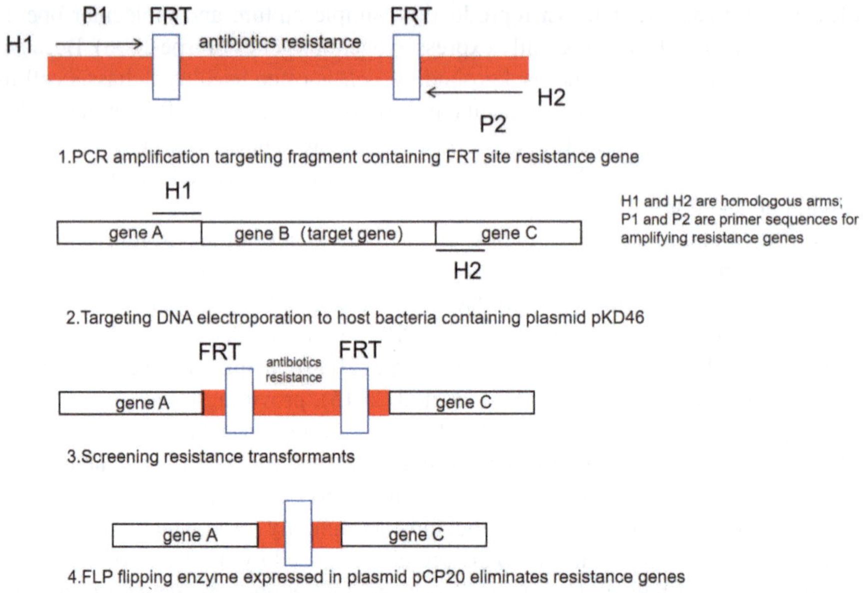

Fig. 3.1 Principle of Red recombination technology (Fan et al. 2018). (Reprinted with permission from *Journal of Nanjing Normal University*, 2018, 41(3):102)

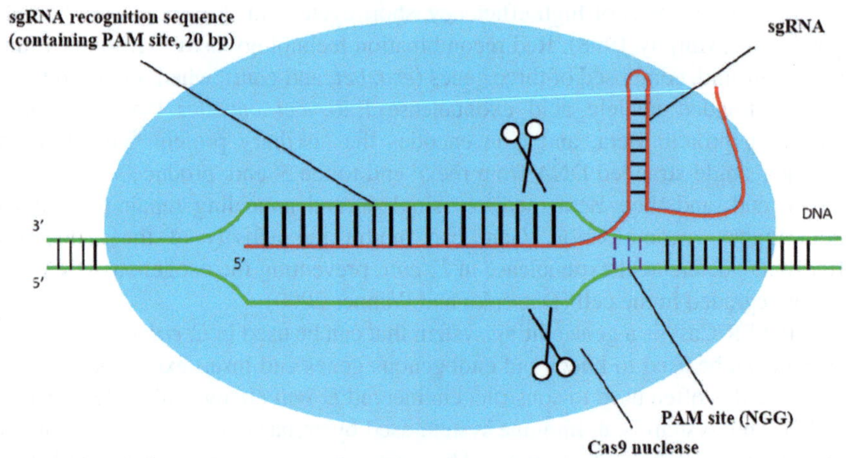

Fig. 3.2 Principle of CRISP/Cas9 technology (see color insert). (Fan et al. 2018)

specific sites. Through the endogenous DNA repair system of *E. coli*, gene knockout or insertion can be efficiently achieved.

2. *Halophilic Microorganisms*

Halophilic microorganisms are a type of microorganism that can grow in environments with high salt concentrations, and they are widely found in the domains of archaea, bacteria, and eukaryotes. Depending on the optimal salt concentration for growth, halophilic microorganisms can be divided into moderate halophiles and extreme halophiles. Halophilic microorganisms have various advantages as chassis cells. Since other bacteria find it difficult to survive in their high-salt growth environments, open fermentation without sterilization can be achieved in industrial production. This not only reduces the cost of equipment, raw materials, and labor but also shortens the production cycle and improves production efficiency. Since the discovery of biodegradable plastic PHA in halophilic microorganisms in 1972, halophilic microorganisms have received much attention in the field of biosynthesis of biobased macromolecular materials (Mengying et al. 2019). For example, the halophilic microorganism *Halomonas* has shown excellent characteristics in the production of PHA (Mengying et al. 2019; Ouyang et al. 2017).

Halophilic microorganisms are a common type of chassis cell in the production of biobased macromolecular material, PHA. In order to produce PHA with different properties and obtain strains with better characteristics, tools suitable for molecular operations of halophilic microorganisms have also been developed.

Homologous recombination induced by suicide plasmids is a mature gene knockout technology in halophilic microorganisms (Fig. 3.3). This system consists of a suicide plasmid marked with the *pyrF* (orotidine-5′-phosphate decarboxylase) gene and a nutritionally defective host lacking the *pyrF* gene (Tao et al. 2017). By transferring the suicide plasmid containing the homologous arm of the target gene into the nutritionally defective host, positive selection of homologous recombination strains is performed using a medium without uracil, followed by counter-selection of *pyrF* gene-deleted recombinant strains using 5-fluoroorotic acid. The target gene is then analyzed by PCR and Southern blotting to complete gene knockout. However, this method is inefficient and time-consuming (editing a single gene in halophilic microorganisms takes more than a month), which severely limits the construction efficiency of engineered strains of halophilic microorganisms.

CRISPR/Cas9 is a gene editing tool that can theoretically be applied to many prokaryotic and eukaryotic cells. There is now a CRISPR/Cas9 system suitable for the halophilic microorganism *Halomonas* (*Halomonas* sp.), which consists of a low-copy plasmid pSEVA321 that can express the Cas9 gene and a high-copy plasmid pSEVA241 that can express single-guide RNA (sgRNA) and contains a 500–1000 bp homologous arm of the donor DNA (Qin et al. 2018). By developing CRISPR/Cas9 technology suitable for halophilic microorganisms, it is possible to edit the genes of halophilic microorganisms in a short time and with high efficiency.

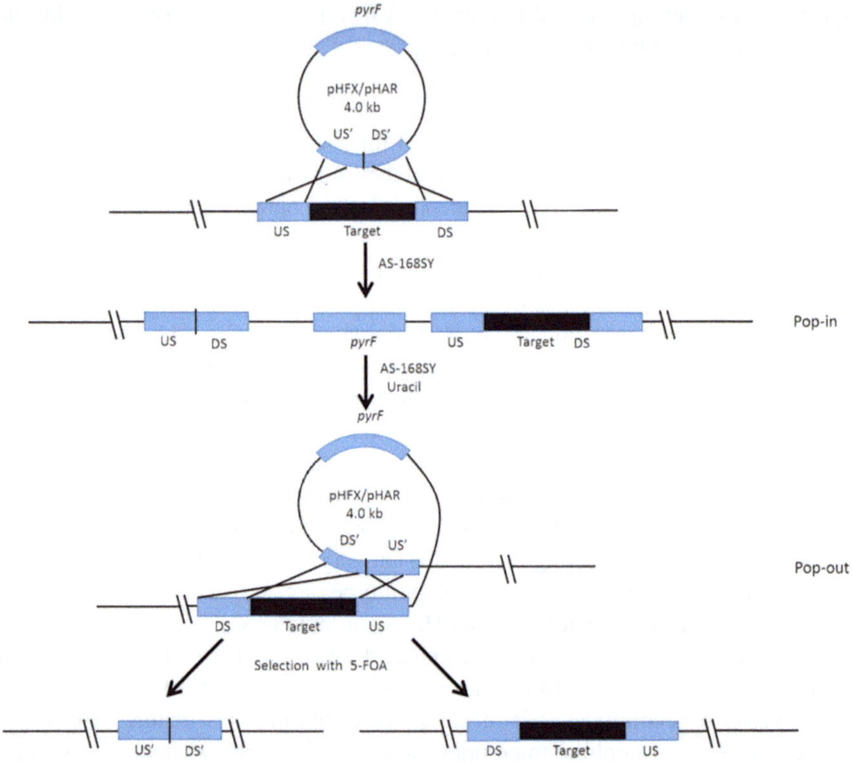

Fig. 3.3 Schematic diagram of the gene knockout system based on *pyrF* (see color insert) (Liu et al. 2011a). (Reprinted with permission from *J Genet Genomics*, 2011, 38(6): 261–269© 2019 Elsevier Ltd)
US/US' upstream fragment of the target gene to be knocked out, *DS/DS'* downstream fragment of the target gene to be knocked *out; Target, target gene*, *5-FOA* 5-fluoroorotic acid, *pyrF* gene encoding orotidine-5′-phosphate decarboxylase

3. *Alcaligenes Eutrophus*

Alcaligenes eutrophus is a Gram-negative bacterium isolated from soil, belonging to the genus *Alcaligenes*. As early as the 1980s, the British ICI company used *A. eutrophus* to successfully ferment the production of biodegradable plastic poly-3-hydroxybutyrate-*co*-3-hydroxyvalerate (PHBV) with glucose and propionic acid as raw materials. Today, *A. eutrophus* has been widely used in the production of biodegradable plastics, PHB and PHBV, and is an excellent chassis cell that can be used for the direct production of biobased macromolecular materials (Wang et al. 2013a).

As a chassis cell capable of producing biobased macromolecular materials, *A. eutrophus* also has a variety of effective gene editing tools applicable to it.

The use of suicide plasmids is a traditional method for constructing *A. eutrophus* mutants. This method uses suicide plasmids with selection markers (resistance to antibiotics or nutritional deficiency) as molecular tools. The principle is similar to the suicide plasmid system used in halophilic microorganisms, but it also has disadvantages such as low efficiency and long time cycle (Saeed et al. 2019).

The gene knockout strategy based on movable group II introns is a more efficient gene knockout method in *A. eutrophus*. This technology calculates the best insertion point of the intron through a computer algorithm and then completes the targeted gene knockout by inserting the intron into a special point of the target gene through the ribonucleoprotein-mediated DNA integration mechanism (retrohoming) (Ewering et al. 2006). When constructing this gene knockout system in *A. eutrophus*, the broad-host-range plasmid pBBR1MCS2 carrying this knockout system is used, and the IPTG-inducible *tac* promoter is used to achieve a higher knockout efficiency.

4. *Lactic Acid Bacteria*

Lactic acid bacteria are a general term for a type of Gram-positive bacteria that can produce a large amount of lactic acid by fermenting carbohydrates. They belong to the order Lactobacillales and the family Lactobacillaceae. These bacteria are diverse and widely distributed in nature, and are widely used in industries such as food, medicine, and others. In the production process of biobased polymers, lactic acid bacteria are mainly used as producers of the PLA monomer—lactic acid. They are the chassis cells that can be used when combining biological and chemical methods to produce biobased macromolecular materials PLA. Depending on the differences in end products of fermentation, lactic acid bacteria can be divided into homolactic and heterolactic fermentation lactic acid bacteria. Homolactic fermentation lactic acid bacteria account for more than 90% of the types of lactic acid bacteria, and the main component of the end product of their fermentation is acetic acid, and they have aldehyde dehydrogenase. In the process of lactic acid fermentation, glucose is degraded into pyruvic acid in these lactic acid bacteria, and then pyruvic acid is reduced to lactic acid under the catalysis of lactic acid dehydrogenase. In the process of homolactic fermentation, one molecule of glucose will be converted into two molecules of lactic acid, and the theoretical value of carbon atom utilization rate can reach 100%; therefore, this type of lactic acid bacteria is an ideal strain for fermentative production of lactic acid. However, in heterolactic fermentation lactic acid bacteria, when glucose is consumed to fermentatively produce lactic acid, there will also be by-products such as ethanol, acetic acid, and CO_2 generated (Myoung et al. 2010), so they are not ideal strains for lactic acid production.

5. *Rhizopus oryzae*

Rhizopus oryzae is a type of mold that can be used to produce lactic acid. It belongs to the subphylum Zygomycota, class Zygomycetes, order Mucorales, family Mucoraceae, and genus Rhizopus. It is widely used in the brewing industry and other fields, and has the ability to directly utilize starch. The L-lactic acid produced by it has a higher optical purity than the L-lactic acid produced by lactic acid bacteria (Haiyan et al. 2006). In addition, *R. oryzae* has developed mycelia, which can form mycelial clumps during the fermentation process for easy separation from the fermentation broth. However, too large mycelial clumps can also cause difficulties in the transfer of O_2 and nutrients and increase the resistance to stirring, thereby reducing the production rate, yield, and stability of the production process. The use of cell immobilization technology can increase the concentration of *R. oryzae* during the lactic acid fermentation process, thereby increasing the yield of lactic acid and improving production efficiency. In addition, it has been reported that the yield and productivity of lactic acid can be increased by mutating to change the morphology of *R. oryzae* (Xiaoming et al. 2008).

6. *Aspergillus terreus*

Aspergillus terreus is a filamentous fungus with high production value, belonging to the class of Ascomycetes, order Eurotiales, family Trichocomaceae, and genus *Aspergillus*. It has the ability to tolerate high osmotic pressure and is widely present in various habitats. Today, *A. terreus* is widely used in the fields of bioengineering and medicine. In the field of biobased macromolecular materials, *A. terreus* can be used to ferment the production of itaconic acid (Fig. 3.4), providing monomers for the synthesis of itaconic acid epoxy resin. As early as the 1940s, Lockwood and others studied the factors affecting the fermentation production of itaconic acid by *A. terreus*, laying the foundation for the industrial production of itaconic acid by *A. terreus* (Xiaolong and Yongqian 2019). Today, deep fermentation of *A. terreus*

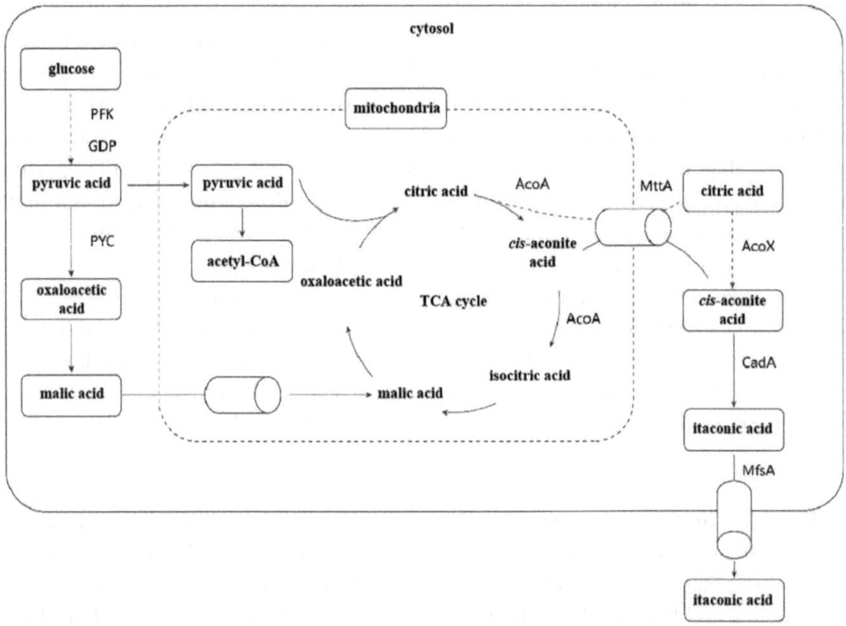

Fig. 3.4 Biological synthesis pathway of itaconic acid in *A. terreus*. (An et al. 2011). *PFK* 6-phosphofructokinase, *GDP* glyceraldehyde 3-phosphate dehydrogenase, *PYC* pyruvate carboxylase, *AcoA* aconitase, *AcoX* acyl-CoA oxidase, *MttA* tricarboxylic acid transporter, *CadA* cis-aconitate decarboxylase, *MfsA* itaconic acid secretion protein

has become the main production method of itaconic acid (Lockwood et al. 1945; Kanamasa et al. 2008).

In order to modify the genetic traits of filamentous fungi such as *A. terreus*, genetic modification elements and gene manipulation tools suitable for *A. terreus* have been developed.

Endogenous promoters of *A. terreus* and promoters from other filamentous fungi, such as P*trpC* , P*adhA* (Aspergillus nidulans), and P*mgpd* (Monascus purpureus), can drive the expression of target genes in *A. terreus*, regulating the metabolic pathway of *A. terreus* (An et al. 2011). These promoters are commonly used genetic elements in the genetic modification of *A. terreus*.

Filamentous fungi such as *A. terreus* have a low homologous recombination efficiency due to their own DNA repair characteristics, making the genetic modification of *A. terreus* difficult. By inactivating the nonhomologous end joining pathway used for double-strand DNA repair in *A. terreus*, the efficiency and specificity of exogenous gene homologous recombination can be improved. The most commonly used homologous recombination system in *A. terreus* is the nutritional transformation system based on the bidirectional selective marker $pyrG_{An}$ established in the *ku80-/pyrG-* double mutant, where the deletion of the *ku80* gene can block the nonhomologous end joining pathway, improve the efficiency of homologous recombination, and the deletion of the *pyrG* gene can form a uracil nutritional deficiency strain, improving the screening marker for homologous recombination (Fig. 3.4) (Xuenian et al. 2020).

7. *Klebsiella* sp.

Klebsiella (*Klebsiella* sp.) is a Gram-negative bacterium, belonging to the phylum Proteobacteria, family Enterobacteriaceae, and genus *Klebsiella.* It has a thick capsule, and most of the bacteria in *Klebsiella* have flagella. In the field of biobased macromolecules, *Klebsiella* is a chassis cell that can be used for the production of biobased macromolecular materials through combined biological and chemical methods. It can ferment glycerol as the sole carbon source and energy source to produce biobased polyethylene terephthalate monomer 1,3-propanediol (Xuenian et al. 2020; Huang et al. 2016) (Fig. 3.5). Under anaerobic conditions, *Klebsiella* mainly metabolizes glycerol through the dihydroxyacetone pathway (DHA pathway), which includes glycerol dehydratase (GDHt), 1,3-propanediol oxidoreductase (PDOR), glycerol dehydrogenase (GDH), and dihydroxyacetone kinase (DHAK). Under aerobic conditions, glycerol fermentation in *Klebsiella* is carried out through the glycerol-3-phosphate system (GLP system), which includes glycerol kinase (GK), aerobic glycerol-3-phosphate (G3P) dehydrogenase, and anaerobic glycerol-3-phosphate (G3P) dehydrogenase. *Klebsiella* generally uses glycerol to produce 1,3-propanediol under anaerobic or microaerobic conditions (Menzel et al. 1997).

In order to achieve the modification of *Klebsiella* and obtain chassis cells with better characteristics, various biological methods related to editing *Klebsiella*'s genes have also been developed.

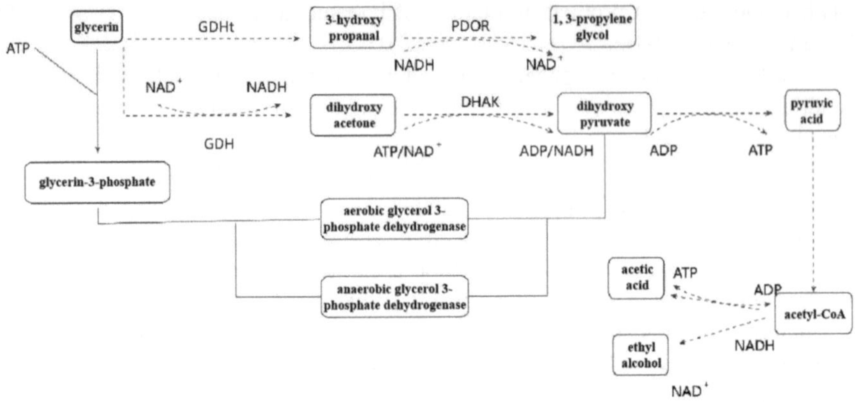

Fig. 3.5 Divergent pathways of glycerol in *Klebsiella* cells. (Jianfeng et al. 2001). *GDHt* glycerol dehydratase, *PDOR* 1,3-propanediol oxidoreductase, *GDH* glycerol dehydrogenase, *DHAK* dihydroxyacetone kinase, *NADH/NAD⁺* reduced form of nicotinamide adenine dinucleotide/nicotinamide adenine dinucleotide, *ATP* adenosine triphosphate, *ADP* adenosine diphosphate; solid lines represent the GLP metabolic pathway, and dashed lines represent the DHA metabolic pathway

Electroporation is a method often used to introduce exogenous genes into *E. coli*, and this method can also be used in *Klebsiella*. Because *Klebsiella* has a thick capsule, this capsule makes it difficult for the bacteria to recover and penetrate the surface, so compared with *E. coli*, the electroporation efficiency of *Klebsiella* is very low. By using a capsule-deficient strain of *Klebsiella* and culturing it on solid medium to change its bacterial morphology, the transformation efficiency of electroporation can be greatly improved (Zeng et al. 2015).

The Red recombination technique commonly used in *E. coli* can also be used for gene modification in *Klebsiella* (Zhang and Xiu 2010; Yan 2007). For example, using the Red recombination technique to knock out the lactate dehydrogenase gene (*ldhA*) in *Klebsiella* can effectively increase the yield of 1,3-propanediol (Lǐfei 2015).

CRISPR/Cas9 can also be applied in *Klebsiella*. In order to use CRISPR/Cas9 to edit genes in *Klebsiella* normally, it is first necessary to construct a CRISPR functional plasmid that can be expressed in *Klebsiella*. Now, there are many examples proving that CRISPR/Cas9 technology can achieve efficient gene editing in *Klebsiella* sp. (Sun et al. 2019; Mcconville et al. 2021).

8. *Corynebacterium glutamicum*

Corynebacterium glutamicum is a Gram-positive bacterium often used in industry to produce amino acids and organic acids, among other products. It belongs to the order Actinomycetales and the genus *Corynebacterium*. *C. glutamicum* has the advantages of being safe and easy to culture, making it a common chassis cell. In the field of biobased macromolecules, *C. glutamicum* can be used to synthesize the monomer pentamethylene diamine for polyamide (PA56) fibers (Buschke et al. 2015).

In order to improve the yield of products, the performance of *C. glutamicum* is being improved, and gene editing techniques suitable for *C. glutamicum* are being continuously developed.

Suicide plasmid-mediated homologous recombination is a mature gene editing technique in *C. glutamicum*. Depending on the mode of homologous recombination, suicide plasmid-mediated homologous recombination can be divided into single-exchange homologous recombination and double-exchange homologous recombination. In the former, resistance genes are generally used as markers, which may leave residues in the genome and potentially have adverse effects on engineered strains. In contrast, double-exchange homologous recombination typically uses auxotrophic strains and corresponding genes as selection markers, following the same mechanism as the suicide plasmid-induced homologous recombination system in halophilic microorganisms (Tan et al. 2012).

The Cre/loxP site-specific recombination system, which introduces specific *loxP* sites into both ends of the homologous sequence and uses the Cre recombinase to recognize these sites to complete homologous recombination, has improved the efficiency of gene manipulation in *C. glutamicum* (Milin et al. 2019; Juanjuan et al. 2020).

The RecT-mediated single-strand recombination system does not rely on the endogenous RecA recombination system of *C. glutamicum*, but directly uses the RecT recombination system from the Rac prophage. This system is simple to operate, can perform gene manipulation at any point in the genome, and greatly improves efficiency, making previously difficult gene manipulations possible (Juanjuan et al. 2020; Stephan et al. 2013).

CRISPR/Cas9 can also be used for gene editing in *C. glutamicum* (Yan 2019; Liu et al. 2017a; Peng et al. 2017). This system has the advantages of short cycles and easy screening, providing great convenience for gene manipulation in *C. glutamicum*. Because *C. glutamicum* lacks a nonhomologous end joining repair mechanism, the introduction of the RecT recombinase from the Rac prophage improves the efficiency of homologous repair, effectively enhancing the editing efficiency of CRISPR/Cas9 in *C. glutamicum*. Today, the CRISPR/Cas9 system in *C. glutamicum* is quite mature, and other CRISPR systems suitable for this bacterium have also been developed to some extent (Mcconville et al. 2021; Junwei et al. 2020; Jiang et al. 2017; Lu 2019).

3.1.2 Metabolic Pathway Modification and Application Related to the Synthesis of Biobased Materials

Metabolic engineering uses multi-gene recombination technology to purposefully modify cellular metabolic pathways, thereby improving the metabolic characteristics of cells and achieving the goal of efficient production of specific target products (Xueming et al. 2015). There are generally two common ways to modify strains in metabolic engineering: one is to introduce pathways that do not exist in the engineered bacteria for heterologous expression, thereby achieving the purpose of producing a certain heterologous product; this process is called the process from nothing to something; the second is for products with very low yield under natural conditions, by introducing efficient pathways or modifying the original metabolic flow distribution to improve product yield, this process is called the process from something to better. In fact, the cell modification from nothing to something also needs to go through a process from something to better to further improve the product conversion rate, thereby reducing the cost of product production (Qianqian et al. 2017).

E. coli is the first host bacterium used for recombinant protein production, with many advantages (Terpe 2006). The level of expression of exogenous gene products is much higher than other gene expression systems, so in the case where the production capacity of the original production bacterium is not ideal, *E. coli* is often used as the first choice of chassis cells for the artificial synthesis pathway of the target product. There are now many examples of using *E. coli* as chassis cells to produce biobased macromolecular material monomers or polymers, and studying the metabolic characteristics of *E. coli* for the modification and application of metabolic pathways related to the synthesis of biobased materials is of great significance.

1. *Metabolic Characteristics of Different E. coli Hosts to Carbon Sources*

As the most widely used carbon source in the culture medium, glucose and glycerol affect the physiological characteristics of *E. coli* to a certain extent (Hadicke and Klamt 2017) (Fig. 3.6). Glucose enters the cell through the phosphotransferase system (PTS) pathway, and at the same time, phosphoenolpyruvate is converted to pyruvate (Dharmadi et al. 2006). *E. coli* decomposes glucose into triose through three pathways, including the glycolysis pathway, Entner–Doudoroff pathway (ED pathway), and pentose phosphate pathway (PPP) (El-Mansi 2019). *E. coli* strains BL21 and JM109 use glucose as a carbon source. When the carbon source metabolic flow passes through pyruvate, if this carbon source metabolic flow exceeds the processing capacity of the tricarboxylic acid cycle, it will produce acetic acid, and the accumulation of acetic acid has medium and strain specificity (Mey et al. 2007). For the above two *E. coli* strains, the BL21 strain has advantages as a host cell expressing recombinant proteins, while the acetic acid concentration produced by JM109 is higher than that of BL21; but when the cell density increases, the acetic acid accumulation rate of the two strains both decreases (Shiloach et al. 1996). Glycerol is converted into dihydroxyacetone phosphate and enters the glycolysis pathway under the catalysis of glycerol dehydrogenase (GDH) and dihydroxyacetone kinase (DHAK) (Yu et al. 2019). When glycerol is used as a carbon source to convert to phosphoenolpyruvic acid or pyruvic acid, the redox equivalent it produces is twice that of glucose. The conversion of glycerol to succinic acid or ethanol is a process of redox balance.

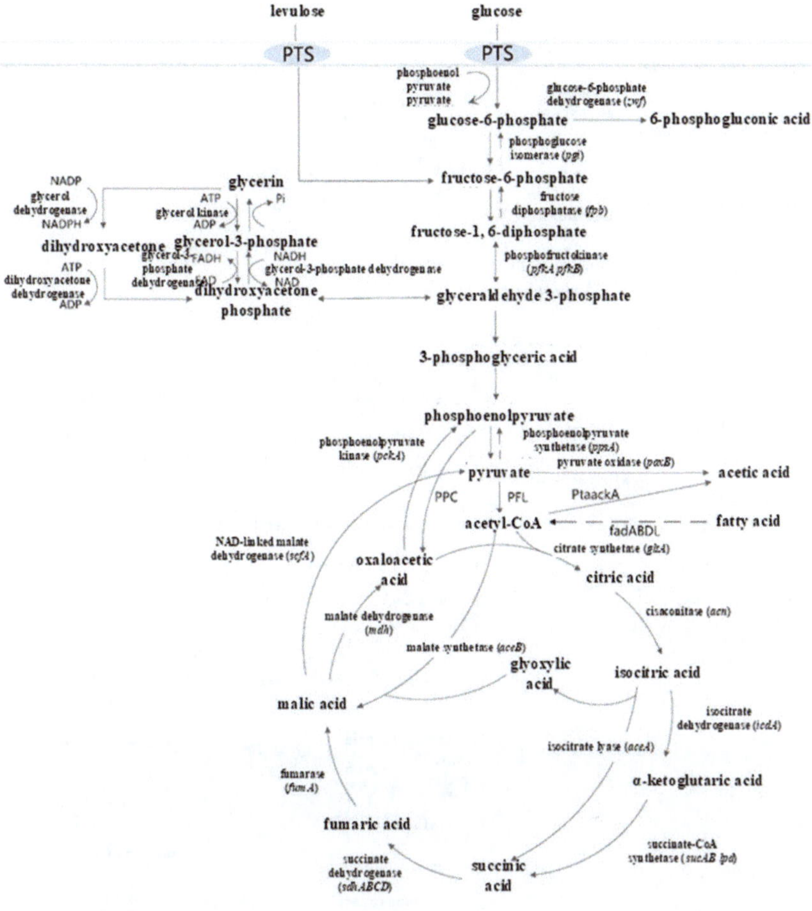

Fig. 3.6 Schematic diagram of the central metabolic pathway under aerobic fermentation conditions in *E. coli*. (Hadicke and Klamt 2017)

2. *Metabolic Remodeling and Application of the Glycolysis Pathway*

The glycolysis pathway, also known as the EMP pathway, is a series of reactions that degrade glucose and glycogen into pyruvic acid and are accompanied by the generation of ATP. It is a common glucose degradation pathway that exists in all biological organisms. The glycolysis pathway can proceed under both anaerobic and aerobic conditions and is a common metabolic pathway for glucose to undergo aerobic or anaerobic decomposition.

Many researchers have increased the synthesis flux of acetyl-CoA in *E. coli* by regulating the expression of key enzymes in glycolysis, thereby increasing the yield of the corresponding target products. Figure 3.7 (Chen et al. 2019) shows the metabolic pathways related to acetyl-CoA in *E. coli* and the various enzymes involved in the reactions. The current metabolic engineering strategies to increase the flux of

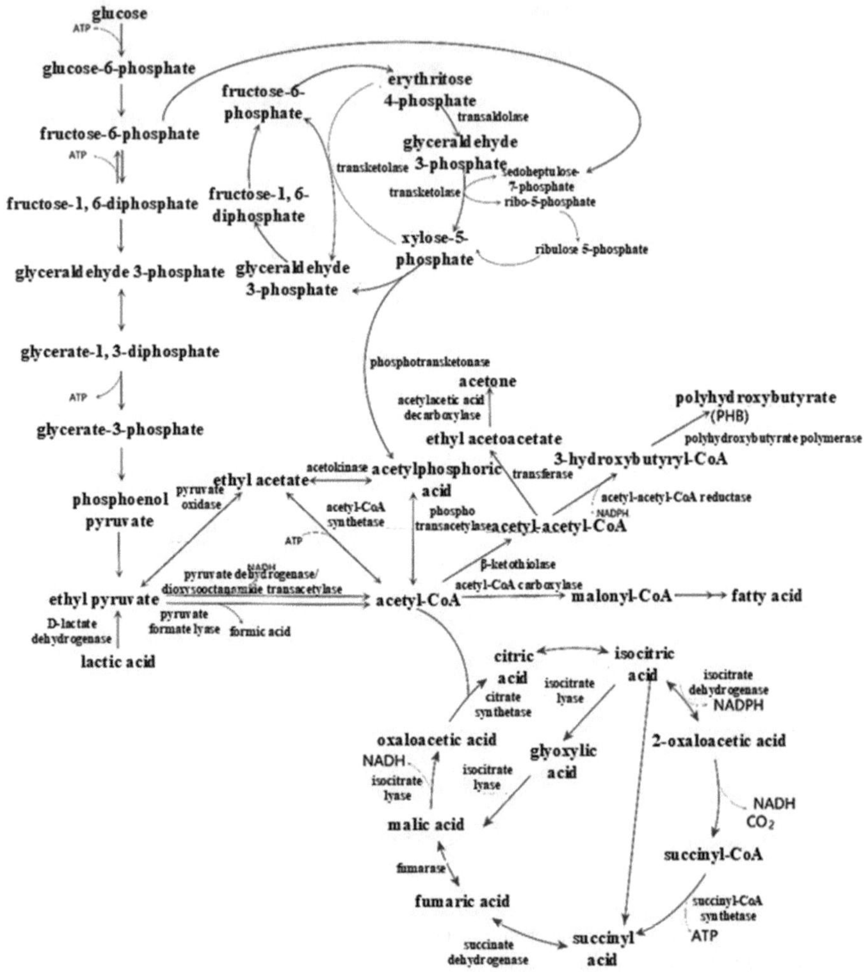

Fig. 3.7 *E. coli* acetyl coenzyme A-related metabolic pathways and enzymes. (Chen et al. 2019)

acetyl-CoA can be divided into five categories: metabolic regulation of the acetic acid pathway, metabolic regulation of acetyl-CoA synthesis from pyruvic acid, metabolic regulation of the central carbon metabolism pathway, β-oxidation synthesis of acetyl-CoA pathway metabolic regulation, and new pathways for acetyl-CoA synthesis (Fig. 3.7). The synthesis of biobased materials mainly involves the regulation and modification of the central carbon metabolism pathway.

Han et al. (2001) performed proteomic analysis on high-yield PHB *E. coli* strains and found that the expression levels of fructose bisphosphate aldolase (FBA) and phosphoglycerate isomerase (TPIA) in the glycolysis pathway were upregulated. FBA catalyzes the conversion of fructose-1,6-bisphosphate into dihydroxyacetone phosphate and glyceraldehyde-3-phosphate, and TPIA converts dihydroxyacetone phosphate into glyceraldehyde-3-phosphate. Both of these enzymes can catalyze the synthesis of glyceraldehyde-3-phosphate. Increasing the expression levels of FBA and TPIA is beneficial to the synthesis of glyceraldehyde-3-phosphate, thereby increasing the synthesis flux of acetyl-CoA and ultimately increasing the yield of PHB. According to the above proteomic analysis results, Lee et al. (2013) overexpressed *fba* and *tpiA* in *E. coli* cells, which increased the synthesis flux of acetyl-CoA and increased the yield of PHB from 0.4 to 1.6 g/L.

In the glycolysis pathway of *E. coli*, the 3-phosphoglyceraldehyde dehydrogenase (GAPDH) encoded by *gapA* is the main source of NADH production. Under anaerobic conditions, the glycolysis pathway of *E. coli* still maintains a relatively high flux, which can meet the needs of NADH-dependent reactions (Bastian et al. 2011), while the lower flux of the PP pathway limits the supply of NADPH. Some researchers have replaced the *gapA* gene of *E. coli* with the *gapC* gene encoding $NADP^+$-dependent GAPDH from *Clostridium acetobutylicum*, which enhanced the synthesis of NADPH in cells, and the production of NADPH-dependent products, lycopene and caproic acid increased by 150% and 95%, respectively (Martínez et al. 2008). Wang et al. (2013b) also replaced the *gapA* gene of *E. coli* with the *gapB* gene encoding $NADP^+$-dependent GAPDH from *Bacillus subtilis,* which also increased the production of three NADPH-dependent compounds: lycopene, caproic acid, and 3-chloropropionic acid.

3. *Metabolic Engineering and Application of the Pentose Phosphate Pathway*

The pentose phosphate pathway (PPP) is a way of glucose oxidative decomposition. Since this pathway starts with glucose-6-phosphate (G-6-P), it is also called the hexose phosphate shunt. This pathway occurs in the cytoplasm and can be divided into two stages. The first stage starts with the dehydrogenation of G-6-P to form 6-phosphoglucono-δ-lactone, which is then hydrolyzed to form 6-phosphogluconic acid, and then oxidized and decarboxylated to form ribulose-5-phosphate. NADP is the electron acceptor in all the above oxidation reactions. The second stage is when ribulose-5-phosphate undergoes a series of ketone and aldehyde transfer reactions, and finally forms glyceraldehyde-3-phosphate and fructose-6-phosphate through intermediates such as erythrose-4-phosphate, ribose-5-phosphate, and sedoheptulose-7-phosphate, which two can reenter the glycolytic pathway for metabolism (Fig. 3.7).

The PPP is the main metabolic pathway for the synthesis of NADPH. Glucose-6-phosphate dehydrogenase (G6PDH) and 6-phosphogluconate dehydrogenase (6PGDH) are the key enzymes for the PPP to reduce $NADP^+$ to synthesize NADPH. Overexpression of G6PDH and 6PGDH can effectively increase the flux of the PPP and the level of NADPH in cells (Lim et al. 2002).

Regulating the flux of the PPP can control the ratio of NADPH/$NADP^+$ and thus control the intracellular level of acetyl-CoA. The production of NADPH is mainly related to two key enzymes of the PPP, one is glucose-6-phosphate dehydrogenase (G6PD, encoded by *zwf* gene). Another is 6-phosphogluconate dehydrogenase (6-PGDH, encoded by *gnd* gene) (Wang et al. 2013c). Overexpression of the *zwf* gene in *E. coli* can increase the NADPH/$NADP^+$ ratio by six times; overexpression of the *gnd* gene can increase NADPH/NADP$^+$ by two times. So, overexpression of these two genes can increase the synthesis flux of acetyl-CoA. Song et al. (2006) overexpressed the *talA* gene (encoding transaldolase) in *E. coli*, increasing NADPH/NADP$^+$ by 1.62 times and ultimately doubling the yield of PHB. Jung et al. (2004) overexpressed the *tktA* gene (encoding transketolase), directing the metabolic flow of the non-oxidative PPP toward the production of glyceraldehyde-3-phosphate, increasing the flux of acetyl-CoA synthesis in *E. coli*, and increasing the supply of NADPH, ultimately increasing the yield of PHB by 1.7 times.

The *pgi* gene encodes phosphoglucose isomerase, which catalyzes glucose-6-phosphate into EMP. Knocking out *pgi* will block glucose-6-phosphate from entering the EMP pathway, and more glucose-6-phosphate will enter the PPP, increasing the carbon flux of the PPP. Studies have shown that knocking out the *pgi* gene in *E. coli* can increase the biotransformation rates of colorless anthocyanins and catechins that depend on NADPH by four times and two times, respectively (Kim et al. 2011). In addition, the knockout of *pgi* also increased the thymidine synthesis ability of *E. coli* by four times, not only because NADPH can drive UDP to transform into dUDP but also because it promotes the transformation of tetrahydrofolate into dihydrofolate. The generated dihydrofolate can be converted into thymidine together with dTMP (Chemler et al. 2010). The knockout of the *pfkA* and *pfkB* genes encoding phosphofructokinase (PFK) can promote one of the PPP products, fructose-6-phosphate, to produce glucose-6-phosphate through the gluconeogenesis pathway. The additional generation of glucose-6-phosphate will further increase the carbon flux of the PPP and the level of NADPH. In *E. coli*, after knocking out *pfkA* and *pfkB*, the YRPG (product molar amount/sugar consumption molar amount) of NADPH-dependent 3-methylhydroxybutyrate (MHB) increased by 44% compared to when *pgi* was knocked out (Lee et al. 2010; Siedler et al. 2011).

4. *Metabolic Modification and Application of the 2-Keto-3-Deoxy-6--Phosphogluconate (KDPG) Pathway*

The 2-keto-3-deoxy-6-phosphogluconate (ED) pathway is an important competitive pathway of the TCA cycle. Glucose, through the ED pathway, generates pyruvate in just four reaction steps, where the key reaction steps in the pathway are catalyzed by 6-phosphogluconate dehydratase and 2-keto-3-deoxy-6--phosphogluconate aldolase, catalyzing 6-phosphogluconate to produce 3-phosphoglyceraldehyde and pyruvate, which can then enter the EMP pathway and PPP.

In *E. coli*, when glucose is used as a carbon source, the ED pathway is in a silent state. Zhang et al. (2014) activated the ED pathway by overexpressing the *edd* gene (encoding 6-phosphogluconate dehydratase) and the *eda* gene (encoding 2-keto-3-deoxy-6-phosphogluconate aldolase) in *E. coli*, and further enhanced the metabolic flow of the ED pathway by overexpressing the *zwf* gene, strengthening the supply of pyruvate and NADPH. Although no increase in PHB production was observed, the accumulation of pyruvate significantly increased. Subsequently, by further overexpressing the pyruvate dehydrogenase complex gene, the concentration of acetyl-CoA was increased by more than three times, effectively improving the ability of *E. coli* to produce PHB.

5. *Metabolic Modification and Application of the Citric Acid Cycle*

The citric acid cycle (also known as the tricarboxylic acid cycle, TCA) is a cyclic system used to oxidize the acetyl group in acetyl-CoA into CO_2 and reducing equivalents through enzymatic reactions. The first step of this cycle is the condensation of acetyl-CoA with oxaloacetate to form citrate. The reactant acetyl-CoA (connected by a molecule of coenzyme A and an acetyl) is a common intermediate product of carbohydrate, fat, and amino acid metabolism. After entering the cycle, it will be decomposed, eventually generating the product CO_2 and producing H, which is transferred to coenzyme I—nicotinamide adenine dinucleotide (NAD^+) and flavin adenine dinucleotide (FAD), transforming them into $NADH^+$, $H^{+,}$ and $FADH_2$. $NADH^+$, $H^{+,}$ and $FADH_2$ carry H into the respiratory chain, where the respiratory chain transfers electrons to O_2 to produce H_2O, while coupling oxidative phosphorylation to produce ATP, providing energy.

E. coli uses glucose as a carbon source, and when fermenting under sufficient O_2 conditions, the cell metabolism is faster, and most of the metabolic flow of acetyl-CoA will enter the TCA cycle due to the high affinity of citrate synthase (CS) (K_m = 0.07 mmol/L). Reducing the flux of the TCA cycle can increase the accumulation of acetyl-CoA, thereby increasing the production of the target product. Fowler et al. (2009) knocked out the succinate dehydrogenase (SDH) gene and citrate lyase (CITE) gene of *E. coli*, blocking the metabolic flow of acetyl-CoA entering the TCA cycle, and finally increased the yield of acetoacetyl-CoA from 0.95 to 2.60 nmol/mg. Xu et al. (2011) knocked out the fumarase (FUM) and succinyl-CoA synthetase (SUC) genes in *E. coli*, down-regulating the flux of the TCA cycle; at the same time, by overexpressing phosphoglycerate kinase (PGK), glyceraldehyde-3-phosphate dehydrogenase (GAPD) and PDH, the metabolic flow of acetyl-CoA in *E. coli* was significantly increased; combined with the overexpression of the acetyl-CoA carboxylase (ACC) gene, the yield of acetoacetyl-CoA was increased nearly four times without affecting cell growth, and the yield of downstream product flavanone was increased by 5.6 times.

6. *Modification and Application of Pyruvate Metabolism*

In *E. coli*, pyruvate is an important metabolic intermediate in the glycolysis pathway, which can be used for the synthesis of lactic acid, alanine, acetate ester, and acetyl-CoA, as well as for the synthesis of ethylene glycol, 2,3-butanediol, and isobutanol (Atsumi et al. 2008; Nielsen et al. 2010). Currently, there are three main ways to modify the pyruvate pathway: one is to reduce the activity of the pyruvate dehydrogenase complex (PDHC), which consists of three subunits, including the EI subunit (pyruvate dehydrogenase) encoded by the *aceE* gene, the E2 subunit (dihydrohpoamide transacetylase) encoded by the *aceF* gene, and the E3 subunit (dihydrolipoamide dehydrogenase) encoded by the *IpdA* gene. Song et al. (Canhui and Weiguo 2013) used the Red recombination system to construct a nutrition-deficient *E. coli* strain MG1655. After knocking out the *aceE* gene in PDHC *of E. coli* strain

MG1655, the flow of pyruvate to the TCA cycle was blocked, thereby promoting the accumulation of pyruvate. The second is to reduce the activity of the α-ketoglutarate dehydrogenase complex (AKGDH), which also consists of three subunits, including the E1 subunit (α-ketoglutarate decarboxylase, KCBX) encoded by the *sucA* gene, the E2 subunit (ipoyl transsuccinylase, LTS) encoded by the *sucB* gene, and the E3 subunit encoded by the *IpdA* gene. Causey et al. (2003, 2004) knocked out the *atpFH*, *adhE* , and *sucA* genes of *E. coli* to reduce its ATP synthesis, as well as reduce its cell growth and CO_2 production rate, thereby enhancing its accumulation from glucose to pyruvate. The third is to reduce the consumption of pyruvate by reducing the metabolic pathway of pyruvate. *E. coli* strain YYC202 is a successfully modified nutritional defect strain used for pyruvate accumulation. In this strain, the genes encoding pyruvate dehydrogenase, phosphoenolpyruvate synthase, pyruvate formate lyase, and pyruvate oxidase are mutated, thereby removing the downstream metabolic pathway that can consume pyruvate (Zelić et al. 2010; Zhu et al. 2007). In summary, there are currently two main strategies to promote the accumulation of pyruvate in *E. coli*: one is to increase the synthesis speed of pyruvate and reduce feedback inhibition regulation; the other is to remove the metabolic pathway of pyruvate and reduce the consumption of pyruvate.

7. *Metabolic Modification and Application of the Glyoxylate Cycle*

The main control point of the glyoxylate cycle is the activation and deactivation of isocitrate dehydrogenase (IDH). When *E. coli* uses acetate as a carbon source, about 75% of IDH is in a deactivated state, which can lead to an increase in the concentration of isocitrate, thereby enabling isocitrate lyase to function (Kurata and Sugimoto 2018). In addition, when the TCA cycle flux is low and acetate is present, the isocitrate lyase of the intracellular glyoxylate cycle branch will also be activated. Isocitrate lyase regulatory factor (ICLR) can regulate the expression level of some genes in the TCA cycle, directing the carbon metabolic flow toward the glyoxylate branch, thereby changing the synthesis flux of acetyl-CoA. ICLR inhibits the expression of the *aceBAK* operon, composed of the three enzyme genes in the glyoxylate branch: malate synthase (MS), isocitrate lyase (ICL), and isocitrate dehydrogenase kinase/phosphatase (ACEK). After knocking out the *iclR* gene, the glyoxylate branch was activated, the metabolic flow of acetic acid and PEP synthesizing acetyl-CoA was increased, and the yield of 3-hydroxypropionate (3-HP) with acetyl-CoA as a precursor was increased by 1.9 times (Liu et al. 2017b).

The general metabolic pathway for anaerobic fermentation synthesis of succinic acid by *E. coli* is each molecule of glucose generates two molecules of NADH, and two molecules of phosphoenolpyruvate (PEP) are involved, with each PEP molecule undergoing reduction via the TCA arm to consume two molecules of NADH and produce succinic acid. Due to the limitation of NADH, the theoretical yield of succinic acid is only 1 mol/mol glucose. The acetaldehyde pathway can also produce succinic acid and helps to reduce the consumption of the cofactor NADH. Knocking out the *iclR* gene (Gui et al. 1996) that suppresses the acetaldehyde pathway operon (aceBAK) can activate the acetaldehyde cycle while blocking the overflow metabolic pathway of succinic acid synthesis to acetic acid, lactic acid, and ethanol. The succinic acid yield of the recombinant *E. coli* strain SBS550MG obtained through the above modifications reached 1.6 mol/mol glucose, with an average consumption of 1.25 mol NADH per mole of succinic acid (Sánchez et al. 2005). Zhu et al. (2013) found that under anaerobic conditions, knocking out the competitive pathway of *E. coli* succinic acid can activate its acetaldehyde pathway, where the knockout of acetic acid synthesis-related genes *ackA* (encoding acetate kinase) and *pta* (encoding phosphate acetyltransferase) plays a key role in activating the acetaldehyde pathway. When *ackA-pta* and other competitive pathways were knocked out in *E. coli*, the flux of the acetaldehyde cycle under anaerobic conditions increased from 0% to 31%, and the fermentation concentration of succinic acid also increased by five times. Therefore, when *E. coli* synthesizes succinic acid under anaerobic conditions, the activation of the acetaldehyde pathway can effectively reduce the consumption of NADH, thereby increasing the yield of succinic acid.

3.2 Biosynthesis of Polyhydroxyalkanoates

Polyhydroxyalkanoates (PHAs) are a general term for high molecular weight polyesters completely synthesized by microorganisms (Chen and Patel 2012). PHAs can be divided into short-chain-length (SCL) PHAs and medium-chain-length (MCL) PHAs based on the number of carbon atoms in their monomers. SCL-PHA monomers usually contain 3 to 5 carbon atoms, such as polyhydroxybutyrate (PHB) and polyhydroxyvalerate (PHV), while MCL-PHA monomers generally contain 6 to 14 carbon atoms, such as polyhydroxyhexanoate (PHHx) and polyhydroxyoctanoate (PHO). Since the polymerization of different monomers requires PHA synthases with different substrate specificities, most natural microorganisms can only synthesize one type of PHA. PHAs composed of short-chain and medium-chain monomers can combine the excellent properties of SCL-PHAs and MCL-PHAs, thus attracting more and more attention from academia and industry. Materials synthesized from PHAs can acquire new functions through chemical modification of part of the PHA structure (side chains containing phenyl rings, halogens, unsaturated bonds, etc.).They can also be block copolymerized, allowing high polymers with different properties to be chemically linked together to obtain various materials with different properties, the properties of which can be controlled by adjusting the ratio of different blocks (Yang et al. 2021). Due to the diversity of PHA monomers and the significant differences in chain length between them, the properties (such as hardness and softness) of different PHA materials vary greatly. PHA has biodegradability and biocompatibility and is therefore considered an environmentally friendly material that can help solve the increasingly serious environmental pollution problem. The production of PHA consists of multiple stages, including synthesis, extraction, processing, and the formation of the final product, involving various fields such as fermentation engineering, material purification, and processing technology. The rapid development of synthetic biology has greatly promoted the discovery of various PHA-producing strains and the development of their production strategies. This section mainly introduces the development history, types, and main production strategies of PHA.

3.2.1 Diversity and Biosynthesis of PHA

1. *Diversity of PHA*

PHA has been developed for over 100 years since it was first discovered in microorganisms in nature. As early as the beginning of the twentieth century, researchers first observed lipid-like inclusions in bacteria, and then found similar lipid-like inclusions {poly(3-hydroxybutyrate) [P(3HB)]} in *B. megaterium*, so P(3HB) is considered the first member of PHA (Raza et al. 2018). Based on this discovery, researchers have studied strains of *Bacillus* such as *B. cereus* and *B. megaterium*, and found that microorganisms of this genus can synthesize and degrade PHA using different substrates (Lemoigne 1926a). Further research found that PHA can serve as an energy storage material in bacteria, and their accumulation or degradation of PHA is easily affected by environmental changes. Currently, over 160 different hydroxy fatty acids (HA) have been identified that can serve as PHA monomers, and plastics synthesized from different HAs exhibit a variety of characteristics ranging from thermoplastics to elastomers (Macrae and Wilkinson 1958; Steinbüchel and Valentin 1995).

HA has a variety of side chains, and the length of their carbon chains varies, so the PHA polymerized from HA as monomers is also diverse. The majority of PHA monomers are 3-hydroxy fatty acids (3HA), but fatty acids with hydroxy in other positions can also serve as monomers to synthesize PHA. For example, recombinant *Pseudomonas putida* can synthesize PHA containing 5-hydroxyhexanoic acid, 4-hydroxyheptanoic acid, and 4-hydroxyoctanoic acid (Chen and Wu 2005). Li et al. found that PHA can also be synthesized using HA with unsaturated double bond side chains as monomers (Valentin et al. 1991). PHA with unsaturated double bonds can bring new high-value applications after subsequent chemical modification and processing (such as grafting other polymers with special functions onto the linear backbone of PHA).

The way different monomers aggregate is another important factor affecting the diversity of PHA. PHA composed of only one type of monomer is called a homopolymer, while PHA formed by the aggregation of two or more types of monomers is called a copolymer. The copolymer PHA synthesized by natural microorganisms is generally a random copolymer, and due to the influence of the metabolic background of the microorganisms, some monomers cannot synthesize the corresponding homopolymers. For example, when long-chain fatty acids are used as substrates, microorganisms will reduce the carbon chain length of fatty acids through fatty acid metabolism, resulting in the PHA synthesized by microorganisms being a random copolymer formed by monomers of different carbon chain lengths. However, with the development of molecular biology and genetics, the process of microbial synthesis of PHA has become controllable, and scientists have begun to try to use recombinant microorganisms to produce new types of PHA that natural microorganisms cannot synthesize. Such as knocking out the genes of related enzymes in the fatty acid metabolism pathway of *Pseudomonas entomophila*, it can use a long-chain fatty acid of a certain carbon chain length as a substrate to synthesize a homopolymer PHA of the corresponding carbon chain length (Li et al. 2014a). Now the synthesis process of copolymer PHA has also become controllable, which has led to the emergence of a new type of polymer—block copolymers (Chung et al. 2011; Tripathi et al. 2012). Block copolymers can theoretically greatly increase the diversity of PHA. According to the composition and order of the blocks, it can be divided into blocks between homopolymers of multiple monomers, blocks between homopolymers and random copolymers, blocks between homopolymers and block copolymers, blocks between random copolymers and block copolymers, etc., which makes the diversity of PHA infinitely broadened. The molecular weight of PHA can range from tens of thousands to tens of millions of daltons, which is also a manifestation of the diversity of PHA. Even in the same microbial cell, the molecular weight of the synthesized PHA is only distributed within a certain range, so the current detection methods give the average molecular weight of all PHA molecules. The controllability of the PHA synthesis process makes the molecular weight of PHA also controllable, such as mutating the N-terminus of the PHA synthase gene to increase the PHA polymerization ability (Tappel et al. 2012), using methanol to induce the termination of the PHA synthesis chain (Zheng et al. 2006), or adding polyethylene glycol to the culture medium to regulate the molecular weight of PHA (Ashby et al. 2015). If the activity of key enzymes in the PHA synthesis process can be precisely controlled by molecular biology means, it will be possible to achieve more precise control of the PHA molecular weight.

2. *Biosynthesis of PHA*

PHA can be divided into short-chain-length (SCL) PHA and medium-chain-length (MCL) PHA.

The monomers of SCL-PHA usually contain 3 to 5 carbon atoms, and its representative examples are P(3HB) and P(3HB-*co*-3HV). P(3HB) is one of the most widely studied PHAs. In its biosynthetic pathway, β-ketothiolase (PhaA) condenses two acetyl-CoA into acetoacetyl-CoA, which is then reduced to (*R*)-3-hydroxybutyryl-CoA by NADPH-dependent acetoacetyl-CoA reductase (PhaB), and finally, (*R*)-3-hydroxybutyryl-CoA is polymerized into P(3HB) under the action of PHA synthase (PhaC). The PHA biosynthetic operon *phaCAB* was first identified in *Cupriavidus necator* (Ashby et al. 1997). It is reported that several microbes can produce P(3HB), including certain strains of *C. necator*, *Rhodospirillum rubrum*, *Beijerinckia indica*, *Azotobacter vinelandii*, *Caryophanon latum*, *Allochromatium vinosum*, *Pseudomonas* sp., *Alcaligenes latus*, and *Bacillus* sp. (Schubert et al. 1988).

The monomers of MCL-PHA found in *Pseudomonads* sp. usually contain 6 to 14 carbon atoms. MCL-HA-CoA can be produced through fatty acid metabolism, such as by using the β-oxidation or de novo fatty acid biosynthesis pathway (Choi et al.

2020). The fatty acid metabolism of the β-oxidation pathway generates (*R*)-3HA-CoA of different carbon chain lengths, which is then polymerized into MCL-PHA under the action of PhaC. Acetyl-CoA can be converted into (*R*)-3HA-CoA through the de novo fatty acid biosynthesis pathway, thus enabling the synthesis of MCL-PHA using inexpensive carbon sources. Many studies have successfully constructed the biosynthetic pathway of MCL-PHA in different strains of *P. aeruginosa* and *P. putida* (Schubert et al. 1988).

As early as the early 1990s, researchers discovered PHA composed of SCL and MCL monomers (mainly 3HB, 3HV, and 3HHx) in bacteria such as *Rhodospirillum rubrum* (Andin et al. 2017), *Rhodocyclus gelatinosus* (Brandl et al. 1989) and *Rhodococcus* sp. (Liebergesell et al. 1991) grown on media with hexanoic acid as a carbon source. In addition, it was found that when fatty acids with more than 12 carbon atoms, such as trilaurin, coconut oil, and olive oil, were used as carbon sources, P(3HB-*co*-3HHx) could also be produced in two *Aeromonas* strains. The 3HHx component in the P(3HB-*co*-3HHx) produced by these two strains would vary with the concentration of the carbon source and the culture conditions (Haywood et al. 1991). After studying the PHA biosynthesis gene of *Aeromonas caviae*, it was found that the key enzyme involved in the PHA biosynthesis of this bacterium is PhaC, and PhaC enzyme has a high substrate specificity for 3HB-CoA and 3HHx-CoA. In addition, it was also found in this bacterium that (*R*)-specific enoyl-CoA hydratase (PhaJ) is responsible for using the corresponding enoyl-CoA to produce 3HB-CoA and 3HHx-CoA (Kobayashi et al. 1994). Currently, genes of several enoyl-CoA hydratases have been identified in strains of *P. aeruginosa*, *P. putida*, and *Escherichia*. These enoyl-CoA hydratases are all involved in the biosynthesis of PHA in the fatty acid β-oxidation pathway (Fukui et al. 2001).

PHA is usually produced in the form of amorphous copolymers because the producing strains will simultaneously generate various HA-CoA monomers and mix them in the cytoplasm. In order to diversify the material characteristics of PHA, block PHA can be synthesized by controlling the order of intracellular generation of each monomer. For example, *C. necator* can produce P(3HB) and P(3HV) [P(3HB)-*b*-P(3HV)] block copolymers by controlling the feeding order of two substrates (fructose and valeric acid), where the P(3HB) segment is mainly formed during the utilization of fructose, while valeric acid in *C. necator* is mainly used to transform into the precursor of 3HV (Park and Lee 2003).

Recombinant *E. coli* is an important chassis cell for the production of block PHA copolymers. Based on the construction and regulation of synthetic pathways, and combined with the use of synthetic biology tools, it is now possible to use recombinant *E. coli* to produce P(3HB)-*b*-P(3HP). In recombinant *E. coli*, the synthesis of 3HB and 3HP has independent regulation systems, which can be regulated separately by different inducers [L-arabinose and isopropyl β-D-1-thiogalactopyranoside (IPTG)] (Pederson et al. 2006). Studies have shown that biosynthetic block copolymers exhibit different material properties from random copolymers composed of similar monomers. For example, compared with P(3HB-*co*-4HB) random copolymers and block copolymers of P(HB) and P(4HB), P(3HB)-*b*-P(4HB) block copolymers exhibit better physical properties, especially in terms of yield strength and tensile strength (Wang et al. 2013d).

3.2.2 Main Strategies for PHA Biosynthesis

1. *Metabolic Engineering Control Strategy*

Metabolic engineering of microorganisms and the introduction of new metabolic pathways can achieve the goal of synthesizing PHA using low-cost substrates. However, to increase the yield of PHA, it is necessary to optimize by regulating the expression of genes in the metabolic pathway and controlling the entire metabolic network of the cell. By adjusting the transcriptional strength of the gene promoter in the metabolic pathway or the translational strength of the ribosome binding site (RBS), the strength of the PHA synthesis metabolic flow can be regulated. However, the synthesis pathway of PHA is not independent of the metabolic network of the microorganism itself. Blindly overexpressing the PHA synthesis pathway does not necessarily increase the production of PHA and may even affect other metabolic pathways in the cell, which will inhibit cell growth and PHA accumulation. Systems biology can start from the entire metabolic network of the cell, systematically analyze the impact of each metabolic pathway on the target product, and optimize the synthesis of PHA by controlling other metabolic pathways. For example, compared to glucose, *P. putida* prefers to use fatty acids as substrates. After knocking out the glucose dehydrogenase gene *gcd*, the expression of glucose metabolism and PHA synthesis-related genes in the recombinant *P. putida* strain was upregulated, and the gene expression of other metabolic pathways was almost unaffected. The ability of the recombinant strain to synthesize PHA also increased by 100% (Hu et al. 2011). In the recombinant strain with the *gcd* gene knocked out, overexpression of the *acoA* gene encoding the subunit of pyruvate dehydrogenase increased the synthesis ability of PHA by 121% compared to the wild-type *P. putida* (Poblete-Castro et al. 2013). The β-oxidation pathway is one of the three main pathways for PHA synthesis. Using fatty acids as substrates to provide short-chain and medium-chain monomers is a common method for producing PHA with this pathway (Borrero-de Acuña et al. 2014; Andreessen et al. 2010; Sánchez et al. 2003; Kroumova et al. 2002; Cavalheiro et al. 2012). Modifying the β-oxidation pathway is of great significance for promoting the synthesis of MCL-PHA or short-medium-long chain copolymer PHA (SCL-*co*-MCL-PHA). In 1994, researchers first discovered that *A. caviae* can use alkanes and oils to produce random copolymers of poly-3-hydroxybutyrate-3-hydroxyhexanoate (PHBHHx) (Doi et al. 1989). β-oxidation is the main pathway in this process, in which acyl-CoA dehydrogenase (*fadE*) catalyzes the reaction of acyl-CoA to enoyl-CoA, which is the rate-limiting step in the β-oxidation process. Studies have shown that overexpression of *fadE* in *E. coli* can increase the amount of enoyl-CoA, and then expressing *phaJ* and *phaC* can make the recombinant strain accumulate more PHA (Kobayashi et al. 1994). However, in the β-oxidation process, cells may convert most of the fatty acids into acetyl-CoA for cell growth, thereby wasting expensive fatty acids on acetyl-CoA that can be produced with cheap substrates such as glucose (Lu et al. 2010; Valentin and Dennis 1997; Li et al. 2010). Due to the presence of β-oxidation, the efficiency of converting fatty acids into PHA is low, resulting in high production costs for MCL-PHA. Therefore, modifying the β-oxidation pathway is also an important strategy to improve PHA production efficiency. In *P. putida* or *P. entomophila*, knocking out the genes encoding 3-hydroxyacyl-CoA dehydrogenase (*fadB*) and 3-ketoacyl-CoA thiolase (*fadA*) in the β-oxidation pathway can convert most fatty acids into 3-hydroxyacyl-CoA for PHA synthesis, instead of being oxidized to acetyl-CoA, thereby significantly improving the conversion rate from substrate to MCL-PHA (Chung et al. 2011; Wang et al. 2009; Chen and Jiang 2017). It has been reported that partial knockout of the β-oxidation pathway or inhibition of β-oxidation with propionic acid can result in PHA synthesis monomers of the same or longer chain as the added fatty acid substrate (Liu et al. 2010). In addition, the ratio of various fatty acids in the substrate can also affect the composition of PHA (Stephan et al. 2013). Using the partially knocked out β-oxidation *P. putida* KT2442 strain as a controllable PHA production platform, the monomer ratio of PHA can be precisely regulated by adding a predetermined ratio of fatty acids, and random and block copolymers of poly(3-hydroxybutyrate-3-hydroxyhexanoate) [P(3HB-*co*-3HHx)] can be synthesized, the monomer composition and material properties of which are quite stable (Qi et al. 1998; Tripathi et al. 2013). Similarly, by partially knocking out the β-oxidation pathway and introducing a pathway for direct synthesis of PHB from acetyl-CoA, the recombinant *P. entomophila* LAC32 strain has been successfully developed into a SCL-*co*-MCL PHA synthesis platform, which can synthesize a variety of new PHAs such as poly(3-hydroxybutyrate-3-hydroxydecanoate) [P(3HB-*co*-3HD)], poly(3-hydroxybutyrate-3-hydroxydodecanoate) [P(3HB-*co*-3HDD)] (Tripathi et al. 2012). In addition, fatty acid oxidation mutants of *Pseudomonas* sp. strains can also produce medium-chain PHAs such as homopolymers, random copolymers, or block copolymers of 3-hydroxyhexanoate (3HHx), 3-hydroxyoctanoate (3HO), 3-hydroxydecanoate (3HD), and 3-hydroxydodecanoate (3HDD) (Li et al. 2018). Although many studies have shown that modifying the β-oxidation pathway can improve the yield of biosynthetic PHA, it is worth noting that β-oxidation-defective strains cultured only with fatty acids may grow slowly or accumulate only a small amount of MCL-PHA (Chen et al. 2015; Green et al. 2002).

Synthetic biology technology is also one of the important means to regulate the PHA synthesis pathway. For example, CRISPR technology can achieve gene knockout, replacement, and other editing purposes, and its derivative technology—CRISPRi can—regulate the metabolic pathway by inhibiting the expression of target genes, which has been used to regulate the synthesis of PHA. Lü et al. used CRISPRi technology to inhibit the expression of the *sad* gene encoding succinic semialdehyde dehydrogenase in a recombinant *E. coli* strain containing the metabolic pathway for synthesizing P3HB4HB from glucose, thereby regulating the metabolic flow of 4HB synthesis (Ward and O'Connor 2005). They designed different sgRNAs to inhibit different expression levels of the *sad* gene, thereby making the metabolic flow intensity toward 4HB synthesis different, and finally synthesized different P3HB4HB with 4HB molar ratio content from 1% to 9% (Ward and O'Connor 2005). On this basis, further inhibition of the expression of the succinyl-CoA synthetase gene (*sucC* and *sucD*) and succinic dehydrogenase gene (*sdhA* and *sdhB*) can further enhance the metabolic flow of 4HB synthesis, expanding the regulation range of the composition ratio of 4HB in P3HB4HB from 1.4% to18.4% (Ward and O'Connor 2005). If these genes are knocked out one by one according to traditional molecular biology methods, it will not only be time-consuming and laborious, but also severely affect the growth of the strain due to the knockout of some essential genes. The use of CRISPRi technology can achieve simultaneous inhibition of multiple target genes, including essential genes, and different degrees of expression of target genes can be inhibited by designing different sgRNAs, allowing for fine-tuning of metabolic pathways.

2. *Key Enzymes in the PHA Synthesis Pathway and Their Modification*

So far, many enzymes related to PHA biosynthesis and degradation have been identified, such as PHA synthase (PhaC), β-ketothiolase (PhaA), NADPH-dependent acetoacetyl-CoA reductase (PhaB), 3HA-ACP:CoA transacylase (PhaG), enoyl-CoA hydratase (PhaJ); enzymes involved in fatty acid β-oxidation, such as 3-ketoacyl-CoA thiolase (FadA), 3HA-CoA dehydrogenase (FadB), acyl-CoA synthetase (FadD) and acyl-CoA dehydrogenase (FadE); de novo fatty acid biosynthesis pathway enzymes, such as β-ketoacyl-ACP reductase (FabG), 3HA-ACP dehydratase (FabA), β-ketoacyl-ACP synthase I (FabB), β-ketoacyl-ACP synthase II (FabF), β-ketoacyl-ACP synthase III (FabH), enoyl-ACP reductase (FabI), malonyl-CoA:ACP-transacylase (FabD), PHA depolymerase (PhaZ), transcription regulator (PhaD); several PHA granule-associated proteins (PhaF, PhaI, PhaP, and PhaM) (Schubert et al. 1988).

Among these enzymes, PhaC plays an important role in the pathway of polymerizing HA-CoA for PHA production. According to substrate specificity and enzyme subunit type, PhaC can be divided into four types (I, II, III, and IV). Type I PhaC, containing a single subunit enzyme (for example, PhaC found in *C. necator* and *A. latus*), exists in the form of a dimer, which can polymerize SCL-HA-CoA. Class III and IV PhaC can also use SCL-HA-CoA as a substrate, but they are composed of different subunits. Class III PhaC, represented by *A. vinosum* PhaC, is composed of two different subunits, PhaC and PhaE. Class IV PhaC, found in *Bacillus,* is composed of PhaC and PhaR subunits. Class II PhaC found in *Pseudomonas* (including single-subunit enzymes, encoded by *phaC1* and *phaC2*) can use MCL-HA-CoA as a substrate to synthesize PHA. The identified PhaCs include Class I: *A. caviae* PhaC, Class II: *Pseudomonas* sp. 61-3, and *Pseudomonas* sp. MBEL 6-19 PhaC, Class III: *Thiocapsa pfennigii* PhaC, and Class IV: *B. cereus* PhaC (Schubert et al. 1988). In addition to PhaC, PHA granules also encapsulate various proteins, such as Phasin, PhaZ, and PhaM. Phasin is an amphipathic protein and is the most abundant protein on the surface of PHA granules. Identified Phasins include PhaP1-5 from *C. necator*, PhaF and PhaI from *P. putida*, and PhaP from *B. megaterium.* Phasin can change the number, size, and distribution of PHA granules and can also regulate the accumulation of PHA. Similar to Phasin, PhaM affects the morphology of PHA granules and activates the biosynthesis of PHA by forming a PhaC/PhaM complex. Studies on the *phaM* deletion mutant of *C. necator* found that PhaM can affect the size and distribution of PHA granules during cell division, because this enzyme can bind both PHA granules and genomic DNA (Lv et al. 2015).

3. *Development and Modification of Production Strains*

Since PhaC can determine the general characteristics of the produced PHA (including monomer type and composition, production rate, molecular weight, and polydispersity), researchers have been committed to designing more active PhaC with specificity for different substrates in order to regulate the monomer composition of PHA (Pfeiffer et al. 2011). The crystal structure of PhaC was only recently resolved, and previous methods of PhaC modification mainly relied on random mutation and screening or structure prediction based on sequence homology (Schubert et al. 1988). Using these methods, many important results have been discovered, such as the synthesis of P(3HB-*co*-3HHx) by *A. caviae* PhaC mutants obtained through random mutation and screening. These mutations were then introduced into *E. coli* that produces P(3HB). The *E. coli* strain JM109 was used, and JM109 was allowed to grow on a medium containing the lipophilic fuel—Nile red, and then further visually screened for mutants with high PHA content (high Nile red content). The mutated genes of the selected mutants were reintroduced into *E. coli* strain LS5218, which can use sodium dodecyl sulfate as the sole carbon source to produce P(3HB-*co*-3HHx). It was found that single amino acid substitutions at Asn149Ser and Asp171Gly could promote the production of P(3HB) and P(3HB-*co*-3HHx), and increase the content of 3HHx component (Sim et al. 1997). In addition, it was found that the double mutant of *A. caviae* PhaC with Asn149Ser and Asp171Gly mutations could combine 3HO monomer with 3HHx from octanoic acid (Kichise et al. 2002). Through random mutagenesis and screening, the amino acid residues Glu130, Ser325, Ser477, and Gln481 in *Pseudomonas* sp. 61-3 PhaC1 were identified as important residues that can change substrate specificity and activity. Compared with the wild-type strain, the P(3HB) yield of the *Pseudomonas* sp. 61-3 mutant strain with Glu130Asp, Ser325Thr, Ser477Gly, and Gln481Lys quadruple mutations increased by 400 times (Tsuge et al. 2007). Various microbial strains have been developed for the production of PHA. The most typical case is the recombinant *E. coli* with the *phaCAB* operon from the *C. necator* strain. Compared with the natural *C. necator*, this recombinant *E. coli* strain has a higher production efficiency and P(3HB) production rate. This recombinant *E. coli* strain can produce ultra-high molecular weight P(3HB) (molecular weight of 20 MDa). In addition to bacteria, yeasts and plants have also been reported to be used for the production of PHA (Schubert et al. 1988). To reduce the production cost of PHA, we can start by saving costs in various stages of the fermentation process. The cost control in the biological fermentation process involves multiple stages, including substrate cost, water consumption, energy consumption brought by sterilization, inefficient production caused by batch fermentation, fixed asset costs required for stainless steel equipment, and labor costs. Although some of these issues can be partially resolved through strain modification, it is still a very large and complex project. Therefore, finding a new PHA-producing strain that can partially solve the above problems as a chassis strain, and then solving the remaining problems through strain modification, may significantly reduce the production cost of PHA. Most of the above problems are caused by the sterilization process of fermentation; for example, sterilization requires high-temperature and high-pressure steam, and the preparation of steam requires a large amount of energy; the equipment needs to withstand high temperature and high pressure (requires stainless steel); to prevent bacterial contamination, the fermentation process has to be divided into multiple batches; sterilization-related operations are complex, time-consuming, and labor-intensive, which will eventually be included in labor costs. Therefore, a new chassis strain that does not require sterilization and will not be contaminated by other miscellaneous bacteria can greatly reduce the production cost of PHA. Extremophiles, due to their unique growth conditions, find it difficult for other miscellaneous bacteria to survive in their suitable growth conditions; therefore, they meet the above conditions and are ideal new PHA-producing bacteria.

Halophilic microorganisms, a type of extremophile, have been used for non-sterilized fermentation research of PHA. The growth conditions required for these microorganisms are high concentrations of salt (NaCl), many halophilic microorganisms also prefer alkaline conditions, under this double pressure, ordinary bacteria cannot grow, therefore, using halophilic microorganisms for the fermentation production of PHA can carry out continuous fermentation without sterilization, greatly reducing the related costs brought by the sterilization process (energy consumption, equipment, labor costs, etc.) (Matsumoto et al. 2006). In addition, the growth of halophilic microorganisms requires a salt-containing medium, which means that seawater can be used to replace the freshwater in the fermentation medium, thus not needing to occupy the increasingly scarce water resources. This seawater-based fermentation production technology is called blue water biotechnology.

Representative strains of halophilic microorganisms that can synthesize PHA include *Halomonas boliviensis*, *Halomonas* sp. TD01, and *H. campaniensis* LS21 of the genus *Halomonas*, and *Haloferax mediterranei* of the genus *Haloferax*. These strains can ferment and produce PHA without sterilization, and there will be no contamination by miscellaneous bacteria. For example, *Halomonas* sp. strain TD01 continuously fermented for 14 days under non-sterilized conditions, and when the cell dry weight reached 40 g/L, the PHB content in the cells reached more than 70% (Yin et al. 2015a); and *H. campaniensis* strain LS21 continuously fermented for 65 days under non-sterilized conditions without any contamination by miscellaneous bacteria, the cell dry weight can reach 70 g/L, and the PHB content reached 74% (Tan et al. 2011). Roughly estimated, after using blue water biotechnology, the production cost of PHA can be reduced by nearly half (Yue et al. 2014).

Synthetic biology and metabolic engineering methods can also be applied to the modification of halophilic microorganisms. In recent years, more and more halophilic microorganisms have obtained whole genome sequences and analysis information, making the modification of halophilic microorganisms more targeted. Yin et al. carried out metabolic engineering modifications on *Halomonas* sp. TD01 and its derivative strains (Li et al. 2014b). They expressed heterologous genes in strain TD01 and its derivative strains and constructed a metabolic pathway for the synthesis of PHBV with glucose as the only carbon source (Li et al. 2014b). Tan et al. overexpressed the gene of cell division inhibitor MinCD in *Halomonas* sp. strain TD01, found that it can change the morphology of cells of the strain, elongate them to nearly a hundred micrometers, effectively increasing the space for PHA accumulation in cells of strain TD01, and increasing the PHA content from 69% of the wild-type strain (control) to 82% of the mutant strain (Yin et al. 2015b). In addition, it has been reported that in *H. mediterranei*, knocking out the *pyrF* gene (orotidine-5′-phosphate decarboxylase), and using 5-fluoroorotic acid and uracil as selection pressures can greatly improve the efficiency of gene recombination, and this efficient gene knockout technology can also be applied to the genome modification of other halophilic microorganisms (Tan et al. 2014). As the technology for modifying halophilic microorganisms matures, all PHA synthesis metabolic pathways may be transferred to this chassis microorganism, and then modified by synthetic biology means, making it possible to achieve the goal of low-cost production of all types of PHA through blue water biotechnology. In summary, integrating metabolic engineering, synthetic biology, morphological engineering, and other modification means with blue water biotechnology can effectively improve the competitiveness of microbial PHA production and obtain diverse PHA materials.

4. *Constructing Recombinant Pathways and Using Cheap Raw Materials to Produce PHA*

The synthesis of a certain PHA requires a compound of the corresponding structure as a substrate. For example, the synthesis of medium and long chain PHAs requires medium and long chain fatty acids as substrates, while the synthesis of PHAs containing 4HB monomers requires butyric acid, 1,4-butanediol, γ-butyrolactone, etc. as substrates. However, these substrates themselves are too expensive, greatly increasing the production cost of PHA. Using cheap substrates such as sugar compounds to directly synthesize various PHAs is a way to reduce production costs. When natural microorganisms use sugars as substrates, the synthesized PHA is mainly P(3HB), which is also the first PHA found in nature (Liu et al. 2011a). Some microorganisms can synthesize copolymers of 3-hydroxybutyric acid and 3-hydroxyvaleric acid (PHBV) when using sugars as substrates (Lemoigne 1926b). But so far, there are no reports of natural microorganisms synthesizing other types of PHAs with sugar as the only carbon source. A large number of studies have shown that using synthetic biology and metabolic engineering technology, new metabolic pathways can be constructed in microorganisms to achieve the synthesis of nonnatural compounds. For example, Wang et al. (Han et al. 2013) constructed two metabolic pathways for the synthesis of 3-hydroxybutyric acid and 3-hydroxyhexanoic acid copolymers (PHBHHx) from glucose in *E. coli*. In addition, two different metabolic pathways for the synthesis of PHBV from glucose have been constructed in *E. coli* by Wang et al. (Han et al. 2013) and the combination of these two metabolic pathways can yield PHBV with different monomer compositions. In addition to the reported metabolic pathways for the synthesis of P3HB4HB (copolymer of 3-hydroxybutyric acid and 4-hydroxybutyric acid) (Wang et al. 2014a) and P4HB (poly-4-hydroxybutyrate) (Li et al. 2010) from glucose, as well as the pathway that natural microorganisms existed for synthesizing PHB from glucose, all commercially available fourth generation PHAs can theoretically be produced using glucose as a substrate. In addition to the commercialized PHAs, other types of PHAs have also achieved the construction of metabolic pathways for synthesis using glucose as a substrate, such as Meng et al. (Zhou et al. 2012) have constructed metabolic pathways for the synthesis of poly-3-hydroxypropionic acid (P3HP), copolymers of 3-hydroxybutyric acid and 3-hydroxypropionic acid (P3HB3HP) from glucose in recombinant *E. coli* strain. Zhuang et al. (Meng et al. 2015) have constructed a trans-fatty acid oxidation pathway to synthesize PHAs containing medium-chain monomers in recombinant *E. coli* strain. The monomer lactic acid of polylactic acid can also be introduced into the synthesis process of PHA through a new metabolic pathway, forming a new type of PHA containing lactic acid monomers (Zhuang et al. 2014). To date, almost all common PHAs have been realized to be synthesized with glucose as the sole carbon source (Park et al. 2013).

In order to reduce the production cost of PHA, various types of biological waste have also begun to be used as raw materials for the production of PHA in recent years, including agricultural waste, food processing waste, dairy waste, biodiesel waste, and synthetic gas (Chen et al. 2015). As shown in Table 3.1, PHA can now be

Table 3.1 Research on the production of PHA using different biological wastes

Raw material	Producing strain	Fermentation scale/type	Polymer type and concentration	References
Lignin hydrolysate	*Paraburkholderia sacchari*	Shake Flask	Poly-β-hydroxybutyrate 34.5 g/L	Bhatia et al. (2021)
Agave tequila bagasse hydrolysate	*Burkholderia sacchari*	Shake flask	Poly-β-hydroxybutyrate 24 g/L	Dietrich et al. (2020)
Waste fish oil	*Salinivibrio species* M318	Fed-batch (10L fermenter)	Poly-β-hydroxybutyrate 69.1 g/L	González-García et al. (2019)
Waste frying oil	*Halomonas hydrothermalis*	Shake flask	Poly(3-hydroxybutyrate-*co*-3-hydroxyvalerate) 50.5 g/L	Van Thuoc et al. (2019)
Organic fraction of municipal solid waste and sewage sludge	Mixed microbial culture	Fed-batch (380L fermenter)	Polyhydroxyalkanoates 0.36 ± 0.05 g/L	Pernicova et al. (2019)
Whole/raw whey	*Bacillus megaterium Ti3*	Shake flask	Poly-β-hydroxybutyrate 2.20 ± 0.11 g/L	Valentino et al. (2019)
Cheese whey hydrolysate	*Halomonas halophila*	Shake flask	Poly-β-hydroxybutyrate 3.26 g/L	Israni et al. (2020)
Synthetic gas	*Rhodospirillum rubrum*	Batch feeding (3.6L fermenter)	Poly-β-hydroxybutyrate 5.32 ± 0.09 g/L	Kucera et al. (2018)

produced using various types of biological waste through various microorganisms. The production of PHA using biological waste not only reduces costs but also transforms the biological waste produced by human activities into bioplastics, which play a protective role for the environment. Although it is not yet possible to produce PHA on a large scale using biological waste, it is believed that with the continuous development of science and technology, the vision of producing PHA on a large scale at low cost with biological waste will eventually be realized.

PHA, due to its good biodegradability and biocompatibility, as well as its variety and performance, has received widespread attention. As more and more PHAs with new functions are developed, the application of PHA will further expand. However, production costs and thermodynamic performance are still the biggest challenges for large-scale production and commercialization of PHA. Through synthetic biology and systems biology, metabolic engineering, and next-generation biotechnology, extreme microorganisms can be used as low-cost production platforms for PHA, greatly simplifying the biological manufacturing process of PHA through open continuous fermentation without sterilization. At the same time, metabolic engineering and synthetic biology techniques can be used to modify the production strains of PHA, directing more metabolic flow toward PHA synthesis, which can improve the conversion rate of substrates to PHA. Morphological and auto-flocculation modifications of the strains can simplify the downstream purification process of PHA, significantly reducing the production cost of PHA. In addition, it is necessary to continuously develop new monomers (and optimal monomer ratios), new structures, and processing techniques that can improve the performance of PHA, so that the thermodynamic performance of PHA can approach or even surpass petroleum-based plastics, further promoting the application and promotion of PHA.

3.3 Biological Synthesis of Polylactic Acid

Due to many shortcomings of the traditional poly(lactate) [PLA] chemical synthesis method, such as multiple process units, high production costs, and the use of metal catalysts and organic solvents that can easily cause environmental pollution, researchers hope to synthesize PLA using a green and environmentally friendly biological method.

Many studies have developed a completely biological process for producing PLA and PLA copolymers from renewable resources by constructing PHA biosynthetic pathways and new metabolic pathways in recombinant bacteria. This can be achieved by introducing propionyl coenzyme A transferase and engineered PHA synthase, and combining the pathways of systems metabolic engineering. Since polyesters are synthesized in cells through PHA synthase polymerizing various HA-CoA produced in different metabolic pathways, and they accumulate in cells as various granules, copolymers containing lactic acid synthesized through the polyester biosynthetic pathway can also be synthesized and accumulated in cells (Karmann et al. 2019). PHA synthase is a key enzyme for the specific polymerization of various monomer substrates, and *E. coli* engineering strains carrying PHA synthase can convert lactic acid into PLA without extraction and purification. Therefore, lactate-polymerizing enzyme (LPE) can replace metal catalysts to establish the biological synthesis process of PLA. In addition to being mild, this biological synthesis method has several other advantages (Steinbüchel and Füchtenbusch 1998):

First, there is no need to use high-purity monomers. The biological method does not require the use of high-purity monomers to limit side reactions during the polymerization process and avoid toxicity brought by the use of metal catalysts, as in the chemical method. This is because PLA copolymers are catalyzed by engineered PHA synthase, using various metabolic intermediates in microorganisms as substrates for synthesis within cells.

Second, various PLA copolymers with new material properties can be synthesized. Since the original metabolic pathways or heterologous metabolic pathways in microorganisms can provide at least 150 types of hydroxy carboxylic acids as monomers, they can be synthesized into various PLA copolymers with new material properties through LPE in microorganisms.

Third, ultra-high molecular weight PLA and PLA copolymers can be synthesized and accumulated in cells. In the chemical method, the viscosity of the chemical polymer increases with the progress of the reaction, and high viscosity can lead to uneven mixing, thus limiting the molecular weight of the synthesized PLA. However, this problem does not exist in the biological method.

Fourth, the properties of the synthesized materials are easy to control. Since PHA synthase only accepts R-configured HA-CoAs, the monomers of PLA and its copolymers synthesized by the biological method have a single optical activity, making their material properties easier to control.

Fifth, there are no product processing problems caused by unreacted residual monomers. In the chemical polymerization reaction of PLA, the remaining unreacted propylene glycol monomer in the reaction system can cause processing problems for PLA polymers. However, there are no unreacted residual monomers in the polymers containing lactate ester synthesized by microbial fermentation, so there will be no subsequent product processing problems caused by residual monomers.

Song et al. designed a new metabolic pathway in *C.glutamicum* to produce monomer substrates—lactoyl coenzyme A (LA-CoA) and 3-hydroxybutyryl coenzyme A (3HB-CoA) for LPE-catalyzed copolymerization to produce PLA (Park et al. 2012). LA-CoA is synthesized by lactate dehydrogenase (LDH) and propionyl coenzyme A transferase, and 3HB-CoA is provided by β-ketone thiolase (PhaA) and NADPH-dependent acetoacetyl coenzyme A reductase (PhaB), eventually producing poly(lactic acid-co-3-hydroxybutyric acid) [P(LA-co-3HB)] with a high lactic acid component (96.8 mol%). The newly designed *C.glutamicum* strain has the potential to become a new application platform for the production of food-grade and biomedical-grade PLA and its copolyesters.

The heat-resistant *Pseudomonas* sp. SG4502 strain isolated by Tajima et al. could accumulate PHA even at 55 °C, so it can be used as a source of heat-resistant PHA synthase (Song et al. 2012). Two genes encoding PHA synthase (PhaC1SG and PhaC2SG) from the strain were identified and cloned. Two mutations (Ser324Thr and Gln480Lys, corresponding to the mutations in LPE of *Pseudomonas* sp. 61-3 strain) were introduced into PhaC1SG to evaluate the potential of the resulting protein as a "heat-resistant LPE." The mutated PhaC1SG[PhaC1SG(STQK)] synthesized P(LA-co-3HB) in a high-temperature reaction system in vitro. This heat-resistant LPE can be used to synthesize various types of lactic acid-based copolymers. With such engineered enzymes and engineered microorganisms, PLA homopolymers and copolymers can be synthesized directly from biomass fermentation without the need for extraction and purification of lactic acid, thus developing a greener PLA synthesis process.

Taguchi et al. constructed a recombinant *E. coli* with a coenzyme A transferase gene that produces LA-CoA. The production of LA-CoA was confirmed by capillary electrophoresis/MS analysis. They introduced an engineered PHA synthase gene based on the principle of substrate similarity, and obtained the copolyester P (94 mol% 3HB-co-6 mol% LA). Its average molecular weight was found to be 1.9×10^5 by gel permeation chromatography, gas chromatography/MS, and NMR. This established a one-step microbial metabolic process for the synthesis of LA-based polyester for the first time (Tajima et al. 2012). The microbial one-step fermentation method for producing PLA is currently the optimal PLA synthesis method, which can control the composition of the polymer by metabolizing lactic acid monomers in different proportions in a one-step process. The enantioselectivity of LDH and LPE enzymes for lactic acid enantiomers is a key factor in the synthesis of lactic acid-based polyesters. LPE can incorporate LA components into the polymer. Yamada M et al. achieved fine control of the proportion of LA components in P (3HB-co-LA) by saturating mutation at the 392nd position (F392X) of LPE (Taguchi et al. 2008). Among the 19 saturated mutants, 17 mutants produced P (3HB-co-LA) with various LA components (16–45 mol%). The F392S mutant showed the highest LA component, 45 mol%, and the polymer content increased to 62 wt%, while the LA component in the polymer produced by the unsaturated mutant strain under the same culture conditions was 26 mol%, and the polymer content was 44 wt%. The F392S mutant under anaerobic conditions can promote the production of LA, leading to a further increase in the LA component, up to 62 mol% (Yamada et al. 2010). This indicates that the mutation at the 392nd position in LPE greatly promotes the fine-tuning of the LA component in the copolymer, which is beneficial to enhance the thermal performance of the copolymer. Yang et al. constructed a metabolic engineering strain of *E. coli* that produces PLA copolymers by expressing propionyl coenzyme A transferase evolved from *C.propionicum* and PHA synthase from *Pseudomonas* sp. MBEL 6-19 strain (Aixia et al. 2020). This strain not only produces PLA, but also PLA copolymers—poly (3-hydroxybutyrate-co-lactic acid) [P (3HB-co-LA)] (Yamada et al. 2009; Yang et al. 2010). This strain can not only use glucose to produce PLA homopolymers with a mass fraction of up to 11 wt%, but also use glucose and 3HB to produce P (3HB-co-LA) copolymers containing 55–86 mol% lactate with a mass fraction of up to 57 wt% (Yamada et al. 2009; Yang et al. 2010). Jung et al. introduced β-ketothiolase and acetoacetyl coenzyme A reductase from *C. necator* into *E. coli*, and were able to prepare P(3HB-co-LA) copolymers with a lactic acid content of up to 70 mol% and a mass fraction of 46 wt% just from glucose (Yang et al. 2010).

Utsunomia et al. reported that an engineered *E. coli* strain can produce D-lactic acid-based oligomers (D-LAO) in one step and autonomously secrete them out of the cell. This is due to their discovery that the strain can autonomously secrete a small amount of oligosaccharide—lactate-co-3-hydroxybutyrate [LA-co-3HB] outside the cell due to the specific expression of D-LPE, without the need to introduce exogenous secretory proteins (Jung et al. 2010). In addition, in order to increase the yield of D-LAO, they tried to use chain transfer (CT) reagents to increase the frequency of LPE chain transfer reactions. They found that adding diethylene glycol to the culture medium was particularly effective. Using 20 g/L of glucose to produce D-LAO, the highest yield can reach 8.3 ± 1.5 g/L, which is 57% of the theoretical maximum yield (Utsunomia et al. 2017). In this one-step secretion system, the lactic acid synthesized inside the cell can be directly converted into a polymer in one step without any downstream processes. Therefore, it is believed that lactic acid bacteria can also serve as hosts for cloning and expressing LPE genes, allowing lactic acid to be directly converted into PLA without the need to use neutralizers to maintain the pH value during the fermentation process. Genetic engineering, metabolic engineering, and systems biology can be used in combination to construct such engineered strains of lactic acid bacteria.

Shozui et al. (2011) used a recombinant *E. coli* strain LS5218 expressing LPE to produce a ternary polyester composed of 96 mol% lactic acid, 1 mol% 3-hydroxybutyrate, and 3 mol% 3-hydroxyvalerate, namely poly(lactic acid-3-hydroxybutyrate-3-hydroxyvalerate) [P(LA-co-3HB-co-3HV)]. This strain grows on glucose and is fed with valerate as a monomer precursor. The glass transition and melting temperatures of the ternary polyester are close to those of chemically synthesized PLLA with similar molecular weights. In addition, due to the strict enantiomer specificity of LPE, the blend with ternary polyester, which is entirely composed of (R)-LA (D-LA), and PLLA forms a stereocomplex with a higher melting point (201.9 °C). The results show that bio-polylactic acid produced by microbial processes has thermal properties similar to those of chemically synthesized PLA.

Although the biosynthesis of PLA has many advantages compared with typical chemical methods, there are several problems that need to be solved. For example, chain termination during polymer synthesis, which limits the molecular weight of PLA; the low production rate in bacteria, which results in low PLA content in cells; in addition, research has found that as the proportion of lactic acid increases, the molecular weight of lactic acid copolymers decreases, which also limits the application of lactic acid copolymers. Therefore, screening new PHA synthases with strong substrate specificity and increasing the biosynthesis of PLA through metabolic engineering, cell morphology transformation, and enzyme fusion are urgent problems to be solved in the biosynthesis of PLA by fermentation.

Although the one-step fermentation synthesis of PLA is mild in reaction conditions and possesses other advantages (such as there being no unreacted residual monomers in the synthesized polymer, and it can synthesize polymers with fine structures in one step from cheap raw materials, which is more attractive than chemical polymerization synthesis of PLA), this method is still in the laboratory research stage, and how to increase the biosynthesis of PLA is still a problem to be further studied. The solution to this problem will greatly promote its industrialization process in the future.

3.4 Chitosan and Its Biosynthesis

3.4.1 Introduction to Chitosan

Chitin is a straight-chain polysaccharide formed by N-acetyl-2-amino-D-glucose through β-1,4 glycosidic bonds, and is a natural high molecular compound. Chitosan is the deacetylated product of chitin. Chitin is the second largest natural high molecular polysaccharide synthesized on earth after cellulose, with an annual biosynthesis of about 10^{10}–10^{11} ton, which is a renewable resource synthesized by a large number of living organisms (Shahbaz 2020). The molecular formula of chitin is $(C_8H_{15}NO_6)n$; its chemical structure is very similar to cellulose (as shown in Fig. 3.8), the difference is that the substituent at the C_2 position is different. The substituent at the C_2 position of cellulose is hydroxyl (–OH), the substituent at the C_2 position of chitin is acetyl amino (–$NHCOCH_3$), and the substituent at the C_2 position of chitosan, as the deacetylated product of chitin, is amino (–NH_2), so chitin and chitosan are often referred to as animal cellulose. In nature, chitin is widely found in the skeletons of some insect organisms, crustaceans (such as shrimp, crab), and mollusks, and also in the cell walls of fungi (such as yeast and green algae). Chitin is generally insoluble in water and common solvents, which limits its application, but chitosan has better solubility and a wider range of applications. In fact, the presence of amino groups on the main chain of chitosan provides it with an important structural feature, which can easily chemically modify it into other derivatives, and it has good biodegradability, biocompatibility, non-toxicity and antibacterial properties (Ahmad et al. 2020; Satitsri and Muanprasat 2020; Elieh Ali Komi et al. 2018).

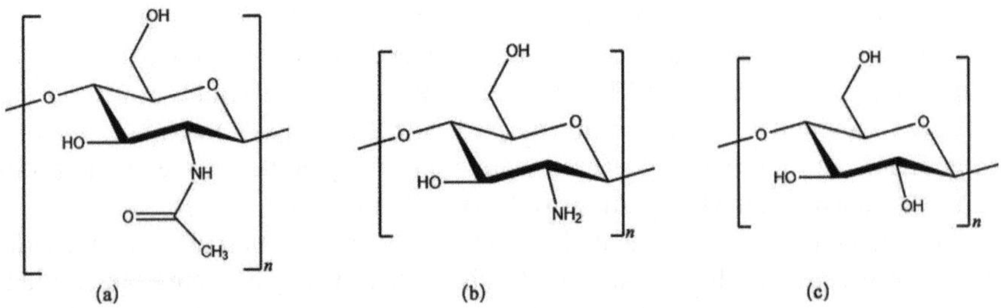

Fig. 3.8 Molecular structure of chitin (**a**), chitosan (**b**), and cellulose (**c**)

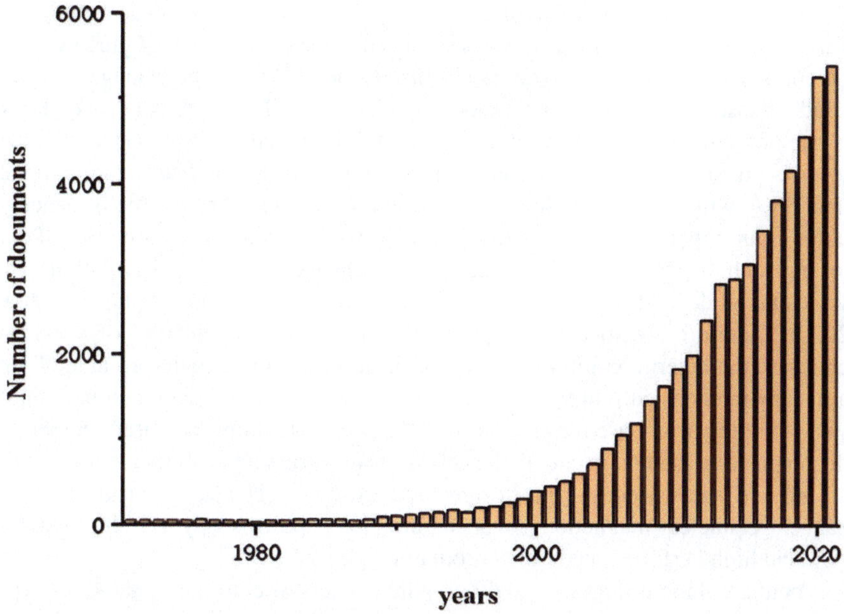

Fig. 3.9 Number of documents related to chitin and its derivatives retrieved by PubMed

Chitin is a renewable resource with great potential. As shown in Fig. 3.9, according to the data statistics obtained by searching for chitin as a keyword in the PubMed database, the attention and importance of chitin is continuously increasing, especially in obtaining high-quality chitin and its derivatives and developing applications have become the current research and development hotspot. Chitin and its derivatives are widely used in textiles, food, agriculture, biomedicine, cosmetics, and other fields. In the field of biomedicine, it has been applied to tissue engineering, drug delivery, diagnosis, molecular imaging, antibacterial treatments, and wound healing (Weiwei et al. 2007b), and its economic value and social benefits are very significant.

1. *Structure and Properties*

Among the physicochemical properties of chitin, the degree of deacetylation (deacetylation degree, DD) and the degree of polymerization (DP) are very important. DP). The differences in deacetylation degree and degree of polymerization (affecting molecular weight) mainly depend on the source of chitosan and its production process conditions. The molecular weight of chitosan determines its viscosity and degradation rate, which can be measured by a viscometer, light scattering method, and gel permeation chromatography. The degree of deacetylation of chitosan affects various functional characteristics, including solubility, chemical reactivity, and biological activity.

The degree of deacetylation of chitin and chitosan is the percentage of the number of glucose units that have been deacetylated to the total number of glucose units. Of course, the degree of acetylation can also be defined as the percentage of acetylated glucose units to the total number of glucose units. The degree of deacetylation is one of the important structural parameters of chitin and chitosan. Generally, chitosan is prepared from natural chitin sources with a degree of deacetylation higher than 55%, which can be dissolved in dilute acid solution. The degree of deacetylation is an important physicochemical property of chitosan. The properties of chitosan, such as solubility, flexibility, ion exchange capacity, flocculation, and viscosity, all depend on the degree of deacetylation to a certain extent (Peng et al. 2020). Therefore, whether it is the production, research, or application of chitin and chitosan, the determination of the degree of deacetylation is indispensable. There are many methods currently used to determine the degree of deacetylation of chitosan, including acid–base titration, colloid titration, potentiometric titration, thermal decomposition method in chemical analysis, and instrument analysis methods (such as infrared spectroscopy, near-infrared spectroscopy, ^{1}H nuclear magnetic resonance, ^{13}C nuclear magnetic resonance, ultraviolet spectroscopy, gel chromatography, and high performance liquid chromatography).

There are three different crystallographic isomers of chitosan in nature: α-type, β-type, and γ-type, which are determined by the different arrangement of molecular chains in the crystal cell. Among the three crystal structures, α-chitosan is the most stable one, mainly found in the cell walls of fungi and bacteria, the limbs and shells of lobsters and crabs, and the shells of common crayfish, etc. (Younes and Rinaudo 2015). β-chitosan is relatively rare and exists in the beak of squid or cuttlefish. γ-chitosan exists in a mixture of α-type and β-type. Under appropriate conditions, β-chitosan and γ-chitosan can also be converted into stable α-chitosan. β-chitosan treated with 6 mol/L hydrochloric acid, and γ-chitosan treated with saturated LiSCN solution, will both become α-chitosan, but this process is irreversible (Moussian 2019).

2. *Applications*

The derivatives of chitin mainly include deacetylated products such as chitosan, oligosaccharides, and monosaccharides with different degrees of polymerization obtained from the degradation of chitin and chitosan, as well as chemically modified derivatives of these substances, such as chitooligosaccharides, chitosan oligosaccharides, N-acetyl glucosamine, glucose hydrochloride, and carboxymethyl chitosan. Chitin and its derivatives, due to their special molecular structure, biocompatibility, biodegradability, and other unique functional characteristics, have attracted widespread attention from the scientific and industrial communities since the late 1970s. The following are some key areas of application of chitin and its derivatives.

(1) Biology and Medicine

The cationic properties and structure of chitin and its derivatives give them unique physicochemical properties, such as pH sensitivity, low immunogenicity, biocompatibility, low toxicity, and biodegradability. These properties have led to their use in wound healing drugs and materials, drug delivery, tissue engineering, and special medical foods. In addition, chitin and its derivatives have other biological properties, such as anti-tumor (Salah et al. 2013), antibacterial (Martins et al. 2014), and antioxidant activities (Ngo and Kim 2014). Chitin and its derivatives have low immunogenicity and generally do not cause immune rejection in the body, demonstrating good biocompatibility. They also have good biodegradability and can generally be degraded into oligosaccharides, N-acetyl glucosamine, and glucosamine in the body, with good safety performance and no toxic side effects (Cheung et al. 2015).

(2) Food Industry

In the food industry, chitin and its derivatives have strong antioxidant and antibacterial abilities (Diegelmann et al. 1996), which can improve the safety, quality, and shelf life of food. Chitin, as a functional food additive, can prevent diet-related diseases on the one hand, and achieve the effect of food therapy and disease relief on the other. It is reported that chitin can prevent microbial food spoilage, extend shelf life, and be made into biodegradable films and food packaging (Friedman and Juneja 2010). China has national standards for chitin (GB1886.312-2020) and chitosan (GB29941-2013) as food additives, and chitooligosaccharides are listed as new food ingredients with industry standards (QB/T 5503-2020). Chitin has also been approved by the US Food and Drug Administration (FDA) as a recognized food additive, dietary fiber (low cholesterol effect), and functional food ingredient. Japan and South Korea have also approved chitin as a food additive.

(3) Agriculture

In the field of agriculture, chitin and its derivatives can be used as seed coating agents to prevent pests, promote plant resistance to harmful microorganisms, act as plant immune inducers to resist cold, drought, salinity, and other stress conditions (Yang et al. 2021), and can also inhibit fungi, viruses, bacterial diseases, and nematodes, so they can also be used as alternatives to biopesticides (Malerba and Cerana 2019). Products on the market include pesticide formulations, bioinsecticides, biofungicides, biofertilizers, fertilizer protectants, disease and stress resistance agents, antifreeze agents, etc. (Friedman and Juneja 2010). Chitosan, chitooligosaccharides, etc. can activate many defense responses in plants, so they are often used as a powerful elicitor for plant disease prevention and control (El Hadrami et al. 2010).

(4) Cosmetics

Chitin and its derivatives can be used as active carriers for cosmetics and other daily necessities, applied to skin care, hair care, and oral care (Jimtaisong and Saewan 2014); they can also be made into shampoo, coloring products, hair gel, toner, cream and detergent, makeup products, deodorant products, microcapsules of active agents, and dental care products (Aranaz et al. 2018). Most chitosan products have a high molecular weight and cannot penetrate the skin, which is an important advantage, making them suitable for skin care. Chitin also has a wide range of applications in skin care because it can be well-tolerated by the skin. It is generally used to provide hydration and prevent dehydration.

3.4.2 Preparation Method of Chitosan

Although chitin is widely present in nature, its main sources are crustaceans, lower fungi, and algae. The former are abundant in water bodies, with the highest content in shrimp and crabs, and the raw materials are easy to accumulate. The dry weight content of chitin in shrimp and crab shells is 10–30% (Yanyan et al. 2015), which has industrial extraction value. So far, the main way to produce chitin on the market is also obtained from crustaceans, mainly crabs and shrimp.

There are mainly two difficulties in obtaining chitin. One is that although chitin exists in crustaceans, the chitin in it forms a complex network structure with proteins and calcium carbonate attached to proteins. The interaction force formed between chitin and protein is strong, and some proteins also participate in the structure of the polysaccharide–protein complex. The second is that in the process of producing chitin, there are also some fats and astaxanthin-based carotenoids and their esters. Only by overcoming these two difficulties and removing impurities can the purity required for food, medicine, or materials applications be achieved. Therefore, the most important steps in producing chitin are the removal of protein and $CaCO_3$, and at the same time, fats and pigments should be removed in the above removal process.

The traditional methods for extracting chitin include the acid–base method, the enzymatic method, and the microbial fermentation method. A very promising method is to use synthetic biology technology to artificially synthesize chitin.

1. *Acid–Base Method*

The traditional process of producing chitin mainly includes three parts: removing protein, removing inorganic salts, and removing pigments. Because there is a chemical bond between chitin and protein, removing protein is the most difficult step among the three. Alkali solution is generally used to remove protein, commonly NaOH, sometimes Na_2CO_3, $NaHCO_3$, KOH, K_2CO_3, $Ca(OH)_2$, Na_2SO_3 are also used. The process of protein removal should be analyzed according to specific situations, striving to avoid excess and not allowing protein residues. This is because if it is excessive, it will cause the deacetylation and hydrolysis of chitin, thereby affecting the final quality of chitin; if there are protein residues, the purity is not high enough, which will affect its application in biomedicine and food, and the application range will be greatly reduced. In addition, some people are allergic to crustacean proteins. The second step is to remove inorganic salts, mainly to remove the residual $CaCO_3$ in chitin. Desalination is generally treated with acid, and commonly used reagents are HCl, sometimes HNO_3, H_2SO_4, CH_3COOH, and HCOOH. The desalination step is relatively easy to complete because $CaCO_3$ is easily acid-soluble. The third step is to remove the pigment. The most popular chitin decolorizing agent in the industry is to use light irradiation and $KMnO_4$ oxidation decolorization. The duration of light exposure should be analyzed according to specific situations. If the sunlight exposure decolorization time is too long, the product will be yellowish (Xuexiao et al. 2014). If $KMnO_4$ is used for decolorization, $NaHSO_3$ and other reducing agents need to be added for reduction treatment. Unreacted $KMnO_4$ will cause environmental pollution. In recent years, hydrogen peroxide is also a commonly used industrial oxidizing decolorizer, which is favored by people due to its low price and pollution-free decomposition products (Le-Ping et al. 2002). The determination of all process flows is not immutable, and it needs to be tested repeatedly to determine the process flow that is most suitable for the product.

From the perspective of the past process of producing chitin using the acid–base method, although the process flow is ever-changing, compared with the process flow of enzymatic method and microbial fermentation method, the advantage of the acid–base method is that the process flow is simple, suitable for large-scale preparation, and the process flow cycle is short. However, the acid–base method also has many disadvantages, mainly because the acid–base method uses a large amount of acid and alkali in the process of extracting chitin, and requires different temperature treatments, which consumes energy and causes serious environmental pollution. The subsequent purification process of chitin is complicated, which increases the production cost. The use of acid and alkali will also cause a decrease in the molecular weight and degree of acetylation of chitin, thereby affecting the quality of chitin products.

2. *Enzymatic Method*

The enzymatic method for extracting chitin is to use commercial protease or protease separated from the fermentation broth to hydrolyze the protein in the shell, and then remove the ash with acid, microwave, and other physical methods to prepare chitin. Proteases are mostly alkaline proteases, as well as pepsin, papain, trypsin, etc., which are conducive to separating proteins from the shells of crustaceans and can be used for the extraction of chitin. This method has the advantages of mild conditions, environmental friendliness, and high product purity (Xiang-Yu et al. 2018).

Angel et al. (Valdez-Pena et al. 2010) treated shrimp heads with alkaline protease, resulting in a dry weight loss rate of 61% and a residual protein content of 275 mg/g. Subsequently, lactate desalting under 400 W microwave radiation resulted in an ash content of 0.2% only. The yield of chitin recovered by the enzymatic and physicochemical combined treatment was 22%. Laila et al. (Manni et al. 2010) extracted protease from the fermentation broth of *B. cereus* for the treatment of waste shrimp shells, and then removed ash with 1.25 M HCl. The final yield of chitin was 16.55 ± 1.5%, the ash content was only 0.433 ± 0.05%, and the protein content was 10.78 ± 0.2%. Islem et al. (Younes et al. 2012) used different microbial crude proteases to extract chitin from shrimp shells. The final yield of chitin was 18.5 ± 2.3%, the protein removal rate was 88 ± 5%, and the ash content was 1.9%.

Compared with the acid–base method, the enzymatic and chemical methods for preparing chitin have milder reaction conditions, can better protect chitin from being reduced in molecular weight and deacetylation, but the enzymatic method takes a longer time and cannot completely remove protein. The residual protein can affect applications in biomedicine and food. Also, the price of commercial enzymes is high. If large-scale production of chitin is needed, a large amount of protease is required, which makes the cost of producing chitin high. However, in the production process of chitin, enzymatic catalysis as a new protein removal method has great application potential (Lopes et al. 2017).

3. *Biological Fermentation Method*

The microbial fermentation method for preparing chitin refers to using waste materials from shrimp and crab shells as raw materials, inoculating microorganisms, and the microorganisms will produce organic acids during growth. The organic acids can dissolve salts such as calcium carbonate in the raw materials of shrimp and crab, thereby achieving the purpose of desalting. Some proteases are also produced during microbial growth, thereby achieving the purpose of removing protein and then obtaining chitin with relatively high purity. Many types of external microorganisms are needed in this method. Generally, strains that produce more acid and protein are employed, such as lactic acid bacteria, *Bacillus*, and *P. aeruginosa.* Sometimes, inoculating a single strain does not achieve the ideal desalting and deproteinization effect, so the method of mixed-strain fermentation is often used.

Flores-Albino et al. (2012) used *Lactobacillus* to ferment crab shells, and after response surface optimization, the desalting rate reached 88%, the deproteinization rate was 56%, and the final chitosan yield was 34%. Jen-KuoYang et al. (2000) used a strain of protease-producing *B. subtilis* to deproteinize untreated shrimp shells, crab shells, and lobster shell waste. The results showed that the protein removal rates were 88%, 67%, and 83%, respectively. In contrast, the protein removal rates of acid-treated waste were 76%, 62%, and 56%, respectively. It can be seen that the protein removal rate of acid-treated waste is far less than that of the fermentation method. Ghorbel-Bellaaj et al. (2011) optimized the fermentation conditions of *P. aeruginosa* A2 strain on shrimp shells using response surface analysis. Under the optimized fermentation conditions, the desalination rate can reach 96%, and the deproteinization rate can reach 89%. The efficiency of mixed fermentation is much higher than that of a single strain, but the research on the process flow of mixed fermentation is much more difficult.

The microbial fermentation method for preparing chitin has the advantages of being environmentally friendly and low cost. However, so far, there have been no reports of industrial production of chitin using microbial fermentation methods, and it is basically at the laboratory level. This requires researchers to continuously innovate and research more effective, environmentally friendly, and low-cost processes for producing chitin based on this method.

3.4.3 Biosynthesis of Chitin

The traditional process for scalable production of chitin is the chemical method, namely the acid–base method. This process has a strong dependence on raw material resources, serious environmental pollution, and the product is not suitable for people who are allergic to seafood or vegetarians. The synthesis of chitin in organisms using synthetic biology can avoid these problems to some extent, so the preparation of chitin through the biosynthetic pathway has great research potential. If synthetic biology is used to synthesize chitin, it can be considered from two aspects. One is that the microbial chassis itself contains chitin, such as *Aspergillus niger* and brewing yeast (Roberts et al. 1983). Their cell wall contains chitin, which can be overproduced by genetic engineering methods, and then chitin can be produced in a green way through separation and purification. The second is to introduce enzymes that synthesize chitin into the cell. The known enzymes that can synthesize chitin are chitinase and glycosyltransferase, but because the degree of polymerization of chitin is relatively high, it is difficult to produce in the cell, so first synthesize chito-oligosaccharides in the cell, which can also produce chitin and its derivatives in a green way.

Most of the research on the synthesis pathway of chitin focuses on crustaceans and fungi. According to the metabolic pathway in the KEGG database, the synthesis pathway of these two types of organisms from fructose-6-phosphate to chitin is roughly the same, as shown in Fig. 3.10.

Referring to the chitin synthesis pathway of *Komagataella phaffii* (yeast) and *Penaeus vannamei* (Pacific white shrimp) in the KEGG database, the synthesis process from fructose-6-phosphate to chitin can be summarized in five steps: ① Glutamine:fructose-6-phosphate transaminase (GFAT) catalyzes glutamine and fructose-6-phosphate to produce glucosamine-6-phosphate (the first rate-limiting step of this process). ② Glucosamine phosphate acetyltransferase (GNA1) catalyzes the acetylation of GlcN-6-P to GlcNAc-6-P. ③ The conversion of GlcNAc-6-P to GlcNAc-1-P by phosphoacetyl glucosamine mutase (AGM1). ④ The activation step of GlcNAc, that is, UDP-GlcNAc pyrophosphorylase (UAP) catalyzes GlcNAc-1-P

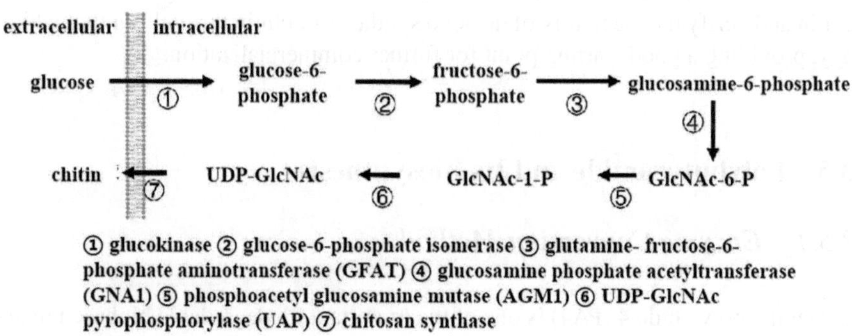

Fig. 3.10 Chitin synthesis pathway in the KEGG database

to produce UDP-GlcNAc. ⑤ The process of connecting the precursor UDP-GlcNAc to the chitin chain by chitin synthase (CHS), and the types and functions of chitin synthase in different species are not the same.

The biotechnological preparation of chitin synthesis is still in its infancy. There are very few research results. Based on the two approaches mentioned above, one example is given for readers' reference.

Our research team (Zhao Liming's team) has conducted a lot of exploration in the biological synthesis of chitin. After cloning the functional genes related to chitin synthesis, using *Pichia pastoris* as the chassis, two strains, GS-1.10 and GS-2.6, were constructed. In strain GS-1.10, glutamine:fructose-6-phosphate aminotransferase (GFAT), UDP-N-acetylglucosamine pyrophosphorylase (UAP), and chitin synthase 1 (CHS1) were overexpressed, and the chitin content increased by 44.3%. In strain GS-2.6, GFAT, glucosamine-6-phosphate N-acetyltransferase (GNA1), phosphoacetylglucosamine mutase (AGM1), UAP and CHS1 were overexpressed, and the chitin content increased by 244.4%. On this basis, the team overexpressed another chitin synthase 3 (CHS3) to obtain strain GS-3.10, the chitin content of this strain was further increased, the content in the cell wall was 162.4 μg/mg, which is 4.43 times higher than the original strain GS115, and finally, through the optimization of fermentation conditions, the final yield of chitin reached 2.23 g/L (Zhang et al. 2022) .

Liu Long's research team at Jiangnan University used combinatorial pathway engineering to achieve de novo production of chitin oligosaccharides in *B. subtilis*. A chitin oligosaccharide synthesis pathway in *B. subtilis* was constructed by this team. First, N-acetylglucosamine glycosyltransferase NodC was expressed in *B. subtilis*, resulting in a shake flask yield of 30 mg/L of chitin oligosaccharides with a degree of polymerization less than 6. Then, six different sources of N-acetylglucosamine glycosyltransferase NodC were introduced, among which the chitin oligosaccharide yield produced by *Rhizobium nodCM* was the highest (560 mg/L). Finally, through the design of the de novo pathway and salvage pathway of UDP-GlcNAc, the biosynthesis of chitin oligosaccharides was further promoted. In a 3L fermenter, the yield of chitin oligosaccharides reached 4.82 g/L, which is the highest level reported so far (Ling et al. 2022).

The above two cases provide new ideas for the artificial biological synthesis of chitin and verify the feasibility of de novo synthesis of chitin through synthetic biology, providing a good starting point for further commercialization.

3.5 Polybutyramide and Its Biosynthesis

3.5.1 Enzyme Conversion Method

Currently, polyamide 4 (PA4) is one of the important biodegradable biobased materials. In its biosynthetic pathway, glutamate decarboxylase (GAD) decarboxylates glutamate to form γ-aminobutyric acid (GABA), which is activated by β-alanine coenzyme A transferase and then spontaneously cyclizes. Therefore, the urgent problem to be solved in the synthesis of PA4 is the selection and modification of the PA4 polymerase. Although there has been no reported achievement of one-step synthesis of PA4 through enzymatic approaches or microbial cell factory under laboratory or industrial conditions, several research groups have demonstrated that certain polyglutamate synthetases and membrane enzyme complexes—such as the pgsB enzyme system used for the production of poly-γ-glutamic acid—exhibit potential for directly polymerizing γ-aminobutyric acid by enzymatic catalysis in vitro. In purified pgsB enzyme system utilizing GABA as the substrate, insoluble white precipitates were observed during the enzymatic reaction in vitro. Unlike the hydrophilicity of butyramide, the physical property of PA4 is a white powder that is insoluble in water. From the perspective of the molecular polymerization mechanism, the most suitable substrate for the polymerization of glutamate synthetase pgsb is L-glutamate, which differs from GABA in spatial structure by one carboxyl group. Therefore, if the affinity of this enzyme for GABA can be improved by subsequent 3D result simulation and enzyme molecular orientation modification, it may fill the gap in the enzymatic synthesis of PA4. From the perspective of the reaction process of PA4 polymerization, the key enzyme system in this process is the amide bond polymerase. The biocatalyst that is both applicable and catalytically active in the amide polymerization process is immobilized Antarctic *Pseudomonas* lipase B (CALB). CALB is stable at high temperatures and accepts a variety of substrates. CALB has a wide specificity for straight-chain fatty acids from butyric acid to stearic acid. In addition, CALB shows high activity for carboxylic acids shorter than hexanoic acid. Although there are no reports on the in vitro cyclization of any length of amino carboxylic acid catalyzed by CALB. But using CALB as a catalyst, ε-caprolactam can be successfully synthesized from 6-aminohexanoic acid. Amino carboxylic acids that are too short do not show the polymerization and cyclization reactions of their own internal amide bonds in the catalytic process, possibly due to the length of the substrate itself, but the high efficiency of amide polymerization makes CALB still a potential amide polymerase for direct polymerization between monomers (Stavila and Loos 2013).

3.5.2 Metabolic Engineering Control Strategy and Chassis Cell Screening for PA4 Synthesis

De novo synthesis of PA4 starts with the TCA cycle of engineered bacteria, catalyzed by glutamate decarboxylase GadB, which catalyzes the glutamate in the bacterial cell, removes a terminal carboxyl group, and generates γ-aminobutyric acid. Then, various coenzyme transferases, such as ACT (coenzyme A transferase) and PCT (propionyl coenzyme A transferase), have good catalytic activity. By transferring coenzyme A to the end of GABA, the activation energy barrier is reduced, and the spontaneous cyclization reaction is realized, thus forming butyramide. By coupling this intracellular reaction with an enzymatic method and constructing a resting cell catalysis and enzyme catalysis composite system, it is expected to realize the de novo synthesis of PA4 polymer from glucose. However, at the same time, it is necessary to metabolically engineer the chassis cells, such as introducing new metabolic pathways. The initial α-decarboxylation reaction in PA4 synthesis is irreversible, caused by pyridoxal 5′-phosphate (PLP) catalyzed by dependent GAD enzymes. PLP is the active form of vitamin B6 and is a coenzyme in many enzymatic reactions. Its supply is crucial for GAD activity. Therefore, to increase the relative yield and precursor accumulation of PA4, the metabolic pathway branches can be altered to block the branch consumption of precursors in PA4 synthesis, thereby increasing the accumulation of precursors. For example, in some microbes, the PA4 precursor GABA is degraded by γ-aminobutyric acid transaminase (encoded by *gabT*), which forms a GABA shunt with succinic semialdehyde dehydrogenase (encoded by *gabD*) and glutamate decarboxylase. By knocking out genes and modifying them to form defective host cells of *Bifidobacteria*, *gabT* has not been found in its genome, but *gabD* exists. This indicates that *Bifidobacteria* do not have the genetic library necessary for GABA degradation to achieve the goal. In addition, due to the high activity and biosafety (GRAS) issues of endogenous GAD, lactic acid bacteria (LAB) are one of the main hosts for the production of PA4 precursors. However, adding L-glutamate or MSG is an essential method to increase the GABA yield in wild-type and engineered LAB strains. Most of the consumed glucose is responsible for cell growth and energy supply, and the metabolic flux used for PA4 synthesis is relatively low. Therefore, it is necessary to enhance the supply of L-glutamate by reconnecting the metabolic pathways in LAB, reducing the metabolic flux toward unwanted by-products during competitive biochemical pathways. Currently, the genome-scale metabolic model (GSMM) provides a useful framework for systematically analyzing the metabolic functions of strains, which will be used for metabolic system engineering in the laboratory to develop advanced strain improvement strategies. Since PA4 itself is a large polymer synthesized by the secondary metabolites of cells, it will have a significant impact on the host cells. Only by reasonably regulating the ratio of material and energy metabolism flow can the growth and production of chassis microbes be balanced. At the same time, by updating high-expression promoters and replacing efficient replicons, the precursor accumulation speed can be greatly enhanced, and the precursor amount for the polymerization reaction can be increased, thereby increasing the polymerization efficiency and reducing the requirements for enzyme activity. So far, PA4 monomers can be biosynthesized by various microbial strains, mainly including strains in *Lactobacillus*, *Glutamicibacter*, *B. subtilis*, *Streptococcus*, *Bifidobacterium*, *Enterococcus*, and *Streptococcus*, especially *Lactobacillus*, *Streptococcus*, and *Enterococcus* show significant GABA titers and have highly active endogenous GAD.

1. *Glutamicibacter*

As the main production bacteria for the amino acid fermentation industry worldwide and the main production strain of glutamate for nearly 40 years, using

Glutamicibacter as the chassis to construct PA4 synthesis cells is a potential choice, but it requires the development of modification tools. Starting with *Glutamicibacter* strain CZ04, using the pK19 mobsacB vector for reverse selection, and by introducing the superoxide dismutase gene promoter (Psod) in front of the genes encoding phosphoenolpyruvate carboxylase (PPC) and pyruvate carboxylase (PYC), the recombinant strains CZ05 and CZ06 with upregulated PPC and PYC were obtained by Li et al. (Wang et al. 2019). The recombinant strain CZ07 with both upregulated PPC and PYC was also obtained with a similar method. The introduction of a strong promoter has minimal impact on the growth of engineered *Glutamicibacter* strain but markedly enhances their carboxylation metabolism, thereby promoting the synthesis and accumulation of organic acids (Jun et al. 2021). This study also developed a complete genome editing operating system, including a homologous recombination-mediated gene editing system, CRISPR-Cas9, and the latest CRISPR-Cpf1/dCpf1 technologies. In 2016, Timothy Lu's team at MIT successfully applied the dCas9-based CRISPRi system to *Glutamicibacter*, bringing about a major reform for low-efficiency recombination methods. The team led by Xu Zhi-Nan at Zhejiang University designed the RE-CRISPR system in *Glutamicibacter*, which can achieve dual functions of genome editing and transcription inhibition. Recently, Zheng Ping and Sun Ji-Bin's team at the Tianjin Institute of Industrial Biotechnology, Chinese Academy of Sciences, developed a multi-gene expression regulation technology based on CRISPR-dCpf1, achieving rapid expression regulation of multiple target genes (Li et al. 2020). The establishment of these technical systems has promoted the optimization and modification of the chassis cells of *Glutamicibacter*, successfully achieving the metabolic control of various energies and substances in the bacterium, making it an ideal carrier for PA4 production.

2. *Bacillus subtilis*

B. subtilis is also a good chassis microorganism for the production of polyglutamic acid, GABA, and polybutyramide. Industrial *B. subtilis* usually uses glycerol, citrate, and glutamate as carbon sources for synthesis. Engineered *B. subtilis* strains that use glucose as the sole carbon source for polybutyramide synthesis have been constructed. By replacing the natural promoter of the polybutyramide synthase operon, using VEG or the xylose-inducible promoter P_{xyl} can enhance the production capacity of *B. subtilis*. As a good host for the production of glutamate, *B. subtilis* possesses all the enzyme systems and metabolic pathways except for polybutyramide, and after many years of iteration and commercialization, it has a good application prospect (Halmschlag et al. 2020).

3. *Recombinant Methylobacterium*

Methanol has recently attracted interest as a potential raw material for biotechnological processes. In recent years, a large number of yeast fermentation industries have used methanol as a carbon source for its many advantages over traditional carbon sources, such as availability, lower cost, chemical purity, and no need to compete with the food industry. BLAST search shows that *Methylobacterium* does not have the glutamate decarboxylase gene in its genome. By using recombinant

Methylobacterium and adding a PLP regeneration pathway to assist in GABA synthesis, the need for additional expensive PLP is avoided, and fermentation production costs are reduced. In addition, natural high L-glutamate production, pyridoxal kinase, and a lack of GABA degradation pathway make Methylobacterium a potential host cell for PA4 production.

4. *Saccharomyces cerevisiae*

Saccharomyces cerevisiae is a microbial cell factory that synthesizes a variety of natural products. By properly utilizing the endogenous metabolic pathways within the *S. cerevisiae* chassis cells, high-value-added biopharmaceuticals, food health products, and fine chemical products can be produced. Therefore, how to finely regulate and optimize the metabolic flow within *S. cerevisiae* cells is a key issue for achieving high yield, high productivity, and high conversion of target chemicals. Acetyl-CoA is a key group in the PA4 metabolic pathway and is also a basic precursor for the synthesis of many natural products. Fine-tuning the synthesis of acetyl-CoA is an important strategy for achieving a high yield of target compounds. Optimizing the synthesis of specific chain length fatty acids and fatty acid derivatives through the fatty acid synthesis pathway, strengthening the metabolic engineering strategy of acetyl-CoA accumulation in *S. cerevisiae*, and reconstructing the metabolic engineering modification strategy of de novo synthesis of amino acid derivatives from the fatty acid pathway make the *S. cerevisiae* chassis cells have unique advantages in the production of natural products.

3.5.3 Conclusion

With the development of the biobased materials industry in recent years, researchers have focused on the biosynthesis of PA4 through enzymatic or intracellular direct synthesis. Compared with the chemical method of PA4 production, the biological method of producing PA4 is considered an inevitable trend. The many commercial enzymes discovered with amide bond polymerization catalytic activity support the extremely high feasibility of biological preparation in theory and meet the requirements of low-cost green manufacturing. Although the prices of pure glucose and MSG are not very high, the overall cost of PA4 production can still be further reduced by changing the chassis host cells and substrates, such as using fermentable sugars derived from lignocellulose (glucose and xylose), methanol, and other low-cost and renewable substrates. In addition, directly synthesizing intracellular PA4 from raw material substrates through microbial methods and the typical fermentation engineering optimization strategies listed as follows are all optional strategies. The typical fermentation engineering optimization strategies include ① developing pretreatment/detoxification technologies with less inhibitor formation, ② constructing robust engineering strains with high tolerance to inhibitors, ③ enhancing co-fermentation of glucose/xylose/methanol, and ④ optimizing fermentation processes and developing new fermentation systems, developing artificial consortia for consolidated bioprocessing (CBP). If this process is realized, the large-scale, low-cost, green biological manufacturing of PA4 will become a reality. With the good biological properties and biodegradability of PA4, it will be used on a large scale to alleviate the domestic white pollution problem.

3.6 γ-Polyglutamic Acid

Polyglutamic acid (PGA) is a biodegradable, water-soluble, nontoxic to the environment and human body, polyanionic biopolymer material formed by the amide linkage of glutamic acid. It is widely used in agricultural production, food, medicine, and other fields. According to the type of glutamic acid carboxyl group (–COOH) involved in the formation of the amide bond, PGA can be divided into α-polyglutamic acid (α-PGA) and γ-polyglutamic acid (γ-PGA) (Fig. 3.11). α-PGA is generally chemically synthesized, while naturally occurring PGA is mostly γ-PGA. Depending on whether the glutamic acid in γ-PGA is D-configuration or L-configuration, γ-PGA can be divided into γ-D-PGA (homopolymer composed of D-glutamic acid), γ-L-PGA (homopolymer composed of L-glutamic acid), and γ-D/L-PGA (copolymer composed of D-glutamic acid and L-glutamic acid in different proportions). The relative molecular weight of γ-PGA varies with the strain, composition of the culture medium, and culture conditions, usually within the range of 10–10,000 kDa. Different relative molecular weights of γ-PGA have different characteristics and functions, and also have different application ranges (Ogunleye et al. 2015; Lei and Xia 2011; Min et al. 2021; Balogun-Agbaje et al. 2021; Mengdi 2021; Poo et al. 2010). Low molecular weight γ-PGA can be used as a food additive and antifreeze (Min et al. 2021), a transdermal absorber in cosmetics, a drug carrier, etc. (Balogun-Agbaje et al. 2021). High molecular weight γ-PGA can be added to cosmetics and agriculture to effectively prevent water loss due to its film-forming properties (Mengdi 2021); it can also effectively bind heavy metal ions and can be used as an environmentally friendly flocculant (Poo et al. 2010).

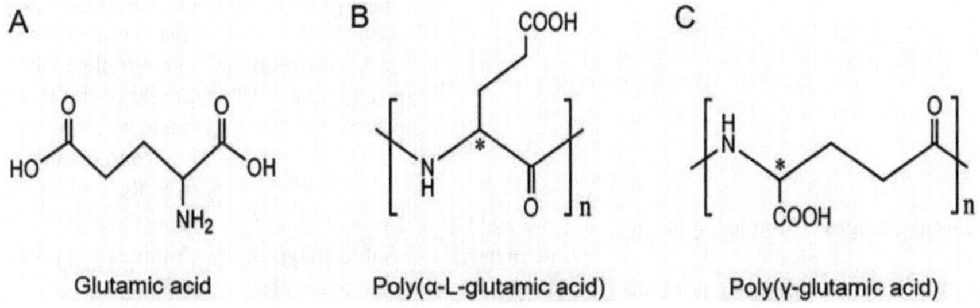

Fig. 3.11 Structural diagrams of glutamic acid monomer, α-polyglutamic acid (α-PGA), and γ-polyglutamic acid (γ-PGA) (Jin and Ning 2005) (**a**) Glutamic acid monomer, (**b**) α-polyglutamic acid (α-PGA), (**c**) γ-polyglutamic acid (γ-PGA)

3.6.1 Preparation Methods of γ-PGA

The preparation methods of γ-PGA include chemical synthesis, extraction, enzyme transformation, and microbial fermentation, each with its own advantages and disadvantages (Table 3.2). With the progress of genetic engineering, synthetic biology, and enzymology technology, the advantages of using genetically modified and/or transformed microorganisms for fermentation or enzyme transformation to prepare γ-PGA are becoming increasingly apparent, but there is still a certain gap in large-scale application.

1. *Extraction Method*

The commonly used raw material for the extraction method is natto, a traditional Japanese food. Natto is a soy product made by fermenting soybeans with *B. subtilis* var. *natto*. It has a certain stickiness, a strong smell, and a slightly sweet taste and not only retains the nutritional value of soybeans but also has a variety of bioactive substances produced during fermentation. The main component of the sticky colloid in natto is γ-PGA. The steps to extract γ-PGA from natto are as follows: cook the natto first, soak it in deionized water, and when γ-PGA is completely dissolved in water, add different concentrations of organic solvents (usually ethanol) to precipitate and extract γ-PGA from the water. Although this method is simple and easy to operate, the content of γ-PGA in the raw material natto is low, and the content of γ-PGA in different varieties and batches of natto varies greatly. The extraction

Table 3.2 Comparison of different preparation methods of γ-PGA (Ogunleye et al. 2015; Lei and Xia 2011; Poo et al. 2010; Jin and Ning 2005; Xiaoli et al. 2012; Xueliang et al. 2022; Peng et al. 2019; He et al. 2020)

	Chemical synthesis	Extraction	Enzyme transformation	Microbial fermentation
Advantages	Peptide synthesis method Dimer condensation method	Easy to operate	Simple process route High yield Few by-products Short production cycle Easy to control the production process	Mild production conditions Simple extraction process Low production cost Large molecular weight of the product Environmentally friendly
Disadvantages	Complex synthesis route Many by-products Low yield Use of toxic gases, not environmentally friendly	Low yield of raw materials Many by-products High cost	Limited known usable enzymes Low molecular weight of the product	Long time Complex metabolic pathway Low overall yield of the product

method is mostly used to prepare crude γ-PGA products. Due to the many by-products (or impurities), it is not suitable for large-scale production.

2. *Chemical Synthesis Method*

There are two types of chemical synthesis methods, the traditional peptide synthesis method and the dimer condensation method. γ-PGA is similar to a peptide polymer, and the peptide synthesis method can be used to connect glutamic acid one by one through group protection, activation, oxidative coupling, and group deprotection. Sanda et al. have improved the traditional peptide method to form the dimer condensation method (Sanda et al. 2001). α-Methyl glutamic acid is produced first by the reaction of D-glutamic acid, L-glutamic acid, or D/L-glutamic acid racemate. After the latter forms a glutamic acid dimer, it reacts with the condensing agent 1,3-dimethylaminopropyl-3-ethyl carbodiimide hydrochloride and 1-hydroxypropyltripyrrole hydrate in N,N-dimethylformamide to form polyglutamic acid methyl ester, which is finally hydrolyzed by alkali to obtain γ-PGA. Overall, the chemical synthesis method has disadvantages such as complicated operations, many by-products, and low yield, making it unsuitable for industrial applications.

3. *Enzyme Conversion Method*

The enzyme conversion method uses enzymatic reactions to link glutamic acid monomers into a γ-PGA polymer. It has advantages such as a simple process route, mild reaction conditions, short cycle, high yield, high purity, few impurities, and can be produced on a large scale. Glutamic acid transpeptidase (GTP) is the only known key enzyme that can complete this process (Ogawa et al. 1991; Shih and Van 2001), but the content and activity of GTP in microbial cells are low. In addition, the degree of polymerization and molecular weight of the products of the enzyme conversion method are low, and the molecular weight of γ-PGA determines its application in different fields. If the molecular weight does not meet the requirements, the application value of the corresponding γ-PGA will also decrease. Therefore, the development and modification of enzymes and the optimization of reaction conditions in the enzyme conversion method need to be improved.

4. *Microbial Fermentation Method*

The microbial fermentation method refers to the extraction of γ-PGA from the fermentation broth after fermenting in a shaking flask or fermenter. The microbial fermentation method is currently the most commonly used method for industrial production of γ-PGA, including batch fermentation, continuous fermentation, liquid–solid two-phase fermentation, stirred tank reactor self-circulation fermentation, and solid fermentation. Its characteristics are mild conditions, a simple process, and large-scale production.

The factors affecting the production of γ-PGA by microorganisms mainly include internal factors (characteristics of the strain itself) and external factors (composition of the culture medium and fermentation method). For internal factors, the strains currently used for fermentation are mainly *B. subtilis*, *B. licheniformis*, and *B. anthracis*. High-yield γ-PGA strains can be obtained by strain mutation and selection, or construction of engineered strains. For external factors, the composition of the culture medium and the culture conditions can be optimized by statistical methods (such as PB design, CCD design, orthogonal experiment, and response surface analysis) to provide an environment conducive to the production of γ-PGA, maximizing the production of γ-PGA by the strain.

3.6.2 Microorganisms Synthesizing γ-PGA

Several microbial taxa—such as Bacteria (*Bacillus*, *Clostridium*), Archaea, and Eukaryotes— are known to synthesize γ-PGA, among which *Bacillus* is the most common producer. The known strains that synthesize γ-PGA in *Bacillus* include *B. anthracis*, *B. subtilis*, *B. subtilis* var. *natto*, *B. licheniformis*, *B. brevis*, *B. thermotolerans*, *B. anmoliquefaciens*, *B. siamese* (Yahui et al. 2021), *B. velezensis* (Lingtian et al. 2020), *B. methylotrophicus* (Senlin et al. 2021), etc., and the Archaea include *Planococcus halophilus*, *Sporosarcina halophile*, *Natrialba asiatica*, *Natronococcus occultus,* etc., and the only representative of eukaryotes is *Cnidaria.* In addition, Kocianova et al. found that *Staphylococcus epidermidis* (da Silva et al. 2020) can also synthesize γ-PGA.

Based on whether exogenous glutamate needs to be added during the fermentation process, the γ-PGA-producing strains are divided into glutamate-dependent type (Type I, which requires the addition of exogenous glutamate to produce γ-PGA) and non-glutamate-dependent type (Type II, which can produce γ-PGA without glutamate). Usually, the γ-PGA yield of non-glutamate-dependent strains is much lower than that of glutamate-dependent strains, which is not suitable for fermentation production. Therefore, the research on the preparation of γ-PGA in China mainly focuses on glutamate-dependent strains. Since the addition of glutamate will increase the cost, when using glutamate-dependent strains, it is necessary to optimize the medium to improve the utilization rate of glutamate.

3.6.3 γ-PGA Biosynthetic Enzyme System and Regulatory Proteins

The synthesis of γ-PGA is through a non-ribosomal dependent peptide synthesis pathway, which is a process mediated by biological membranes, requiring ATP as energy, glutamate as substrates, and metal ions as cofactors. The γ-PGA synthesis mechanism first proposed by Troy et al. (1973) in *B. licheniformis* is as follows: First, ATP activates L-glutamate. Next, AMP formed by ATP removing pyrophosphate (ppi) binds with glutamate at the γ position to form glutamyl-γ-AMP. Then, glutamyl-γ-AMP binds with a -SH enzyme or receptor (such as some thiol, temporarily represented by X), and isomerizes to glutamyl-X. Finally, glutamyl is connected to the γ-PGA fragment, and X is removed, completing the extension of the γ-PGA fragment. Ashiuchi et al. (2004) suggest another mechanism according to their findings during the process of *B. subtilis* synthesizing γ-PGA in vitro. They find that the hydrolysis of ATP forms ADP instead of AMP when *B. subtilis* synthesizes γ-PGA in vitro. The mechanism suggested by Ashiuchi et al. is as follows: Firstly, ATP is hydrolyzed by ATP hydrolase into ADP and Pi; then the phosphate group binds to the C-terminus of the small molecule γ-PGA. Second, the amino terminus of D- or L-glutamic acid undergoes a nucleophilic attack on the C-terminus phosphorylated small molecule γ-PGA, generating Pi. Finally, the action of γ-PGA synthase, the γ-PGA chain is extended. However, it is still uncertain which mechanism is correct. Different microorganisms have different mechanisms for synthesizing γ-PGA due to their biological characteristics and metabolic regulation. According to the state of γ-PGA synthesized by microorganisms, it is divided into bound and free types. The genes of γ-PGA synthase are divided into two types, namely the cap (capsule) system and the pgs (polyglutamate synthase) system (Mingfeng et al. 2011). If the γ-PGA synthesized by the strain is bound to the cell wall of the bacterium as part of the capsule structure, it is called the bound type, and the gene of the γ-PGA synthase of this bacterium is called the cap system gene, represented by *B. anthracis*. If the synthesized γ-PGA is released into the external environment of the bacterium, the gene of this bacterium is called the pgs system gene, represented by *B. subtilis*. The cap system gene is located on the plasmid, composed of *capB*, *capC*, and *capA*. The pgs system gene exists in the pgs operon in the genome, containing *pgsB*, *pgsC*, *pgsA*, and *pgsE,* four genes. The similarity of the related genes of the two is very high (over 50%). Research on γ-PGA synthase and its expression regulation is mainly focused on the Pgs enzyme system (Fig. 3.12). PgsB is a γ-amide linker, located on the plasma membrane through a nutmeg anchor, it has two molecular weights (35 kDa and 45 kDa), both are active, and they are more stable when combined; PgsC is embedded in the cell membrane

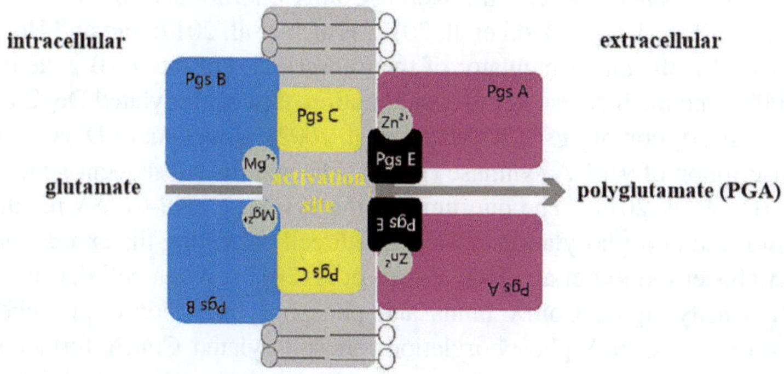

Fig. 3.12 Schematic diagram of the cellular localization of polyglutamic acid synthase complex (see color insert). (Zhong et al. 2010)

through 4 transmembrane regions and multiple nutmeg anchors, there is an amidation site, it may work with PgsB to play the role of ATPase, or it may be responsible for connecting PgsB and PgsA to form a functional complex; PgsA binds to the plasma membrane through an N-terminal transmembrane region and a nutmeg anchor; it has multiple phosphorylation sites, its main responsibility is to transport the synthesized γ-PGA to the outside of the cell (Zhong et al. 2010; Ashiuchi 2013). The three components PgsB, PgsC, and PgsA can only synthesize γ-PGA when they are stably bound to the cell membrane and assisted by ATP; if the cell membrane only contains PgsB, PgsC, PgsA, PgsBC, PgsBA, or PgsCA protein, the cell cannot synthesize long-chain γ-PGA. PgsE is a 6.5 kD small protein; it can bind metal ions (such as Zn^{2+}), can promote *B. subtilis* to synthesize γ-PGA (Wenjuan et al. 2012), and will also affect the product γ-PGA (Ashiuchi et al. 2013). Ken-Ichi et al. analyzed the conditions for *B. subtilis* to produce γ-PGA and found that when PgsE is present, the molecular weight of γ-PGA (about 2,900,000 g/mol) is 60 times that of γ-PGA (about 47,000 g/mol) when PgsE is absent (Fujita et al. 2021).

Both D-glutamic acid and L-glutamic acid can serve as precursor molecules for γ-PGA synthesis. The ratio of D-glutamic acid to L-glutamic acid in γ-PGA produced with different bacterial strains and culture conditions varies. In γ-PGA produced by *B. subtilis* var. *natto*, 50–80% is D-glutamic acid, while in γ-PGA produced by *B. subtilis*, the proportion of D-glutamic acid is 35–75%. In γ-PGA produced by *B. licheniformis*, the proportion of D-glutamic acid increases with the extension of culture time. When overexpressing the *pgs* BCA gene cluster from *B. licheniformis* in *E. coli* and *C. glutamicum* to produce γ-PGA, the γ-PGA is 100% L-glutamic acid (Halmschlag et al. 2019). Halmschlag et al. found that the polyglutamic acid synthase and glutamic acid racemase from different bacterial strains have a great impact on the yield, molecular weight, and D-/L-glutamic acid ratio in γ-PGA, so γ-PGA can be reasonably designed by the combination of these enzymes (Halmschlag et al. 2019).

The transcription and expression levels of γ-PGA synthase also have a great impact on the ability of microbial strains to produce γ-PGA. It has been reported that the transcription of γ-PGA synthase (PgsBCAE) is regulated by the DegS-DegU two-component system, the ComP-ComA quorum sensing system, DegQ, SwrAA, etc. (Fig. 3.13) (Xiaoli et al. 2012; Hsueh et al. 2017; Tao and Hongsheng 2015). DegU is the direct regulator of the transcription of the *pgs*B gene (Ohsawa et al. 2009), and the increase in the concentration of phosphorylated DegU can activate the transcription of *pgsBCA* (Mader et al. 2002). Knockout of DegQ will block the transcription of γ-PGA synthase (PgsBCAE) and the subsequent generation of γ-PGA (Do et al. 2011). The quorum-sensing system ComP-ComA regulates the expression and phosphorylation of DegQ, indirectly affecting the expression of the *pgsBCA* cluster (Xiaoli et al. 2012; Hsueh et al. 2017). When cell density is high, the cell density signal ComX binds and phosphorylates ComP, phosphorylated ComP promotes ComA phosphorylation, phosphorylated ComA further induces DegQ expression and phosphorylation, after DegQ is phosphorylated, it promotes the phosphorylation of the two-component system DegS-DegU, phosphorylated DegU (DegU-Pi) binds to the promoter region of the *pgsBCA* cluster, inducing the

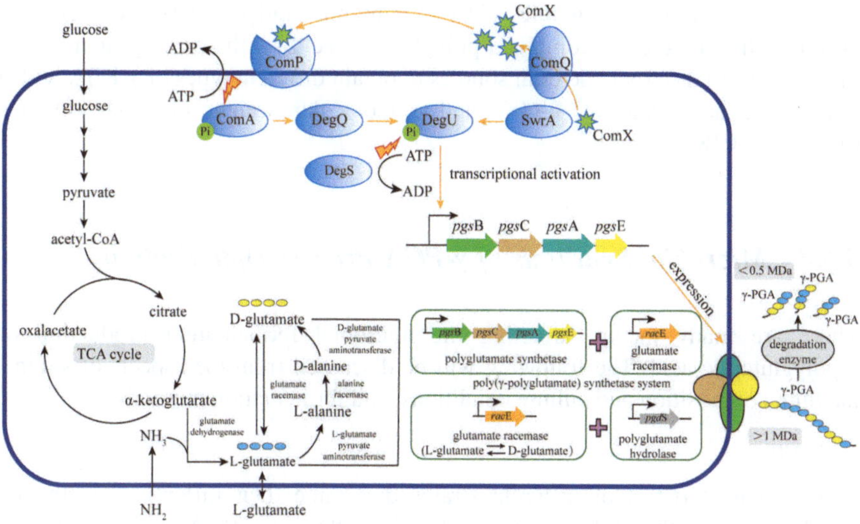

Fig. 3.13 Biosynthesis and regulation of γ-PGA (see color insert). (Xiaoli et al. 2012)

transcription and expression of the *pgsBCA* gene cluster (Xiaoli et al. 2012; Hsueh et al. 2017). Changes in the ComP encoding gene *comP* will affect the ability of bacterial strains to produce γ-PGA. Nagai et al. transposed the insertion sequence IS4Bsu1 into the *comP* gene of the strain *B. subtilis* var. *natto* NAF4, and the ability of this strain to produce γ-PGA spontaneously disappeared (Nagai et al. 2000). The essential protein of aggregation, SwrA, in *Bacillus* also has a synergistic effect on the phosphorylation of DegU and the transcriptional regulation of *pgsBCAE* gene cluster. Osera et al. find that the presence of SwrAA alone or phosphorylated DegU has only a weak effect on the transcription of the *pgsBCAE* gene cluster and γ-PGA production. Only when SwrAA and phosphorylated DegU coexist can the transcription of the *pgsBCAE* gene cluster be fully activated (Kimura et al. 2009). The insertion sequence IS4Bsu1 transposes into the *swrA* gene of the strain *B. subtilis*, also causing the strain to lose its ability to produce γ-PGA (Kimura et al. 2011).

Under different fermentation conditions, the ability of different strains to produce γ-PGA, the ratio of D-/L-glutamic acid in γ-PGA, and the size of the molecular weight of γ-PGA produced are different. In addition to being related to the expression and activity of the above-mentioned γ-PGA synthase, it may also be related to the γ-PGA depolymerase YwtD and D-/L-glutamic acid racemase in a strain. When the strain is in a state of starvation, it will initiate the expression of γ-PGA depolymerase YwtD, degrading γ-PGA to provide the nitrogen source needed for bacterial growth. Xu et al. find that in *B. subtilis* IFO16449 strain, γ-PGA depolymerase YwtD only degrades the γ-glutamyl bond formed by D- and L-glutamic acid in γ-PGA, and can degrade the γ-PGA produced by the strain into two products with different molecular weights: high molecular weight product (490 kD, composed entirely of L-glutamic acid) and low molecular weight product

(11 kD, the ratio of D-glutamic acid to L-glutamic acid is 80:20) (Xinwei et al. 2004). Halmschlag et al. combined polyglutamic acid synthase and glutamic acid racemase from different bacterial sources to obtain different types of γ-PGA with a D-glutamic acid ratio of 4–60% and a molecular weight of 40–8,500 kDa (Xueliang 2018).

3.6.4 Microbial Synthesis of γ-PGA Process Optimization

There are many factors affecting the microbial fermentation production of γ-polyglutamic acid. The following will be discussed from the aspects of strains, medium composition and culture conditions, and fermentation methods.

1. *Strain*

The characteristics of different strains determine their different growth and metabolisms, so the ability to ferment and synthesize γ-PGA is also different. Improving the production capacity of the strain is an important prerequisite for the production of polyglutamic acid. In response to the problem of low γ-PGA yield in natural strains, people use mutation and genetic engineering methods (gene cloning, knockout, transcription, and expression) to construct genetically engineered bacteria and improve their strain performance.

Mutation is to use physical factors (ultraviolet rays, γ rays, lasers, etc.), chemical mutagens (nitrosoguanidine, diethyl sulfate, colchicine, etc.) and other means to act on the bacteria, causing them to produce a large number of mutants in a short time, and then artificially select high-yield mutants that can be stably inherited. Zhang Rui et al. used the ultraviolet-nitrosoguanidine compound mutation method to select a mutant strain that can be stably inherited, with a γ-PGA yield of 24.2 g/L, which is 31.5% higher than that of the starting strain (Xinwei et al. 2004). Chen Yongzhu et al. used ultraviolet light and nitrosoguanidine to induce mutations in *B. licheniformis* ATCC9945A strain and selected a high-yield mutant strain Zγ-29 that can stably inherit, which increased the yield of γ-PGA by four times (Yongzhu 2005). Chen Shuangxi et al. used atmospheric pressure room temperature plasma (ARTP) to induce mutations in *Bacillus subtilis* HD11strain and obtained a high-yield strain HNCL1266. The yield of γ-PGA in this mutant strain was 26 g/L, which was 30% higher than that of the original strain HD11 (Shuangxi et al. 2015). Li Nan obtained the strain S004-50-01 after inducing mutations in the strain *B. subtilis* var. *natto* S004. The yield of γ-PGA increased from 4.25 g/L of the original strain to 10.0 g/L, and the utilization rate of glutamic acid increased from 18.2% to 42.8% (Nan 2006). Liu Dandan et al. used ARTP technology to induce mutations in *B. subtilis* var. *natto*, and obtained a high-yield strain with a 22% increase in γ-PGA yield through casein plate transparent circle quick screening (Dandan et al. 2021).

With the deepening understanding of the biosynthesis mechanism of γ-PGA and the development of genetic engineering technology, the application of genetic engineering technology in the breeding and selection of γ-PGA-producing microorganisms is also constantly expanding. Su et al. integrated the bacterial hemoglobin gene (*vgb*) into the chromosome of *B. subtilis* by homologous recombination, enhancing the oxygen absorption capacity of the mutant strain, successfully overcoming the problem of insufficient oxygen dissolution caused by increased viscosity during fermentation, and increasing the fermentation cell concentration by 1.26 times, γ-PGA yield increased to 60.5 g/L (Su et al. 2010). The introduction of the *vgb* gene also successfully increased the γ-PGA production capacity of *B. amyloliquefaciens* (Hui 2012), *B. subtilis* NX-2 (Bao et al. 2016), *B. licheniformis* (Cai et al. 2018), *C. glutamicum* (Xu et al. 2021)· and others. Yeh et al. integrated the high-efficiency synthetic expression control sequence (SECS) into the upstream of the *ywsC* gene of the non-γ-PGA synthesis strain *B. subtilis* DB430 and obtained a recombinant strain *B. subtilis* PGA6-2 that can produce 28 g/L of γ-PGA in a medium without additional glutamic acid and ammonium chloride (Yeh et al. 2010). Gao et al. improved the glutamic acid metabolism related network in *B. amyloliquefaciens* (knocking out *fadR*, *lysC*, *aspB*, *pckA*, *proAB*, *rocG*, and *gudB* genes) and inserted a strong promoter P_{C2up} in front of the *icd* gene, as well as inserted regulating genes, *pgi* and *gndA*, to promote the intracellular NADPH levels in this strain, resulting in a 2.5-fold increase in γ-PGA production (Gao et al. 2019).

2. *Composition of Culture Medium*

After selecting the appropriate strain, the composition and ratio of the culture medium are key factors affecting the growth and metabolism of the strain. Currently, the culture medium used for γ-PGA synthesis is mostly synthetic medium, so the optimization of key components of the culture medium such as carbon source, nitrogen source, metal ions, and inorganic salts is crucial for the industrial preparation of γ-PGA.

The most suitable carbon sources for *Bacillus* strains to ferment and synthesize γ-PGA are often citric acid, glycerol, fructose, starch, glucose, and maltose. Some strains prefer organic nitrogen sources (such as soy milk, peptone, yeast extract, corn slurry, corn slurry powder, soybean cake powder, and peanut cake powder) as the optimal nitrogen sources, while others prefer inorganic nitrogen sources (such as ammonium chloride, ammonium sulfate, ammonium nitrate, and urea) as the optimal nitrogen sources. Different strains are suitable for different carbon and nitrogen sources. The optimal carbon sources of strain *B. licheniformis* WBL-3 are citric acid and glycerol, as well as glucose and glycerol are the optimal carbon sources of strain *B. licheniformis* ATCC 9945a. *B. subtilis* prefers citric acid and glucose as the optimal carbon sources. Strain *B. subtilis* NX-2 prefers starch and maltose as the optimal carbon sources. For the optimal nitrogen source, when strain *B. licheniformis* ATCC 9945a uses soybean milk as the nitrogen source, γ-PGA yield reaches its maximum (35 g/L). The yield of γ-PGA of strain *B. licheniformis* CGMCC3336 increased by 17% when using yeast extract as a nitrogen source, compared to when using ammonium nitrate as a nitrogen source (Junnan et al. 2018).

Metal ions (such as K^{+}, Mn^{2+}, Fe^{3+}, Mg^{2+}, Zn^{2+}, and Ca^{2+}) are very important for the synthesis amount, molecular weight, and the ratio of D-/L-glutamic acid in γ-PGA of *Bacillus* strains (Stark and von Stockar 2003). Low concentrations of

Mn^{2+} are beneficial for the growth of *B. subtilis*, while high concentrations of Mn^{2+} will inhibit the growth of *B. subtilis*. Gross et al. find that as the concentration of Mn^{2+} gradually increases, the yield of γ-PGA shows a trend of first increasing and then decreasing. Cromwick et al. also find that the concentration of Mn^{2+} ions in the culture medium can regulate the ratio of D-type or L-type glutamic acid in γ-PGA (Peng et al. 2019).

The addition of some salts (such as NaCl) can also affect the synthesis of γ-PGA. Studies have found that the addition of an appropriate concentration of NaCl can reduce the viscosity of the fermentation broth in the later stages of fermentation, playing a defoaming role. However, if the concentration of NaCl is too high, it will cause the osmotic pressure in the fermentation broth to be too high, leading to cell dehydration and death, thereby affecting the yield of γ-PGA (Wei et al. 2010; Shuiyang and Dan 2012).

3. *Culture Conditions*

Aerobic microorganisms need to provide sufficient dissolved oxygen to the cells during growth. When the oxygen supply is insufficient, the normal physiological metabolism of aerobic microorganisms under anaerobic conditions will be affected. Low rotation speed and uneven mass transfer result in insufficient dissolved oxygen. Whereas high rotation speed, providing sufficient dissolved oxygen, also results in too large a shear force generated on the cells, which will damage the cell wall and cell membrane, is not conducive to cell growth, and then affects the production of the product. In addition, optimizing the culture rotation speed is also very important for fermentation preparation. Cromwick et al. increased the ventilation volume in the fermentation of strain *B. licheniformis* ATCC 9945a by increasing the stirring speed (250–800 rpm) and aeration rate (0.5–2.0 L/min), which led to the cell dry weight doubling and the γ-PGA yield increasing nearly three times (Cromwick et al. 1996).

4. *Fermentation Method*

Fermentation methods can be divided into liquid fermentation and solid fermentation, and liquid fermentation is mostly used in industrial production. The yield and molecular weight of γ-PGA are optimized by adjusting the pH, temperature, ion strength of the culture medium, and inoculation amount during the culture process. Strain *B. licheniformis* ATCC 9945a produces the maximum yield of γ-PGA when the pH is maintained at 6.5, and when the pH is less than 5.5 or greater than 7.4, its γ-PGA yield significantly decreases (Shuiyang and Dan 2012). Zhao et al. (Xiaoxing 2017) explore the effects of initial pH and temperature on the production of γ-PGA in strain *B. amyloliquefaciens* YP-2 by liquid fermentation. The results showed that when the initial pH was 7–7.5 and the fermentation temperature was 37 °C, the γ-PGA yield was the highest (38.39 g/L) (Xiaoxing 2017). Although liquid fermentation has the advantages of easy process control, stable yield, and adjustable product molecular weight, it still has some shortcomings. For example, the viscosity of the fermentation broth will continuously increase with the increase of γ-PGA concentration. A large amount of foam will be produced during the fermentation process, which is difficult to control and increases the probability of contamination. The preservatives added to prevent the spoilage of liquid γ-PGA may affect the quality of γ-PGA itself.

Solid fermentation uses agricultural and industrial solid waste as substrates, successfully avoiding the shortcomings brought by liquid fermentation, and the cost is lower. However, it has the disadvantages of uneven fermentation, large fermentation scale, and difficulty in online monitoring and process control. So far, the residues from cow manure compost, monosodium glutamate, and vinegar production have been studied for the preparation of γ-PGA, which not only obtains the target product, but also improves the utilization rate of waste resources. Wu et al. have applied for a patent for this achievement of using glutamic acid fermentation waste cells to ferment and produce polyglutamic acid (Linhui et al. 2018). Zhang et al. optimized the process of preparing γ-PGA from MSG factory wastewater with strain *B. subtilis* 168, resulting in the yield of γ-PGA reaching 53.51 ± 0.92 g/L (Yanli 2018). Han et al. (Wenjing et al. 2019) also explored the feasibility of fermenting γ-PGA from MSG by-products. By controlling the acid production rate during fermentation, the yield of γ-PGA reached a maximum of 57.8 g/L, realizing the reuse of by-products and reducing production costs (Wenjing et al. 2019). In 2021, Zhang et al. explored the feasibility of preparing γ-polyglutamic acid from poplar wood chip enzymatic hydrolysate. After optimizing the culture conditions, the amount of glucose added to the medium was reduced; however, the yield of γ-PGA could reach 30.87 g/L, which is an increase of 5.11% compared to the pure addition of glucose (Zhang et al. 2021a).

5. *Conclusion*

γ-PGA has the advantages of water solubility, biodegradability, edibility, multipurpose use, and environmental friendliness, and has a wide range of application prospects in the fields of medicine, food, plastics, water treatment, and agriculture. The diversity of γ-PGA affects the application potential of γ-PGA. It is very important to prepare γ-PGA with a specific stereoisomer composition and specific molecular weight according to demand. Although there have been some successful cases in recent years with the efforts of researchers, it is far from meeting market demand. The road to exploring strategies for low-cost, efficient preparation of γ-PGA with specific properties is still long. Fully elucidating the molecular regulation mechanism of γ-PGA biosynthesis, combining synthetic biology, metabolic engineering, and other technologies to modify γ-PGA preparation strains, regulating the expression of γ-PGA synthesis enzymes and hydrolases in strains, optimizing fermentation culture conditions of strains, etc., are indispensable steps to obtain specific γ-PGA and to widen the use of γ-PGA, which are also the key directions of γ-PGA research and development.

3.7 ε-Polylysine

ε-Polylysine (ε-PL) is generally a polymer formed by the α-amino and ε-carboxyl groups of 25–35 L-lysine residues connected by amide bonds (Fig. 3.14), first isolated and purified from the fermentation broth of *Streptomyces albulus* by Japanese scholars Shima and Sakai (1977). ε-PL is a white or light yellow powder, slightly bitter, unstable, highly hygroscopic, very soluble in water, insoluble in organic solvents, and mostly exists in the form of hydrobromide. ε-PL has the characteristics of broad-spectrum antibacterial activity, strong bactericidal ability, safety and non-toxicity, good water solubility and thermal stability, high temperature resistance, and is relatively stable in acidic and alkaline environments. It also has excellent properties such as biodegradability and edibility, and has been approved as a food preservative by countries such as the United States, Japan, South Korea, and China. In addition, ε-PL can also be used as a gene carrier, weight loss health product, drug carrier, new type of water-absorbing material, chip, and bioelectronic coating agent (Chheda and Vernekar 2015; Tuersuntuoheti et al. 2019; Chen et al. 2021; Patil and Kandasubramanian 2021; Aixia et al. 2020; Yanyun et al. 2019).

The methods for synthesizing ε-polylysine include chemical synthesis, biosynthesis, and enzymatic methods. Among them, the most ideal method is to ferment *S. albulus* to produce ε-polylysine. In this method, *S. albulus* first converts the substrate glucose into pyruvic acid, which combines with CO_2 under the catalysis of pyruvate carboxylase to produce oxaloacetic acid. Oxaloacetic acid generates lysine under the action of various enzymes. Subsequently, lysine condenses to form ε-polylysine under the action of L-lysine polymerase (Aixia et al. 2020). The biosynthetic pathway of ε-polylysine is shown in Fig. 3.15.

3.7.1 ε-PL High-Yield Bacteria Breeding

1. *Traditional Mutation Breeding*

Traditional random mutation (UV mutation or chemical mutation, etc.) followed by screening is one of the main methods for ε-PL high-yield bacteria breeding. The

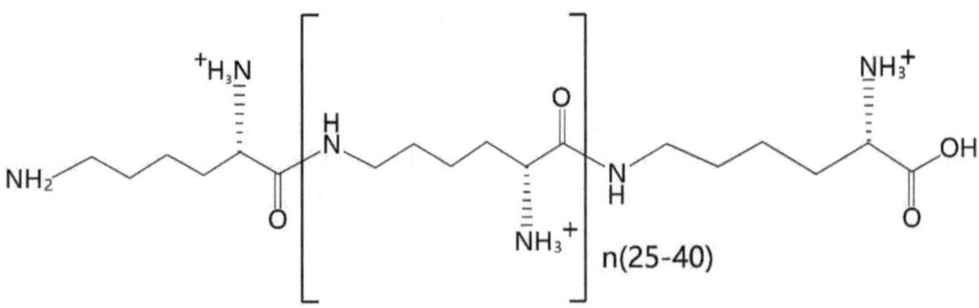

Fig. 3.14 Structure of ε-PL. (Lei and Xia 2011)

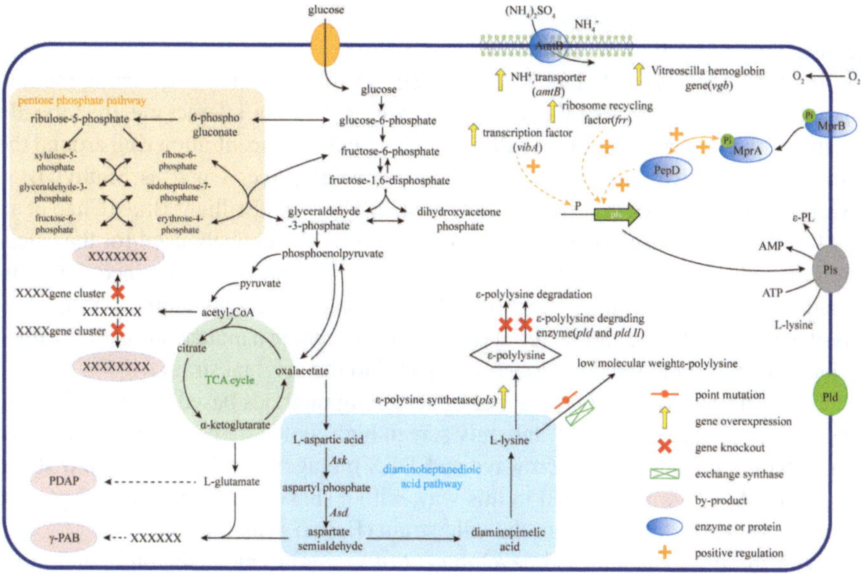

Fig. 3.15 ε-PL biosynthetic pathway (see color insert)

most widely used screening method is the "AEC+Gly" resistance screening method by Japanese scholar Hiraki (1998). The principle is glycine (Gly) is a structural analog of D-alanine, an important component of cell wall synthesis. The Gly added to the culture medium, if absorbed by the cell wall and used for the synthesis of peptidoglycan, will affect the cross-linking degree of peptidoglycan, increase the permeability of the cell wall, and thus promote more lysine structural analogs (AEC) into the cell.The screened AEC-resistant strains can relieve the feedback inhibition of lysine on aspartate kinase (ASK), thereby increasing the precursor of ε-PL synthesis.

Regarding the screening of ε-PL high-yield bacterial strains, the earliest and most studied internationally are mainly Japanese scholars. As early as 1998, the ε-PL shake flask yield of the high-yield mutant strain they screened reached 2.11 g/L, which was the highest in the world at that time (Hiraki et al. 2003; Weiwei et al. 2007a). Many domestic scholars also use AEC as a "sieve" for resistance screening of strains. Chen et al. (Weiwei et al. 2007a) screened AEC-resistant mutants after mutating *Kitasatospora* with diethyl sulfate (DES), and the ε-PL shake flask yield of the mutant strain was two times higher than that of the wild-type strain, reaching 1.17 g/L. Yu (Weiwei et al. 2007b) used physical (UV) and chemical (nitrous acid) methods to mutate *S. albulus*, and the ε-PL yield of the AEC-resistant strain *S. albulus* UN2-71 was 1.64 g/L, which was 1.42 times that of the starting strain. Hong Zong (2012) obtained a strain with an ε-PL yield of 1.6 g/L by ARTP mutation and AEC resistance screening method, which was 31.6% higher than that of the starting strain. Although many experimental results show that AEC can be used as a good "sieve" for screening high-yield strains, this method also has many problems, such as the high price of AEC (1000 yuan/g) and large usage, time-consuming, and laborious screening.

In 2002, Japanese scholars Nishikawa and Ogawa invented a screening method based on the principle that the acidic dye Poly R-478 can interact electrostatically with positively charged ε-PL, resulting in a noticeable concentration ring around the colony (Nishikawa and Ogawa 2002). Subsequently, dyes (such as Methyl Blue (Geng et al. 2014), Methyl Orange (Liu et al. 2017c), and Brilliant Blue (Chao et al. 2006), which are based on similar principles) were successively used for the selection of ε-PL production strains. Zhang et al. (Geng et al. 2014) obtained the high-yield ε-PL *S. albus* strain Z-18 with the Brilliant Blue screening method. Yang et al. used the Brilliant Blue antibacterial ring method to screen a mutant strain of *S. albus*, which increased the yield by 20.9% compared to the original strain.

In addition to the above two types of screening methods based on "sieve," some researchers do not use "sieve"and only screen high-yield strains through significant phenotypic changes after microwave and DES mutagenesis. The high-yield strain *S. albulus* SF-21 was obtained in this way, with a shake flask yield of 0.85 g/L, an increase of 41.3% compared to the wild strain (Haitao et al. 2007).

With the improvement and development of screening methods, more and more high-yield ε-PL strains have been discovered. So far, high-yield ε-PL strains are mainly concentrated in the genus *Streptomyces*, among which the most productive and most studied is *S. albulus*. Others include *Kitasatospora*, *S. ahygroscopicus*, *S. griseofuscus*, and *S. diastatochromogenes* (Aixia et al. 2020). In addition, strains with ε-PL production capabilities such as *B. licheniformis*, *B. cereus*, *Lactobacillus delbrueckii strain*, *Lactococcus lactis strain*, and *C. variabile* have also been screened (Aixia et al. 2020).

The advantage of traditional breeding methods is that mutation breeding can be carried out without understanding the genetic background and gene information of the strain, which is simple to operate and effective. The disadvantage is that it is too random, time-consuming, labor-intensive, and can easily lead to a waste of resources.

2. *Genome Rearrangement Breeding*

Hopwood et al. proposed the concept of protoplast fusion in 1979. Stemmer et al. proposed a new breeding method called DNA shuffling (genome shuffling, genome rearrangement, genome reorganization) in 1998. The idea of this method is to promote the recombination or genome rearrangement of the chromosomes of the protoplasts of the positive mutants after obtaining the positive mutants by traditional mutation breeding, and to concentrate the positive mutant genes together to obtain the fusion mutant strains with greatly improved target traits (Kejia and Shuyi 2020; Qianqian et al. 2019). This method has been widely used in the improvement of industrial microbial strains (Yin et al. 2016; Wang et al. 2017; Magocha et al. 2018). Yin et al. have increased the flask yield and the tank yield of glutathione in *S. cerevisiae* by 32 times and 33 times, respectively, with only two rounds of genome rearrangement (Yin et al. 2016). However, if traditional mutagenesis techniques were used, it would take at least 2 years and require screening of more than 30,000 mutant strains to achieve this effect. This technique has also been successfully used for the screening of high-yield ε-PL strains (Wang et al. 2016; Li et al. 2013; Zhou et al. 2015). Li et al. obtained the mutant strain *Streptomyces* sp. FEEL-1 using genome rearrangement and protoplast fusion technology, with a fermentation yield of 24.5 g/L, an increase of 81% compared to the wild-type strain (Li et al. 2013). Zhou et al. used genome rearrangement technology to screen ε-PL-tolerant bacteria and obtained the mutant recombinant strain *Streptomyces* sp. F4-22, with a yield of 39.96 g/L, an increase of 32.7% compared to the wild-type strain (Zhou et al. 2015).

Compared to traditional mutagenesis breeding, genome rearrangement breeding is simpler, less time-consuming, and more efficient, and it does not require prior understanding of the full genome sequence data of the strain, the metabolic regulation information related to the target product, and the biosynthesis information.

3. *Genetic Engineering Breeding*

In addition to random mutation breeding, rational breeding methods (such as metabolic engineering breeding, and genetic engineering breeding) are also one of the main methods for breeding high-yield ε-PL strains. Rational breeding has strong targeting and significant effects, mainly using metabolic engineering means to overexpress key enzymes in the metabolic pathway of the target product, knock out or reduce enzymes that decompose intermediate products, or optimize the heterologous expression of enzymes by changing codon bias. This method requires a deep understanding of the metabolic regulation mechanism and biosynthesis mechanism of the target product, so that key regulatory proteins or genes encoding key enzymes can be selected and overexpressed or knocked out accordingly.

Polylysine synthetase (Pls) is a key enzyme in the ε-PL synthesis pathway, and overexpression of Pls is expected to increase the yield of ε-PL (Aixia et al. 2020). Qin et al. used the constitutive promoter and ribosome binding site of *Streptomyces* to achieve overexpression of the Pls gene, and the fermentation yield of ε-PL in the overexpressed strain was more than twice that of the original strain (Aixia et al. 2020; Jiayang et al. 2019). Ask enzyme is a key enzyme in the synthesis of the substrate lysine in the bacterium and is subject to feedback inhibition by L-Lys. Based on the traditional mutagenesis breeding AEC (L-Lys analog) resistance screening principle, Hamano et al. constructed a high-yield ε-PL strain rAsk (M68V) that can relieve the feedback inhibition of L-Lys in strain *S. albulus* IFO14147 (Hamano et al. 2007). To solve the problem of oxygen limitation during the fermentation process of ε-PL in this bacterium, Xu et al. integrated the hemoglobin gene expression frame from *Vitreoscilla* into the chromosome of strain *S. albulus* PD-1; the ε-PL yield of the integrated strain significantly increased (Xu et al. 2015a). Subsequently, this research team also introduced the ammonium ion transport protein gene into *S. albulus*. In strain PD-1, the utilization rate of nitrogen sources of the recombinant strain was increased, and the yield of ε-PL was increased from 22.7 to 35.7 g/L (Xu et al. 2018). It is worth noting that not all operations on the expected genes can increase the yield of ε-PL. Yamanaka et al. knocked out the ε-PL degradation enzyme (Plds) gene of the *S. albulus* CRM001 strain, but the yield of ε-PL did not increase (Yamanaka et al. 2020). With the development of high-throughput sequencing technology and the deepening research on the synthesis mechanism and regulation system of ε-PL, the pace of constructing high-yield ε-PL strains by genetic engineering will become faster and faster.

3.7.2 Fermentation Production of ε-PL

1. *Optimization of Cultivation Conditions*

Since the first strain of *S. albulus* producing ε-PL was screened out in 1977, researchers have begun to optimize the fermentation cultivation conditions of the strain. Shima et al. found that the culture medium with glycerol or glucose as the carbon source, and ammonium sulfate and yeast powder as the nitrogen source, was more suitable for the strain to ferment and produce ε-PL, but the yield was only 0.3 g/L (Weiwei et al. 2007b). The research group continued to explore the effect of adding amino acids to the culture medium on the yield of ε-PL and found that proline could significantly increase the yield of ε-PL in the strain (Shima and Sakai 1981). Liao et al. found that when arginine, alanine, and glycine were added to the culture medium, the ε-PL yield of *S. graminearus* significantly increased (Lijuan et al. 2011). Among the many components of the culture medium, the carbon source is crucial for improving the yield of ε-PL. Research by Indian and Chinese scholars almost simultaneously proves that glycerol is a better carbon source for fermenting ε-PL, and its ε-PL yield is 2.3–2.8 times that of glucose as the carbon source (Bankar and Singhal 2011; Chen et al. 2012a). Based on this, Chen et al. further increased the yield of ε-PL by using the "glucose + glycerol" mixed carbon source fermentation method (Chen et al. 2012b). Other scholars have developed and applied cheap carbon sources (sugar cane molasses) and cheap nitrogen sources to explore the feasibility of reducing the fermentation cost of ε-PL. The results show that when the strain *S. albulus* PD-1 ferments ε-PL in a ton-level fermenter, the yield can reach 20.6 g/L (Xiao et al. 2014). In addition to carbon and nitrogen sources, adding different metal ions to the culture medium will also affect the yield of ε-PL. Research by Kobayashi et al. shows that metal ions (Fe^{2+}, Mn^{2+}, and Co^{2+}) can promote the fermentation synthesis of ε-PL by *Kitasatospora kifunense*, and the promoting effect of Fe^{2+} is the strongest (Kobayashi and Nishikawa 2007). Wang et al. also found that Fe^{2+} could increase the ability of *S. diastatochromogenes* to synthesize ε-PL (Wang et al. 2011).

2. *pH Regulation*

pH plays a decisive role in the synthesis and secretion process of ε-PL. In 2001, Kahar's research found that adopting a "high first, low later" pH regulation strategy is beneficial for quickly increasing the fermentation yield of ε-PL producing bacteria (Kahar et al. 2001). The specific operation of this pH regulation strategy is as follows: during the early stage of bacterial culture, the pH of the culture medium is not controlled, allowing it to naturally drop to 5.0 to promote rapid bacterial growth;

in the later stage, the pH is maintained at the optimal fermentation pH (around 4.0) until the end of fermentation, promoting the bacteria to quickly synthesize and secrete the target product ε-PL. By combining this two-stage pH regulation with a substrate pulse flow adding strategy (maintaining glucose concentration below 10 g/L, relieving substrate inhibition), the ε-PL yield reached 48.3 g/L after 192 h of fermentation (Kahar et al. 2001). Shih et al. combined a two-stage pH control process (pH 6.8–4.0) with a pulse substrate flow adding strategy, increasing the ε-PL yield by 2.58 times (Shih and Shen 2006). Chen et al. used a "low first, high later" two-stage pH regulation strategy to culture the *Streptomyces sp*. M-Z18 strain, achieving an ε-PL yield of 30.11 g/L, which is 20% higher than that in constant fermentation at pH 3.8 (Chen et al. 2011). In 2015, Ren et al. further improved the two-stage pH regulation strategy to form a "pH shock" strategy, that is, controlling the pH to 5.0 during the mycelium culture stage, then allowing the pH to naturally drop to the lowest point and maintaining it for 12 h for pH shock, and finally adjusting the pH back and maintaining it at pH4.0 to ensure rapid synthesis and secretion of ε-PL by the strain until the end of fermentation (Ren et al. 2015). Using this strategy in *Streptomyces* sp. M-Z18 strain for 192 h of fermentation, the ε-PL yield and space–time yield were 57.4 g/L and 6.84 g/L/d, respectively. When *S. albulus* FEEL-1 mutant strain R6 was cultured using the pH shock method, a yield of 70.3 g/L of ε-PL was obtained, which is the highest ε-PL yield reported so far (Wang et al. 2019). Pan et al. conducted a comparative transcriptome analysis of the bacterial cells at different stages of pH shock culture, revealing the potential mechanism of the pH shock culture method to increase ε-PL yield (Pan et al. 2019).

3. *The Impact of Dissolved Oxygen*

Dissolved oxygen (DO) is also one of the important parameters affecting ε-PL biosynthesis, mycelial biomass, and substrate utilization rate during ε-PL fermentation. With the increase in cell density, the increase in fermentation fluid viscosity, and the excessive consumption of oxygen, the supply of oxygen is often limited. In 2011, Bankar and Singhal proposed a two-stage DO control strategy to increase ε-PL yield based on the growth kinetics of *S. noursei* NRRL 5126 strain under different ventilation, stirring, and dissolved oxygen levels (Bankar and Singhal 2011). In 2015, Xu et al. used a similar strategy to increase the ε-PL yield of *S. albulus* PD-1 strain by 21.8% (Xu et al. 2015b). To avoid the problem of increased energy consumption required for high-speed stirring and aeration, Xu et al. attempted to culture *S. albulus* PD-1strain with the addition of an oxygen carrier, and the results showed that when 0.5% dodecane was added during the culture process and the dissolved oxygen was maintained above 32%, both the biomass and the yield of ε-PL were increased (Xu et al. 2015c).

4. *Other*

(1) Application of Cell Immobilization Technology

During the fermentation of ε-PL-producing strains, as the biomass increases, the viscosity of the fermentation broth gradually increases, and the liquid mass transfer ability gradually decreases. Although these problems can be solved by changing the stirring speed, the characteristics of the stirring paddle, and the aeration volume, too high a speed will break up the mycelium and adversely affect the morphology of the bacteria and the yield of ε-PL. In addition, because the fermentation time of ε-PL is long (at least 8 days), too high speed or aeration volume will lead to an increase in energy consumption, thereby increasing the cost of fermentation. In 2010, Zhang et al. immobilized *Beijerinckia* MY 5-36 strain with loofah to ferment ε-PL in a fermenter. After 120 h of fermentation, the yield of ε-PL and the biomass reached 1.51 and 1.1 times that of the control group, respectively. In addition to increasing the yield of ε-PL and the biomass per batch, the immobilized cells left over from the previous fermentation can be used as the "seed broth" for the next fermentation. These immobilized cells can be reused five times, with a total fermentation time of up to 526 h, greatly reducing the production cost of ε-PL (Zhang et al. 2010).

(2) Application of In Situ Extraction Technology (ISPR)

As the fermentation time extends, the harmful toxic substances produced in the fermentation of ε-PL will gradually accumulate, affecting the activity of the bacteria and the production efficiency. The use of in situ extraction technology can effectively separate the bacteria and harmful substances, thereby improving production efficiency (Stark and von Stockar 2003). After screening, Liu et al. selected the resin D152 with strong adsorption of ε-PL to in situ adsorb ε-PL in fermentation broth, relieving its inhibitory effect on the producing strain. The yields of ε-PL in shake flask and fermenter culture were 3.62 and 6.22 times higher than those of the control group, respectively (Liu et al. 2011b). The combination of immobilization technology and in situ extraction technology (ISPR) was used to culture *S. ahygroscopicus* GIM8 strain. The ε-PL levels produced under continuous ISPR and intermittent ISPR (three rounds of ISPR during the culture process) at the shake flask level were 5.88 and 2.6 times higher than those of the control group, respectively, and the ε-PL production capacity in the 5L fermenter was 2.89 and 2.74 times that of the control group, respectively (Liu et al. 2021).

(3) Application of Airlift Bioreactor (ABR)

In order to reduce energy consumption and lower costs in fermentation production, ABR technology has also been applied to the fermentation production of ε-PL. Research by Kahar et al. shows that the energy consumption of ABR is only 70% of that of traditional stirred fermenters, and reducing stirring, avoiding leakage of intracellular nucleic acids and other substances, is also beneficial to the extraction and purification of ε-PL in downstream products (Kahar et al. 2002).

3.7.3 Summary

In the past five years, significant breakthroughs have been made in the selection and fermentation process of domestic research on ε-PL producing strains. The yield and volumetric production efficiency of ε-PL have reached or exceeded the world's advanced level. But except for a few studies (Pan et al. 2019; Zeng et al. 2019), most research still remains at the stage of strain selection and process optimization, with insufficient exploration into the mechanisms of high-yield strains. The use of comparative genomics, comparative transcriptomics, and other methods to analyze the high-yield mechanisms of strains provides theoretical support for further enhancing the production capacity of strains through genetic engineering and metabolic engineering. This is a prerequisite for the efficient preparation of ε-PL and one of the future research directions.

L-lysine is the precursor of ε-PL synthesis, and ε-PL synthase is the key enzyme that converts L-lysine to ε-PL. How to increase the concentration of L-lysine in cells and enhance the activity of ε-PL synthase is also one of the key research directions for the construction of high-yield strains. Wu et al. studied in *B. licheniformis* the ways to increase the concentration of L-lysine in cells by enhancing the synthesis of L-lysine, reducing the degradation of L-lysine, increasing the transport of L-lysine from outside to inside the cell, and reducing the excretion of L-lysine (Wu et al. 2019). The strategies of this study can be imitated to construct high-yield ε-PL strains. Overexpression of genes and directed evolution of enzymes are effective methods to enhance the activity of target enzymes in strains. Bai et al. tested the strength of over 200 promoters and over 200 ribosome binding sites in *Streptomyces* (Bai et al. 2015). The appropriate combination of promoters and ribosome binding sites will provide effective help in enhancing the expression level and enzyme activity of ε-PL synthase.

Although domestic biotechnology companies, such as Zhejiang Xinyinxing and Jiangsu Yiming, have successively established industrial production lines for ε-PL, there are still problems such as insufficient development of product usage and homogenization of products. At present, the main use of ε-PL is as a food preservative, but it has good application potential in new antibacterial materials, drug carriers, and nucleic acid carriers (Bai et al. 2015), which may be the main growth point for the future application of ε-PL.

3.8 Polymalic Acid

Polymalic acid (PMA) is a high molecular weight polymer formed by ester linkage of malic acid monomers. Due to the excellent biological properties of polymalic acid, it has a wide range of applications in biomedicine, food packaging, cosmetics, and fragrances. Polymalic acid is a natural polymer, a biodegradable amphiphilic polyester, with good biocompatibility and complete degradability similar to polylactic acid. The molecular structure contains abundant carboxyl groups, which can endow polymalic acid with special properties, including good water solubility. Polymalic acid can be enzymatically or naturally degraded into small-molecule malic acid in aqueous solution. L-malic acid, as a flavor enhancer and acidulant, is widely used in food additives, beverages, and cosmetics. After entering the body, L-malic acid participates in metabolism and is absorbed by the body, demonstrating good biological safety. At the same time, malic acid can also be used as a raw material for the production of other chemicals. The current global production of malic acid is 80,000–100,000 tons per year, with an annual market demand of 200,000 tons. In the food and pharmaceutical fields, the annual demand growth rate for malic acid is about 4%, indicating a stable and sustainable market demand. Due to the large amount of other organic acids contained in the biological products of malic acid, the separation and purification are difficult and the purification cost is high. However, the separation of polymalic acid can be achieved through simple ultrafiltration or precipitation, and its downstream process is simple and low-cost. Therefore, the process of producing polymalic acid through microbial production and then degrading it into malic acid is more suitable for the production of L-malic acid. The production methods of polymalic acid include chemical synthesis and microbial synthesis. The products of chemical synthesis have low molecular weight, harsh reaction conditions, and are difficult to separate and purify, while the products obtained from microbial fermentation production of polymalic acid have the advantages of high molecular weight and high product purity. Microorganisms that can currently synthesize polymalic acid include *Aureobasidium pullulans*, oxidized black yeast, *Penicillium*, and *Physarum polycephalum*. The screening of high-yield strains is the focus of preliminary research, mainly focusing on the strain screening of *A. pullulans*. By controlling the fermentation conditions, the yield can be increased and the production of by-products can be reduced. For malic acid and polymalic acid, the downstream separation, extraction, and purification processes are also the main factors affecting their purity and economy. In recent years, researchers have also conducted related research in these areas.

3.8.1 Microbial Synthesis

Polymalic acid mainly has three structures: α-type, β-type, and γ-type (a mixture of α and β, as shown in Fig. 3.16). The polymalic acid in the body is β-type, so the polymalic acid synthesized by organisms has absorbability and biodegradability. Compared with the chemical method of preparing polymalic acid, the microbial fermentation method directly converts components such as glucose and starch in renewable resources into polymalic acid by microorganisms. After the downstream process of separation and purification, pure polymalic acid can be obtained.

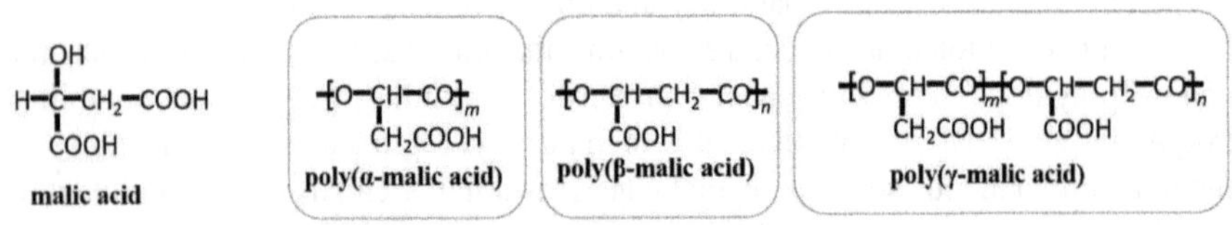

Fig. 3.16 Structure of malic acid and polymalic acid. (Taibo et al. 2020)

The advantages and disadvantages of polymalic acid prepared by chemical and microbial methods are significantly different. The chemical production of polymalic acid includes direct condensation and ring-opening polymerization. The former has a simple reaction and high yield, but the product has a lower molecular weight; the latter has a higher molecular weight, but the reaction steps are more, so the yield is not high. The chemical method will produce three products, and due to the high cost of multi-step reactions and the difficulty of separation and purification, it is not conducive to commercial applications. The microbial fermentation method for preparing polymalic acid has significant advantages and promising prospects. On the one hand, the microbial fermentation method only prepares β-polymalic acid; the product has high purity and high molecular weight, which is conducive to later development and utilization, and the microbial fermentation method for preparing poly malic acid uses renewable resources as raw materials, the source of raw materials is wide, and the resource recycling is realized. On the other hand, the microbial fermentation method has mild reaction conditions and less environmental pollution (Taibo et al. 2020).

The monomer of β-polymalic acid is L-malic acid, L-malic acid as an intermediate metabolite of the TCA cycle, participates in the body's transformation and can be widely used in the food and other fields. L-malic acid as a precursor, after transformation, is used to produce other valuable chemicals. Due to the transformation of microorganisms, the process of producing malic acid and polymalic acid has realized the full utilization of renewable resources, and the yield can be improved, while the production of by-products can be reduced through strategies such as microbial strain selection, metabolic regulation, and genetic modification. By optimizing the downstream process, the product purity is improved, and its economic cost is significantly reduced.

1. *Fermentation Strains*

In 1969, Shimada et al. discovered that polymalic acid is produced from *Penicillium cyclopium*, which is an acidic polymer compound that can inhibit protease activity. In 1989, polymalic acid was also found in the metabolic products of slime molds, and since then, strains including *P. cyclopium*, *A. pullulans*, and *P. polycephalum* have been found to be able to ferment and produce polymalic acid. Among them, *A. pullulans* has a stable morphology, is easy to control, and its ability to synthesize polymalic acid is higher than that of other strains.

The main strains for producing polymalic acid include molds, yeasts, and bacteria, which can also be used to produce polymalic acid. The widely studied *A. pullulans* mainly uses glucose as a carbon source to ferment and produce polymalic acid, with a yield of about 30–60 g/L. The addition of calcium carbonate in the process of fermenting and producing polymalic acid by *A. pullulans* is of great significance. The *P. polycephalum* is a gelatinous mold, and the current production of polymalic acid by fermenting with *P. polycephalum* uses protoplast cell fermentation method, and related research is mostly focused on synthesis mechanism and regulation (Bin and Haifeng 2018). Compared to *A. pullulans*, *P. polycephalum* is easier to synthesize and secrete polymalic acid with a molecular weight of

3–200 kDa. Currently, the research focus on the production of polymalic acid has begun to shift to yeast strains. The oxidative black yeast uses mannitol as the main carbon source, ammonium nitrate as the nitrogen source, adds corn soaking, uses succinic acid as the seed, and adds calcium carbonate to form the culture medium, which can be used to produce polymalic acid (Bin and Haifeng 2018).

2. *High-Yield Strain Screening*

Currently, the production strains of polymalic acid have problems such as long fermentation cycles, high production costs, and low yields per unit time. Therefore, obtaining high-yield strains through natural selection or genetic modification is an effective way to increase yield.

Polymalic acid is one of the main metabolic products of *A. pullulans*. *A. pullulans* is a common screening strain (Bin and Haifeng 2018; Madi et al. 1997) , widely present in nature, and its morphology is affected by Ca^{2+} (Madi et al. 1997), pH, aeration (Madi et al. 2015) and other factors; it has good adaptability to extreme environments (Botic et al. 2014). The research team of Zou Xiang at Southwest University used the *A. pullulans* CCTCC M2012223 strain as the starting strain for high-throughput screening. The mutant strain AH-21 obtained has a fast cell growth rate. After 76 h of fermentation, the yield of polymalic acid reached 41.9 g/L, and after 180 h of fermentation, it reached 127.2 g/L, with an acid production rate of 0.71 g/L and a sugar-acid conversion rate of 0.51 g/g glucose. This mutant strain has good polymalic acid fermentation performance and can be used for industrial production (Li 2017). Lv et al. screened 21 bacteria strains for polymalic acid production from the natural environment, with the highest yield reaching 43.08 ± 0.36 g/L (Leilei et al. 2020).

3.8.2 Optimization of Polymalic Acid Fermentation

After screening for dominant strains, various fermentation conditions also need to be optimized during the fermentation process to increase yield, including optimization of culture medium, inoculation amount of fermentation, temperature, time, product extraction purification methods, etc. (Shan et al. 2017).

1. *Culture Medium Optimization*

When using *A. pullulans* to produce polymalic acid, the components of the fermentation culture medium include glucose, ammonium succinate, calcium carbonate, manganese sulfate, magnesium sulfate, and other inorganic salts. After optimizing the above culture medium components, the fermentation yield of polymalic acid can be doubled.

2. *Fermentation Condition Optimization*

The inoculation amount, temperature, stirring speed, and culture time are very important for the yield of polymalic acid. The optimal growth temperature for the polymalic acid production strain is 25–30 °C, the optimal acid production temperature is 25 °C, and the optimal inoculation amount is 8%. In addition, the repeated batch fermentation method using corncob hydrolysate and immobilized fiber bed reactor technology can increase the concentration of initial hydrolyzed sugar, thereby increasing the yield of polymalic acid.

3. *Product Extraction and Purification*

The fermentation process of polymalic acid will produce many by-product polysaccharides. In order to extract polymalic acid, it is necessary to remove the pullulan polysaccharides in the fermentation broth. The principle of extraction is to reduce the use of organic solvents and reduce the impact on the environment. Generally, a two-step ultrafiltration membrane method (Yicheng et al. 2015), ion exchange method (Hao et al. 2010), etc. are used to extract and purify the polymalic acid produced by *A. pullulans* during fermentation.

4. *Polymerization Pathway*

L-Malic acid is an intermediate metabolite of the TCA cycle, and the synthesis of malic acid can be regulated by controlling the TCA cycle or rTCA cycle. During the synthesis of polymalic acid, glucose first undergoes glycolysis to produce pyruvate. The carboxylation pathway of pyruvic acid and the acetaldehyde pathway are the main pathways of malic acid production. The TCA cycle is weaker in the late stage of fermentation, and the high-yield strains have stronger polymalic acid synthesis ability than the starting strains. In a medium containing $CaCO_3$, pyruvic acid produces malic acid through the TCA cycle, while in a medium without $CaCO_3$, pyruvic acid is first carboxylated to oxaloacetic acid and then reduced to malic acid. Malic acid is polymerized to form polymalic acid under the action of polymerase (Wang et al. 2014b). By adjusting the fermentation temperature in stages, the pathway of *A. pullulans* accumulating polymalic acid can be regulated, which can increase the yield of polymalic acid.

5. *Metabolic Engineering and Metabolic Regulation Pathways of Malic Acid and Polymalic Acid* (Lee et al. 2006)

The black yeast *A. pullulans* has a large genome, can secrete a large amount of hydrolytic enzymes to degrade plant materials, and has various metabolic pathways and biosynthetic pathways. It can produce many extracellular metabolites, including PMA polymers prepared from low-value biomass. The annual demand for PMA and its monomer L-malic acid is large, however, the current chemical methods for producing PMA and L-malic acid have limitations and cannot achieve environmental protection and sustainable development. The malic acid production of yeast and filamentous fungi is high, but the strains producing malic acid have not yet achieved effective commercialization. The difficulty lies in it being hard to separate malic acid and other organic acids in downstream processes. However, malic acid can be produced through PMA, and the process of concentrating and purifying PMA through simple ultrafiltration or ethanol precipitation is simple and convenient, which can replace the production of malic acid. In addition, the PMA fermentation requires fewer neutralizing agents, will not be inhibited by the product, but compared to malic acid, the yield and production rate of PMA are still relatively low. Some by-products may cause low yield and difficulties in regeneration and purification, so it is necessary to improve the property of the production strain through metabolic regulation strategies.

Therefore, it requires an in-depth understanding of strains, such as *A. pullulans*, including carbon metabolism and nitrogen metabolism, especially the role of the TOR signaling pathway in carbon metabolism and nitrogen metabolism, as well as the cell growth regulation process in the PMA biosynthesis process. The classic CRISPR-Cas9 can also be used for gene editing of *A. pullulans* strains that produce PMA. In addition, the enzyme molecules in the process of malic acid polymerization to form PMA can also be developed and studied to optimize the production of malic acid.

3.8.3 Application

Due to the good water solubility, biocompatibility, and biodegradability of polymalic acid, it has a wide range of applications in food, biomedicine, and environmental protection.

1. *Drug Carrier*

Large molecules that can serve as drug carriers usually need to have excellent characteristics such as non-toxicity, biodegradability, non-immunogenicity, and controlled release. Polymalic acid has many of these advantages, as well as good adsorption, which makes it easy to encapsulate drugs. Therefore, it has been widely developed and researched for use in the field of drug carriers. By covalently binding polymalic acid with antisense nucleic acids to form a nanocomposite, a multi-targeted drug delivery system can be constructed. This system carries multiple antisense oligonucleotide drugs and antibodies concentrated at the tumor site, significantly reducing toxicity in other areas (Lee et al. 2006; Fujita et al. 2006; Fujita et al. 2007; Ljubimova et al. 2008a, b). Polymalic acid is easily degraded in a water environment. To increase its liposolubility, methylation can reduce the solubility of polymalic acid, making it easier to disrupt the lipid bilayer, thereby facilitating entry into tumor cells such as brain cancer and breast cancer (Portilla-Arias et al. 2010; Inoue et al. 2011, 2012).

2. *Tumor Diagnosis and Treatment*

Due to the modifiability of polymalic acid, it can be directly coupled with other polymers or bioactive molecules or further coupled with ligands through connectors, for example, monoclonal antibodies, transferrin receptors, etc. Coupling with fluorescent molecules or imaging agents can achieve tumor imaging and analysis of drug distribution in the body. The nanostructure formed by polymalic acid can pass through the blood-brain barrier, thereby enhancing the imaging and treatment effects of brain tumors. Its small physical size, multifunctional coupling, and superior water solubility make polymalic acid a carrier for various types of drugs, playing a role in tumor diagnosis and treatment (Zhang et al. 2021b).

3. *Biomaterials*

Polymalic acid dialysis bags have good colloid osmotic pressure and can serve as ideal dialysis materials. Due to its good biocompatibility and non-immunogenicity, it is also used in the field of biomaterials to make medical bandages and surgical sutures. Polymalic acid itself has no healing effect on wounds, but after binding with heparin-cross-linked growth factors, it can stimulate bone repair and speed up wound healing. High polymer compounds with added polymalic acid can have multiple functions such as degradability and conductivity (Li et al. 2009).

Hydrogels and films based on polymalic acid developed as smart materials can be used in the medical field and have the effect of promoting cell growth. Polymalic acid with extended carboxyl chains or RGD modifications has good blood compatibility. The former's linked platelets have good cell affinity and can be used for cell proliferation, while the latter's material can inhibit platelet adhesion and thrombus formation after growing a monolayer of endothelial cells. These functions make polymalic acid have good potential application value in the field of tissue engineering. In addition, there are polymeric acid materials with shape memory functions, and their shape transition temperature can be conveniently controlled by adjusting the curing time, and their shape can be permanently maintained by heating coupling (Zou et al. 2020).

3.8.4 Future Prospects

Polylactic acid, as an important biobased product, has various applications in the field of biomaterials. At present, the research on the biological preparation of polylactic acid is not rich enough, and most of the research is concentrated at the laboratory stage, with not many successful cases of large-scale commercial production. Therefore, it is necessary to increase efforts in strain selection and optimization of fermentation conditions, and at the same time, in-depth research on fermentation pathways and mechanisms should be carried out to seek breakthroughs.

References

Ahmad SI, Ahmad R, Khan MS et al (2020) Chitin and its derivatives: structural properties and biomedical applications. Int J Biol Macromol 164:526–539

Aixia W, Xiuwen W, Jiayang Q et al (2020) Biosynthesis of ε-polylysine and its application in the medical field. J Binzhou Med College 43(3):6

An L, Luijk NV, Beek MT, Caspers M et al (2011) A clone-based transcriptomics approach for the identification of genes relevant for itaconic acid production in *Aspergillus*. Fungal Genet Biol 48(6):602–611

Andin N, Longieras A, Veronese T et al (2017) Improving carbon and energy distribution by coupling growth and medium chain length polyhydroxyalkanoate production from fatty acids by *Pseudomonas putida* KT2440. Biotechnol Bioprocess Eng 22:308–318

Andreessen B, Lange AB, Robenek H et al (2010) Conversion of glycerol to poly(3-Hydroxypropionate) in recombinant *Escherichia coli*. Appl Environ Microbiol 76(2):622–626

Aranaz I, Acosta N, Civera C et al (2018) Cosmetics and cosmeceutical applications of chitin, chitosan and their derivatives. Polymers (Basel) 10(2)

Ashby RD, Shi FY, Gross RA (1997) Use of poly(ethylene glycol) to control the end group structure and molecular weight of poly(3-hydroxybutyrate) formed by *Alcaligenes latus* DSM1122. Tetrahedron 53(45):15209–15223

Ashby RD, Solaiman DK, Strahan GD et al (2015) Methanol-induced chain termination in poly(3-hydroxybutyrate) biopolymers: molecular weight control. Int J Biol Macromol 74:195–201

Ashiuchi M (2013) Biochemical engineering of PGA. Microb Biotechnol 6(6):664–674

Ashiuchi M, Shimanouchi K, Nakamura H et al (2004) Enzymatic synthesis of high-molecular-mass poly-γ-glutamate and regulation of its stereochemistry. Appl Environ Microbiol 70(7):4249–4255

Ashiuchi M, Yamashiro D, Yamamoto K (2013) Bacillus subtilis EdmS(formerly PgsE) participates in the maintenance of episomes. Plasmid 70(2):209–215

Atsumi S, Hanai T, Liao JC (2008) Non-fermentative pathways for synthesis of branched-chain higher alcohols as biofuels. Nature 451(7174):86–89

Bai C, Zhang Y, Zhao X et al (2015) Exploiting a precise design of universal synthetic modular regulatory elements to unlock the microbial natural products in *Streptomyces*. PNAS 111(39):12181–12186

Balogun-Agbaje OA, Odeniyi OA, Odeniyi MA (2021) Drug delivery applications of poly-γ-glutamic acid. Future J Pharm Sci 7(1):1–10

Bankar SB, Singhal RS (2011) Improved poly-ε-lysine biosynthesis using *Streptomyces noursei* NRRL 5126 by controlling dissolved oxygen during fermentation. J Microbiol Biotechnol 21(6):652–658

Bao T, Xiaohai F, Dan Z et al (2016) Enhancing the biosynthesis of γ-polyglutamic acid through the expression of vitreoscilla hemoglobin gene. Biochem Eng J 14(2):1–6

Bastian S, Xiang L, Meyerowitz JT et al (2011) Engineered ketol-acid reductoisomerase and alcohol dehydrogenase enable anaerobic 2-methylpropan-1-ol production at theoretical yield in *Escherichia coli*. Metab Eng 13(3):345–352

Bhatia SK, Otari SV, Jeon JM et al (2021) Biowaste-to-bioplastic (polyhydroxyalkanoates): conversion technologies, strategies, challenges, and perspective. Bioresour Technol 326:124733

Bin L, Haifeng Y (2018) Screening and identification of short stalk mold and the effect of dissolved oxygen on its fermentation. Food Industry Sci Technol 39(6):102–107

Borrero-de Acuña JM, Bielecka A, Häussler S et al (2014) Production of medium chain length polyhydroxyalkanoate in metabolic flux optimized *Pseudomonas putida*. Microb Cell Factories 13:88–102

Botic T, Kralj-Kuncic M, Spepcic K et al (2014) Biological activities of organic extracts of four *Aureobasidium pollulans* varieties isolated from extreme marine and terrestrial habitats. Nat Prod Res 28(12):874–882

Brandl H, Knee EJ Jr, Fuller RC et al (1989) Ability of the phototrophic bacterium *Rhodospirillum rubrum* to produce various poly(β-hydroxyalkanoates): potential sources for biodegradable polyesters. Int J Biol Macromol 11:49–55

Buschke N, Schröder H, Wittmann C (2015) Metabolic engineering of Corynebacterium glutamicum for production of 1,5-diaminopentane from hemicellulose. Biotechnol J 6(3):306–317

Cai D, Chen Y, He P et al (2018) Enhanced production of poly-γ-glutamic acid by improving ATP supply in metabolically engineered *Bacillus licheniformis*. Biotechnol Bioeng 115(10):2541–2553

Canhui S, Weiguo Z (2013) The effect of knocking out the *aceE* gene on the growth and pyruvate metabolism of *E. coli*. Biochem Eng 11(6):15–18

Causey TB, Zhou S, Shanmugam KT et al (2003) Engineering the metabolism of *Escherichia coli* W3110 for the conversion of sugar to redox-neutral and oxidized products: Homoacetate production. PNAS 100(3):825–832

Causey TB, Shanmugam KT, Yomano LP et al (2004) Engineering *Escherichia coli* for efficient conversion of glucose to pyruvate. PNAS 101(8):2235–2240

Cavalheiro JM, Raposo RS, de Almeida MC et al (2012) Effect of cultivation parameters on the production of poly(3-hydroxybutyrate- *co* -4-hydroxybutyrate) and poly(3-hydroxybutyrate-4-hydroxybutyrate-3-hydroxyvalerate) by *Cupriavidus necator* using waste glycerol. Bioresour Technol 111:391–397

Chao Z, Dongrong Z, Wei H et al (2006) A simple and sensitive method for screening ε-PL producing strains from soils. J Shandong Univ 44(11):1104–1107

Chemler JA, Fowler ZL, Mchugh KP et al (2010) Improving NADPH availability for natural product biosynthesis in *Escherichia coli* by metabolic engineering. Metab Eng 12(2):96–104

Chen GQ, Jiang XR (2017) Engineering bacteria for enhanced polyhydroxyalkanoates (PHA) biosynthesis. Synth Syst Biotechnol 2(3):192–197

Chen GQ, Patel MK (2012) Plastics derived from biological sources: present and future: a technical and environmental review. Chem Rev 112(4):2082–2099

Chen GQ, Wu Q (2005) The application of polyhydroxyalkanoates as tissue engineering materials. Biomaterials 26(33):6565–6578

Chen XS, Li S, Liao LJ et al (2011) Production of epsilon-poy-l-lysine using a novel two-stage pH control strategy by *Streptomyces* sp. M-Z18 from glycerol. Bioprocess Biosyst Eng 34(5):561–567

Chen XS, Tang L, Li S et al (2012a) Optimization of medium for enhancement of epsilon-poly-L-lysine production by *Streptomyces* sp M-Z18 with glycerol as carbon source. Bioprocess Biosyst Eng 35:469–475

Chen XS, Ren XD, Dong N et al (2012b) Culture medium containing glucose and glycerol as a mixed carbon source improve epsilon-poly-l-lysine production by *Streptomyces* sp M-Z18. Bioprocess Biosyst Eng 35:469–475

Chen GQ, Hajnal I, Wu H et al (2015) Engineering biosynthesis mechanisms for diversifying polyhydroxyalkanoates. Trends Biotechnol 33(10):565–574

Chen L, Dingyu L, Baowei W et al (2019) Progress in the study of acetyl CoA metabolism regulation and its application in *Escherichia coli*. Chem Industry Progr 38(9):4218–4226

Chen S, Huang S, Li Y et al (2021) Recent advances in epsilon-poly-L-lysine and L-lysine-based dendrimer synthesis, modification, and biomedical applications. Front Chem 9:169

Cheung RC, Ng TB, Wong JH et al (2015) Chitosan: an update on potential biomedical and pharmaceutical applications. Mar Drugs 13(8):5156–5186

Chheda AH, Vernekar MR (2015) A natural preservative ε-poly-L-lysine: fermentative production and applications in food industry. Int Food Res J 22(1):23–30

Choi SY, Rhie MN, Kim HT et al (2020) Metabolic engineering for the synthesis of polyesters: a 100-year journey from polyhydroxyalkanoates to non-natural microbial polyesters. Metab Eng 58:47–81

Chung AL, Jin HL, Huang LJ et al (2011) Biosynthesis and characterization of poly(3-hydroxydodecanoate) by β-oxidation inhibited mutant of *Pseudomonas entomophila* L48. Biomacromolecules 12(10):3559–3566

Cromwick AM, Birrer GA, Gross RA (1996) Effects of pH and aeration on γ-poly (glutamic acid) formation by *Bacillus licheniformis* in controlled batch fermentor cultures. Biotechnol Bioeng 50(2):222–227

da Silva FRG, Campos ACA, Souza IS et al (2020) Production of poly-γ-glutamic acid (γ-PGA) by clinical isolates of *Staphylococcus Epidermidis*. Open Microbiol J 14(1):30–37

Dandan L, Yipeng Z, Li W et al (2021) Establishment of a rapid screening method for ARTP-induced high-yield γ-polyglutamic acid strains. J Anhui Univ Technol 36(1):7

Datsenko KA, Wanner BL (2000) One-step inactivation of chromosomal genes in *Escherichia coli* K-12 using PCR products. PNAS 97(12):6640–6645

Datta S, Costantino N, Court DL (2006) A set of recombineering plasmids for gram-negative bacteria. Gene 379:109–115

Dharmadi Y, Murarka A, Gonzalez R (2006) Anaerobic fermentation of glycerol by *Escherichia coli*: a new platform for metabolic engineering. Biotechnol Bioeng 94:821–829

Diegelmann RF, Dunn JD, Lindblad WJ et al (1996) Analysis of the effects of chitosan on inflammation, angiogenesis, fibroplasia, and collagen deposition in polyvinyl alcohol sponge implants in rat wounds. Wound Repair Regen 4(1):48–52

Dietrich K, Oliveira-Filho ER, Dumont M-J et al (2020) Increasing PHB production with an industrially scalable hardwood hydrolysate as a carbon source. Ind Crop Prod 154:112703

Do TH, Suzuki Y, Abe N et al (2011) Mutations suppressing the loss of DegQ function in *Bacillus subtilis* (natto) poly-γ-glutamate synthesis. Appl Environ Microbiol 77(23):8249–8258

Doi Y, Segawa A, Kunioka M (1989) Biodegradable poly(3-hydroxybutyrate-*co*-4-hydroxybutyrate) produced from γ-butyrolactone and butyric acid by *Alcaligenes eutrophus*. Polym Commun 30(6):169–171

El Hadrami A, Adam LR, El Hadrami I et al (2010) Chitosan in plant protection. Mar Drugs 8(4):968–987

Elieh Ali Komi D, Sharma L, Dela Cruz CS (2018) Chitin and its effects on inflammatory and immune responses. Clin Rev Allergy Immunol 54(2):213–223

El-Mansi M (2019) Contrasting effects of isocitrate dehydrogenase deletion on fluxes through enzymes of central metabolism in *Escherichia coli*. FEMS Microbiol Lett 366(15):187

Ewering C, Heuser F, Benölken JK et al (2006) Metabolic engineering of strains of *Ralstonia eutropha* and *Pseudomonas putida* for biotechnological production of 2-methylcitric acid. Metab Eng 8(6):587–602

Fan X, Huixian W, Jiawei H et al (2018) Research progress on gene editing methods—taking the gene knockout method of *E. coli* as an example. J Nanjing Normal Univ (Natural Sci Ed) 41(3):102–108

Flores-Albino B, Arias L, Gómez J et al (2012) Chitin and L(+)-lactic acid production from crab (*Callinectes bellicosus*) wastes by fermentation of *Lactobacillus* sp. B2 using sugar cane molasses as carbon source. Bioprocess Biosyst Eng 35(7):1193–1200

Fowler ZL, Gikandi WW, Koffas MAG (2009) Increased malonyl coenzyme A biosynthesis by tuning the *Escherichia coli* metabolic network and its application to flavanone production. Appl Environ Microbiol 75(18):5831–5839

Friedman M, Juneja VK (2010) Review of antimicrobial and antioxidative activities of chitosans in food. J Food Prot 73(9):1737–1761

Fujita M, Khazenzon NM, Ljubimov AV et al (2006) Inhibition of laminin-8 in vivo using a novel poly(malic acid)-based carrier reduces glioma angiogenesis. Angiogenesis 9(4):183–191

Fujita M, Lee BS, Khazenzon NM et al (2007) Brain tumor tandem targeting using a combination of monoclonal antibodies attached to biopoly(β-L-malic acid). J Control Release 122(3):356–363

Fujita KI, Tomiyama T, Inoi T et al (2021) Effect of pgsE expression on the molecular weight of poly (γ-glutamic acid) in fermentative production. Polym J 53(2):409–414

Fukui T, Kichise T, Iwata T et al (2001) Characterization of 13 kDa granule-associated protein in *Aeromonas caviae* and biosynthesis of polyhydroxyalkanoates with altered molar composition by recombinant bacteria. Biomacromolecules 2:148–153

Gao W, He Y, Zhang F et al (2019) Metabolic engineering of Bacillus amyloliquefaciens LL 3 for enhanced poly-γ-glutamic acid synthesis. Microbial Biotechnol 12(5):932–945

Geng W, Yang C, Gu Y et al (2014) Cloning of ε-poly-L-lysine (ε-PL) synthetase gene from a newly isolated ε-PL-producing Streptomyces albulus NK 660 and its heterologous expression in *Streptomyces lividans*. Microb Biotechnol 7(2):155–164

Ghorbel-Bellaaj O, Hmidet N, Jellouli K et al (2011) Shrimp waste fermentation with *Pseudomonas aeruginosa* A2: optimization of chitin extraction conditions through Plackett-Burman and response surface methodology approaches. Int J Biol Macromol 48(4):596–602

González-García Y, Grieve J, Meza-Contreras JC et al (2019) Tequila agave bagasse hydrolysate for the production of polyhydroxybutyrate by *Burkholderia sacchari*. Bioengineering (Basel, Switzerland) 6(4):115–127

Green PR, Kemper J, Schechtman L et al (2002) Formation of short chain length/medium chain length polyhydroxyalkanoate copolymers by fatty acid β-oxidation inhibited *Ralstonia eutropha*. Biomacromolecules 3(1):208–213

Gui L, Sunnarborg A, Pan B et al (1996) Autoregulation of iclR, the gene encoding the repressor of the glyoxylate bypass operon. J Bacteriol 178(1):321–324

Hadicke O, Klamt S (2017) EColiCore2: a reference network model of the central metabolism of *Escherichia coli* and relationships to its genome scale parent model. Sci Rep 7:39647

Haitao Z, Yan L, Jie O et al (2007) Induced breeding of ε-polylysine-producing bacterial mutants. Food Sci 28(9):398–401

Haiyan W, Ming L, Huajun W et al (2006) Microbial metabolic engineering in lactic acid production. J Process Eng 3:178–182

Halmschlag B, Steurer X, Putri SP et al (2019) Tailor-made poly-γ-glutamic acid production. Metab eng 55:239–248

Halmschlag B, Putri SP, Fukusaki E et al (2020) Poly-γ-glutamic acid production by *Bacillus subtilis* 168 using glucose as the sole carbon source: a metabolomic analysis. J Biosci Bioeng 130(3):272–282

Hamano Y, Nicchu I, Shimizu T et al (2007) epsilon-Poly-L-lysine producer, *Streptomyces albulus*, has feedback-inhibition resistant aspartokinase. Appl Microbiol Biotechnol 76(4):873–882

Han MJ, Yoon SS, Lee SY (2001) Proteome analysis of metabolically engineered *Escherichia coli* producing poly(3-hydroxybutyrate). J Bacteriol 183(1):301–308

Han J, Hou J, Zhang F et al (2013) Multiple propionyl coenzyme A-supplying pathways for production of the bioplastic poly(3-hydroxybutyrate- *co* -3-hydroxyvalerate) in *Haloferax mediterranei*. Appl Environ Microbiol 79(9):2922–2931

Hao W, Yifeng Z, Yuanai C et al (2010) Extraction of polymalic acid from fermentation broth by ion exchange method. Ion Exchange Adsorp 27(3):257–263

Haywood GW, Anderson AJ, Williams DR et al (1991) Accumulation of a poly(hydroxyalkanoate) copolymer containing primarily 3-hydroxyvalerate from simple carbohydrate substrates by *Rhodococcus* sp. NCIMB 40126. Int J Biol Macromol 13:83–88

He Y, Weiguang L, Juanqin Z et al (2020) Research progress on γ-polyglutamic acid. Anhui Agric Sci 48(18):18–22

Hiraki J, Masakazu H, Hiroshi M et al (1998) Improved ε-Poly-L-Lysine production of an S-(2-Aminoethyl)-L-cysteine resistant mutant of Streptomyces albulus. Seibutsu Kogakkaishi 76:487–493

Hiraki J, Ichikawa T, Ninomiya S et al (2003) Use of ADME studies to confirm the safety of epsilon-polylysine as a preservative in food. Regul Toxicol Pharmacol 37(2):328–340

Hsueh YH, Huang KY, Kunene SC et al (2017) Poly-γ-glutamic acid synthesis, gene regulation, phylogenetic relationships, and role in fermentation. Int J Mol Sci 18(12):2644

Hu D, Chung AL, Wu LP et al (2011) Biosynthesis and characterization of polyhydroxyalkanoate block copolymer P3HB- *b* -P4HB. Biomacromolecules 12:3166–3173

Huang X, Mei C, Li J et al (2016) Establishing an efficient gene-targeting system in an itaconic-acid producing *Aspergillus terreus* strain. Biotechnol Lett 38(9):1603–1610

Hui J (2012) Expression of vitreoscilla hemoglobin in γ-PGA synthesis bacterium B.amyloliquefaciens LL3. Nankai University

Inoue S, Ding H, Portilla-Arias J et al (2011) Polymalic acid-based nanobiopolymer provides efficient systemic breast cancer treatment by inhibiting both HER2/neu receptor synthesis and activity. Cancer Res 71(4):1454

Inoue S, Pati R, Portilla-Arias J et al (2012) Nanobiopolymer for direct targeting and inhibition of EGFR expression in triple negative breast cancer. PLoS ONE 7(2):e31070

Israni N, Venkatachalam P, Gajaraj B et al (2020) Whey valorization for sustainable polyhydroxyalkanoate production by *Bacillus megaterium*: production, characterization and in vitro biocompatibility evaluation. J Environ Manag 255:109884

Jianfeng W, Zhilong X, Haijun L et al (2001) Research on the microaerobic fermentation of 1,3-propanediol by *Klebsiella pneumoniae*. Modern Chem Industry 5:28–31

Jiang W, Bikard D, Cox D et al (2013) RNA-guided editing of bacterial genomes using CRISPR-Cas systems. Nat Biotechnol 31(3):233–239
Jiang Y, Qian F, Yang J et al (2017) CRISPR-Cpf1 assisted genome editing of *Corynebacterium glutamicum*. Nat Commun 8:15179
Jiayang Q, Aixia W, Yubin X et al (2019) A genetically engineered strain of *Streptomyces albus* and its application in the production of ε-polylysine. China, 201911198884.5
Jimtaisong A, Saewan N (2014) Utilization of carboxymethyl chitosan in cosmetics. Int J Cosmet Sci 36(1):12–21
Jin H, Ning C (2005) Properties and production methods of γ-polyglutamic acid. Amino Acids Biol Resour 26(3):4–8
Juanjuan Y, Xiaoyu M, Xiaorui W et al (2020) Research progress in gene editing of Corynebacterium glutamicum. J Biol Eng 5:820–828
Jun C, Yang C, Ding-Qi W et al (2021) The effect of enhancing the carboxylation pathway of glutamic acid rod-shaped bacteria on the yield of organic acids. J Wuhan Univ Sci Technol 44(2):112–118
Jung YM, Lee JN, Shin HD et al (2004) Role of *tktA* gene in pentose phosphate pathway on odd-ball biosynthesis of poly-β-hydroxybutyrate in transformant *Escherichia coli* harboring phbCAB operon. J Biosci Bioeng 98(3):224–227
Jung YK, Kim TY, Park SJ et al (2010) Metabolic engineering of *Escherichia coli* for the production of polylactic acid and its copolymers. Biotechnol Bioeng 105(1):161–171
Junnan F, Juan L, Lishan X et al (2018) Progress in microbial fermentation production of γ-polyglutamic acid. J Appl Environ Biol 24(5):1041–1049
Junwei L, Ye L, Wang Y et al (2020) Condition optimization of base editing in Corynebacterium glutamicum. J Biol Eng 36(1):143–151
Kahar P, Iwata T, Hiraki J et al (2001) Enhancement of ε-polyysine production by *Streptomyces albulus* strain 410 using pH control. J Biosci Bioeng 91(2):190–194
Kahar P, Kobayashi K, Iwata T et al (2002) Production of epsilon-polylysine in an airlift bioreactor (ABR). J Biosci Bioeng 93(3):274–280
Kanamasa S, Dwiarti L, Okabe M et al (2008) Cloning and functional characterization of the cis-aconitic acid decarboxylase (CAD) gene from *Aspergillus terreus*. Appl Microbiol Biotechnol 80(2):223–229
Karmann S, Panke S, Zinn M (2019) Fed-batch cultivations of *Rhodospirillum rubrum* under multiple nutrient-limited growth conditions on syngas as a novel option to produce poly(3-hydroxybutyrate) (PHB). Front Bioeng Biotechnol 7:59–69
Kejia W, Shuyi Q (2020) Research progress on the application of genome rearrangement technology in microbial strain breeding. Food Industry Sci Technol 41(3):6
Kichise T, Taguchi S, Doi Y (2002) Enhanced accumulation and changed monomer composition in Polyhydroxyalkanoate(PHA) Ccopolyester by in vitro evolution of *Aeromonas caviae* PHA synthase. Appl Environ Microbiol 68:2411–2419
Kim YM, Cho HS, Jung GY et al (2011) Engineering the pentose phosphate pathway to improve hydrogen yield in recombinant *Escherichia coli*. Biotechnol Bioeng 108(12):2941–2946
Kimura K, Tran LSP, Do TH et al (2009) Expression of the pgsB encoding the poly-gamma-DL-glutamate synthetase of *Bacillus subtilis* (natto). Biosci Biotechnol Biochem 73(5):1149–1155
Kimura K, Tran LSP, Funane K (2011) Loss of poly-γ-glutamic acid synthesis of *bacillus subtilis* (*natto*) due to IS4Bsu1 translocation to *swrA* gene. Food Sci Technol Res 17(5):447–451
Kobayashi K, Nishikawa M (2007) Promotion of ε-poly-l-lysine roduction by iron in Kitasatosporakifunense. World J Microbiol Biotechnol 23(7):1033–1036
Kobayashi G, Shiotani T, Shima Y et al (1994) Biosynthesis and characterization of poly(3-hydroxybutyrate- *co* -3-hydroxyhexanoate) from oils and fats by *Aeromonas* sp. OL-338 and *Aeromonas* sp. FA-440. Biodegrad Plastics Polym 12:410–416
Kroumova AB, Wagner GJ, Davies HM (2002) Biochemical observations on medium-chain-length polyhydroxyalkanoate biosynthesis and accumulation in *Pseudomonas mendocina*. Arch Biochem Biophys 405(1):95–103

Kucera D, Pernicová I, Kovalcik A et al (2018) Characterization of the promising poly(3-hydroxybutyrate) producing halophilic bacterium *Halomonas halophila*. Bioresour Technol 256:552–556

Kurata H, Sugimoto Y (2018) Improved kinetic model of *Escherichia coli* central carbon metabolism in batch and continuous cultures. J Biosci Bioeng 125(2):251–257

Lee BS, Fujita M, Khazenzon NM et al (2006) Polycefin, a new prototype of a multifunctional nanoconjugate based on poly(β-L-malic acid) for drug delivery. Bioconjug Chem 17(2)

Lee HC, Kim JS, Jang W et al (2010) High NADPH/$NADP^+$ ratio improves thymidine production by a metabolically engineered *Escherichia coli* strain. J Biotechnol 149(1-2):24–32

Lee SH, Kang K, Kim EY et al (2013) Metabolic engineering of *Escherichia coli* for enhanced biosynthesis of poly(3-hydroxybutyrate) based on proteome analysis. Biotechnol Lett 35(10):1631–1637

Lei J, Xia M (2011) Improvement of extraction methods for γ-polyglutamic acid. Modern Chem Industry S1:267–270

Leilei L, Danfeng Z, Linjie C et al (2020) Screening and identification of high-yield polymalic acid bacteria. Food Industry Science and Technology

Lemoigne M (1926a) Products of dehydration and polymerization of beta-oxybutyric acid. Finanz-Rundschau Ertragsteuerrecht 91(1):449–454

Lemoigne M (1926b) Products of dehydration and of polymerization of β-hydroxybutyric acid. Bull Soc Chem Biol 8:770–782

Le-Ping G, Yu-Min D, Hua-Tang Y (2002) Study on the decolorization conditions and molecular weight of hydrogen peroxide chitosan. J Wuhan Univ Sci Ed 48(4):4

Li H (2017) High-throughput screening and fermentation characteristics of polymalic acid-producing bacteria. Southwest University, Chongqing

Li SP, Yan YH, Zhang QS et al (2009) Biodegradable conductive biomedical polymer materials. ZL200810197694.7

Li ZJ, Shi ZY, Jian J et al (2010) Production of poly(3-hydroxybutyrate- *co* -4-hydroxybutyrate) from unrelated carbon sources by metabolically engineered *Escherichia coli*. Metab Eng 12(4):352–359

Li S, Chen X, Dong C et al (2013) Combining genome shuffling and interspecific hybridization among *Streptomyces* improved ε-poly-L-lysine production. Appl Biochem Biotechnol 169(1):338–350

Li S, Cai L, Wu L et al (2014a) Microbial synthesis of functional homo-, random, and block polyhydroxyalkanoates by β-oxidation deleted *Pseudomonas entomophila*. Biomacromolecules 15(6):2310–2319

Li T, Chen XB, Chen JC et al (2014b) Open and continuous fermentation: products, conditions and bioprocess economy. Biotechnol J:1503–1511

Li M, Chen X, Che X et al (2018) Engineering *Pseudomonas entomophila* for synthesis of copolymers with defined fractions of 3-hydroxybutyrate and medium-chain-length 3-hydroxyalkanoates. Metab Eng 52:253–262

Li M, Chen J, Wang Y et al (2020) Efficient multiplex gene repression by CRISPR-dCpf1 in *Corynebacterium glutamicum*. Front Bioeng Biotechnol 8:357

Liebergesell M, Hustede E, Timm A et al (1991) Formation of poly(3-hydroxyalkanoates) by phototrophic and chemolithotrophic bacteria. Arch Microbiol 155(5):415–421

Lìfei C (2015) Strain improvement of high-yield 1,3-propanediol Klebsiella pneumoniae. Master's thesis, Qilu University of Technology

Lijuan L, Fulin Z, Xusheng C et al (2011) The effect of amino acids on the biosynthesis of ε-polylysine by Streptomyces griseus. Industrial Microbiol 04:43–49

Lim SJ, Jung YM, Shin HD et al (2002) Amplification of the NADPH-related genes zwf and gnd for the oddball biosynthesis of PHB in an *E. coli* transformant harboring a cloned phbCAB operon. J Biosci Bioeng 93(6):543–549

Lin Y, Lin Z, Huang C et al (2015) Metabolic engineering of *Escherichia coli* using CRISPR-Cas9 meditated genome editing. Metab Eng

Ling M, Wu Y, Tian R et al (2022) Combinatorial pathway engineering of *Bacillus subtilis* for production of structurally defined and homogeneous chitooligosaccharides. Metab Eng 70:55–66

Lingtian W, Yunlan Z, Mengrong C et al (2020) A strain of *Bacillus velezensis* and its application in the co-production of microbial polysaccharides and γ-polyglutamic acid. CN202010649356.3

Linhui W, Xu L, Jiaojiao L et al (2018) A method for producing polyglutamic acid by fermenting glutamic acid fermentation waste mycelium, CN108841882A

Liu H, Lu T (2015) Autonomous production of 1,4-butanediol via a de novo biosynthesis pathway in engineered *Escherichia coli*. Metab Eng 29:135–141

Liu Q, Luo G, Zhou XR et al (2010) Biosynthesis of poly(3-hydroxydecanoate) and 3-hydroxydodecanoate dominating polyhydroxyalkanoates by β-oxidation pathway inhibited *Pseudomonas putida*. Metab Eng 13(1):11–17

Liu H, Jing H, Liu X et al (2011a) Development of *pyrF* -based gene knockout systems for genome-wide manipulation of the archaea *Haloferax mediterranei* and *Haloarcula hispanica*. J Genet Genomics 38(6):261–269

Liu SR, Wu QP, Zhang JM et al (2011b) Production of epsilon-poly-L-lysine by Streptomyces sp using resin-based, in situ product removal. Biotechnol Lett 33(8):1581–1585

Liu J, Wang Y, Lu Y et al (2017a) Development of a CRISPR/Cas9 genome editing toolbox for *Corynebacterium glutamicum*. Microb Cell Factories 16(1):205

Liu M, Ding Y, Chen H et al (2017b) Improving the production of acetylCoA-derived chemicals in *Escherichia coli* BL21(DE3) through *iclR* and *arcA* deletion. BMC Microbiol 17(1):2–9

Liu YJ, Chen XS, Zhao JJ et al (2017c) Development of microtiter plate culture method for rapid screening of ε-poly-L-lysine-producing strains. Appl Biochem Biotechnol 183(4):1209–1223

Liu SR, Yang XJ, Sun DF (2021) Enhanced production of ε-poly-L-lysine by immobilized Streptomyces ahygroscopicus through repeated-batch or fed-batch fermentation with in situ product removal. Bioprocess Biosyst Eng:1–12

Ljubimova JY, Fujita M, Khazenzon NM et al (2008a) Nanoconjugate based on polymalic acid for tumor targeting. Chem Biol Interact 171(2):195–203

Ljubimova JY, Fujita M, Ljubimov AV et al (2008b) Poly(malic acid) nanoconjugates containing various antibodies and oligonucleotides for multi-targeting drug delivery. Nanomedicine 3(2):247–265

Lockwood LB, Raper KB, Moyer AJ et al (1945) The production and characterization of ultravioletnduced mutations in *Aspergillus terreus*. III. Biochemical characteristics of the mutations. Am J Bot 32(4):214–217

Lopes C, Antelo LT, Franco-Uría A et al (2017) Chitin production from crustacean biomass: Sustainability assessment of chemical and enzymatic processes. J Clean Prod 172(PT.4):4140–4151

Lu L (2019) Research on CRISPR/Cpf1 system in genome editing of Corynebacterium glutamicum ATCC 14067. Master's thesis, South China University of Technology

Lu X, Zhang J, Wu Q et al (2010) Enhanced production of poly(3-hydroxybutyrate-*co* -3-hydroxyhexanoate) via manipulating the fatty acid β-oxidation pathway in *E. coli*. FEMS Microbiol Lett 221(1):97–101

Lv L, Ren YL, Chen JC et al (2015) Application of CRISPRi for prokaryotic metabolic engineering involving multiple genes, a case study: controllable P(3HB- *co* -4HB) biosynthesis. Metab Eng 29:160–168

Macrae RM, Wilkinson JF (1958) Poly-beta-hydroxybutyrate metabolism in washed suspensions of *Bacillus cereus* and *Bacillus megaterium*. J Microbiol 19(1):210–222

Mader U, Antelmann H, Buder T (2002) *Bacillus subtilis* functional genomics: genome-wide analysis of the DegS-DegU regulon by transcriptomics and proteomics. Mol Gen Genomics 268(4):455–467

Madi N, McNeil B, Harvey L (1997) Effect of exogenous calcium on morphological development and biopolymer synthesis in the fungus *Aureobasidium pullulans*. Enzyme Microb Technol 21(2):102–107

Madi N, McNeil B, Harvey L (2015) Influence of culture pH and aeration on ethanol production and pullulan molecular weight by *Aureobasidium pullulans*. J Chem Technol Biotechnol 65(4):343–350

Magocha TA, Zabed H, Yang M et al (2018) Improvement of industrially important microbial strains by genome shuffling: current status and future prospects. Bioresour Technol 6:1–14

Malerba M, Cerana R (2019) Recent applications of chitin- and chitosan-based polymers in plants. Polymers (Basel) 11(5)

Manni L, Ghorbel-bellaaj O, Jellouli K et al (2010) Extraction and characterization of chitin, chitosan, and protein hydrolysates prepared from shrimp waste by treatment with crude protease from *Bacillus cereus* SV1. Appl Biochem Biotechnol 162(2):345–357

Martínez I, Zhu J, Lin H et al (2008) Replacing *Escherichia coli* NAD-dependent glyceraldehyde 3-phosphate dehydrogenase (GAPDH) with a NADP-dependent enzyme from Clostridium acetobutylicum facilitates NADPH dependent pathways. Metab Eng 10(6):352–359

Martins AF, Facchi SP, Follmann HD et al (2014) Antimicrobial activity of chitosan derivatives containing N-quaternized moieties in its backbone: a review. Int J Mol Sci 15(11):20800–20832

Matsumoto K, Aoki E, Takase K et al (2006) In vivo and in vitro characterization of Ser477X mutations in polyhydroxyalkanoate (PHA) synthase 1 from *Pseudomonas* sp. 61-3: effects of beneficial mutations on enzymatic activity, substrate specificity, and molecular weight of PHA. Biomacromolecules 7:2436–2442

Mcconville TH, Giddins MJ, Uhlemann AC (2021) An efficient and versatile CRISPR-Cas9 system for genetic manipulation of multi-drug resistant *Klebsiella pneumoniae*. STAR Protocols 2(1):100373

Meng DC, Wang Y, Wu LP et al (2015) Production of poly(3-hydroxypropionate) and poly(3-hydroxybutyrate- *co* -3-hydroxypropionate) from glucose by engineering *Escherichia coli*. Metab Eng 29:189–195

Mengdi W (2021) Synthesis methods of polyglutamic acid and its applications in agriculture. China Salt Industry 395(20):48–50

Mengying Z, Yahui L, Yuanlong Z et al (2019) Research progress on the biosynthesis of polyhydroxyalkanoates (PHAs) by halophilic bacteria. Biotechnol Bull 35(6):172–177

Menzel K, Zeng A, Deckwer W (1997) High concentration and productivity of 1,3-propanediol from continuous fermentation of glycerol by *Klebsiella pneumoniae*. Enzym Microb Technol 20(2):82–86

Mey M, Maeseneire S, Soetaert W et al (2007) Minimizing acetate formation in *E. coli* fermentations. J Ind Microbiol Biotechnol 34(11):689–700

Milin Z, Baojun K, Hui Z et al (2019) Comparison of CRISPR-Cpf1 and *Cre/loxP* gene knockout technologies in *Corynebacterium glutamicum*. Microbiol Bull 46(2):65–78

Min P, Yingqing Z, Ting W et al (2021) Research progress on the functional impact of polyglutamic acid on food. China Food Addit 7:138–142

Mingfeng C, Yinghong J, Hui X et al (2011) Microbial synthesis of γ-polyglutamic acid, related genes and application prospects. Microbiol Bull 38(3):388–395

Moussian B (2019) Chitin: structure, chemistry and biology. Adv Exp Med Biol 1142:5–18

Murphy KC (1998) Use of bacteriophage λ recombination functions to promote gene replacement in *Escherichia coli*. J Bacteriol 180(8):2063–2071

Myoung PJ, Yu-Sin J, Yong KT et al (2010) Development of a gene knockout system for *Ralstonia eutropha* H16 based on the broad-host-range vector expressing a mobile group II intron. Fems Metab Eng 2:193–200

Nagai T, Phan Tran LS, Inatsu Y et al (2000) A new IS 4 family insertion sequence, IS 4Bsu 1, responsible for genetic instability of poly-γ-glutamic acid production in *Bacillus subtilis*. J Bacteriol 182(9):2387–2392

Nan L (2006) Screening and optimization of γ-PGA producing strain S004-50-01. Food Ferment Industries 6(31):1–6

Ngo DH, Kim SK (2014) Antioxidant effects of chitin, chitosan, and their derivatives. Adv Food Nutr Res 73:15–31

Nielsen DR, Yoon SH, Yuan CJ et al (2010) Engineering Acetoin and meso-2,3-butanediol biosynthesis in *E. coli*. Biotechnol J 5(3):274–284

Nishikawa AM, Ogawa K (2002) Distribution of microbes producing antimicrobial epsilon-poly-L-lysine polymers in soil microflora determined by a novel method. Appl Environ Microbiol 68(7):3575–3581

Ogawa Y, Hosoyama H, Hamano M et al (1991) Purification and properties of γ-Glutamyltranspeptidase from *Bacillus subtilis* (natta). Agric Biol Chem 55(12):2971–2977

Ogunleye A, Bhat A, Irorere VU et al (2015) Poly-γ-glutamic acid: production, properties and applications. Microbiology 161(1):1–17

Ohsawa T, Tsukahara K, Ogura M (2009) *Bacillus subtilis* response regulator DegU is a direct activator of pgsB transcription involved in γ-poly-glutamic acid synthesis. Biosci Biotechnol Biochem 73(9):2096–2102

Ouyang P, Wang H, Hajnal I et al (2017) Increasing oxygen availability for improving poly(3-hydroxybutyrate) production by *Halomonas*. Metab Eng 45:20–31

Pan L, Chen X, Wang K et al (2019) Understanding high ε-poly-l-lysine production by *Streptomyces albulus* using pH shock strategy in the level of transcriptomics. J Ind Microbiol Biotechnol 46(12):1781–1792

Park SJ, Lee SY (2003) Identification and characterization of a new enoyl coenzyme a hydratase involved in biosynthesis of medium-chain-length polyhydroxyalkanoates in recombinant *Escherichia coli*. J Bacteriol 185:5391–5397

Park SJ, Lee SY, Kim TW et al (2012) Biosynthesis of lactate-containing polyesters by metabolically engineered bacteria. Biotechnol J 7(2):199–212

Park SJ, Jang YA, Lee H et al (2013) Metabolic engineering of *Ralstonia eutropha* for the biosynthesis of 2-hydroxyacid-containing polyhydroxyalkanoates. Metab Eng 20:20–28

Patil NA, Kandasubramanian B (2021) Functionalized polylysine biomaterialls for advanced medical applications: a review. Eur Polym J 146:110248

Pederson EN, McChalicher CWJ, Srienc F (2006) Bacterial synthesis of PHA block copolymers. Biomacromolecules 7:1904–1911

Peng F, Wang X, Sun Y et al (2017) Efficient gene editing in *Corynebacterium glutamicum* using the CRISPR/Cas9 system. Microb Cell Factories 16(1):201

Peng G, Kun W, Yahui C et al (2019) Synthesis and application of γ-polyglutamic acid. J Xuchang Univ 38(5):92–95

Peng D, Jinlu Z, Bingrong L et al (2020) Chemical modification and application of chitin and chitosan. Polym Bull 7:17

Pernicova I, Kucera D, Nebesarova J et al (2019) Production of polyhydroxyalkanoates on waste frying oil employing selected *Halomonas* strains. Bioresour Technol 292:122028

Pfeiffer D, Wahl A, Jendrossek D (2011) Identification of a multifunctional protein, PhaM, that determines number, surface to volume ratio, subcellular localization and distribution to daughter cells of poly(3-hydroxybutyrate), PHB, granules in *Ralstonia eutropha* H16. Mol Microbiol 81:936–951

Poblete-Castro I, Binger D, Rodrigues A et al (2013) In-silico-driven metabolic engineering of *Pseudomonas putida* for enhanced production of poly-hydroxyalkanoates. Metab Eng 15:113–123

Poo H, Park C, Kwak MS et al (2010) New biological functions and applications of high-molecular-mass poly-γ-glutamic acid. Chem Biodivers 7(6):1555–1562

Portilla-Arias J, Pati R, Hu J et al (2010) Nanoconjugate platforms development based in poly(β-L-malic acid) methyl esters for tumor drug delivery. J Nanomater:825363

Przystałowska H, Zeyland J, PowałOwska DS et al (2015) 1,3-Propanediol production by new recombinant *Escherichia coli* containing genes from pathogenic bacteria. Microbiol Res 171:1–7

Qi QS, Steinbüchel A, Rehm BHA (1998) Metabolic routing towards polyhydroxyalkanoic acid synthesis in recombinant *Escherichia coli* (*fadR*): inhibition of fatty acid β-oxidation by acrylic acid. FEMS Microbiol Lett 167(1):89–94

Qianqian Y, Feiran L, Hao L et al (2017) Discovery of new strategies for microbial metabolic engineering modification from metabolic network analysis. Chem Industry Progr 36(12):4592–4600

Qianqian T, Yaliang L, Shunchang W et al (2019) Research progress on microbial genome rearrangement technology. J Chifeng Univ (Natural Sci Ed) 35(10):18–19

Qin Q, Ling C, Zhao Y et al (2018) CRISPR/Cas9 editing genome of extremophile *Halomonas* spp. Metab Eng 47:219–229

Raza ZA, Abid S, Banat IM (2018) Polyhydroxyalkanoates: characteristics, production, recent developments and applications. Int Biodeterior Biodegradation 126:45–56

Ren XD, Chen XS, Zeng X et al (2015) Acidic pH shock induced overproduction of epsilon-poly-l-lysine in fed-batch fermentation by *Streptomyces* sp M-Z18 from agro-industrial by-products. Bioprocess Biosyst Eng 38(6):1113–1125

Roberts RL, Bowers B, Slater ML et al (1983) Chitin synthesis and localization in cell division cycle mutants of *Saccharomyces cerevisiae*. Mol Cell Biol 3(5):922–930

Saeed KA, Eribo BE, Ayorin De FO et al (2019) Characterization of copolymer hydroxybutyrate/hydroxyvalerate from *Saponified Vernonia*, soybean, and "Spent" frying oils. J AOAC Int 4:4

Salah R, Michaud P, Mati F et al (2013) Anticancer activity of chemically prepared shrimp low molecular weight chitin evaluation with the human monocyte leukaemia cell line, THP-1. Int J Biol Macromol 52:333–339

Sánchez RJ, Schripsema J, Da Silva LF et al (2003) Medium-chain-length polyhydroxyalkanoic acids (PHA mcl) produced by *Pseudomonas putida* IPT046 from renewable sources. Eur Polym J 39(7):1385–1394

Sánchez AM, Bennett GN, San KY (2005) Novel pathway engineering design of the anaerobic central metabolic pathway in *Escherichia coli* to increase succinate yield and productivity. Metab Eng 7(3):229–239

Sanda F, Fnjiyama T, Endo T (2001) Chemical synthesis of poly-γ-glutamic acid by condensation of γ -glutamic acid dimer: synthesis and reaction of poly-γ-glutamic acid methyester. Polym Sci 39(5):732–741

Satitsri S, Muanprasat C (2020) Chitin and chitosan derivatives as biomaterial resources for biological and biomedical applications. Molecules 25(24)

Schubert P, Steinbüchel A, Schlegel HG (1988) Cloning of the *Alcaligenes eutrophus* genes for synthesis of poly-beta-hydroxybutyric acid (PHB) and synthesis of PHB in *Escherichia coli*. J Biotechnol 170:5837–5847

Senlin W, Wen L, Yan W et al (2021) A strain of methylotrophic *Bacillus*, method for fermenting γ-polyglutamic acid and its application. CN202110869912.2

Shahbaz U (2020) Chitin, characteristic, sources, and biomedical application. Curr Pharm Biotechnol 21(14):1433–1443

Shan C, Yanling H, Le Z et al (2017) Determination study of polymalic acid production by microbial fermentation. Agric Products Process 5:51–53

Shih IL, Shen MH (2006) Application of response surface methodology to optimize production of poly-epsilon-lysine by *Streptomyces albulus* IFO 1417. Enzym Microb Technol 39(1):15–21

Shih L, Van YT (2001) The production of poly-(γ-glutamic acid) from microorganisms and its various applications. Bioresour Technol 79(3):207–225

Shiloach J, Kaufman J, Guillard AS et al (1996) Effect of glucose supply strategy on acetate accumulation, growth, and recombinant protein production by *Escherichia coli* BL21(λDE3) and *Escherichia coli* JM109. Biotechnol Bioeng 49:421–428

Shima S, Sakai H (1977) Polylysin produced by *Streptomyces*. Agric Biol Chem 41(9):1807–1809

Shima S, Sakai H (1981) Poly-L-lysine produced by *Streptomyces*. Part II. Taxonomy and fermentation studies. Agric Biol Chem 45:2497–2502

Shozui F, Ken'ichiro M, Motohashi R et al (2011) Biosynthesis of a lactate (LA)-based polyester with a 96 mol% LA fraction and its application to stereocomplex formation. Polym Degrad Stab 96(4):499–504

Shuangxi C, Erchao Z, Lele Z, Qihao X (2015) Atmospheric pressure room temperature plasma mutation breeding of γ-polyglutamic acid producing strains. Chin J Pharm Industry 46(9):960–964

Shuiyang Z, Dan Z (2012) Method for producing γ-polyglutamic acid by adding NaCl during fermentation, CN102533885A

Siedler S, Bringer S, Bott M (2011) Increased NADPH availability in *Escherichia coli* : improvement of the product per glucose ratio in reductive whole-cell biotransformation. Appl Microbiol Biotechnol 92(5):929

Sim SJ, Snell KD, Hogan SA et al (1997) PHA synthase activity controls the molecular weight and polydispersity of polyhydroxybutyrate in vivo. Nat Biotechnol 15:63–67

Solaiman D, Foglia A (2002) Synthesis of poly(hydroxyalkanoates) by *Escherichia coli* expressing mutated and chimeric PHA synthase genes. Biotechnol Lett 24:1011–1016

Song BG, Kim TK, Jung YM et al (2006) Modulation of talA gene in pentose phosphate pathway for overproduction of poly- β -hydroxybutyrate in transformant *Escherichia coli* harboring phbCAB operon. J Biosci Bioeng 102(3):237–240

Song Y, Matsumoto K, Yamada M et al (2012) Corynebacterium glutamicum as an endotoxin-free platform strain for lactate-based polyester production. Appl Microbiol Biotechnol 93(5):1917–1925

Stark D, von Stockar U (2003) In situ product removal (ISPR) in whole cell biotechnology during the last twenty years. Adv Biochem Eng Biotechnol 80:149–175

Stavila E, Loos K (2013) Synthesis of lactams using enzyme-catalyzed aminolysis. Tetrahedron Lett 54(5):370–372

Steinbüchel A, Füchtenbusch B (1998) Bacterial and other biological systems for polyester production. Trends Biotechnol 16(10):419–427

Steinbüchel A, Valentin HE (1995) Diversity of bacterial polyhydroxyalkanoic acids. FEMS Microbiol Lett 128(3):219–228

Stephan B, Solvej S, Jan M et al (2013) Recombineering in *Corynebacterium glutamicum* combined with optical nanosensors: a general strategy for fast producer strain generation. Nucleic Acids Res 41(12):6360–6369

Su Y, Li X, Liu Q et al (2010) Improved poly-γ-glutamic acid production by chromosomal integration of the Vitreoscilla hemoglobin gene (vgb) in Bacillus subtilis. Bioresour Technol 101(12):4733–4736

Sun Q, Wang Y, Shen L et al (2019) Application of CRISPR/Cas9-based genome editing in studying the mechanism of pandrug resistance in *Klebsiella pneumoniae*. Antimicrob Agents Ch 63(7):e00113–e00119

Taguchi S, Yamada M, Matsumoto K et al (2008) A microbial factory for lactate-based polyesters using a lactate-polymerizing enzyme. PNAS 105(45):17323–17327

Taibo H, Kai Y, Weiqiang Z et al (2020) Research on the preparation of poly(malic acid) by fermentation of budding short stalk mold. Biochem Eng 6:134–139

Tajima K, Han X, Satoh Y et al (2012) In vitro synthesis of polyhydroxyalkanoate (PHA) incorporating lactate (LA) with a block sequence by using a newly engineered thermostable PHA synthase from *Pseudomonas* sp. SG4502 with acquired LA-polymerizing activity. Appl Microbiol Biotechnol 94(2):365–376

Tan D, Xue YS, Aibaidula G et al (2011) Unsterile and continuous production of polyhydroxybutyrate by *Halomonas* TD01. Bioresour Technol 102(17):8130–8136

Tan Y, Xu D, Ye L et al (2012) Construction of a novel sacB-based system for marker-free gene deletion in *Corynebacterium glutamicum*. Plasmid 67(1):44–52

Tan D, Wu Q, Chen JC et al (2014) Engineering *Halomonas* TD01 for the low-cost production of polyhydroxyalkanoates. Metab Eng 26:34–47

Tao Y, Hongsheng X (2015) Progress in the study of microbial synthesis of γ-polyglutamic acid, related genes, synthesis mechanism and fermentation. Biotechnol Bull 31(3):25–34

Tao W, Lv L, Chen GQ (2017) Engineering Halomonas species TD01 for enhanced polyhydroxyalkanoates synthesis via CRISPRi. Microb Cell Factories 16(1):48

Tappel RC, Kucharski JM, Mastroianni JM et al (2012) Biosynthesis of Poly[(R)-3-hydroxyalkanoate] copolymers with controlled repeating unit compositions and physical properties. Biomacromolecules 13(9):2964–2972

Terpe K (2006) Overview of bacterial expression systems for heterologous protein production: from molecular and biochemical fundamentals to commercial systems. Appl Microbiol Biotechnol 72(2):211–222

Tripathi L, Wu LP, Chen JC et al (2012) Synthesis of diblock copolymer poly-3-hydroxybutyrate-block-poly-3-hydroxyhexanoate [PHB- *b* -PHHx] by a β-oxidation weakened *Pseudomonas putida* KT2442. Microb Cell Factories 11(1):44–54

Tripathi L, Wu LP, Dechuan M et al (2013) *Pseudomonas putida* KT2442 as a platform for the biosynthesis of polyhydroxyalkanoates with adjustable monomer contents and compositions. Bioresour Technol 142:225–231

Troy FA (1973) Chemistry and biosynthesis of the poly (γ-d-glutamyl) capsule in *Bacillus licheniformis*: I. Properties of the membrane-mediated biosynthetic reaction. J Biol Chem 248(1):305–315

Tsuge T, Watanabe S, Shimada D et al (2007) Combination of N149S and D171G mutations in *Aeromonas caviae* polyhydroxyalkanoate synthase and impact on polyhydroxyalkanoate biosynthesis. FEMS Microbiol Lett 277:217–222

Tuersuntuoheti T, Wang Z, Wang Z et al (2019) Review of the application of ε-poly-L-lysine in improving food quality and preservation. J Food Preserv 42(10):e14153

Utsunomia C, Matsumoto K, Taguchi S (2017) Microbial secretion of D-lactate-based oligomers. ACS Sustain Chem Eng 5(3):2360–2367

Valdez-Pena AU, Espinoza-perez JD, Sandoval-fabian GC et al (2010) Screening of industrial enzymes for deproteinization of shrimp head for chitin recovery. Food Sci Biotechnol 19(2):553–557

Valentin HE, Dennis D (1997) Production of poly(3-hydroxybutyrate- *co* -4-hydroxybutyrate) in recombinant *Escherichia coli* grown on glucose. J Biotechnol 58(1):33–38

Valentin HE, Schönebaum A, Steinbüchel A (1991) Identification of 5-hydroxyhexanoic acid, 4-hydroxyheptanoic acid and 4-hydroxyoctanoic acid as new constituents of bacterial polyhydroxyalkanoic acids. Appl Bacteria Arch Microbiol 155(5):415–421

Valentino F, Moretto G, Lorini L et al (2019) Pilot-scale polyhydroxyalkanoate production from combined treatment of organic fraction of municipal solid waste and sewage sludge. Ind Eng Chem Res 58(27):12149–12158

Van Thuoc D, My DN, Loan TT et al (2019) Utilization of waste fish oil and glycerol as carbon sources for polyhydroxyalkanoate production by *Salinivibrio* sp. M318. Int J Biol Macromol 141:885–892

Wang HH, Li XT, Chen GQ (2009) Production and characterization of homopolymer polyhydroxyheptanoate (P3HHp) by a *fadBA* knockout mutant *Pseudomonas putida* KTOY06 derived from *P. Putida* KT2442. Process Biochem 44(1):106–111

Wang G, Jia S, Wang T et al (2011) Effect of ferrous ion on epsilon-poly-L-lysine biosynthesis by Streptomyces diastatochromogenes CGMCC3145. Curr Microbiol 62(3):1062–1067

Wang Y, Chen R, Cai JY et al (2013a) Biosynthesis and thermal properties of PHBV produced from Levulinic acid by *Ralstonia eutropha*. PLoS One 8(4)

Wang Y, San KY, Bennett GN (2013b) Improvement of NADPH bioavailability in *Escherichia coli* by replacing NAD(+)-dependent glyceraldehyde-3-phosphate dehydrogenase GapA with NADP (+)-dependent GapB from Bacillus subtilis and addition of NAD kinase. J Ind Microbiol Biotechnol 40(12):1449–1460

Wang Y, San KY, Bennett GN (2013c) Improvement of NADPH bioavailability in *Escherichia coli* through the use of phosphofructokinase deficient strains. Appl Microbiol Biotechnol 97(15):6883–6893

Wang Q, Yang P, Xian M et al (2013d) Biosynthesis of poly(3-hydroxypropionate-*co* -3-hydroxybutyrate) with fully controllable structures from glycerol. Bioresour Technol 142:741–744

Wang Q, Liu XL, Qi QS (2014a) Biosynthesis of poly(3-hydroxybutyrate- *co* -3-hydroxyvalerate) from glucose with elevated 3-hydroxyvalerate fraction via combined citramalate and threonine pathway in *Escherichia coli*. Appl Microbiol Biotechnol 98(9):3923–3931

Wang Y, Quan Y, Song C (2014b) Progress in microbial synthesis and application of polymalic acid. Chin J Biotechnol 30(9):1331–1340
Wang L, Chen X, Wu G et al (2016) Genome shuffling and gentamicin-resistance to improve ε-poly-l-lysine productivity of *Streptomyces albulus* W-156. Appl Biochem Biotechnol 180(8):1601–1617
Wang Y, Zhang G, Zhao X et al (2017) Genome shuffling improved the nucleosides production in *Cordyceps kyushuensis*. J Biotechnol 260(1):42
Wang L, Li S, Zhao J et al (2019) Efficiently activated ε-poly-l-lysine production by multiple antibiotic-resistance mutations and acidic pH shock optimization in *Streptomyces albulus*. MicrobiologyOpen 8(5):e00728
Ward PG, O'Connor KE (2005) Bacterial synthesis of polyhydroxyalkanoates containing aromatic and aliphatic monomers by *Pseudomonas putida* CA-3. Int J Biol Macromol 35(3/4):127–133
Wei X, Ji Z, Chen S (2010) Isolation of halotolerant *Bacillus licheniformis* WX-02 and regulatory effects of sodium chloride on yield and molecular sizes of poly-γ-glutamic acid. Appl Biochem Biotechnol 160(5):1332–1340
Weiwei C, Hongyang Z, Hong X (2007a) Breeding of mass-producing ε-polylysine mutant and its batch fermentation. Industrial Microbiol 37(2):28–30
Weiwei C, Hongyang Z, Hong X (2007b) Research on the selection and batch fermentation of high-yield ε-polylysine strains. Industrial Microbiol 37(2):28–30
Wenjing H, Yingchao L, Guanghao Z et al (2019) Study on the conditions for fermentation production of polyglutamic acid using monosodium glutamate and by-products. Food Ferment Technol 55(3):39–42
Wenjuan Y, Wenjun F, Xiaole X et al (2012) Bioinformatics analysis of the structure of PgsBCA, the γ-polyglutamic acid synthase system. J Nantong Univ (Natural Sci Ed) 11(2):41–46
Wu F, Cai D, Li L et al (2019) Modular metabolic engineering of lysine supply for enhanced production of bacitracin in *Bacillus licheniformis*. Appl Microbiol Biotechnol 103(21–22):8799–8812
Xiang-Yu S, Qi-Feng W, Xiu-Lian R (2018) Progress in the extraction process and application of chitosan/chitin from shrimp and crab shells. Food Res Dev 39(22):6
Xiao J, Xu ZX, Xu H et al (2014) Economical production of poly(epsilon-L-lysine) and poly(L-diaminopropionic acid) using cane molasses and hydrolysate of streptomyces cells by *Sreptomyces albulus* PD-1. Bioresour Technol 164:241–247
Xiaoli Y, Zhiqiang F, Xiaohong H et al (2012) Synthesis of a new γ-polyglutamic acid water-absorbing resin. Synthetic Chemistry
Xiaolong S, Yongqian F (2019) Morphological mutant strain of Rhizopus with high yield of L-lactic acid screened by compound mutation and carbon metabolic flux analysis. Jiangsu Agric Sci 47(01):294–299
Xiaoming S, Xiaoya W, Zhen W et al (2008) Research progress on the fermentation production of lactic acid by immobilized Rhizopus. Chem Industry Progr 2:206–208
Xiaoxing Z (2017) pH regulation and variable temperature fermentation of Bacillus amyloliquefaciens YP-2 for γ-polyglutamic acid production. Henan Agricultural University, Zhengzhou
Xinwei X, Changqing S, Jialong W et al (2004) Biosynthesis of γ-polyglutamic acid and its related genes. Chin J Biotechnol 24(8):38–41, 47
Xu P, Ranganathan S, Fowler ZL et al (2011) Genome-scale metabolic network modeling results in minimal interventions that cooperatively force carbon flux towards malonyl-CoA. Metab Eng 13(5):578–587
Xu ZX, Cao CH, Sun ZZ et al (2015a) Construction of a genetic system for *Streptomyces albulus* PD-1 and improving poly(epsilon-L-lysine) production through expression of vitreoscilla hemoglobin. J Microbiol Biotechnol 25(11):1819
Xu Z, Feng X, Sun Z et al (2015b) Economic process to co-produce poly(ε-l-lysine) and poly (l-diaminopropionic acid) by a pH and dissolved oxygen control strategy. Bioresour Technol 187:70–76
Xu Z, Bo F, Xia J et al (2015c) Effects of oxygen-vectors on the synthesis of epsilon-poly-lysine and the metabolic characterization of *Streptomyces albulus* PD-1. Biochem Eng J 94:58–64

Xu D, Yao H, Cao C et al (2018) Enhancement of epsilon-poly-L-lysine production by overexpressing the ammonium transporter gene in *Streptomyces albulus* PD-1. Bioprocess Biosyst Eng 41(9):1337–1345

Xu G, Wang J, Gu L et al (2021) Functional characterization of CapBCA in controlling poly-γ-glutamic acid synthesis in corynebacterium glutamicum

Xueliang Z (2018) Regulation of γ-polyglutamic acid molecular weight and its adsorption of heavy metal ions. Henan University, Kaifeng

Xueliang Z, Wenya L, Guang L et al (2022) Adsorption properties of γ-polyglutamic acid hydrogel for Cd^2+,Pb^2+. 13

Xueming Z, Tao C, Zhiwen W (2015) Metabolic engineering. Higher Education Press, Beijing

Xuenian H, Shen T, Xuefeng L (2020) Progress and prospects in the synthetic biology research of the industrial filamentous fungus Aspergillus terreus. Synth Biol 2:187–211

Xuexiao C, Qixing J, Yanshun X et al (2014) Study on the decolorization technology of chitin from Antarctic krill shell. J Zhengzhou Univ Light Ind 29(3):2095-476X(2014)03-0021-04

Yahui C, Qing W, Wenyu W et al (2021) Optimization of fermentation medium for γ-polyglutamic acid production by *Bacillus siamensis* LW-1. Food Industry Sci Technol 42(16):163–170

Yamada M, Matsumoto K, Nakai T et al (2009) Microbial production of lactate-enriched poly[(R)-lactate-co-(R)-3-hydroxybutyrate]. Biomacromolecules 10(4):677–681

Yamada M, Matsumoto, Ken'ichiro et al (2010) Adjustable mutations in lactate (LA)-polymerizing enzyme for the microbial production of LA-based polyesters with tailor-made monomer composition. Biomacromolecules 11(3):815–819

Yamanaka K, Hamano Y, Oikawa T (2020) Enhancement of metabolic flux toward ε-poly-l-lysine biosynthesis by targeted inactivation of concomitant polyene macrolide biosynthesis in *Streptomyces albulus*. J Biosci Bioeng

Yan Z (2007) Research on gene transformation technology of Klebsiella pneumoniae. Master's thesis, Shanghai Jiaotong University

Yan X (2019) Exploration of gene knockout and recombination engineering mediated point mutation methods in Corynebacterium glutamicum ATCC 13032. Master's thesis, Nanjing Normal University

Yang J, Shih II, Tzeng Y et al (2000) Production and purification of protease from a *Bacillus subtilis* that can deproteinize crustacean wastes. Enzym Microb Technol 26(5-6):406–413

Yang TH, Kim TW, Kang HO et al (2010) Biosynthesis of polylactic acid and its copolymers using evolved propionate CoA transferase and PHA synthase. Biotechnol Bioeng 105(1):150–160

Yang L, Yang Z, Liqiang F et al (2021) Study on the cold resistance of N-acetylglucosamine in rice. Southern Agric J 52(6):11

Yanli Z (2018) Cultivation of Bacillus subtilis to produce γ-polyglutamic acid using monosodium glutamate wastewater and preliminary characterization. J Ecol Environ 27(10):1949–1957

Yanyan L, Shanshan Z, Guodong R (2015) Progress in the extraction of chitin from crustaceans, insects, and fungi. Food Res Dev 36(7):5

Yanyun F, Haijuan G, Hailiang L et al (2019) Research progress on the biological preservative polylysine. Agric Products Process 477(4):57–62

Yeh CM, Wang JP, Lo SC et al (2010) Chromosomal integration of a synthetic expression control sequence achieves poly-γ-glutamate production in a *Bacillus subtilis* strain. Biotechnol Prog 26(4):1001–1007

Yicheng B, Hongyan L, Hao W et al (2015) Two-step ultrafiltration method for separating and extracting polymalic acid from fermentation broth. Membr Sci Technol 35(1):97–102

Yin J, Chen JC, Wu Q et al (2015a) Halophiles, coming stars for industrial biotechnology. Biotechnol Adv 33:1433–1442

Yin J, Wang H, Fu XZ et al (2015b) Effects of chromosomal gene copy number and locations on polyhydroxyalkanoate synthesis by *Escherichia coli* and *Halomonas* sp. Appl Microbiol Biotechnol 99(13):5523–5534

Yin H, Ma Y, Deng Y et al (2016) Genome shuffling of Saccharomyces cerevisiae for enhanced glutathione yield and relative gene expression analysis using fluorescent quantitation reverse transcription polymerase chain reaction. J Microbiol Methods 127:188–192

Yongzhu C (2005) Mutation breeding of γ-polyglutamic acid producing bacteria and application research of heavy metal adsorption. Sichuan University

Younes I, Rinaudo M (2015) Chitin and chitosan preparation from marine sources. Structure, properties and applications. Mar Drugs 13(3):1133–1174

Younes I, Ghorbel-bellaaj O, Nasri R et al (2012) Chitin and chitosan preparation from shrimp shells using optimized enzymatic deproteinization. Process Biochem 47(12):2032–2039

Yu Y, Zhu X, Xu H et al (2019) Construction of an energy-conserving glycerol utilization pathways for improving anaerobic succinate production in *Escherichia coli*. Metab Eng 56:181–189

Yue HT, Ling C, Yang T et al (2014) A seawater-based open and continuous process for polyhydroxyalkanoates production by recombinant *Halomonas campaniensis* LS21 grown in mixed substrates. Biotechnol Biofuels 7:108–120

Yuhui W (2018) Study on the synthesis of ethylene glycol by *E. coli* transformation of xylose, Master's thesis, Shandong University

Zelić B, Gostović S, Vuorilehto K et al (2010) Process strategies to enhance pyruvate production with recombinant *Escherichia coli*: from repetitive fed-batch to in situ product recovery with fully integrated electrodialysis. Biotechnol Bioeng 85(6):638–646

Zeng AP, Menzel K, Deckwer WD (2015) Kinetic, dynamic, and pathway studies of glycerol metabolism by *Klebsiella pneumoniae* in anaerobic continuous culture: II. Analysis of metabolic rates and pathways under oscillation and steady-state conditions. Biotechnol Bioeng 52(5):561–571

Zeng X, Miao W, Wen B et al (2019) Transcriptional study of the enhanced ε-poly-L-lysine productivity in culture using glucose and glycerol as a mixed carbon source. Bioprocess Biosyst Eng 42(4):555–566

Zhang Q, Xiu Z (2010) Metabolic pathway analysis of glycerol metabolism in *Klebsiella pneumoniae* incorporating oxygen regulatory system. Biotechnol Prog 25(1):103–115

Zhang Y, Feng X, Xu H et al (2010) Epsilon-poly-L-lysine production by immobilized cells of Kitasatospora sp. MY 5-36 in repeated fed-batch cultures. Bioresour Technol 101(14):5523–5527

Zhang Y, Lin Z, Liu Q et al (2014) Engineering of Serine-Deamination pathway, Entner-Doudoroff pathway and pyruvate dehydrogenase complex to improve poly(3-hydroxybutyrate) production in *Escherichia coli*. Microb Cell Factories 13(1):172

Zhang C, Ren H, Zhong C (2021a) Preparation of γ-polyglutamic acid from enzymatic hydrolysate of poplar sawdust. Arab J Chem 14(4):103095

Zhang J, Chen D, Liang G, Xu W et al (2021b) Biosynthetic polymalic acid as a delivery nanoplatform for translational cancer medicine. Trends Biochem Sci 46(3):213–224

Zhang X, Zhang C, Zhou M et al (2022) Enhanced bioproduction of chitin in engineered *Pichia pastoris*. Food Biosci:47

Zheng Z, Li M, Xue XJ et al (2006) Mutation on N-terminus of polyhydroxybutyrate synthase of *Ralstonia eutropha* enhanced PHB accumulation. Appl Microbiol Biotechnol 72(5):896–905

Zhong Z, Jianguang W, Lequan Q et al (2010) Research progress on microbial polyglutamic acid (γ-PGA) synthase and synthesis mechanism. Biotechnol Bull 6:52–56

Zhou XY, Yuan XX, Shi ZY et al (2012) Hyperproduction of poly(4-hydroxybutyrate) from glucose by recombinant *Escherichia coli*. Microb Cell Factories 11:54–61

Zhou YP, Ren XD, Wang L et al (2015) Enhancement of ε-poly-lysine production in ε-poly-lysine-tolerant *Streptomyces* sp. by genome shuffling. Bioprocess Biosyst Eng 38(9):1705–1713

Zhu Y, Eiteman MA, Dewitt K et al (2007) Homolactate fermentation by metabolically engineered *Escherichia coli* strains. Appl Environ Microbiol 73(2):456–464

Zhu LW, Li XH, Zhang L et al (2013) Activation of glyoxylate pathway without the activation of its related gene in succinate-producing engineered *Escherichia coli*. Metab Eng 20:9–19

Zhuang Q, Wang Q, Liang Q et al (2014) Synthesis of polyhydroxyalkanoates from glucose that contain medium-chain-length monomers via the reversed fatty acid β-oxidation cycle in *Escherichia coli*. Metab Eng 24:78–86

Zong H, Zhan Y, Li X et al (2012) A new mutation breeding method for Streptomyces albulus by an atmospheric and room temperature plasma. Afr J Microbiol Res 6:3154–3158

Zou X, Li S, Wang P et al (2020) Sustainable production and biomedical application of polymalic acid from renewable biomass and food processing wastes. Crit Rev Biotechnol 41(2):216–228

Chapter 4
Engineered Living Materials

Jiaofang Huang, Xiaojuan Zhu, Xumeng Huang, Ning Jiang, Zihan Qi, Yumei Hu, and Liming Zhao

4.1 Concept and Composition of Engineered Living Materials

Engineered Living Materials (ELMs) consist of living matter or cell communities that are embedded in self-regenerating matrices made of either their own natural components or artificial scaffolds. The living cells form or assemble the material itself (with responsive function or/and scaffold function), or regulate the functional performance of the material in some way. The development of engineered living materials will push the boundaries and frontiers of synthetic biology, materials engineering, nanotechnology, biomaterials, and artificial intelligence, and guide the emergence of new fields. The self-replication, self-repair, and environmental response of living systems, as well as the ability to interact with inorganic and/or organic substances, enable biological systems to develop engineered living materials with special properties (Fig. 4.1) (An et al. 2023).

J. Huang (✉)
State Key Laboratory of Bioreactor Engineering, East China University of Science and Technology, Shanghai, China

College of Life Science, Jiangxi Normal University, Nanchang, China
e-mail: huangjf@ecust.edu.cn

X. Zhu · X. Huang · N. Jiang · Z. Qi · Y. Hu
State Key Laboratory of Bioreactor Engineering, East China University of Science and Technology, Shanghai, China

L. Zhao
State Key Laboratory of Bioreactor Engineering, East China University of Science and Technology, Shanghai, China

Key Laboratory of Biobased Material Engineering of China Light Industry, East China University of Science and Technology, Shanghai, China
e-mail: zhaoliming@ecust.edu.cn

L. Zhao (ed.), *Research and Applications of Bio-based Degradable Materials*,
https://doi.org/10.1007/978-981-95-1188-4_4

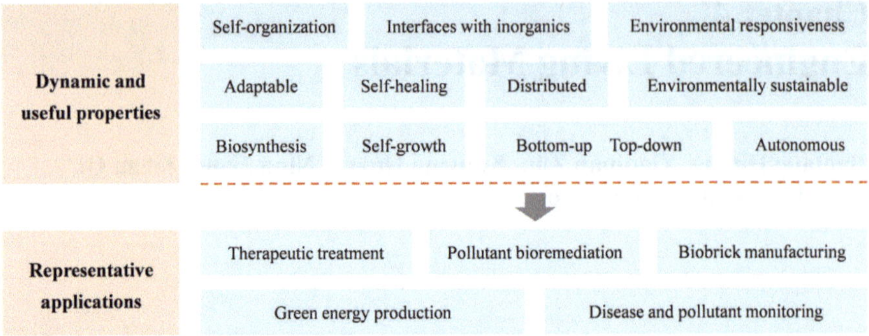

Fig. 4.1 The foundational properties of living systems contribute to the development of beneficial properties for material function and synthesis. (An et al. 2023)

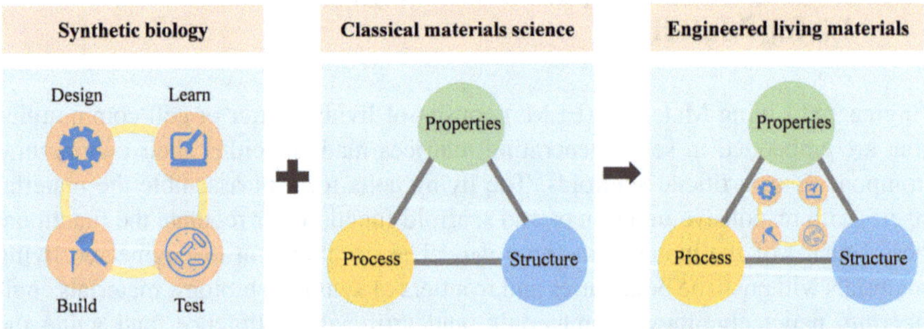

Fig. 4.2 The cross-integration of synthetic biology and traditional materials science. (Srubar 2021)

To design ELMs, it is necessary to integrate synthetic biology and materials science. Synthetic biology mainly involves applications in learning, designing, constructing, and testing, with genetically engineered organisms possessing new or enhanced functions (such as material production, sensing, and response functions). After rational design and planning, ELM researchers combine this with the processes, structures, and properties of natural and engineered materials to achieve the goal of new functional materials (Fig. 4.2) (Srubar 2021).

The existing ELMs field mainly involves the genetic engineering construction of algae, bacteria, fungi, and plants. Living cells serve as material factories, utilizing energy and raw materials from the environment to produce biological polymer objects. They also play a role in guiding the formation or maintenance of necessary materials, as well as integrating other inorganic polymers, particles, or scaffolds into the materials as part of material assembly. This distinction is a key factor that sets ELM apart from other bio-hybrid devices (Fig. 4.3). Engineering techniques at ELMs encompass a variety of methods, including genetic engineering of chassis cells and spatial/mechanical engineering to confine or locate cells. In contrast to most self-assembly biomaterials that utilize purified components, ELMs can effectively guide the formation of materials under complex conditions containing multiple components, such as nutrient feedstock or metabolic waste. When necessary,

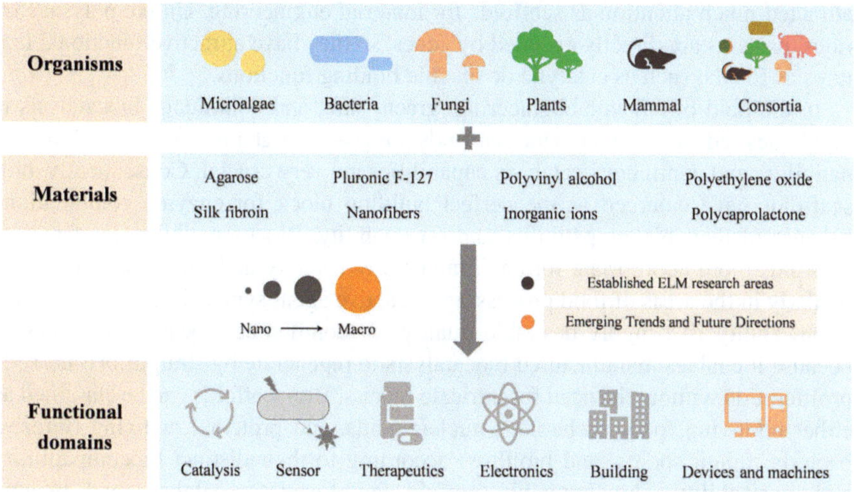

Fig. 4.3 Composition and research field of engineered living materials. (Srubar 2021; Rodrigo-Navarro et al. 2021)

ELMs can be treated at specific times to remove or eliminate living cells, thereby preserving the assembled material without concern for the ongoing maintenance of the cells or potential biosecurity threats.

4.2 Application Research of Engineered Living Materials

The fundamental principle of ELMs is that engineered living cells actively participate in material synthesis, self-organization into higher-order structures, and maintenance of these structures to respond to environmental stimuli. The research begins with bacteria, as they have rapid reproduction and many of them can be genetically controlled. During the process of cell biofilm formation, cells release extracellular polysaccharides, proteins, eDNA, etc., which assemble into complex spatial structures outside the cell, known as the extracellular matrix (ECM). Extracellular polysaccharides are the most abundant substances in the ECM, composed of various carbohydrate monomers with diverse and complex structures. Functionally, polysaccharides serve as a "molecular glue," allowing bacteria to nonspecifically adhere to surfaces and each other, and can also protect cells from harsh environments. As an engineering platform, extracellular polysaccharides have advantages due to their high content in the ECM, ease of separation, and high stability. However, compared to proteins and nucleic acids, it is more difficult to structurally modify polysaccharides using genetic engineering techniques, because the biosynthetic pathways of different species lack standardization. Extracellular proteins are also a common component of bacterial biofilm matrices, in forms such as functional amyloid proteins, pili, and flagella. Especially, functional amyloid proteins, which have recently attracted much attention as scaffolds for material engineering. Unlike polysaccharides, proteins are directly encoded by genes, so they have attractive functional features as ELMs, such as catalytic or specific binding functions.

In the field of enzyme engineering, green, safe, and continuous biocatalysis is highly desired. As such, enzyme scaffolds possessing high biosafety, good biocompatibility, and continuous catalysis capabilities are very crucial. Consequently, bio-scaffolds have emerged as the perfect building block for enzyme combinations because of their biocompatibility and renewability. Biocompatibility renders bio-scaffolds more appropriate for environmentally friendly and safe production, particularly in the areas of food processing, bioactive agent synthesis, and diagnostics. Renewability is a highly desirable quality for continuous biocatalysis processes because it enables manufactured biocatalysts to regenerate by straightforward self-proliferation without the need for intricate repairs. Bio-scaffolds can be classified as either nonliving (polysaccharides, nucleic acids, and proteins) or living (viruses, bacteria, fungi, spores, and biofilms) according to their distinct biocompatibility and renewability. The living bio-scaffolds employed for catalysis and enzyme assembly will be the subject of the following.

4.2.1 Research on the Application of Bacterial Cellulose-Based Living Materials

The high structural complexity of bacterial cell surface polysaccharides, such as lipopolysaccharides and capsular polysaccharides, opens up possibilities for the development of a wide range of functional living materials. Yi et al. (2009) used an intracellular biosynthesis system to modify rhamnose, a common bacterial polysaccharide. By integrating the GDP-rhamnose pathway of *Bacteroides fragilis* into *E. coli*, replacing the native *E. coli* GDP-rhamnose de novo synthesis pathway, various nonnatural rhamnoses were integrated. Polysaccharides were synthesized through metabolic engineering using a range of chemical functional groups, including azides, alkynes, ketones, and amino acid groups. Overall, metabolic sugar engineering has found wider applications in the field of chemical biology. This method of manufacturing polysaccharide polymers with nonstandard chemical components using metabolic engineering is attractive for the development of ELMs. However, before this strategy becomes more feasible for scalable material manufacturing, issues related to the poor processing efficiency of modified sugars may need to be resolved.

Among the various extracellular polysaccharides produced by bacteria, cellulose (such as xanthan gum, dextran, alginates, and other extracellular polysaccharides) has received the most attention in recent years. The molecular structural units of bacterial cellulose (BC) and natural cellulose derived from plants or algae are the same; however, the properties of BC are distinct, with bacterial cellulose fibers having high crystallinity, high tensile strength, ultra-fine network structure, strong water retention capacity, high biocompatibility, adaptability, and good biodegradability, and no lignin or other byproducts. The high purity and excellent performance of bacterial cellulose fibers make them widely used in medical materials, the food industry, paper industry, high-end audio equipment diaphragms, and other fields. People have discovered substances containing bacterial cellulose as early as ancient times, such as the gelatinous bacterial film formed on the surface of the fermentation liquid during the vinegar brewing process recorded in 《Qi Min Yao Shu》. In 1976, Brown (R.M. Brown) and his collaborators first described the movement of acetic acid bacteria in the process of cellulose biosynthesis. Bacterial cellulose has been a worthy research area for a long time. Recently, bacterial cellulose has become the focus of attention in the field of synthetic biology and materials science. On the one hand, the naturally produced cellulose-producing strains, such as *Gluconobacter xylinum*, are relatively easy to culture under laboratory conditions; on the other hand, compared with the pellicle produced by other bacteria, the biofilm produced by *G. xylinum* is more robust mechanically and suitable for various physical operations, including drying. Studies on bacterial cellulose as a material are continually being advanced; for example, genetic tools are being developed to control the material's mechanical and functional properties, and modern manufacturing technologies are being used to shape the material into a variety of useful forms. People can enhance the yield of cellulose and other biopolymers or alter their chemical composition by modifying or optimizing specific biosynthetic pathways of cellulose and employing metabolic engineering techniques. Fang et al. (2015) extracted the biofilm matrix component—curdlan (β-1,3-dextran) from the commercial bacterium *Agrobacterium* sp. Although curdlan and cellulose both utilize the same precursor (UDP-glucose) for biosynthesis, curdlan is easier to degrade in the human body than cellulose due to the β-1.3-linkage between glucose monomers, making it a potential scaffold for tissue engineering. To develop a material that merges the mechanical strength of cellulose with the degradability of curdlan, researchers integrated a single curdlan synthase into the cellulose-producing strain *Gluconobacter xylinum*, thereby achieving simultaneous secretion of curdlan and cellulose to produce a composite biofilm comprising fibers intertwined with a blend of curdlan and cellulose polysaccharides. The surface morphology and hydrophilic characteristics of this composite biofilm are between pure curdlan and pure cellulose materials. Therefore, it has been proven that it has the ability to engineer cell types capable of simultaneously biosynthesizing and assembling multiple biopolymers. Florea et al. (2016) have discovered a new cellulose-producing bacterial strain, *Komagataei bacterrhaeticus* iGEM. This strain contains a gene-encoded cellulose production pathway. Researchers have designed this strain to be compatible with a "toolkit," such as annotated chromosomes, a small plasmid library, and a set of reporter genes, which makes the engineering of this strain more feasible. Researchers have also developed an inducible system based on the acyl homoserine lactone (AHL) quorum-sensing pathway, which can regulate the production of cellulose and has been proven to be valuable in tracking cellulose films over time and space during the bacterial growth phase. Mangayil et al. (2017) increased the cellulose yield of *Komagataeibacter xylinus* four times by recombining the endogenous bacterial cellulose (BC) synthase operon. Compared with the wild-type strain, all four genes in the BC synthase operon are overexpressed, jointly promoting the biosynthesis, regulation, chain crystallization, and export of the polymer, resulting in higher glucose utilization efficiency and thicker cellulose biofilms. In addition to enhancing cellulose production efficiency by modifying bacteria using synthetic biology-related techniques, it is also feasible to optimize the production and performance of bacterial cellulose through microtechnology and materials science techniques. For example, using microfluidic encapsulation and processing techniques, it has been proven that *G. xylinus* can produce cellulose microspheres after being encapsulated in a microfluidic droplet system for several days (Yu et al. 2016). Because modifying the genes encoding bacterial cellulose is challenging, current research primarily focuses on utilizing it as a scaffold material. This involves fixing functional substances on cellulose by editing other protein-encoding genes. Subsequently, the cellulose film formed is used to achieve specific functionalities. Gilbert et al. (2021) designed yeast strains as factories for enzyme expression and secretion. The synthesized secreted enzymes are fused with tags that bind to cellulose and then immobilized within bacterial cellulose. This process allows for the autonomous production of catalytic materials that continue to grow. In addition, engineered yeast can be incorporated into the developing cellulose matrix, forming living materials that can sense and respond to chemical and optical stimuli (Fig. 4.4). This symbiotic culture of bacteria and yeast is a flexible platform for producing engineered living materials made of bacterial cellulose, with potential applications in biosensing and biocatalysis. Engineering biosynthetic pathways provides new opportunities for controlling the chemical composition and material characteristics of ELMs. In addition, genetically engineer *Komagataeibacter rhaeticus* to create a bacterial strain that grows self-pigmenting BC (Walker et al. 2024). In itself, this control is valuable from the perspective of biomanufacturing or new polymers, and it is of special significance for the development of ELMs.

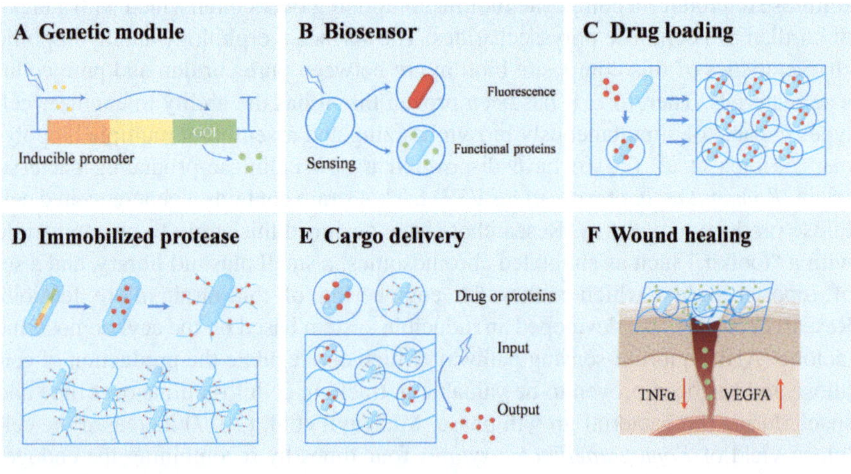

Fig. 4.4 Schematic diagram of the design of functionalized engineered living materials based on polysaccharides cellulose

4.2.2 Research on the Application of Living Materials Based on Biofilm Amyloid Proteins

Protein engineering can regulate polymer sequences, allowing them to possess complex functions, such as specific molecular recognition, organized self-assembly, and catalysis, which occur naturally in biological systems. It is for these reasons that proteins are used as building blocks, such as collagen, keratin, and silk.

Early on, researchers successfully reconstructed the surface of microbial cells into programmable scaffolds to regulate interactions involving S-layer proteins between cells and other elements in the extracellular environment (Messner and Sleytr 1992). The S-layer is an ancient cell encapsulation strategy, found in almost all archaea and many strains of both Gram-positive and Gram-negative bacteria, which assemble into two-dimensional structures relying on proteins (Beveridge 1994). Single-layer lattice-like proteins can form crystal patterns with different symmetries and play roles in membrane barrier functions, bacterial adhesion, and enzyme scaffolding. Fusing heterologous structural domains with the S-layer can make it a useful scaffold for creating novel synthetic nanobiomaterials (Beveridge 1994; Sleytr et al. 2007). In fact, the combination of a rich toolkit with S-layer proteins from sources such as *Lactobacillus* spp. has been utilized in various fields such as vaccine development, heavy metal bioremediation, sensor diagnostics, and cell-free nanobiomaterials (Sleytr et al. 2007). Variants displayed on the bacterial cell surface focus on controlling higher-order interactions that may be of particular interest for ELM development (Michon et al. 2016) . Moraïs et al. (2014) designed a consortium composed of *Lactobacillus plantarum*, and assembled it into an extracellular artificial cellulose-degrading enzyme complex or cellulase. The consortium consists of three strains, two of which are modified to have soluble forms of recombinant cellulase and xylanase, which can be fused to different dockerin domains. The third strain uses a cell wall anchoring scheme to display an assembly scaffold. The consortium can degrade cellulose biomass with an efficiency almost as high as that of the corresponding soluble enzyme, and it has stronger stability. This strategy of dividing tasks among different engineered strains may be an effective method to avoid overburdening the metabolism of any single cell type.

In recent years, significant progress has been made in the research of biofilm amyloid proteins as scaffolds for engineered living materials (Fig. 4.5). The most extensively studied functional amyloid protein is the Curli system produced by *E. coli*, which is also characterized in *Salmonella*, *Citrobacter*, and *Enterobacter spp.* (Dueholm et al. 2012). The Curli fibers of Escherichia coli are formed by the extracellular self-assembly of CsgA, a 13 kDa protein, which is secreted in the form of unstructured monomers and then aggregates extracellularly. Several research groups have used Curli fibers to create highly designable functional materials. Chen et al. demonstrated that this overall strategy can achieve in situ assembly of extracellular fibers during cell growth (Chen et al. 2014).

In addition, the nanomechanical properties of Curli fibers can be adjusted by controlling the relative concentrations of CsgA and another protein in the Curli

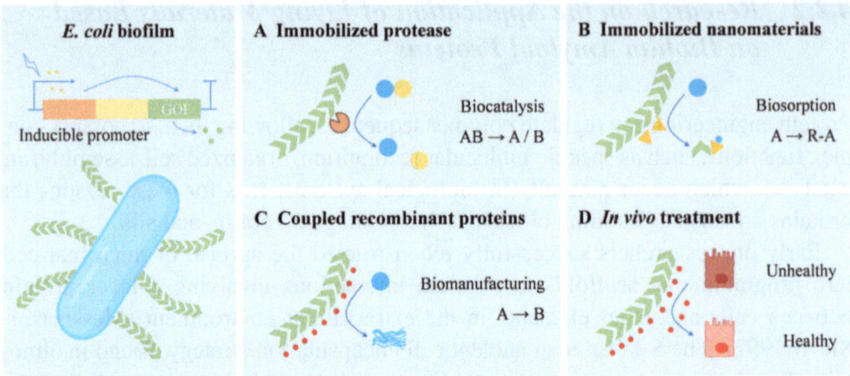

Fig. 4.5 Research on *Escherichia coli* biofilm amyloid proteins as scaffolds for engineered living materials

system, CsgB, as well as protein structures, such as fused linkers, etc. (Abdelwahab et al. 2017). Nguyen et al. developed a biofilm-integrated nanofiber display technology (BIND) (Nguyen et al. 2014). Botyanszki et al. (2015) used a popular and efficient engineering two-component "SpyTag/SpyCatcher" covalent bioconjugation technology in subsequent research to fix enzymes on biofilms. Zhong's group (Wang et al. 2019) used biofilm-anchored gold nanoparticles to reduce nitroaromatic compounds, such as the pollutant p-nitrophenol, and used biofilm-anchored mixed $Cd_{0.9}Zn_{0.1}S$ quantum dots and gold nanoparticles to degrade organic dyes in a semi-artificial photosynthesis system and biofilm-anchored CdSeS@ZnS quantum dots for hydrogen production.

In addition to controlling genes with nanoscale characteristics, engineered biofilm living materials have applications in broader areas such as the environment and energy (Dorval Courchesne et al. 2017). To realize the true potential of ELM and make it responsive to environmental signals, appropriate continuous remodeling of the Curli matrix can be carried out, such as responding to and adsorbing heavy metals (Tay et al. 2017). Researchers use the SpyTag+SpyCatcher binding scheme encoded by genes to produce modular extracellular polymers (Ma et al. 2011). Duprey et al. designed a strain of *E. coli* to produce a "bacterial biofilter" in response to increased metal ion concentrations. NicoT is an exogenous metal transporter from *Novosphingobium aromaticivorans*, which can be re-expressed in *E. coli* to enhance cell absorption of nickel and cobalt ions. This method, combined with the "synthetic adhesion operon," triggered the formation of various biofilms. The resulting strain can sense metals in the environment and produce biofilms to adsorb heavy metals (Duprey et al. 2014). However, SpyTag/SpyCatcher has the drawback of inhibiting enzyme function. Therefore, Hao Dong, Jiaofang Huang, and others used phage affinity peptide screening technology to screen the lipase binding peptide (LBP), modified it (LBP2) and co-expressed it with CsgA (CsgALBP2), and developed *E. coli* BL21△CsgA-CsgB-CsgALBP2 (LBP2 functionalized) biofilm as a surface display platform to maximize the catalytic efficiency of lipase (Lip181).

After Lip181 was fixed onto the LBP2 functionalized biofilm material, it showed stronger thermostability, pH stability, and storage stability. Surprisingly, through this fixation strategy, the relative activity of fixed Lip181 increased from 8.43 to 11.33 U/mg. In addition, the highest lactose load on the LBP2 functionalized biofilm material reached 27.90 mg/g of wet biofilm material, which is equivalent to 210.49 mg/g of dry biofilm material, demonstrating its potential for a high enzyme loading surface. In addition, the fixed Lip181 was utilized to hydrolyze phthalate esters, achieving a hydrolysis rate of 100% for dibutyl phthalate. Therefore, in terms of maximizing enzyme performance, lipid fixation mediated by LBP2 is more favorable than the traditional SpyTag/SpyCatcher strategy (Dong et al. 2021).

In addition, Jiaofang Huang and colleagues have constructed and developed a *Bacillus subtilis* biofilm living material platform based on the engineering modification of biofilm proteins TasA and BslA (Fig. 4.6). Compared with previous reports, this platform can display larger and more functional proteins and has successfully achieved functional applications. For example, *B. subtilis* biofilm TasA protein serves as a display scaffold for functional proteins (such as organic phosphorus hydrolase OPH and plastic degradation enzyme MHETase) with environmental remediation function (Huang et al. 2019). Besides, *B. subtilis* biofilm was engineered as a living glue material by displaying mussel foot proteins (Mfps) on TasA and BslA biofilm proteins (Huang et al. 2019; Zhang et al. 2019). Recently, Huang's group developed *B. subtilis* biofilm as an efficient catalyst for toluene degradation using *B. subtilis* biofilm-supported Mn–Ce/zeolite (Mohsin et al. 2024), and developed a novel living material that can detect various heavy metal ions simultaneously and realize the in situ remediation for heavy metal ion pollution (Zhu et al. 2024).

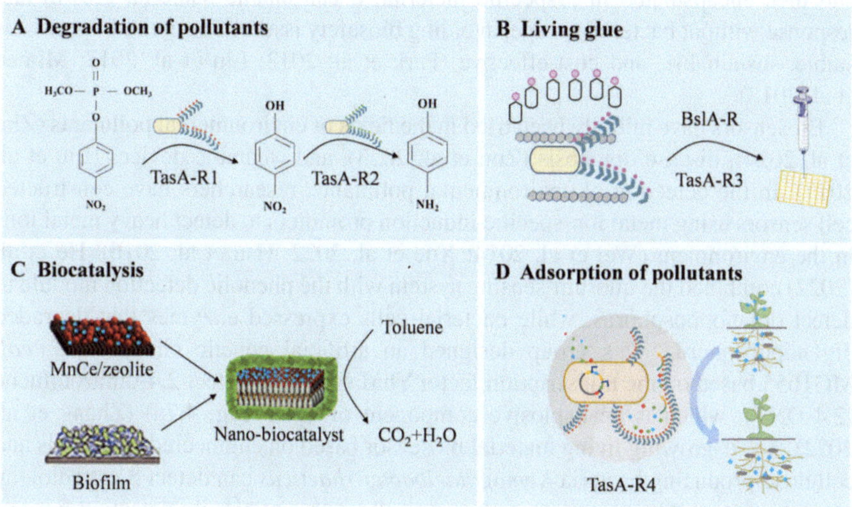

Fig. 4.6 Schematic diagram of *Bacillus subtilis* biofilm as an engineering living material display platform

In summary, the field of biofilm engineering research is endless. In addition to the examples above, it involves various methods and applications, such as from bioremediation schemes to continuous flow biocatalysis systems, etc. (Halan et al. 2012; Singh et al. 2006; Lu et al. 2021). Scientists are focusing on applied research related to the promotion of ELM, aiming to spatiotemporally control the assembly of living cells and materials using engineering tools.

4.2.3 Application Research of Living Materials for Biosensing

The biological system can receive external signals and respond, enabling the activation of specific gene expression under certain stimuli, which is an important prerequisite for developing engineered living sensors in biological systems. These stimuli and responses are generally achieved through promoter switches. For instance, bacterial biosensing can trigger the synthesis of reporter proteins in response to signal input by modifying kinases, transcription factors, repressors, or RNA ribosome switches (Park et al. 2013). Current research generally uses fluorescent proteins or bioluminescent proteins as reporter molecules, which is conducive to detection and quantitative analysis. A single detected signal input can trigger the expression of the reporter gene, and the expression of the reporter gene is a continuous accumulation, achieving the effect of signal amplification (Gui et al. 2017). When using biosensors, researchers must consider biosafety issues. This can be achieved by introducing toxic protein genes and suicide genes to control bacterial reproduction or encapsulating bacteria in a gel to prevent bacterial diffusion into the environment. If the gene circuit encoding a detection system is constructed inside a microorganism and then encapsulated in a specific material, it can enable bacterial growth and response without bacterial escape, avoiding biosafety issues. The entire biosensor is stable, sustainable, and cost-effective (Park et al. 2013; Liu et al. 2017; Mimee et al. 2018).

Biosensors have initially been tried in the fields of environmental pollutants (Zhu et al. 2024), disease diagnosis (Zou et al. 2023), and wearable devices (Liu et al. 2017). In the detection of environmental pollutants, researchers have constructed cell sensors using metal ion-specific induction promoters to detect heavy metal ions in the environment (Wei et al. 2014; Xue et al. 2022; Hsu et al. 2016). He et al. (2022) combined the quorum-sensing system with the phenolic detection module to detect organophosphorus, while bacterial cells expressed enzymes that degraded organophosphorus. Ye's group designed an artificial genetic circuit in *E. coli* MG1655 based on the transcription factor YhaJ system to detect 2,4-dinitrotoluene (2,4-DNT), which is an explosive component of TNT (Fig. 4.7a) (Zhang et al. 2022). A self-growing living material biosensor based on engineered yeast cells and cellulose-producing bacteria *Komagataeibacter rhaeticus* can detect β-estradiol, an estrogenic steroid hormone, through yeast cells with a minimum detection limit of 5 nM (Gilbert et al. 2021).

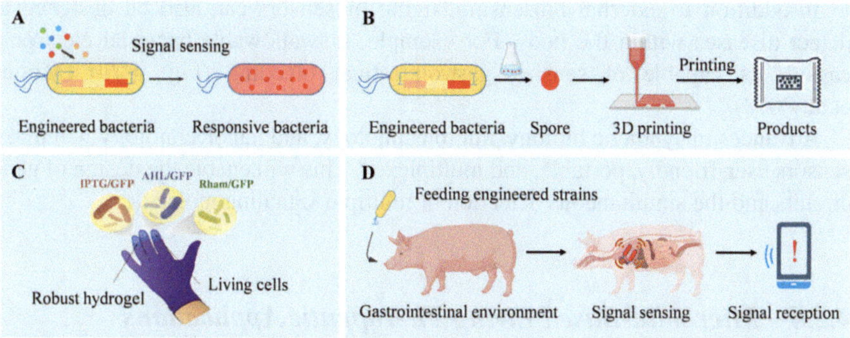

Fig. 4.7 Living materials as biosensors

Mora et al. (2015) detected lactose or galactose in milk by encapsulating transgenic *E. coli* in agarose hydrogel, which is used for the detection of human food regulation. However, due to the inability of *E. coli* to produce spores and the poor mechanical stability of agarose, it cannot be stored at low temperatures, limiting its large-scale production. Fortunately, *B. subtilis* is generally considered a safe strain and can germinate spores. The germinated spores are able to withstand harsh environments such as low temperatures, dryness, detergents, and acids and bases, which is beneficial for their production as a sensor. Schulz-Schönhagen et al. encapsulated *B. subtilis* in polyvinyl alcohol (PVA) hydrogel to detect IPTG in the environment. The dormant spores within these hydrogels can be stored at temperatures between −80 and −20 °C for an extended period (at least 4 weeks), and can also be dried, meeting the industry standard production requirements for biosensors (Schulz-Schönhagen et al. 2019) (Fig. 4.7b).

Wearable biosensors are an important need for future intelligent robots and disease detection. Biosensors achieve noninvasive real-time health monitoring by detecting biochemical substances in body fluids, such as sweat, and can sense body movement through changes in muscle contraction. Compared to pure hydrogel biomaterials, bioactive wearable sensor devices need to be made of flexible materials and have good wear resistance. Additionally, the encapsulated living cells can maintain normal physiological metabolism for sensing and detecting different signal inputs and outputs. For instance, Liu et al. encapsulated engineered *E. coli* in alginate-polyacrylamide hydrogel, allowing the diffusion of nutrients and products. The sensor can be composed of an elastic separator that supports gas exchange, and the device can be stretched to 1.8 fold its original length. When applied to skin or nitrile gloves, it can detect different small molecules. In addition, the authors encapsulated these bacteria in Pluronic F-127 acrylate encapsulation and used it as ink to print a flexible living tattoo that can sense IPTG, rhamnose, and AHL on the skin (Fig. 4.7c) (Liu et al. 2017). In addition, *B. subtilis* spores are printed within the material, which can sense or kill *Staphylococcus aureus*, and this work demonstrates the potential of living materials for scenarios requiring storage or exposure to environmental stress (González et al. 2020).

In addition to external biosensing, living biosensors can also be designed to detect diseases within the body. For example, a swallowable bacterial electronic capsule is capable of sensing gastrointestinal bleeding (Fig. 4.7d) (Mimee et al. 2018).

Advances in synthetic biology, microtechnology, and nanotechnology will make sensors user-friendly, portable, and multiplexed. This will enable the design of gene circuits and the simultaneous detection of multiple signaling molecules.

4.2.4 Microbial-Based Living Therapeutic Applications

The use of biological metabolites for the treatment and prevention of diseases has a long history. Chinese traditional medicine utilizes various plant and microbial metabolites to achieve disease treatment and prevention. Traditional industries employ natural and genetically modified microbes to produce specific metabolites. With advancements in synthetic biology, material science, and computer science, living cells can be combined with biocompatible materials that have low immunogenicity, controlled drug release, and high safety. These living therapeutic materials are used for the treatment and prevention of diseases, aiming to achieve therapeutic effects that cannot be achieved by other treatments (Zhai et al. 2024; Omer et al. 2022; Jiang et al. 2024a, b).

Therefore, biocompatible hydrogels with mechanical robustness and selective permeability are the preferred artificial materials for manufacturing living therapeutic devices. For example, soft biocompatible agarose hydrogels can accommodate genetically modified *E. coli*, which secretes drugs into the culture medium under light. Composite materials enable the regulation of reporter gene production, localization, and dosage release through dynamic control of light metabolic pathways. Similarly, microcapsules based on biopolymers and nanoporous membranes can also serve as cell containers, ensuring material exchange between cells and the surrounding environment, and providing continuous nutrient supply for the encapsulated cells (Dai et al. 2019; Gerber et al. 2012).

Combined living therapies can not only deliver drugs in vivo but also be used to treat pathogenic infections on the skin (Wei et al. 2024). Antibiotic-producing microbes encased in soft, hydrated hydrogels can be used to make medical bandages with long-term or on-demand antibacterial properties. For example, a 3D printed hydrogel patch that fits the shape of a wound, containing *B. subtilis* spores secreting staphylolytic and penicillin. This patch is capable of detecting and eliminating *Staphylococcus aureus* when applied to skin wounds (Fig. 4.8a) (Schulz-Schönhagen et al. 2019). Genetically modified bacteria can also manipulate the behavior of mammalian cells by secreting metabolites (Fig. 4.8b) (Sankaran et al. 2018). Therefore, they can also be used in the field of regenerative medicine. For example, nonpathogenic bacteria *Lactococcus lactis* displaying recombinant human fibronectin can be displayed on the extracellular biofilm. It can be planted on organic

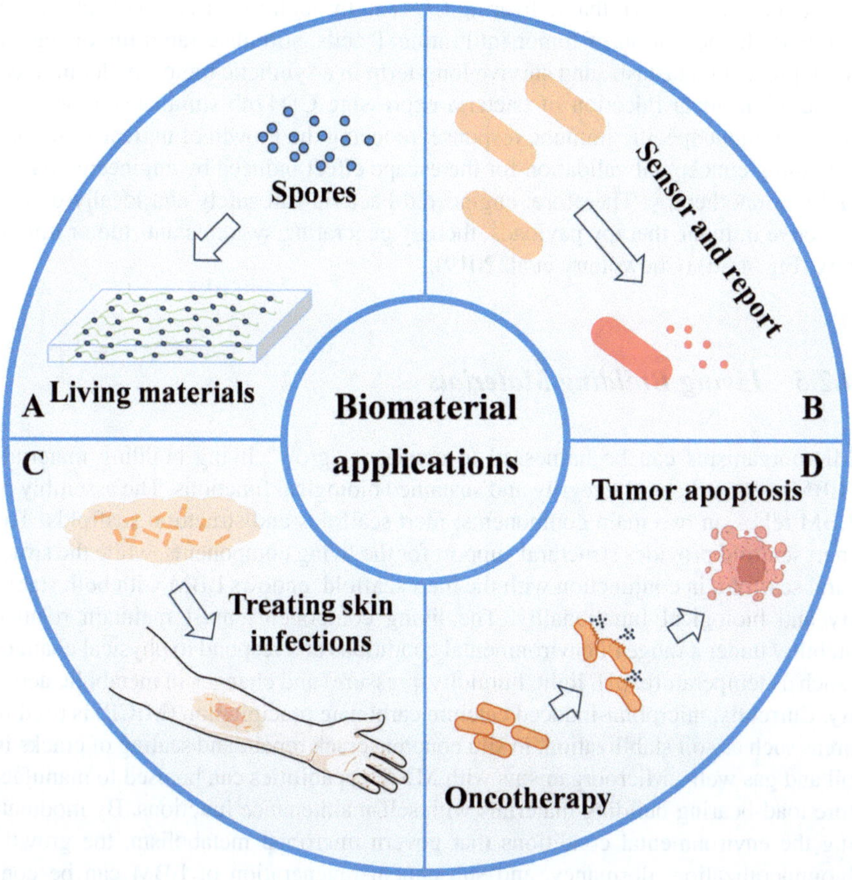

Fig. 4.8 Schematic diagram of therapeutic application of living materials based on microorganisms

or inorganic surfaces to form a "living bio-interface," inducing the differentiation of mesenchymal stem cells (Rodrigo-Navarro et al. 2014; Hay et al. 2018).

Synthetic biology provides the possibility of designing life systems with customized functions by reprogramming gene circuits. Therefore, drug-releasing cells can also be used in other healthcare fields, such as producing low-cost cosmetics or masks with virus decomposition capabilities. In addition to the dynamic characteristics of engineered life systems, artificial synthetic materials can also give mixed composite materials customized characteristics, such as responsiveness. A smart, adaptable gel made from the thermosensitive polymer Pluronic F-127 contains living Bacillus subtilis spores and can be used to treat superficial fungal infections. When the temperature rises to 37 °C, this gel will transition from a liquid state to a hydrogel state (Fig. 4.8c) (Lufton et al. 2018).

Studies have shown that delivering CD47nb through tumor colony bacteria can promote the activation of tumor-infiltrating T cells, stimulate rapid tumor regression, prevent metastasis, and survive long-term in a synthetic tumor model in mice. In addition, local injection of bacteria expressing CD47nb stimulates a systemic tumor-antigen-specific immune response, reducing the growth of untreated tumors, providing conceptual validation for the escape effect induced by engineered bacterial immunotherapy. Therefore, engineered bacteria can safely and locally deliver effective immune therapy payloads, thereby generating systemic anti-tumor immunity (Fig. 4.8d) (Chowdhury et al. 2019).

4.2.5 Living Building Materials

Microorganisms can be harnessed to create or "grow" living building materials (LBM) with structural integrity and sustained biological functions. The assembly of LBM relies on two main components: inert scaffolds and structural scaffolds. The inert scaffold provides structural support for the living components, while the structural scaffold, in conjunction with the inert scaffold, endows LBM with both stability and biological functionality. The living components must maintain relative stability under a range of environmental conditions and respond to physical changes (such as temperature, pH, light, humidity, pressure) and changes in metabolic activity. Currently, microbial-induced calcium carbonate precipitation (MICP) is used in areas such as soil stabilization, in situ concrete crack repair, and sealing of cracks in oil and gas wells. Microorganisms with MICP capabilities can be used to manufacture load-bearing building materials with self-maintenance functions. By modulating the environmental conditions that govern microbial metabolism, the growth, biomineralization, dormancy, and subsequent regeneration of LBM can be controlled on demand. These environmental transformations can regenerate LBMs from a parent inoculum, providing new possibilities for the manufacture, use, and remanufacturing of infrastructure materials (Fig. 4.9). This living building material is made by mixing microbial cells with a calcium-containing nutrient medium, gelatin, and sand, and then using temperature and humidity switches to enable it to regenerate exponentially and then obtain structural integrity through drying (Liu et al. 2024; Kang et al. 2021). After serving as a load-bearing structural material, LBMs can be deconstructed and recycled as a source of new LBMs (Heveran et al. 2020).

Coral reefs, as living building materials, are typical examples of microbial participation in the biomineralization of inorganic nanomaterials. They are essentially calcium carbonate deposits, with enzymes produced by marine organisms catalyzing Ca^{2+} and CO_3^{2-} , to generate calcium carbonate precipitation (Achal et al. 2015; Bertucci et al. 2013) . In a similar way, other naturally occurring and engineered microbes can be studied. They can catalyze the synthesis of organic materials (such as polymers), or coordinate the assembly of non-biological modules in other ways, which can be found in their environment. In other examples, prefabricated

Fig. 4.9 Living building materials (LBMs). (Heveran et al. 2020)

structures composed of synthetic polymers or inorganic materials can serve as scaffolds to guide cell growth and behavior. Here, engineers can utilize a wealth of micro-manufacturing and nano-manufacturing technologies to regulate cell behavior on a range of length scales.

As previously mentioned, research has been conducted to use bacteria-produced calcium carbonate as a binding material for the manufacture of bricks in construction. Microbially induced calcite precipitation (MICP) as an alternative method to improve brick structural performance is being studied. This approach mainly utilizes the ability of certain microbes to secrete urease, which produces CO_2 from urea, to catalyze the formation of calcite form of calcium carbonate. In this process (which takes several days to weeks), calcite acts as a matrix that binds clay or sand particles together to form bricks. Raut and others (Raut et al. 2014) using *Bacillus pasteurii*, the compressive strength of a 28d cured brick is approximately 8 MPa, which is two- to threefolds lower than that of a traditionally fired brick. McBee et al. (2022) used fungi and bacteria along with wood chip fragments to cultivate microbial living bricks, which can support the weight of an adult. The concept of "cultivating" bricks was proposed by the start-up company (bioMASON), which has shown a keen interest in this technology and claims that through genetic engineering, bricks can be created with capabilities such as glowing in the dark, absorbing pollutants, and being used for water quality detection. The ultimate goal of these active calcium carbonate materials is to produce biomaterials, self-healing concrete, and MICP bricks, with the aim of reducing structural maintenance costs, preserving structural integrity, and minimizing environmental pollutants generated during brick production.

4.2.6 Living Materials for Electron Transfer and Energy Conversion

Engineered living conductive materials hold significant potential in a range of applications where traditional conductive polymers are unsuitable, including as improved catalytic coatings for microbial fuel cell electrodes, self-powered sensors for harsh environments, and next-generation active components for bioelectronic devices that interact with the human microbiome. Researchers use the conductivity of microbial biofilms or exogenous electricity-producing strains (*S. oneidensis* or *Geobacter* spp.) to transfer electrons or convert organic matter into electricity. Microalgae or cyanobacteria use solar energy to convert inorganic matter into organic matter, even forming fuel cells, etc. (Tang et al. 2021) Seker et al. (2017) created a bio-inorganic hybrid material for conduction by displaying conductive gold nanoparticles binding domains on CsgA.

4.3 Scale Production and Processing of Living Materials

Compared to biomaterials made from purified biopolymers, living materials self-grow without the need for excessive processing during assembly or manufacturing when close to their final form. Common examples of engineered living materials include biofilms, mycelial materials, and bacterial cellulose. The main challenge is the choice of production methods, which can range from small-batch analytical scale to larger semi-preparative production scale, and then to full industrial scale production (Manjula-Basavanna et al. 2024). Common examples of engineered living materials include biofilms, mycelial materials, and bacterial cellulose. This process can be divided into two stages: the first is the production of the biomaterial itself, usually through in situ biosynthesis, and the second is the processing of the material to obtain the required purity and form.

4.3.1 Scale Production of Living Materials

Mycelium is a new category of self-growing, fibrous, naturally composite materials that can be produced in large quantities and large areas. This material is primarily derived from white rot and brown rot fungi, which are capable of degrading lignin and cellulose, making it one of the most commercially successful living materials that can be prepared on a large scale. Mycelium not only grows naturally but can also be shaped into various forms, such as bricks made from mycelium used in construction.

Recent reports have detailed the physical properties of mycelium: it is not only thermally stable and hydrophobic but also has a high water contact angle and a relatively low water absorption value (Haneef et al. 2017). These characteristics are essential for many applications on both small and large scales. Currently, the developed mycelium materials are natural polymer composites (chitin, cellulose, proteins, etc.), which require less energy for production (self-growth), and these characteristics can be adjusted by modifying their nutrient substrates for actual situations and potential uses. Compared to other multicellular tissues, mycelium grows more rapidly, can be easily shaped using molds, and has the potential to alter its physicochemical properties. Thus, mycelial technology holds great promise for the large-scale production of engineered living materials. Numerous studies have shown that mycelium bio-templates combine the rapid growth potential of fungi with ideal inorganic material characteristics, greatly expanding the range of uses for mycelium materials (Bigall et al. 2008).

Initially, Valentini et al. (2016a) focused on the fields of environmentally friendly packaging and construction materials, using nutrients essential for microbial growth to regulate their development process. They claimed that by altering the mix of fungal species, the material properties of the mycelial structure could be adjusted, resulting in biohybrid materials with unique characteristics that cannot be produced by conventional non-biological processes, thus holding commercial potential (Fig. 4.10). At the same time, the team also adopted a similar process, adding graphene nanosheets during the fermentation process of brewing yeast fungus extract to couple the two, forming a composite film with better mechanical properties. This mycelium can be used to produce leather-like materials, and even used to form bricks for house construction (Valentini et al. 2016b).

In addition to structural materials, this technology also has potential uses in advanced application fields such as bio-integrated electronic circuits. Research has shown (Kozicki et al. 1995) that, on custom-designed integrated test circuits, the rust fungus of cowpea with precise signal recognition ability can grow mycelium, forming bacterial tubes, thereby creating a bio-hybrid integrated system. A promising application for adjusting the functional parameters of mycelial materials is the production of extracellular polymeric substances (EPS) by engineered fungi. Currently, very little is known about the composition and biosynthesis process of EPS from lignin-digesting fungi, only that it is mainly a polysaccharide network, which acts as a fixed point for secreted enzymes or as a structural sheath around the

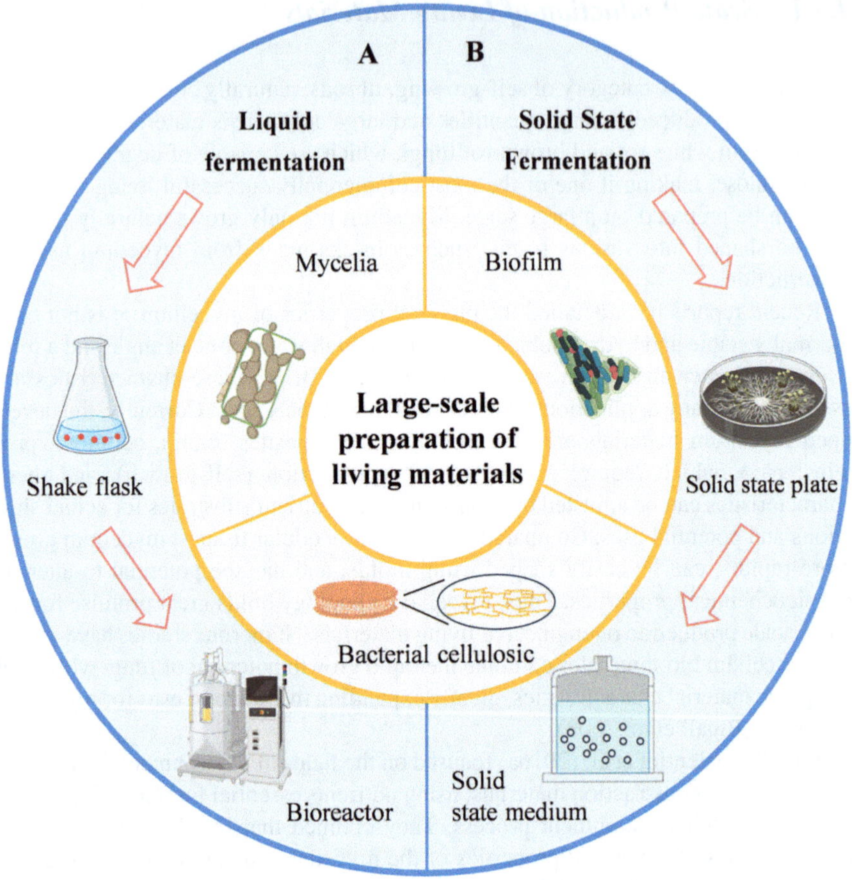

Fig. 4.10 Scale production of living materials

mycelium (Hietala et al. 2009). Most genetic tools available for the modification of filamentous fungi are underdeveloped, making it difficult to use genetic engineering to understand the biological characteristics of these fungi and to fully industrialize them. Therefore, there is a need to develop general methods for genetic manipulation of non-model filamentous fungi, and a CRISPR-Cas9-based system suitable for filamentous fungi has been developed. This system is simple and universal, and can achieve RNA-guided mutation by transforming the target fungus with a single plasmid. The system currently includes four CRISPR-Cas9 vectors, all of which carry common fungal markers, allowing selection in a wide range of fungi, and the prospects for customized filamentous materials are extremely broad (Nodvig et al. 2015).

4.3.2 Scale-Up Production of Bacterial Cellulose Materials

Cellulose is one of the most abundant polymers in nature, usually associated with other polymers (pectin, lignin, arabinoxylan, etc.) in the cell walls of plants, algae, and some bacteria (rhizobia, agrobacteria, etc.). However, only a few bacterial species taxonomically related to Acetobacter (acetic acid bacteria) can secrete cellulose extracellularly.

Bacterial cellulose, obtained from acetic acid bacteria and cellulose-producing bacteria, can be considered a successful example of living materials that can be implemented on a large scale. It possesses a range of unique properties, including high mechanical strength, high water absorption capacity, high crystallinity, and an ultra-fine and highly pure fibrous network structure. Similar to mycelial materials, the process of large-scale preparation primarily changes the functional form of natural materials through physical effects. If the production process can be improved, especially through techniques like deep fermentation, bacterial cellulose materials are expected to become a versatile new commercial product.

Currently, bacterial cellulose materials have been commercialized in various fields. For instance, medical skin dressings made from microbial cellulose materials, building materials created on jute fiber scaffolds, and a wide range of garments produced using large-scale textile materials. In addition, due to its strong mechanical properties, it can also be used to manufacture high-strength paper. Using its acoustic properties, it can be used to prepare diaphragms for high-end audio speakers (Widmaier et al. 2009). Currently, the use of bacterial cellulose in the manufacture of advanced displays (Shah and Brown 2005) and electroactive paper for energy storage devices is being explored (Li et al. 2014).

Currently, commercial research on bacterial cellulose is mainly concentrated in biomedical fields, such as the preparation of burn dressings and tissue engineering scaffolds, led by several companies focusing on the development of bacterial cellulose materials (Wood et al. 2016; Hong et al. 2010) . One major obstacle to the widespread commercial application of this multifunctional material is the high production cost, primarily due to the inefficient cultivation process of bacteria, which may take a week or longer to form a film on a desktop. Therefore, continuous exploration of the synthesis pathway of bacterial cellulose and other microbial chassis cells (such as cyanobacteria) is crucial for increasing the yield of bacterial cellulose and reducing the cost of cultivation (Florea et al. 2016; Qian et al. 2021).

4.3.3 Large-Scale Production of Biofilm Living materials

Biofilm refers to a common survival state of bacteria in nature, where bacterial cells adapt to the environment by adhering to contact surfaces and secreting heterogeneous extracellular matrix, such as polysaccharides, proteins, and nucleic acids. The cells encapsulate themselves within this matrix, forming an aggregated membrane-like structure (Branda et al. 2005; Ma et al. 2012; Stoodley et al. 2002). Biofilms have a three-dimensional structure that is solid, stable, and resistant to destruction. Scholars have conducted research and development on the process of large-scale production of *Escherichia coli* biofilm living materials (Fig. 4.10c).

For ELM materials composed of functional amyloid proteins or cellulose and other bacterial ECMs, large-scale production mainly uses traditional bioreactors, fixed/moving-bed bioreactors, hollow fiber bioreactors, and static bioreactors. The widespread use of bioreactors has accelerated the development of bioactive materials, but there are still difficulties in maintaining the stability and rate of biological processes. The simplest bioreactor is a static bioreactor, which has a relatively low cost, the culture is not disturbed, and is very suitable for the production of in situ assembled floating film materials, such as bacterial cellulose. However, due to the lack of effective mixing of nutrients and waste transfer, the culture usually grows slowly. Through CFD modeling, microbubbles generated by fluid oscillation can provide an effective, low-energy way to increase the high interfacial area of mass transfer and improve the mixing liquid circulation process. This reactor is called an airlift reactor. The use of microbubbles not only increases the surface area to volume ratio but also improves the mixing efficiency by increasing the liquid circulation speed around the drainage pipe. Cultivating *Acetobacter xylinum* used for bacterial cellulose production in this improved airlift reactor, the concentration of bacterial cellulose is about threefold higher than in the traditional bubble tower, which proves that this reactor can significantly increase the yield of bacterial cellulose (Al-Mashhadani et al. 2015; Brophy and Voigt 2014).

Another traditional bioreactor is the continuous stirred tank reactor (CSTR), which is also one of the most common bioreactors in large-scale preparation. The feature of this reactor is that it can continuously stir the culture with an impeller, while adding nutrients at the same speed as removing products, and it has been used to culture self-assembled living materials of suspended cells (Pawar et al. 2015; Tian et al. 2012). However, for interface-assembled ELM materials, a key factor for large-scale production is to increase the available adsorption surface area to maximize the adhesion of cells or ECM in the biofilm, so CSTR is not suitable. To enhance the adsorption surface area, one can achieve this by selecting a support structure, such as glass beads, glass wool, carbon fibers, bone charcoal, porous brick particles, polymer supports, wood chips, and metals, each varying in composition and geometric shape (Junker et al. 2007).

If you want to enable the growth of biofilm living materials on a large-scale interface, you can use a solid carrier reactor. Depending on whether the fixed-bed layer can float, it can be divided into a fixed-bed or a moving-bed bioreactor. Among them, the fixed-bed reactor can be further divided into a packed-bed reactor (PBR) and a trickle-bed reactor (TBR). PBR achieves long-term culture of various cell lines by fixing cells in a suitable substrate that forms the bed, allowing them to adsorb on the surface of the substrate. Both PBR and TBR have been successfully used to produce biofilms in the field of biocatalysis, but they are usually not used in the field of material production. PBR is currently widely used in the development of biological artificial liver support systems and the in vitro expansion research of bone marrow cells (Sen et al. 2017). TBR has been used in wastewater management systems and is now using natural biofilms to degrade pollutants (Hekmat et al. 2004) by the biotransformation method.

4.4 Manufacturing and Processing of Living Materials

4.4.1 3D Printing

3D printing is a technology that manufactures 3D products by adding materials layer by layer using 3D printing equipment based on a pre-designed 3D model, and it has made significant advancements in areas such as construction, manufacturing, molds, and biomedicine (Mirkhalaf et al. 2023; Ortega et al. 2023; Pitta et al. 2023). It has made significant progress in the fields of construction, manufacturing, mold, biomedicine, etc. (Xiongfa et al. 2018). Among them, bioprinting technology, as a type of 3D printing, is a general term for a series of bottom-up, biomimetic additive manufacturing technologies. By using bio-ink to carry live cells directly for structural construction, it can precisely control the spatial distribution of cells and other bioactive components in the scaffold. This spatial programming method allows the development of heterogeneous ELMs of any shape. Nowadays, using biomaterials can achieve bioprinting of scaffolds and cells, and can also print live cells, as well as polymer drug growth factors, to achieve 3D printing at specific locations (Koo and Kim 2016; Kang et al. 2016; Jakus et al. 2016).

Currently, using 3D printing technology, by utilizing the inherent viscoelasticity of bacterial biofilms, a programmable, printable biofunctional material platform can be constructed, which can realize new applications in the fields of nanomanufacturing, biocatalysis, biomedicine, etc. Huang et al. (2019) developed an active functional material platform, using the amyloid-like protein expression and assembly mechanism of *B. subtilis* biofilm, to develop genetically programmable TasA fusion proteins with different structural domain combinations. These fusion proteins can be secreted and self-assembled into extracellular nanostructures with adjustable physicochemical properties around living cells and can be flexibly configured in space (Fig. 4.11). These constructed living materials, while maintaining their natural activity and various cellular capabilities (such as self-regeneration), can also exhibit the nonnatural functions of the combined functional domains, including changed enzyme activity, red fluorescence, and the templated assembly ability of inorganic nanoparticles (NP).

In addition, the programmable group sensing circuits within these spatially arranged cellular materials can be utilized to create dynamic engineering living materials that change with space and time (Kuang et al. 2024). Connel et al. (2013) used a micro 3D printing strategy to encapsulate and position multiple bacterial populations (adjacent, nested, and free-floating colonies) within sealed cavities formed by highly porous cross-linked gelatin. The cell populations grow sealed

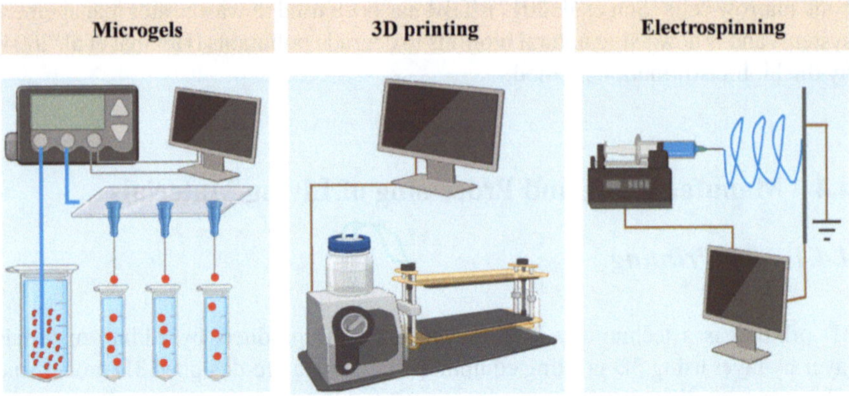

Fig. 4.11 Manufacturing and processing of engineered living materials. Focus on the application of biofilms in areas such as microgels, 3D printing, and electrospinning

within the gelatin and can access peptides, antibiotics, and quorum sensing signals, and can organize within almost any 3D geometric structure. Using gel materials, processing the designed structures into small-scale microchambers allows for the rapid diffusion of dual-related molecules such as group signaling signals and antibiotics. This provides a means to investigate how bacterial social behavior depends on population size and density, container shape, and the flow rate of the surrounding medium. This technique has demonstrated structures containing multiple bacterial chambers; these bacterial chambers have interconnected "corridors" and nested chambers containing different types of bacteria.

Although these 3D-printed microchambers have so far only been used as analytical tools, they have the potential to be used to rearrange specific ELM-generating cells in 3D to create well-defined heterogeneous materials. A recent low-cost alternative has emerged that uses a modified low-cost 3D printer to print engineered *E. coli* into alginate bio-ink by integrating a computer-controlled injection pump with the 3D printer. Although the 3D resolution of the structure is about 1–2 mm, given the low construction cost and low resource requirements, this bacterial bioprinting system should be able to be more widely used to explore 3D-printed ELMs.

Currently, cutting-edge 3D-printing technology can print biological tissues and organs. Organs, as a collection of cells, are closely related to ELM technology and perform physiological functions. By selecting suitable biomaterials and living cells, it is possible to directly simulate real organs and vascular networks outside the body and print them for tissue engineering. Based on the highly integrated characteristics of 3D printing, it can be used to prepare artificial electronic skin, muscles, and even bionic robots. Therefore, many current studies use 3D printing technology to assemble complex organs by inoculating cells in biodegradable substrates. Such as Homan et al. (2016) recently developed a bioprinting method to create 3D artificial kidney proximal tubules in vitro, embedding them in the extracellular matrix and placing them in printable tissue chips (Fig. 4.11b). This method can programmatically define tubule size and geometry (including convolution), and compared to cells grown in a 2D control, the manufactured proximal tubules show significantly enhanced epithelial morphology and functional characteristics, demonstrating strong potential for manufacturing complex ELMs.

4.4.2 Electrospinning

Electrospinning is a processing technique that uses polymer solutions or melts to form jets under the action of a strong electric field, thereby producing continuous fibers. It mostly involves solution spinning with a small amount of melt spinning. This method involves the deformation of charged melts or polymer solutions under the action of electrostatic field forces, solidification through melt cooling or solvent evaporation, and finally obtaining fibrous substances (Ji et al. 2024). Among them, electrospinning is a universal method for the large-scale preparation of uniform and continuous nanostructured materials. It is capable of producing continuous polymer fibers with diameters ranging from microns to nanometers and has been employed to create fiber materials containing living cells. This material is supported by a polymer substrate and can achieve a certain function (Barth et al. 2010; Huang et al. 2003; Lu et al. 2009).

Letnik et al. (2015) developed a new "living composite material" of yeast polymers, encapsulating yeast cells in polymer microtubes made of core-shell electrospun fibers. This electrospun composite material possesses a distinctive structure with a large surface area, enabling unrestricted liquid transfer. This feature prevents cell leakage and creates a conducive environment for sustaining cell viability (Fig. 4.11c). Compared to free cells, electrospun yeast cells can also biodegrade polyphenols and produce ethanol. These findings suggest that electrospun composite materials can provide a more promising platform for active microbial polymer systems that can be used in the water-based field. In addition, composite fibers containing *Sphaerotilus natans* and Nitrobacter can be used for metal chelation and nitrate removal from solutions, respectively. Installing bacteria-containing fibers on a metal wire frame to form a mesh structure can be conveniently used for metal chelation. Subsequently, the mesh can be further cultured for the recovery and regeneration treatment of live bacteria from the fibers.

Electrospinning has been used to encapsulate bacterial cells for drug delivery, as well as to capture live cells in the pores of fibrous matrices. Xie et al. (2016) constructed a plasmid containing a reporter gene encoding green fluorescent protein (GFP), which was then transformed into bacteria by electroporation. By encapsulating engineered bacteria in core-sheath fibers through coaxial electrospinning, the researchers provided a strategy for immobilizing engineered bacteria on electrospun fibers for drug delivery (Fig. 4.11). Eroglu et al. (2012) found that electrospun chitosan nanofiber mats can effectively immobilize microalgal cells, which can extract nitrates from wastewater, thus serving as a durable model system for wastewater treatment.

As a novel processing method for preparing ultrafine fibers, electrospinning is a promising approach that can be further developed to directly encapsulate cells in fibers using robust shell materials. This method captures engineered microbes and their extracellular matrices to manufacture large-scale materials like nets and fabrics, paving the way for the creation of functional wearable devices and clothing in the future.

References

Abdelwahab MT, Kalyoncu E, Onur T et al (2017) Genetically-tunable mechanical properties of bacterial functional amyloid nanofibers. Langmuir 33(17):4337–4345

Achal V, Mukherjee A, Kumari D et al (2015) Biomineralization for sustainable construction—a review of processes and applications. Earth-Sci Rev 148:1–17

Al-Mashhadani MKH, Wilkinson SJ, Zimmerman WB (2015) Airlift bioreactor for biological applications with microbubble mediated transport processes. Chem Eng Sci 137:243–253

An B, Wang Y, Huang Y et al (2023) Engineered living materials for sustainability. Chem Rev 123(5):2349–2419

Barth S, Hernandez-Ramirez F, Holmes JD et al (2010) Synthesis and applications of one-dimensional semiconductors. Prog Mater Sci 55(6):563–627

Bertucci A, Moya A, Tambutté S et al (2013) Carbonic anhydrases in anthozoan corals—a review. Biorg Med Chem 21(6):1437–1450

Beveridge TJ (1994) Bacterial S-layers. Curr Opin Struct Biol 4(2):204–212

Bigall NC, Reitzig M, Naumann W et al (2008) Fungal templates for noble-metal nanoparticles and their application in catalysis. Angew Chem Int Ed Eng 47(41):7876–7879

Botyanszki Z, Tay PK, Nguyen PQ et al (2015) Engineered catalytic biofilms: Site-specific enzyme immobilization onto *E. coli* curli nanofibers. Biotechnol Bioeng 112(10):2016–2024

Branda SS, Vik S, Friedman L et al (2005) Biofilms: the matrix revisited. Trends Microbiol 13(1):20–26

Brophy JA, Voigt CA (2014) Principles of genetic circuit design. Nat Methods 11:508

Chen AY, Deng Z, Billings AN et al (2014) Synthesis and patterning of tunable multiscale materials with engineered cells. Nat Mater 13:515

Chowdhury S, Castro S, Coker C et al (2019) Programmable bacteria induce durable tumor regression and systemic antitumor immunity. Nat Med 25(7):1057–1063

Connell JL, Ritschdorff ET, Whiteley M et al (2013) 3D printing of microscopic bacterial communities. Proc Natl Acad Sci USA 110(46):18380–18385

Dai Z, Lee AJ, Roberts S et al (2019) Versatile biomanufacturing through stimulus-responsive cell-material feedback. Nat Chem Biol 15(10):1017–1024

Dong H, Zhang W, Xuan Q et al (2021) Binding peptide-guided immobilization of lipases with significantly improved catalytic performance using *Escherichia coli* BL21(DE3) biofilms as a platform. ACS Appl Mater Interfaces 13(5):6168–6179

Dorval Courchesne NM, Duraj-Thatte A, Tay PKR et al (2017) Scalable production of genetically engineered nanofibrous macroscopic materials via filtration. ACS Biomater Sci Eng 3(5):733–741

Dueholm MS, Albertsen M, Otzen D et al (2012) Curli functional amyloid systems are phylogenetically widespread and display large diversity in operon and protein structure. PLoS One 7(12):e51274

Duprey A, Chansavang V, Frémion F et al (2014) "NiCo Buster": engineering *E. coli* for fast and efficient capture of cobalt and nickel. J Biol Eng 8(1):19

Eroglu E, Agarwal V, Bradshaw M et al (2012) Nitrate removal from liquid effluents using microalgae immobilized on chitosan nanofiber mats. Green Chem 14(10):2682

Fang J, Kawano S, Tajima K et al (2015) In vivo curdlan/cellulose bionanocomposite synthesis by genetically modified Gluconacetobacter xylinus. Biomacromolecules 16(10):3154–3160

Florea M, Hagemann H, Santosa G et al (2016) Engineering control of bacterial cellulose production using a genetic toolkit and a new cellulose-producing strain. Proc Natl Acad Sci 113(24):E3431–E3440

Gerber LC, Koehler FM, Grass RN et al (2012) Incorporation of penicillin-producing fungi into living materials to provide chemically active and antibiotic-releasing surfaces. Angew Chem Int Ed Eng 51(45):11293–11296

Gilbert C, Tang T-C, Ott W et al (2021) Living materials with programmable functionalities grown from engineered microbial co-cultures. Nat Mater 20(5):691–700

González LM, Mukhitov N, Voigt CA (2020) Resilient living materials built by printing bacterial spores. Nat Chem Biol 16(2):126–133

Gui Q, Lawson T, Shan S et al (2017) The application of whole cell-based biosensors for use in environmental analysis and in medical diagnostics. Sensors (Basel, Switzerland) 17(7):1623

Halan B, Buehler K, Schmid A (2012) Biofilms as living catalysts in continuous chemical syntheses. Trends Biotechnol 30(9):453–465

Haneef M, Ceseracciu L, Canale C et al (2017) Advanced materials from fungal mycelium: fabrication and tuning of physical properties. Sci Rep 7(1):41292

Hay JJ, Rodrigo-Navarro A, Petaroudi M et al (2018) Bacteria-based materials for stem cell engineering. Adv Mater 30(43):1804310

He J, Zhang X, Qian Y et al (2022) An engineered quorum-sensing-based whole-cell biosensor for active degradation of organophosphates. Biosens Bioelectron 206:114085

Hekmat D, Feuchtinger A, Stephan M et al (2004) Biofilm population dynamics in a trickle-bed bioreactor used for the biodegradation of aromatic hydrocarbons from waste gas under transient conditions. Biodegradation 15(2):133–144

Heveran CM, Williams SL, Qiu J et al (2020) Biomineralization and successive regeneration of engineered living building materials. Matter 2(2):481–494

Hietala AM, Nagy NE, Steffenrem A et al (2009) Spatial patterns in hyphal growth and substrate exploitation within norway spruce stems colonized by the pathogenic white-rot fungus *Heterobasidion parviporum*. Appl Environ Microbiol 75(12):4069–4078

Homan KA, Kolesky DB, Skylar-Scott MA et al (2016) Bioprinting of 3D convoluted renal proximal tubules on perfusable chips. Sci Rep 6:34845

Hong SH, Lee J, Wood TK (2010) Engineering global regulator Hha of *Escherichia coli* to control biofilm dispersal. Microb Biotechnol 3(6):717–728

Hsu CY, Chen BK, Hu RH et al (2016) Systematic design of a quorum sensing-based biosensor for enhanced detection of metal ion in *Escherichia Coli*. IEEE Trans Biomed Circuits Syst 10(3):593–601

Huang Z-M, Zhang YZ, Kotaki M et al (2003) A review on polymer nanofibers by electrospinning and their applications in nanocomposites. Compos Sci Technol 63(15):2223–2253

Huang J, Liu S, Zhang C et al (2019) Programmable and printable *Bacillus subtilis* biofilms as engineered living materials. Nat Chem Biol 15(1):34–41

Jakus AE, Rutz AL, Jordan SW et al (2016) Hyperelastic "bone": a highly versatile, growth factor-free, osteoregenerative, scalable, and surgically friendly biomaterial. Sci Transl Med 8(358):358ra127

Ji D, Lin Y, Guo X et al (2024) Electrospinning of nanofibres. Nature Rev Methods Primers 4(1):1

Jiang N, Ding W, Zhu X et al (2024a) Lipid-encapsulated engineered bacteria living materials inhibit COX-2 to enhance DOX toxicity. BioDesign Res 0(ja)

Jiang N, Zhao K, Liu C et al (2024b) Tumor microenvironment responsive multifunctional smart living materials based on engineered bacteria for inducing macrophage polarization to enhance tumor immunotherapy. Chem Eng J 488:150820

Junker LM, Toba FA, Hay AG (2007) Transcription in *Escherichia coli* PHL628 biofilms. FEMS Microbiol Lett 268(2):237–243
Kang H-W, Lee SJ, Ko IK et al (2016) A 3D bioprinting system to produce human-scale tissue constructs with structural integrity. Nat Biotechnol 34(3):312–319
Kang S-Y, Pokhrel A, Bratsch S et al (2021) Engineering *Bacillus subtilis* for the formation of a durable living biocomposite material. Nat Commun 12(1):7133
Koo Y, Kim G (2016) New strategy for enhancing in situ cell viability of cell-printing process via piezoelectric transducer-assisted three-dimensional printing. Biofabrication 8(2):025010
Kozicki MNRR, Whidden TK et al (1995) Directed growth of Uromyces hyphae on integrated circuit substrates. J Vac Sci Technol A 13:1808–1813
Kuang J, Bonanno S, Chang W-T et al (2024) Microbially driven reversible size- and color-changing materials. Matter 7(5):1848–1866
Letnik I, Avrahami R, Rokem JS et al (2015) Living composites of electrospun yeast cells for bioremediation and ethanol production. Biomacromolecules 16(10):3322–3328
Li S, Huang D, Zhang B et al (2014) Flexible supercapacitors based on bacterial cellulose paper electrodes. Adv Energy Mater 4(10):1301655
Liu X, Tang T-C, Tham E et al (2017) Stretchable living materials and devices with hydrogel-elastomer hybrids hosting programmed cells. Proc Natl Acad Sci USA 114(9):2200–2205
Liu Y, Wu M, Kang A et al (2024) Living seed materials made by metal–organic framework-encapsulated *Bacillus Subtilis* spore. Adv Funct Mater 34(11):2309288
Lu X, Wang C, Wei Y (2009) One-dimensional composite nanomaterials: synthesis by electrospinning and their applications. Small 5(21):2349–2370
Lu C, Jiaofang H, Hao D et al (2021) Synthetic biology modified microorganisms and biofilms used for heavy metal pollution detection and repair. Chin J Biotechnol 41(1):62–71
Lufton M, Bustan O, Eylon B-h et al (2018) Living bacteria in thermoresponsive gel for treating fungal infections. Adv Funct Mater 28(40):1801581
Ma Q, Yang Z, Pu M et al (2011) Engineering a novel c-di-GMP-binding protein for biofilm dispersal. Environ Microbiol 13(3):631–642
Ma L, Wang J, Wang S et al (2012) Synthesis of multiple *Pseudomonas aeruginosa* biofilm matrix exopolysaccharides is post-transcriptionally regulated. Environ Microbiol 14(8):1995–2005
Mangayil R, Rajala S, Pammo A et al (2017) Engineering and characterization of bacterial nanocellulose films as low cost and flexible sensor material. ACS Appl Mater Interfaces 9(22):19048–19056
Manjula-Basavanna A, Dural-Thatte AM, Joshi NS (2024) Mechanically tunable, compostable, healable and scalable engineered living materials. bioRxiv:2024.01.31.578234
McBee RM, Lucht M, Mukhitov N et al (2022) Engineering living and regenerative fungal-bacterial biocomposite structures. Nat Mater 21(4):471–478
Messner P, Sleytr UB (1992) Crystalline bacterial cell-surface layers. Adv Microb Physiol 33:213–275
Michon C, Langella P, Eijsink VG et al (2016) Display of recombinant proteins at the surface of lactic acid bacteria: strategies and applications. Microb Cell Factories 15:70
Mimee M, Nadeau P, Hayward A et al (2018) An ingestible bacterial-electronic system to monitor gastrointestinal health. Science 360(6391):915–918
Mirkhalaf M, Men Y, Wang R et al (2023) Personalized 3D printed bone scaffolds: a review. Acta Biomater 156:110–124
Mohsin MZ, Mohsin A, Zaman WQ et al (2024) Efficient toluene degradation using *Bacillus subtilis* biofilm-supported Mn–Ce/zeolite catalysts. Green Synth Catalysis
Mora CA, Herzog AF, Raso RA et al (2015) Programmable living material containing reporter micro-organisms permits quantitative detection of oligosaccharides. Biomaterials 61:1–9
Moraïs S, Shterzer N, Lamed R et al (2014) A combined cell-consortium approach for lignocellulose degradation by specialized *Lactobacillus plantarum* cells. Biotechnol Biofuels 7:112
Nguyen PQ, Botyanszki Z, Tay PKR et al (2014) Programmable biofilm-based materials from engineered curli nanofibres. Nat Commun 5(1):4945

Nodvig CS, Nielsen JB, Kogle ME et al (2015) A CRISPR-Cas9 system for genetic engineering of filamentous fungi. PLoS One 10(17):e0133085

Omer R, Mohsin MZ, Mohsin A et al (2022) Engineered bacteria-based living materials for biotherapeutic applications. Front Bioeng Biotechnol 10:870675

Ortega JS, Corrales-Orovio R, Ralph P et al (2023) Photosynthetic microorganisms for the oxygenation of advanced 3D bioprinted tissues. Acta Biomater 165:180–196

Park M, Tsai SL, Chen W (2013) Microbial biosensors: engineered microorganisms as the sensing machinery. Sensors (Basel, Switzerland) 13(5):5777–5795

Pawar SS, Vongkumpeang T, Grey C et al (2015) Biofilm formation by designed co-cultures of *Caldicellulosiruptor* species as a means to improve hydrogen productivity. Biotechnol Biofuels 8:19

Pitta KC, Panahkhahi S, Putra NE et al (2023) Biomimetic approaches for the design and fabrication of bone-to-soft tissue interfaces. ACS Biomater Sci Eng 9(7):3810–3831

Qian M, Lei H, Villota E et al (2021) High yield production of nanocrystalline cellulose by microwave-assisted dilute-acid pretreatment. Combined with enzymatic hydrolysis. Chem Eng Process 160:108292

Raut SH, Sarode DD, Lele SS (2014) Biocalcification using *B. pasteurii* for strengthening brick masonry civil engineering structures. World J Microbiol Biotechnol 30(1):191–200

Rodrigo-Navarro A, Rico P, Saadeddin A et al (2014) Living biointerfaces based on non-pathogenic bacteria to direct cell differentiation. Sci Rep 4(1):5849

Rodrigo-Navarro A, Sankaran S, Dalby MJ et al (2021) Engineered living biomaterials. Nat Rev Mater 6(12):1175–1190

Sankaran S, Zhao S, Muth C et al (2018) Toward light-regulated living biomaterials. Adv Sci 5(8):1800383

Schulz-Schönhagen K, Lobsiger N, Stark WJ (2019) Continuous production of a shelf-stable living material as a biosensor platform. Adv Mater Technol 4(8):1900266

Seker UO, Chen AY, Citorik RJ et al (2017) Synthetic biogenesis of bacterial amyloid nanomaterials with tunable inorganic-organic interfaces and electrical conductivity. ACS Synth Biol 6(2):266–275

Sen P, Nath A, Bhattacharjee C (2017) Packed-bed bioreactor and its application in dairy, food, and beverage industry. In: Larroche C, Sanromán MÁ, Du G, Pandey A (eds) Current developments in biotechnology and bioengineering. Elsevier, London, pp 235–277

Shah J, Brown RM (2005) Towards electronic paper displays made from microbial cellulose. Appl Microbiol Biotechnol 66(4):352–355

Singh R, Paul D, Jain RK (2006) Biofilms: implications in bioremediation. Trends Microbiol 14(9):389–397

Sleytr UB, Huber C, Ilk N et al (2007) S-layers as a tool kit for nanobiotechnological applications. FEMS Microbiol Lett 267(2):131–144

Srubar WV (2021) Engineered living materials: taxonomies and emerging trends. Trends Biotechnol 39(6):574–583

Stoodley P, Sauer K, Davies DG et al (2002) Biofilms as complex differentiated communities. Ann Rev Microbiol 56:187–209

Tang T-C, An B, Huang Y et al (2021) Materials design by synthetic biology. Nat Rev Mater 6(2):332–350

Tay PKR, Nguyen PQ, Joshi NSA (2017) Synthetic circuit for mercury bioremediation using self-assembling functional amyloids. ACS Synth Biol 6(10):1841–1850

Tian B, Liu J, Dvir T et al (2012) Macroporous nanowire nanoelectronic scaffolds for synthetic tissues. Nat Mater 11:986

Valentini L, Bon SB, Signetti S et al (2016a) Fermentation based carbon nanotube multifunctional bionic composites. Sci Rep 6:27031

Valentini L, Bittolo Bon S, Signetti S et al (2016b) Graphene-based bionic composites with multifunctional and repairing properties. ACS Appl Mater Interfaces 8(12):7607–7612

Walker KT, Li IS, Keane J et al (2024) Self-pigmenting textiles grown from cellulose-producing bacteria with engineered tyrosinase expression. Nat Biotechnol

Wang X, Pu J, Liu Y et al (2019) Immobilization of functional nano-objects in living engineered bacterial biofilms for catalytic applications. Natl Sci Rev 6(5):929–943

Wei W, Liu X, Sun P et al (2014) Simple whole-cell biodetection and bioremediation of heavy metals based on an engineered lead-specific operon. Environ Sci Technol 48(6):3363–3371

Wei Y, Han Z, Mao X (2024) Injectable living probiotic dressing built by droplet-based microfluidics and photo-cross-linking to prevent pathogenic infection and promote wound repair. Adv Healthc Mater 13(4):2302423

Widmaier DM, Tullman-Ercek D, Mirsky EA et al (2009) Engineering the Salmonella type III secretion system to export spider silk monomers. Mol Syst Biol 5:309

Wood TL, Guha R, Tang L et al (2016) Living biofouling-resistant membranes as a model for the beneficial use of engineered biofilms. Proc Natl Acad Sci 113(20):E2802–E2811

Xie S, Tai S, Song H et al (2016) Genetically engineering of *Escherichia coli* and immobilization on electrospun fibers for drug delivery purposes. J Mater Chem B 4(42):6820–6829

Xiongfa J, Hao Z, Liming Z et al (2018) Recent advances in 3D bioprinting for the regeneration of functional cartilage. Regen Med 13(1):73–87

Xue Y, Qiu T, Sun Z et al (2022) Mercury bioremediation by engineered *Pseudomonas putida* KT2440 with adaptationally optimized biosecurity circuit. Environ Microbiol 24(7):3022–3036

Yi W, Liu XW, Li YH et al (2009) Remodeling bacterial polysaccharides by metabolic pathway engineering. Proc Natl Acad Sci USA 106(11):4207–4212

Yu J, Huang TR, Lim ZH et al (2016) Production of hollow bacterial cellulose microspheres using microfluidics to form an injectable porous scaffold for wound healing. Adv Healthc Mater 5(23):2983–2992

Zhai L, Fu L, Wei W et al (2024) Advances of bacterial biomaterials for disease therapy. ACS Synth Biol 13(5):1400–1411

Zhang C, Huang J, Zhang J et al (2019) Engineered *Bacillus subtilis* biofilms as living glues. Mater Today 28:40–48

Zhang Y, Zou ZP, Chen SY et al (2022) Design and optimization of *E. coli* artificial genetic circuits for detection of explosive composition 2,4-dinitrotoluene. Biosens Bioelectron 207:114205

Zhu X, Xiang Q, Chen L et al (2024) Engineered *Bacillus subtilis* Biofilm@Biochar living materials for in-situ sensing and bioremediation of heavy metal ions pollution. J Hazard Mater 465:133119

Zou ZP, Du Y, Fang TT et al (2023) Biomarker-responsive engineered probiotic diagnoses, records, and ameliorates inflammatory bowel disease in mice. Cell Host Microbe 31(2):199–212.e5

Chapter 5
Chemical Synthesis of Biobased Materials

Fangke Zhan and Liming Zhao

With the rapid growth of the world population, the global demand for energy and new materials has sharply increased. Most of today's chemicals and polymers still come from fossil resources; about 8% of petroleum is used to manufacture polymers. However, the production and use of petroleum-based materials often accompany the generation of a large amount of pollutants, and their waste is also increasing. Moreover, most polymers are discarded after use, and the long-term non-degradability of these products poses a great threat to the earth's ecological environment, making the environment on which humans depend for survival increasingly deteriorate. According to statistics, tens of millions of tons of plastic waste will be emitted globally every year. People are increasingly concerned about the adverse environmental and socioeconomic consequences of petrochemicals, and the lifespan of fossil resources is also an important factor that people have to consider. Therefore, it is of extraordinary significance to use natural biomass polymers as research objects, explore their different synthesis techniques, and study the innovative development of corresponding raw materials. In this chapter, we will focus on the introduction of biobased materials and their chemical synthesis methods, as well as the main degradable biobased materials' chemical synthesis routes, and summarize some practical problems in these materials' synthesis and the latest synthesis processes.

F. Zhan
State Key Laboratory of Bioreactor Engineering, East China University of Science and Technology, Shanghai, China

L. Zhao (✉)
State Key Laboratory of Bioreactor Engineering, East China University of Science and Technology, Shanghai, China

Key Laboratory of Biobased Material Engineering of China Light Industry, East China University of Science and Technology, Shanghai, China
e-mail: zhaoliming@ecust.edu.cn

L. Zhao (ed.), *Research and Applications of Bio-based Degradable Materials*,
https://doi.org/10.1007/978-981-95-1188-4_5

5.1 Biobased Polyesters

Polyester is a general term for a class of polymers obtained by the condensation of polyols and polyacids. It has excellent performance and a wide range of uses. It is a common engineering plastic, widely used in laminated resins, hot pressing molding, surface coating, films, rubber, fibers, plasticizers, and other fields. Polyester mainly includes polyester resins and polyester elastomers. Polyester mainly refers to polyethylene terephthalate (PET) and conventionally includes linear thermoplastic resins such as polyarylate (PAR) and polybutylene terephthalate (PBT). This section only summarizes and demonstrates the chemical synthesis process of biodegradable biobased polyesters derived from biomass. The commonly reported biobased degradable polyesters include polylactic acid (PLA), polybutylene succinate (PBS), and poly β-hydroxybutyrate (PHB). Therefore, the main chemical synthesis pathways of the above three materials and their synthesis methods will be summarized and sorted out. The advantages and disadvantages will also be summarized and stated.

5.1.1 Polylactic Acid (PLA)

PLA, also known as polylactic acid, is a polyester polymer obtained by polymerization reaction with lactic acid as the main raw material. It is currently the most researched new type of biodegradable material. In recent years, PLA has won the favor of the global plastics industry with its excellent performance, wide range of applications, significant social benefits, and green economic characteristics. The exploration and application of its related materials in the field of biobased materials are increasing, and its related products are of great significance to human sustainable development. The chemical synthesis methods of PLA include the polylactide ring-opening polymerization method (Fig. 5.1) and the direct condensation method of lactic acid.

1. *Propylene Glycol*

Polylactide is a cyclic dimer of lactic acid, called 3,6-dimethyl-1,4-dioxane-2,5-dione. Polylactide is polymerized from lactic acid, so there are three different optical isomers of the cyclic dimer of lactic acid (Fig. 5.2): L-polylactide, D-polylactide,

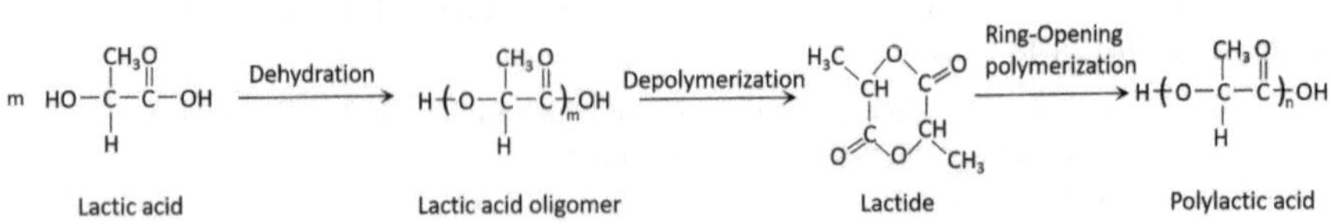

Fig. 5.1 Polylactic acid prepared by ring-opening polymerization of lactide

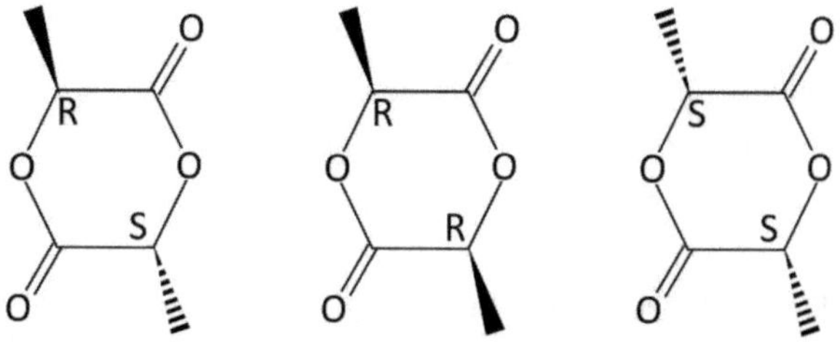

Fig. 5.2 Enantiomers of lactide

Table 5.1 Physical properties of lactide (Bailly 2002; Dorgan et al. 2000)

Physical properties	L-Propylene glycol	D-Propylene glycol	Racemic Propylene glycol	Levorotatory propylene glycol
Molar mass (g/mol)	144.12	144.12	144.12	/
Melting point (°C)	96–97	96	53	125–127
Heat of fusion (J/g)	/	146	128	185
Density (g/ml)	/	1.32–1.38	1.32–1.38	/
Viscosity (mPa·s)	/	2.71 (110 °C)	/	/
Optical rotation (*)	+260	−260	/	/

and meso-polylactide. The polylactide formed by the mixture of L-type and D-type is called racemic polylactide (Table 5.1).

2. *Preparation of Propylene Glycol*

Pelouze first proposed the synthesis method of propylene glycol in 1845. Lactic acid produces insoluble substances in the process of heating, dehydration, and self-esterification of lactic acid; water-insoluble substances are produced; the crystals obtained after distillation are named lactide. Propylene glycol is generally synthesized from lactate or directly from lactic acid, but the use of lactate as a raw material to produce propylene glycol is too expensive, and impurities such as acetic acid are easily mixed in the product, with low yield and high purification cost, and it is rarely used in industry. The use of lactic acid produced by biological fermentation as the raw material for the production of propylene glycol can effectively reduce its production cost (Sinclair 1977), and make the reaction more economical while obtaining a higher yield. The specific process flow of propylene glycol production is shown in Fig. 5.3: First, lactic acid is esterified and dehydrated to form low molecular weight lactic acid, and then propylene glycol is obtained by cracking the polymer under high temperature through a catalyst.

The methods for preparing propylene glycol are generally divided into the normal pressure method and the reduced pressure method. The normal pressure method is to introduce inert gas into the reaction system and use these inert gases to carry

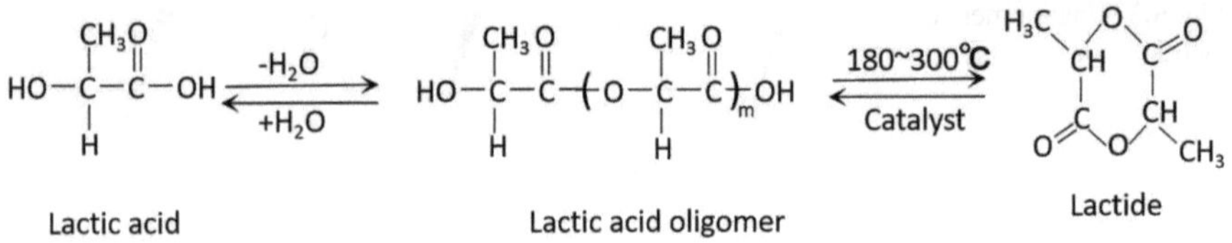

Fig. 5.3 Schematic diagram of propylene glycol preparation. (Wei et al. 2012)

propylene glycol out of the reaction system. Okuyama used N_2 and CO_2 as inert gases to introduce into the reaction system, and continuously carried out the generated propylene glycol from the reaction system by reducing the partial pressure of propylene glycol vapor (Lu et al. 2008). Due to the presence of inert gas, the discoloration and carbonization reactions caused by oxidation are avoided, thereby improving the yield and purity of propylene glycol. Although the normal pressure method has low technical difficulty and simple operation, it has low production efficiency, long dehydration time, and low product yield.

The reduced pressure method is currently the most widely used method in the synthesis process of propylene glycol. This method mainly places the reaction system in a high vacuum environment, quickly evaporates the reaction-generated propylene glycol, and avoids the oxidation effect of O_2, thereby improving the yield of propylene glycol (Chaorong and Huanfeng 2010). Some researchers have found that adding a certain amount of inert solution (such as glycerin and alcohol solution) to the reaction system based on the reduced pressure method can effectively improve the problem of easy carbonization and carbonization of the reaction liquid, and allow lactide to quickly evaporate (Tao and Chaowen 1998; Nan et al. 2005; Hualin et al. 2005; Weiwei 2009). In addition, researchers have also studied the impact of propylene glycol depolymerization on the molecular weight of low molecular weight lactic acid. They believe that when the molecular weight of low molecular weight lactic acid is small, less propylene glycol and more lactic acid are produced by depolymerization; however, when the molecular weight of low molecular weight lactic acid is too large, it exceeds the range suitable for depolymerization, and more residues are produced during depolymerization, resulting in a lower yield of propylene glycol (Jiangtao and Haihong 2009). However, researchers have not yet reached a consensus on the numerical value of the viscosity-average molecular weight of lactic acid oligomers that is most suitable for depolymerization.

3. *Purification of Lactide*

The lactide produced by the high-temperature cracking of lactic acid generally contains impurities such as water, lactic acid, and lactic acid oligomers. These impurities can cause lactide to re-hydrolyze into lactic acid, making it difficult for lactide to polymerize into high molecular weight PLA. Therefore, how to remove the impurities in the crude lactide becomes a key issue in the purification of lactide.

The methods for purifying lactide mainly include solvent recrystallization, melt crystallization, and distillation.

(1) Solvent Recrystallization

Recrystallizing lactide in a mixture of toluene and acetic acid is a common method for preparing lactide in the laboratory. In addition to toluene and acetic acid, ethyl acetate, ethanol, and other solvents can also be used for the recrystallization of lactide (Nan et al. 2003). Li Xia and others (2011) obtained lactide with high yield and optical purity in the laboratory by using ethyl acetate and ethanol as solvents, combined with hydrolysis and recrystallization methods. The optical purity reached 99.5%, and the recrystallization yield reached 44.6%, which meets the standard for preparing PLA monomers.

(2) Melt Crystallization

Melt crystallization is a method of purifying lactide by utilizing the differences in the freezing points of the components or key components of the substance to be separated. It does not require the addition of an external solvent, which has low energy consumption, low operating temperature, high selectivity, less environmental pollution, and can prepare high-purity or ultra-pure products. It is a green purification technology. Stepwise melt crystallization generally consists of two processes: crystallization and sweating. The crude lactide containing impurities is heated and melted in a collection tank, then pumped to the top of the film crystallizer for stepwise cooling crystallization. First, the temperature is reduced to 90 °C within 5 min to induce lactide crystallization; then crystallization cooling is carried out at a condensation rate of 5 °C/min. When the liquid level in the collection tank drops to 1/5–1/4 of the initial amount, the crystallization ends and the mother liquor is discharged; then, the temperature is raised to sweat the formed crystals. When the sweat is 5–10% of the initial amount, the sweating ends and the sweat is discharged; finally, the temperature is raised until all the crystals are melted, and one purification process is completed (Qimei and Chonghui 2015). Typically, following two purification processes, lactide can reach a very high purity, with a mass fraction of up to 99.99%, and no meso-lactide can be detected in the product.

(3) Distillation

Distillation is another important method for purifying lactide. This method involves adding a mixture of lactide, water, lactic acid, and lactic acid oligomers to a vacuum device, and distilling and separating them according to the different residence times of different substances in the distillation tower (Abe et al. 2004). The distillation equipment is easy to scale up and continuous industrial production, but due to the accumulation of impurities that affect the reaction rate in the reaction tower (such as meso-lactide), and the boiling points of lactide and meso-lactide are very close, this method has high requirements for equipment (Zhao et al. 2021).

4. *Synthesis of Polylactic Acid*

As early as 1845, Pelouze obtained lactic acid linear dimer-lactyl lactic acid by dehydrating lactic acid through distillation at high temperature (130 °C). It was not until 68 years later that scientist Nef first obtained low molecular weight PLA (Nef

1914) under low-pressure and high-temperature conditions by directly dehydrating and condensing lactic acid. This is also the origin of the condensation method currently used to prepare PLA. In 1932, Wallace Carothers, a scientist at DuPont, obtained PLA with a molecular weight of several thousand by ring-opening polymerization of lactic acid's cyclic dimer—propylene glycol, based on the work of his predecessors. Although its mechanical properties were still not up to industrial standards, it laid the foundation for Lowe's subsequent research. In 1954, Lowe from the same company further improved this technology and obtained high molecular weight PLA (Lowe 1954) for the first time. This is the precursor of the ring-opening polymerization method commonly used by current PLA production companies. From then on, PLA materials with practical value truly appeared. One of the initial criticisms of PLA was that, unlike other polyester fibers, it did not have good stability in humid environments and would slowly hydrolyze. However, decades later, this feature has become a widely recognized advantage among scientists.

Correspondingly, the industrial production of PLA began in 1932 at DuPont (Shaohui and Tao 2020), but the production cost was too high until 1987, when they developed a new PLA manufacturing process in collaboration with Cargill. In 2001, Cargill and Dow jointly invested in a PLA production line with an annual output of about 140,000 tons. Mitsui Chemicals of Japan followed closely, cooperating with Cargill in 2001, and other early participants included Shimadzu Corporation and Dainippon Ink and Chemicals (Nef 1914; Shaohui and Tao 2020). In addition, Germany's Uhde Inventa-Fischer and BASF, Italy's Snamprogetti, and the Netherlands' Hycail were also participants in the PLA production industry during the same period.

The domestic PLA industry started late, and the Shanghai Institute of Industrial Microbiology and the Jiangsu Institute of Microbiology were the first domestic institutions to use fermentation to produce lactic acid for PLA synthesis. According to data, China's PLA production reached 2 tons/year in 2002 (Shaohui and Tao 2020). Zhejiang Hisun Pharmaceutical Co., Ltd. cooperated with the Changchun Institute of Applied Chemistry of the Chinese Academy of Sciences in 2004 to conduct research on PLA preparation technology, carrying out in-depth cooperative research on lactic acid fermentation production, extraction, and polymerization, optimizing industrial technology, and under the requirements of national environmental protection production and clean production, they built the largest degradable PLA resin industrial production demonstration line in the country, with an output of up to 5000 tons/year (now expanded to an annual output of 15,000 tons). The establishment of this production line marked China's alignment with international standards in PLA production. In 2013, Shanghai Tongjie Liang Biomaterials Co., Ltd. established a 10,000-ton production line for developing different grades of PLA resin in Ma'anshan, Anhui, and the quality of various resins (injection grade, sheet grade, film grade, fiber grade) produced is comparable to domestic and foreign manufacturers. In 2016, COFCO Group invested 850 million yuan to build a 30,000-ton PLA raw material project and a 30,000-ton PLA downstream product project. There is also a 50000-ton PLA production line built by Jiangsu Yunyoucheng Bio-environmental Materials Co., Ltd. in 2012.

PLA, due to its excellent biocompatibility and biodegradability, has been widely used in various fields. Although numerous PLA production lines have been established domestically and abroad, the still immature industrial chain cannot meet the industry demand. Currently, PLA has become an important biodegradable polymer medical material that can be used in the human body, and is widely used in the field of biomedical polymers (Gadda et al. 2014; Nofar et al. 2019; Liu et al. 2020). At the same time, due to the high strength of PLA, high modulus, transparency, and ease of processing, it is also widely used in the field of general plastics. Currently, the mature PLA preparation in the industry all adopts the two-step method of fermentation–chemical polymerization, that is, first obtaining and separating lactic acid through bacterial fermentation of sugars in renewable resources such as corn and sugarcane (Södergård and Stolt 2010), and then using chemical methods to polymerize lactic acid into homopolymers or copolymers. The general process is ① microbial fermentation to produce lactic acid; ② purify lactic acid, then prepare its cyclic dimer (lactide); and ③ lactic acid condensation or lactide ring-opening polymerization to form PLA.

5. *Chemical Synthesis of Polylactic Acid*

The pathways for chemically synthesizing high molecular weight PLA mainly include three types (Fig. 5.4): direct condensation of free lactic acid, azeotropic dehydration condensation, and lactide ring-opening polymerization.

(1) Direct Condensation Method

The direct condensation method of lactic acid refers to a synthesis process in which lactic acid (chemically or fermentation-produced) is heated under vacuum at 50 °C, and in the presence of a dehydrating agent, the hydroxyl and carboxyl groups between lactic acid molecules undergo direct condensation dehydration reaction

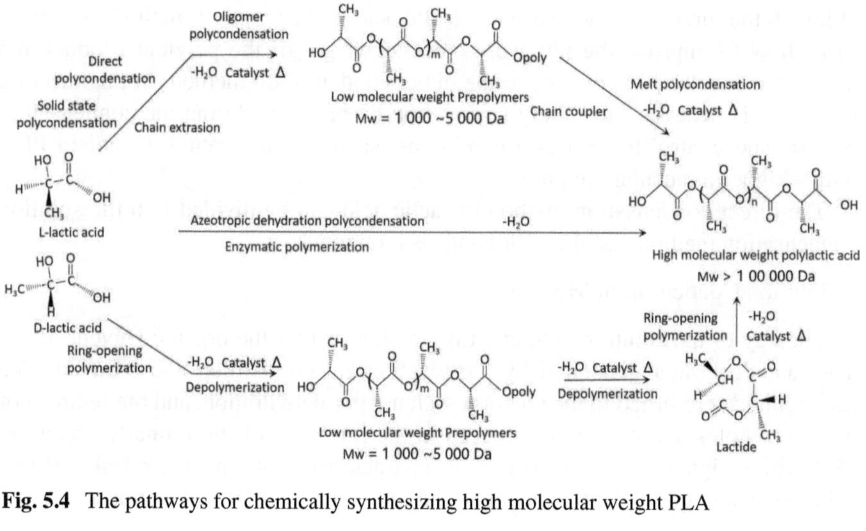

Fig. 5.4 The pathways for chemically synthesizing high molecular weight PLA

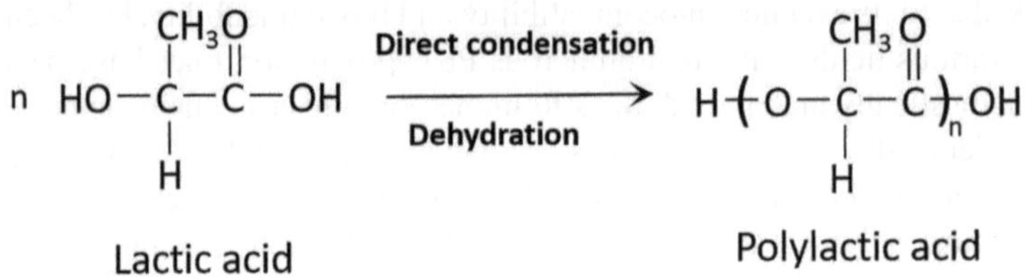

Fig. 5.5 Preparation of polylactic acid by direct condensation method. (Liu et al. 2020; Södergård and Stolt 2010; Auras et al. 2004; Lasprilla et al. 2012; Li et al. 2020)

(Fig. 5.5). This reaction involves a balance between condensation and depolymerization, and it is a reversible process. The entire system contains free lactic acid, free water, PLA, and lactide, but the viscosity of the system will continue to increase as the reaction proceeds, and many distillation by-products (such as water and lactoyl lactic acid) (Liu et al. 2020; Södergård and Stolt 2010; Auras et al. 2004; Lasprilla et al. 2012; Li et al. 2020) will be produced. Compared with the ring-opening polymerization of propylene carbonate, the direct condensation method has abundant raw material sources, a simple process flow, and a higher conversion rate of lactic acid monomers. It does not even require the purification of intermediates, making it cost-effective. The main problem with the direct condensation method is that the coexistence system of free lactic acid, water, oligomers, and propylene carbonate is relatively stable, and the by-products, such as water generated by the reaction, are difficult to remove in the viscous melt. Because it is difficult to remove by-products in the viscous system (even under vacuum conditions), the PLA obtained by this method has a low molecular weight, low mechanical properties, a wide distribution range, and requires a long reaction time. The direct impact of by-product retention makes the reaction difficult to proceed in the forward direction; the relative molecular weight of the produced PLA is less than 5000 Da, and the color is impure when the polymerization temperature is higher than 180 °C. Therefore, although the direct polymerization of lactic acid is the simplest method to prepare PLA, how to improve the relative molecular weight of the polymer product and efficiently decolorize is the key to the direct condensation method. In this reaction system, it is necessary to timely remove the by-products during the condensation process and control the reaction conditions, such as temperature, to obtain PLA with a higher molecular weight.

The direct condensation method of lactic acid can be divided into the solution condensation method and the melt condensation method.

① Solution Condensation Method

The key to the solution condensation method is that the organic solvent, lactic acid, and the water generated by the reaction are co-distilled and refluxed. The reflux liquid is returned to the reaction system after dehydration, and the removal of water promotes the reaction to proceed in the forward direction, finally preparing PLA with a higher molecular weight. Its characteristics are that the reaction is in a solution state, good heat exchange, the polymer will not decompose due to temperature changes, and the cost is lower than the ring-opening method. However, the solution condensation method also has some disadvantages, such as being easily affected by solvent impurities and various side reactions (including racemization and transesterification). In addition, this method also consumes a large amount of organic solvents, which will increase the pressure of environmental pollution.

The organic solvent added in this method needs to meet two conditions: One is that it cannot participate in the reaction, and the other is that it can dissolve the polymer and co-boil with water. In the entire reaction process, the added solvent mainly plays three roles (Keyou et al. 1999): one is to dissolve the lactic acid monomer, so that the growing PLA is dissolved or swollen, promoting the growth of the PLA chain; the second is to reduce the viscosity of the entire reaction system, absorb the heat released by the reaction, and reduce the intensity of the reaction; the third is to form azeotrope with the by-product water in the reaction system, so that the water can be gradually removed with the increase of reflux time. Ajioka et al. (1995) optimized the synthesis process of continuous azeotropic dehydration condensation, and obtained PLA with a molecular weight of more than 300,000 Da through this process, and the obtained PLA has good mechanical properties.

② Melt Condensation Method

The melt condensation method refers to the method of dehydrating lactic acid in a molten state to form PLA. The by-products, such as water generated in this system are discharged by vacuuming or carried by inert gas. Melt polymerization is a relatively economical and easy-to-control process, which is affected by factors such as reaction temperature, reaction time, catalyst, and pressure. Commonly used esterification reaction catalysts include medium-strong acids (such as H_2SO_4 and H_3PO_4), transition metals and their oxides (Sn, Zn, SnO_2, ZnO), transition metal salts ($SnCl_2$, $SnCl_4$), metal organics (stannous octoate, triethylaluminium), etc.

Moon et al. (2000) used L-lactic acid as the raw material to prepare PLLA by melt polycondensation, and studied the effect of different catalyst systems on the reaction. They found that tin oxide and tin chloride can effectively increase the molecular weight of PLA. The study found that by adding protonic acid to the reaction system at 180 °C to activate the Sn(II) base catalyst (SnO and $SnCl_2 \cdot 2H_2O$), a PLLA with a molecular weight of about 100,000 Da was finally obtained. The introduction of protonic acid as a co-catalyst enhanced the activity of the Sn(II) compound, and effectively inhibited the discoloration of the product and the racemization problem during the melt polycondensation process of lactic acid under continuous high temperature.

In recent years, researchers have hoped to increase the molecular weight of PLA by increasing the reaction time, reducing the reaction temperature, using better catalysts, or optimizing new processes. The reactions involving the optimization of these conditions mainly include melt-solid-phase polymerization, chain extension reactions, etc.

Melt-solid-phase polymerization (Moon et al. 2001; Yaoming et al. 2002). This method is to further polymerize the prepolymer at a temperature higher than its Tg (glass transition temperature) and lower than its Tm (melting temperature) to obtain the target polymer. The principle is that the low molecular weight in the amorphous region reacts with the end groups of large molecules to form high polymers, and the molecular chains continue to grow and aggregate in the crystalline region, improving the molecular weight, purity, and crystallinity of PLA. This method avoids the occurrence of side reactions at high temperatures, the process is simple, and the cost is low, but the polycondensation time of this reaction is longer. Moon et al. (2001) catalyzed the polycondensation of L-lactic acid with $SnCl_2$ and p-toluenesulfonic acid at 180 °C for 5 h, obtained a prepolymer with a molecular weight of up to 20,000 Da, then heat treated it at 105 °C for 2 h to crystallize it, and solid-phase polymerized it at 150 °C for about 30 h, to obtain a PLLA with a molecular weight of about 500,000 Da.

Chain extension reaction (Chaoyang and Yaoming 2003) using chain extenders to extend the prepolymer can obtain PLA with a higher molecular weight. Chain extenders refer to bifunctional small organic molecules that can react quickly with the hydroxyl (–OH) or carboxyl (–COOH) groups at the end of the prepolymer chain, causing the PLA molecular weight to multiply in a short time. There are many types of chain extenders, and the chain extenders currently commonly used in PLA chain extension synthesis mainly include diisocyanates, diacyl chlorides, diazirines, dianhydrides, and diepoxides, etc. The chain extension polymerization method has obvious advantages, such as fast reaction speed and short reaction time; but the higher the reactivity of the chain extender, the worse its thermal stability often is. Woo et al. (1995) polycondensed L-lactic acid into a prepolymer with a molecular weight of 11,000 Da, and performed a chain extension reaction on it at below 160 °C with 1,6-hexamethylene diisocyanate (HDI) as the chain extender, to obtain a PLLA with a molecular weight of about 76,000 Da.

In summary, these one-step polymerization processes are relatively economical and easy to control, but they are equilibrium reactions affected by many parameters such as temperature, reaction time, catalyst, and pressure. These factors have a significant impact on the molecular weight of the product PLA. In addition, the water produced during the polymerization process can cause high molecular weight PLA to decompose at high temperatures. Therefore, the PLA produced by a one-step polymerization reaction usually has a lower molecular weight. To obtain high molecular weight PLA, it is necessary to pay attention to controlling reaction kinetics, removing the formed water, and preventing the degradation of PLA chains during the reaction process.

(2) Ring-Opening Polymerization of Lactide (ROP)

Due to the many disadvantages of the direct condensation method of lactic acid (such as the presence of free lactic acid, water and impurities in the system, the obtained polymer has a low molecular weight, the need to remove the generated water, and the reaction needs to be carried out under high vacuum and high temperature, which cause PLA to color and racemize), the ring-opening polymerization method of lactide is currently the commonly used method for preparing high molecular weight PLA. This method is divided into two steps: First, lactide is prepared by the dehydration cyclization of lactic acid; then, high molecular weight PLA is prepared by the ring-opening polymerization of refined lactide.

Lactic acid is a chiral molecule with L and D isomers, and its optical purity is an important factor in determining the physical properties of PLA. Lactic acid can form three forms of PLA: poly L-lactic acid (PLLA), poly D-lactic acid (PDLA), and poly D,L-lactic acid (PDLLA) (Griffith 2000). Among them, PLLA is highly regarded for its excellent biocompatibility and mechanical properties, but it also has disadvantages such as slow degradation rate, high crystallinity, and strong hydrophobicity (Singhvi et al. 2019). D-lactic acid degrades faster, so we can combine L-lactic acid and D-lactic acid monomers to produce PLA to avoid the aforementioned disadvantages of PLLA (Discher and Ahmed 2006; Saeidlou et al. 2012). The ring-opening polymerization of lactide can control the ratio and order of D-lactic acid and L-lactic acid units in the final PLA polymer by controlling the residence time, temperature, and the type and concentration of the catalyst, thereby regulating the molecular weight and molecular chain structure of PLA and its physical and chemical properties. Therefore, this method is currently used for the industrial production of PLA. However, the heavy metal-based catalysts used in this method are very likely to contaminate the product, making the purification process of PLA more complicated and also limiting the application of PLA in the fields of food packaging and biomedicine.

Depending on the catalyst used and the reaction mechanism, the ring-opening polymerization of lactide can be mainly divided into ionic ring-opening polymerization and coordination ring-opening polymerization.

Ionic ring-opening polymerization can be divided into cationic ring-opening polymerization and anionic ring-opening polymerization.

In cationic ring-opening polymerization, the cationic catalysts used for ionic ring-opening polymerization mainly include protonic acids (such as HCl and HBr), Lewis acids (such as $AlCl_3$ and $SnCl_4$), and alkylating agents (such as CF_3SO_3Me), three types of acidic substances (Basko et al. 2015). Cationic polymerization catalysts are electrophilic reagents, which can interact with the oxygen atoms in the lactide monomer, generate intermediate cations, which cause carbonyl oxygen bond breakage, and then initiate the ring-opening reaction of lactide to generate acyl cations, and initiate monomer growth.

In anionic ring-opening polymerization, the anionic catalysts used for propylene carbonate ring-opening polymerization mainly include alkali metals and alkali metal alcohol oxidants, such as sodium alcohol, potassium alcohol, and butyl lithium. The principle of anionic ring-opening polymerization is that the negative ions of the catalyst attack the acyl oxygen bond of propylene carbonate, forming an active center lactone anion, which then inserts into the main chain to trigger the chain growth of propylene carbonate. Its reaction characteristics are high initiation activity and fast speed, but there are many side reactions that easily produce racemization reactions, so it is not suitable for the synthesis of high molecular weight PLA.

Coordination polymerization, also known as Ziegler–Natta polymerization, is one of the most common and effective methods for preparing PLA. The propylene carbonate monomer first forms an activated coordination complex with the organometallic catalyst, and then inserts into the metal-carbon bond to generate a high molecular weight polymer. The initiators of coordination polymerization are mainly organometallic compounds of transition metals and lanthanide metals, such as alkyl (aryl) metals, alkoxy metals, carboxylates, and metal oxides (Singhvi et al. 2019). Alkoxy aluminum is an effective catalyst for lactone ring-opening polymerization, among which $Al(Oi\text{-}Pr)_3$ is widely used in lactic acid polymerization; while carboxylic acid tin compounds are the most widely used catalysts in PLA industrial production, such as octanoic acid tin, its activity is much greater than $Al(Oi\text{-}Pr)_3$. Although coordination polymerization can effectively produce PLA, some uncontrollable side reactions, such as ester alcoholysis (ester exchange) reaction, will also occur in the later stage of its reaction.

The catalyst is a key factor affecting the coordination polymerization reaction, so more and more researchers have begun to explore new catalysts that can effectively catalyze the ring-opening polymerization reaction of propylene carbonate. Because $Pr(OTf)_3$ itself has poor catalytic activity, Köhn et al. (Köhn et al. 2003) first reported a new catalyst—bis(trimethyltriazacyclohexane) trifluoromethanesulfonate praseodymium $[(Me_3TAC)_2Pr(OTf)_3](Cat)$, which can effectively catalyze the ring-opening polymerization reaction of D,L-propylene carbonate in different solvents (tetrahydrofuran, dichloromethane, ethyl acetate, and toluene). Under the conditions of 170 °C and a substrate to catalyst ratio of 1000, a reaction for 18 h, the highest yield of PLA can reach 95%, and the molecular weight can reach 18,000 Da. At present, there are few studies on copper-based catalysts for propylene carbonate polymerization. John et al. (2007) found that under the catalysis of {2-[1-(2,6-diethylphenylamino)ethyl]phenoxy}2Cu(II), the highest molecular weight of PLA produced by the ring-opening polymerization reaction of L-propylene carbonate can reach 26.3×10^3 Da, and the monomer conversion rate can reach 57%. In addition, researchers have found two other copper complexes {2-[1-(2,6-dimethylphenylamino)ethyl]phenoxy}2Cu(II) and {2-[1-(2-dimethylphenylamino)ethyl]phenoxy}2Cu(II). Cu(II) can also catalyze the ring-opening polymerization of L-lactide under solvent-free, molten conditions (160 °C), but they can only catalyze the production of medium molecular weight PLA (molecular weights of 12.0×10^3 Da and 15.9×10^3 Da).

In the commonly used PLA production process, PLA is generally formed by the ring-opening polymerization of lactide catalyzed by heavy metals, and lactide itself is derived from the conversion of fermented lactic acid. The residual traces of heavy metal catalysts not only adversely affect the polymerization reaction, but also severely affect the application of PLA in the medical and food fields. Therefore, it is urgent to develop safe and environmentally friendly new catalysts to replace the commonly used heavy metal catalysts for new PLA polymerization production.

With the deep implementation of green environmental protection concepts, green organic synthesis technology has also made considerable progress. In the new PLA polymerization technology, enzyme-catalyzed polymerization,supercritical CO_2 polymerization, and other organic ring-opening polymerization-related research have received widespread attention. In recent years, enzyme-catalyzed polymerization has become one of the most feasible methods to replace traditional polymer synthesis methods due to its environmentally friendly characteristics. Compared with organic synthesis, enzyme-catalyzed polymerization has many advantages (Jingchang et al. 2017), such as high catalytic activity, mild reaction conditions, fewer by-products, and better control of regioselectivity. In addition, highly specific enzyme-catalyzed reactions can synthesize finely structured polymers from cheap raw materials, while traditional chemical synthesis reactions require high-purity monomers, anhydrous conditions, and high temperatures to reduce the production of by-products.

The production of PLA by enzyme-catalyzed polymerization using lactic acid polymerase (LPE) is an ideal PLA biosynthesis process, which can be used to replace traditional chemical synthesis for PLA production. Therefore, the urgent task for researchers is to discover microorganisms that can produce PLA and excavate the lactic acid polymerization-related enzymes they carry.

Chanfreau and others found a method (Chanfreau et al. 2010) for producing poly L-lactide (PLLA) and poly L-lactide-co-glycolide (PLLGA) by enzyme-catalyzed polymerization in ionic liquid 1-hexyl-3-methylimidazolium hexafluorophosphate using lipase B from Antarctic yeast (*Candida antarctica*) (Novozyme 435). The reaction at 90 °C resulted in a molecular weight of PLLA of 37.89×10^3 g/mol as measured by size exclusion chromatography, with a maximum yield of 63% (Chanfreau et al. 2010).

Whulanza and others studied the role of yeast lipase (CRL) as a biological catalyst for the ring-opening polymerization of L-lactide in the production of metal-free PLA (Whulanza et al. 2018). They tested the effects of different temperatures and CRL concentrations on the ring-opening polymerization production of PLA, and the results showed a close relationship between the formation of PLA and CRL activity (Whulanza et al. 2018). At 90 °C and a CRL concentration of 2%, the activity of CRL in catalyzing the ring-opening polymerization of L-lactide was the highest; and under these conditions, the yield and molecular weight of PLA were also the highest (Badens et al. 2018).

Supercritical fluid is a new type of green and pollution-free solvent that can replace traditional organic solvents, and its use is safer and more environmentally friendly. Using supercritical CO_2 Using supercritical carbon dioxide ($ScCO_2$) as a solvent to synthesize PLA and its copolymers has various advantages such as adjustable process parameters, simple and efficient separation and purification, and mild preparation conditions. It is especially suitable for the preparation of biomedical polyester materials (Fan et al. 2017). Compared with conventional solvents, $ScCO_2$ has the following excellent characteristics: ① The reaction is inert. Because $ScCO_2$ is a chemically stable solvent, it will not participate in the synthesis reaction of PLA. ② Good solubility and swelling performance. Improving the solubility of the solute in the solvent is beneficial to improve the reaction efficiency. ③ Moderate critical conditions. The critical pressure of CO_2 is 7.38 MPa, and the pressure conditions of the polymerization reaction with $ScCO_2$ as the medium are mostly higher than this critical pressure, and higher pressure helps the polymerization reaction (Fan et al. 2017). ④ Simplify the product purification process. Compared with other solvents, when $ScCO_2$ is used as the reaction medium, this polymerization reaction system does not need to undergo additional purification processes (such as dissolution, precipitation, and filtration after the reaction).

Ferrari and others used L-lactide as a raw material, a copolymer monomer with a double bond (2-hydroxyethyl methacrylate) as an initiator, and carried out ring-opening polymerization in $ScCO_2$ to synthesize biomedical PLA macromolecules with a definite average chain length (Ferrari et al. 2014). This study used $ScCO_2$ as a solvent and carried out the polymerization reaction at a temperature of 90–130 °C. The results showed that low oligomers with a narrow molecular weight distribution can be generated at the lowest test temperature of 90 °C, and the adverse effects of secondary reactions on PLA synthesis are significantly reduced at this temperature, thus obtaining crude PLA macromolecules with higher purity.

In recent years, in order to facilitate industrial production, the synthesis of polylactic acid mainly focuses on research in efficient catalytic ring-opening polymerization systems, the synthesis of copolymers with new structures and compositions, etc., to prepare PLA with higher molecular weight (Dorgan et al. 2000). However, the characteristics of PLA depend on the purity of lactide, and an important key point to consider is that residual monomers will cause PLA hydrolysis. At present, the main obstacles to the commercial production of polylactic acid in China are the complexity of the production process, the long process, and the high cost of the product. If the production of polylactic acid is to be industrialized as soon as possible, the core is to reduce production costs, especially the cost of synthesizing polylactic acid. China is the second-largest grain producer in the world. According to China's sustainable development strategy, deep processing of food and production of high-value-added products is the only way to achieve leapfrog economic development. Therefore, using corn and other agricultural products as raw materials, the market potential for producing biodegradable polylactic acid using biotechnology is huge. However, the existing lactic acid varieties are all general-purpose racemic lactic acid, and the quality does not meet the polymerization requirements, so only by increasing research efforts in this area can China's products be aligned with international standards.

5.1.2 Polybutylene Succinate (PBS)

PBS is an important biodegradable synthetic polyester, which is prepared by an efficient condensation reaction of aliphatic dicarboxylic acids and their derivatives and aliphatic diols, and belongs to thermoplastic resins. PBS is a white crystalline polymer with a density of 1.27 g/cm^3, a melting point of 115 °C, a crystallinity of 30–60%, and a crystallization temperature of 75 °C. Its structural formula is shown in Fig. 5.6.

The traditional synthesis methods of PBS mainly include biological fermentation and chemical synthesis. The chemical synthesis method includes direct esterification and transesterification.

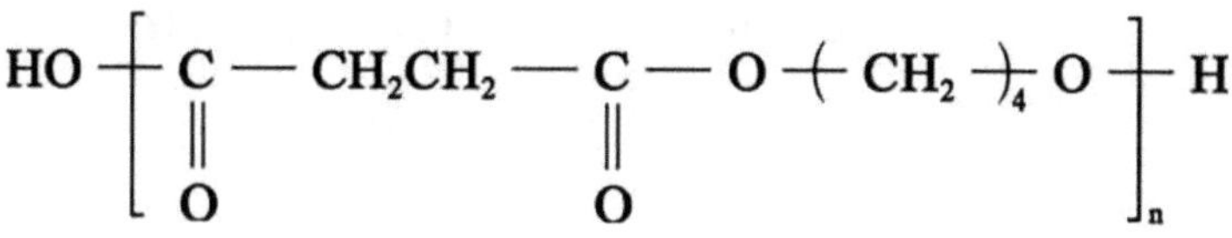

Fig. 5.6 PBS structural formula

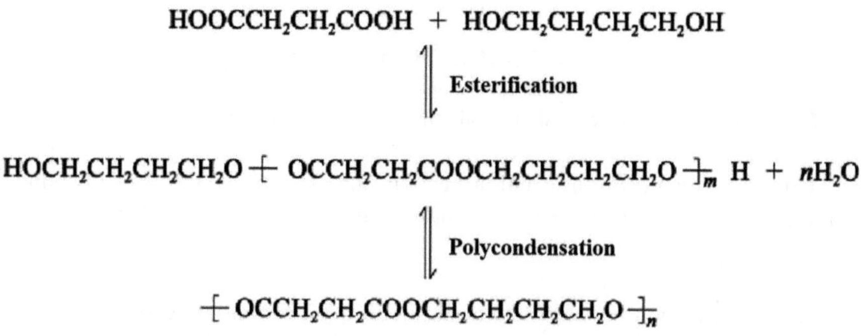

Fig. 5.7 Direct esterification method for preparing PBS polyester. (Changhui and Xia 2008)

1. *Direct Esterification*

Direct esterification is the process of direct polymerization of dicarboxylic acids and alkylene diols. This method is relatively simple and can produce PBS with a higher molecular weight, so it is widely used in industrial production. The method can be described as follows: Under lower reaction temperatures, dicarboxylic acids are esterified with an excess of diols to form carboxyl-terminated prepolymers, and then the diols are removed under high temperature, high vacuum, and the presence of a catalyst to obtain PBS polyester. The reaction equation is shown in Fig. 5.7. The catalysts used in the esterification process are mainly titanates and p-toluenesulfonic acid (Changhui and Xia 2008). The temperature during the polymerization stage is generally controlled at about 160 °C. Temperature control is a very important step because the cyclization of butanediol and high-temperature oxidation will cause the polymer to turn yellow. In addition to using temperature control to prevent the production of yellow esters, the method of cold traps in the reaction process, and the extraction of the by-product tetrahydrofuran can achieve the same effect. Other researchers have also reported reasons that can cause the polyester to turn yellow, such as the presence of $Ti(iOPr)_4$ and Ti(OBu) (Papageorgiou and Bikiaris 2005). Laboratory reactions add antioxidants during polymerization, and polyphosphoric acid is added during the second step of the condensation reaction to prevent side reactions such as etherification and thermal decomposition, which can improve the degree of yellowing of the ester. Currently, the synthesis of high molecular weight PBS is still a challenge, because various catalysts make the product toxic, and at the same time have a certain impact on the environment, so under the action of nontoxic

$$HOCH_2CH_2CH_2CH_2OH + H_3COOCCH_2CH_2COOCH_3$$

$$\rightleftharpoons \text{Transesterification}$$

$$HOCH_2CH_2CH_2CH_2O \left[OCCH_2CH_2COOCH_2CH_2CH_2CH_2O \right]_m H + nCH_3OH$$

$$\rightleftharpoons \text{Polycondensation}$$

$$\left[OCCH_2CH_2COOCH_2CH_2CH_2CH_2O \right]_n$$

Fig. 5.8 Transesterification method for synthesizing PBS polyester. (Jinbo 2013)

and harmless green environmental protection catalysts, the synthesis of PBS under low temperature and short time conditions has become a key research point.

2. *Transesterification*

Transesterification is one of the earliest and most commonly used methods for synthesizing PBS. The reaction equation for synthesizing PBS polyester by transesterification (Jinbo 2013) is shown in Fig. 5.8.

In the process of ester exchange synthesis, PBS is polymerized in a molten state from equal amounts of dimethyl succinate and 1,4-butanediol or an excess of 1,4-butanediol not exceeding 10%. This environment requires the presence of a catalyst. The synthesis steps can be roughly divided into two steps: ester exchange and polymerization. Before the reaction, nitrogen is introduced into the reactor in advance to exclude the air, to avoid oxidation during the esterification process. Then, under a nitrogen atmosphere, the reactor is heated to 150–190 °C, the esterification reaction occurs, and methanol and moisture in the reaction system can be removed. Subsequently, the polymerization reaction occurs at 200 °C under vacuum conditions. The relative molecular weight of the PBS obtained from the above reaction is 59,500 g/mol, and the weight-average relative molecular weight is 104,100 g/mol (Min et al. 2006).

The basic reaction steps included in the two methods of direct esterification and ester exchange are esterification and condensation reactions, with the difference lying in the first step of esterification. The ester exchange method achieves esterification by removing methanol through ester exchange, while the direct esterification method achieves esterification through alcohol-acid dehydration. Due to its low boiling point, methanol is volatile, which makes the reaction tend toward the completion of the ester exchange reaction. The advantage of the ester exchange method among the two methods is that it can effectively avoid the end-capping phenomenon caused by unreasonable reactant ratios. However, this method is more costly, requires the prior acquisition of dimethyl succinate, and the solvent it contains is highly toxic, which increases the potential for environmental issues. The advantage of the direct esterification method is more prominent; its product is water, it is non-polluting to the environment, and it has a lower cost.

In the 1970s, from the first synthesis of polybutylene succinate, the synthesis of PBS has undergone long and sufficient development (Changhui et al. 2008); chemical synthesis methods include ① melt polycondensation method (Yoo and Km 1988): succinic acid and butanediol esterify at a certain temperature, then undergo high temperature and high vacuum polycondensation. ② Solution polymerization method (Ajioka et al. 1998): using different solvents, let the solvent reflux at a constant temperature and time to remove some water, let succinic acid and butanediol esterify, and finally perform condensation reaction at high temperature. ③ Melt-solution combined method: Toluene is used as a solvent, after 24 h of solution esterification at 130–135 °C, toluene is distilled out, and high molecular weight copolymers are obtained by high-temperature and high-vacuum melt polymerization under catalytic reaction. ④ Ester exchange method: Dimethyl succinate and a certain amount of diol undergo ester exchange to remove methanol, and then condense into an ester. ⑤ Prepolymer chain extension method: The use of this method simplifies the operation of the reversible reactions of the direct esterification method and the ester exchange reaction method, which need to continuously exclude small molecular substances (water or methanol). The polyester obtained after the condensation reaction is of low molecular weight, which is due to the fact that during the condensation reaction, especially in the later stages of the reaction, the temperature is often higher than 200 °C. At this temperature, decarboxylation, thermal degradation, thermal oxidation, and other side reactions inevitably occur. The chain extension reaction can choose to use the active groups of the chain extender to react with the terminal hydroxyl groups of the polyester, in order to increase the relative molecular weight of the polyester. At the same time, when synthesizing PBS using the chain extension method, the conditions for the condensation reaction are more relaxed than the direct esterification method. However, the use of the chain extension method to produce PBS will reduce the biosafety of the material and affect its biodegradability (Pingkai 2012).

5.1.3 Poly(2,5-Furandicarboxylate) (PEF)

The chemical structure of PEF is shown in Fig. 5.9. There are two process routes for synthesizing PEF: The first is the ester exchange method using dimethyl 2,5-furandicarboxylate (DMFD) and ethylene glycol (EG) as raw materials; the

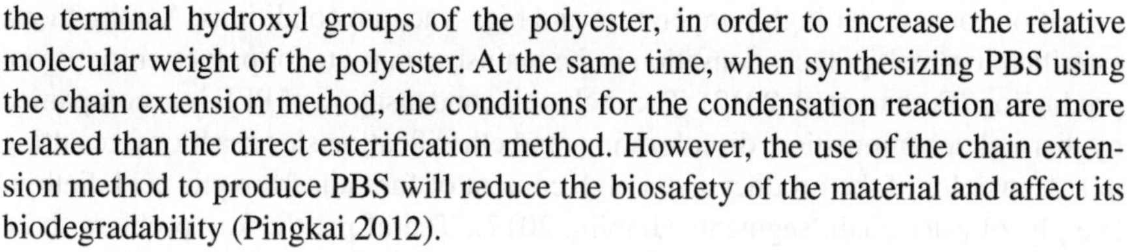

Fig. 5.9 Chemical structure of poly(2,5-furandicarboxylate)

second is the esterification method using FDCA (2,5-furandicarboxylic acid) and EG (ethylene glycol) as raw materials. The main raw materials for synthesizing PEF are DMFD, FDCA, and EG, and the main catalysts are tetrabutyl titanate, antimony trioxide, antimony acetate, and dibutyltin oxide. The polymerization reaction of PEF usually involves solution polycondensation, melt polycondensation, melt-solid-phase polycondensation, and ring-opening polycondensation methods (Xie Hongzhou and Linbo 2019).

1. *Solution Polycondensation Method*

The solution polycondensation method is a method of polycondensation in an inert solvent. Gandini et al. (2009) prepared PEF with a degree of polymerization of 70 using 2,5-furandicarbonyl chloride and ethylene glycol as raw materials, pyridine as a catalyst, and 1,1,2,2-tetrachloroethane as a solvent at room temperature. However, due to the huge challenges in the synthesis of 2,5-furandicarbonyl chloride monomer and the use and recovery of solvents in terms of technology, environmental protection, and energy consumption, and the low molecular weight of the product, this method has stopped at the industrial production of PEF.

2. *Melt Polycondensation Method*

This method includes two technical routes, namely, the esterification or ester exchange reaction of dicarboxylic acid or its diester with diol under normal pressure or pressure to generate the corresponding prepolymer, and then the melt polycondensation route under high-temperature and high-vacuum conditions. Among them, esterification-melt polycondensation is the mainstream route for polyester industrial production (Eerhart et al. 2012). The melt polycondensation of PEF is accompanied by thermal decomposition, discoloration, and etherification side reactions, resulting in a deep color of the product and a high content of furandicarboxylic acid diethylene glycol ester chain segments (Jiaping 2017). Therefore, the key problem to be solved in the synthesis of PEF by the melt polycondensation method is to first select the appropriate polymerization process and catalyst to prepare high molecular weight PEF under suitable process conditions, while strictly suppressing the above side reactions. At present, most PEF synthesis research chooses the above-mentioned ester exchange route; the reason is that the current FDCA refining technology is not mature. In addition, another important factor is that the ester exchange route is conducive to suppressing the occurrence of the above side reactions.

3. *Melt-Solid-Phase Polycondensation Method*

Knoop et al. (2013) synthesized PEF prepolymer with a number average molecular weight of 5000 g/mol using isopropyl titanate as a catalyst through ester exchange-melt polycondensation, and then carried out solid-phase polycondensation reaction under high vacuum at 180 °C for 72 h, the molecular weight of PEF can be increased to 3000 g/mol. Van Berkel et al. (2015) used propyl ester–antimony trioxide as a catalyst, and prepared PEF with an intrinsic viscosity of 0.80 dL/g through ester exchange-melt/solid-phase polycondensation. Hong et al. (2016) By extending the solid-state reaction time to 48 h, the intrinsic viscosity of PEF was successfully increased to 0.72 dL/g. Kasmi et al. (2018) Through melt crystallization technology, a secondary solid-phase condensation was carried out at 205 °C for 6 h, and the final PEF polyester's high molecular weight intrinsic viscosity exceeded 1 dL/g. The advantage of solid-phase condensation is that it can significantly increase the molecular weight of PEF and suppress side reactions, but it also has obvious disadvantages, namely, the reaction speed is low, and the reaction time is too long. This situation inevitably leads to a large reactor, high investment and production costs, and it is difficult to become the first choice of process for PEF industrial production.

4. *Ring-Opening Polymerization*

Ring-opening polymerization refers to the method of obtaining corresponding polymers by ring-opening polymerization of cyclic monomers, which is very common in the polymerization production of polylactic acid. Morales-Huerta et al. (2016) Under the catalysis of triethylene diamine, FDCA and dichlorosulfone were reacted in the polar solvent N,N-dimethylformamide to obtain 2,5-furan dicarboxyl chloride, which was then reacted with ethylene glycol to obtain a cyclic monomer. Then, ring-opening polymerization was carried out under the catalysis of tin oxalate to obtain PEF with a higher molecular weight. Rosenboom et al. (2018) First, under the action of dibutyltin oxide, the ester exchange reaction of furan dicarboxylic acid dimethyl ester and ethylene glycol was used to obtain the prepolymer of PEF, which was then depolymerized. The depolymerized product was ring-opened and polymerized under the action of a plasticizer and a tin catalyst to obtain a high molecular weight colorless PEF polyester. Ring-opening polymerization is indeed beneficial for controlling molecular weight and suppressing discoloration, and can obtain high molecular weight PEF, but the reaction conditions are harsh, the monomer preparation process is complex, the cost is high and not environmentally friendly, and it is difficult to apply to industrial production. In summary, the classic methods of synthesizing polyesters can be used to synthesize PEF, and the most industrially valuable is the esterification/ester exchange-melt condensation method. However, challenges still exist, and the goal of in-depth research on this method is to solve the bottleneck problem of promoting molecular weight growth while suppressing side reactions.

5.1.4 Polyglycolic Acid (PGA)

Polyglycolic acid, also known as polyhydroxyacetic acid, is the simplest linear aliphatic polyester and a rigid, high-crystalline thermoplastic polymer (45–55%). Polyglycolic acid has excellent biodegradability, biocompatibility, heat resistance, gas barrier properties, and mechanical strength. It can also be copolymerized with other biobased materials to improve the degradation rate and improve its bulk properties. It is one of the most widely studied and most used biodegradable materials to date. At present, the methods for synthesizing polyglycolic acid at home and abroad mainly include direct polycondensation of glycolic acid, ring-opening polymerization of glycolate, polycondensation of halogenated acetic acid, polymerization of formaldehyde and carbon monoxide, and suspension polymerization.

1. *Direct Polymerization of Glycolic Acid*

Direct polymerization of glycolic acid refers to the method of using glycolic acid to directly heat and dehydrate into a polymer. Takahashi and Kimura (2000) and others obtained high molecular weight polyglycolic acid by direct condensation melting/solidification. After comparing different catalysts, such as zinc and tin, it was found that in the direct polymerization of glycolic acid body condensation, $Zn(CH_3CO_2)_2{\cdot}2H_2O$ is the ideal catalyst for obtaining high molecular weight polymers, but there are problems such as high relative molecular mass of the polymer and long synthesis time. Wang et al. (2004) synthesized polyglycolic acid from glycolic acid under the action of tin catalyst $SnCl_2$, which has been proven to be used as a biodegradable material for medical fibers. Ayyoob et al. (2017) and others used diphenyl sulfone as a solvent and methanesulfonic acid as a catalyst to successfully synthesize polyglycolic acid under vacuum conditions through a solution, and the intrinsic viscosity of the final product polyglycolic acid was 0.2 dL/g. The intrinsic viscosity of polyglycolic acid synthesized by melt/solid condensation is as high as 0.35 dL/g. Wu Yibin et al. (2017) and others used microwave radiation to prepare polyglycolic acid from glycolic acid or glycolic acid methyl ester. The polyglycolic acid obtained by this method has a high molecular weight. Compared with the traditional melting method, the microwave radiation method heats up quickly and evenly, and can complete the reaction in a shorter time compared with the traditional melting method, improving the reaction efficiency. The direct condensation method of glycolic acid to synthesize polyglycolic acid eliminates the steps of preparing and purifying the intermediate glycolide, with short steps, simple operation, less equipment, less consumption of raw material reagents, and is conducive to reducing the cost of synthesis. It is the simplest method to prepare polyglycolic acid, but the disadvantage is that the reaction time is longer.

2. *Glycolide Ring-Opening Polymerization Method*

Glycolide is the cyclic dimer form of glycolic acid. It is generally synthesized by the cyclization of glycolic acid, and then the product polyglycolic acid is polymerized by the ring-opening of glycolide, as shown in Fig. 5.10.

Various catalysts can be used in the process of glycolide ring-opening polymerization. The commonly used types of catalysts are also zinc-based (such as zinc acetate, zinc acetate dihydrate) and tin-based (such as stannous octoate, tin chloride) and also include bismuth-based (bismuth acetate) catalysts. The most common catalyst abroad is stannous octoate, which is characterized by a faster reaction rate, high yield, and high molecular weight of polyglycolic acid. However, the biggest feature of this catalyst is that the reaction activity is not high; the reaction not only needs to be carried out at a higher temperature and for a longer time, but the catalyst will also remain in the product, and the toxicity cannot be measured. Jing Xiabin

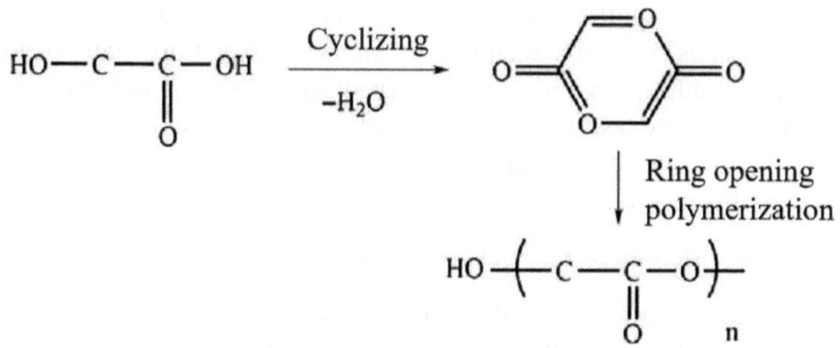

Fig. 5.10 Synthesis of polyglycolic acid by glycolide ring-opening polymerization method

et al. (2001) used calcium ammonia to initiate the ring-opening polymerization reaction of cyclic esters. In an anhydrous and high-purity argon reaction environment, the ring-opening polymerization of cyclic esters such as glycolide was initiated by the solution polymerization method, resulting in high yield and high molecular weight polyglycolic acid products. The purity of the products obtained by this method is also very high. Yamane et al. (2014) Innovative technology has been used to produce high-purity intermediate ethylene glycol, using the uncommon solvent polyalkylene glycol diether to dissolve low molecular weight ethyl acrylate in the synthesis process, followed by distillation. The purpose of distillation is to recover the ethylene glycol produced by the depolymerization reaction along with the solvent, resulting in higher-purity ethylene glycol. This polymerization reaction was initially induced in the molten state, and after reaching appropriate reaction conversion conditions, the polymer precipitated in solid form to promote polymerization, which can prepare high molecular weight polyethylene glycol. Wu Guibao and others (2013) used $SnCl_2 \cdot 2H_2O$ as a catalyst to study the ring-opening polymerization reaction of ethylene glycol under high pressure. The results showed that by precisely controlling the reaction temperature, reaction time, and catalyst dosage, polyethylene glycol with a relative average molecular weight of up to 1.61×10^5 can be obtained. The ring-opening polymerization of ethylene glycol can produce high molecular weight polyethylene glycol, but the purity requirements for ethylene glycol are extremely high. This synthesis route has prominent disadvantages, namely, a long process, low product yield, and high production costs. Because there is no technology for producing high-purity ethylene glycol in China that can be used for efficient, large-scale production of polyethylene glycol, polyethylene glycol is generally produced and sold in small quantities as a very expensive high-value-added product. However, domestic research on this route is gradually deepening. Hu Cuiqiong and others (2004) used ethylene glycol as a monomer and methyl silicone oil as a dispersant to prepare polyethylene glycol with a characteristic viscosity of 0.9 dL/g by the suspension polymerization method under nitrogen protection. Lu Weiliang and others (2013) improved on the above basis, using emulsifier MOA-3 as a dispersant, ethyl acrylate as a monomer, and tin octoate as a catalyst, and also used the suspension synthesis method to synthesize polyethylene glycol with a characteristic viscosity of up to 0.845 dL/g, but the reaction time span was too long. The method of synthesizing polyethylene glycol by suspension polymerization has the advantages of easy heat transfer in the reaction system and simple subsequent processing, but it is not easy to obtain high molecular weight polyethylene glycol products with this method.

3. *Condensation Method of Halogenated Acetic Acid*

This method is to directly condense halogenated acetic acid into polyethylene glycol with active atom-containing end groups, as shown in Fig. 5.11.

Tang and others (2015) dissolved chloroacetic acid in ethyl acetate, then added triethylamine, and obtained ethyl acetate after stirring, refluxing, and distilling. After washing to remove by-products, the remaining insoluble polymer was collected to obtain the product polyethylene glycol. Yu Juan and others (2006) used chloroacetic acid as the raw material, acetone as the solvent, under the action of a certain concentration of triethylamine; the reaction was heated and refluxed at 65 °C for 4 h. The product is washed, dried, reheated, melted, polymerized, and then washed and purified to obtain solid polyvinyl alcohol. The obtained polyvinyl alcohol has good crystallinity, and the melting point has increased. Shi Yuxiang and others (2000) under the action of triethylamine and chloroform, chloroacetic acid is used as the reaction raw material to polymerize polyvinyl alcohol in one step, and the product yield exceeds 80%. The method of one-step polymerization of halogenated acetic acid to produce polyvinyl alcohol has the advantages of simple reaction process, a short production process, easy operation, and low production cost.

4. *Formaldehyde and Carbon Monoxide Polymerization Method*

The formaldehyde and carbon monoxide polymerization method is a method of directly polymerizing formaldehyde and carbon monoxide to form polyvinyl alcohol. The synthesis process is shown in Fig. 5.12.

Göktürk and others (2015) proposed a cheap and effective new technology for synthesizing polyvinyl alcohol by cationic alternating copolymerization of formaldehyde and carbon monoxide, successfully avoiding the cumbersome route that usually requires high-purity ethyl ester. At 170 °C and 5.516 MPa, after reacting for a period of time under the initiation of trifluoromethanesulfonic acid, the yield of polyvinyl alcohol exceeds 90%. It is worth noting that although the reaction achieved a high copolymerization conversion rate, the molecular weight of the polymer is relatively low.

$$\mathrm{X{-}CH_2{-}COOH} \longrightarrow \mathrm{-\!\!\!\!+\,CH_2COO\,+\!\!\!\!-_{n}}$$

Fig. 5.11 Synthesis of polyethylene glycol by halogenated acetic acid condensation method

$$CO + CH_2O \longrightarrow \left[CH_2COO \right]_n$$

Fig. 5.12 Formaldehyde and carbon monoxide polymerization method for synthesizing polyvinyl alcohol

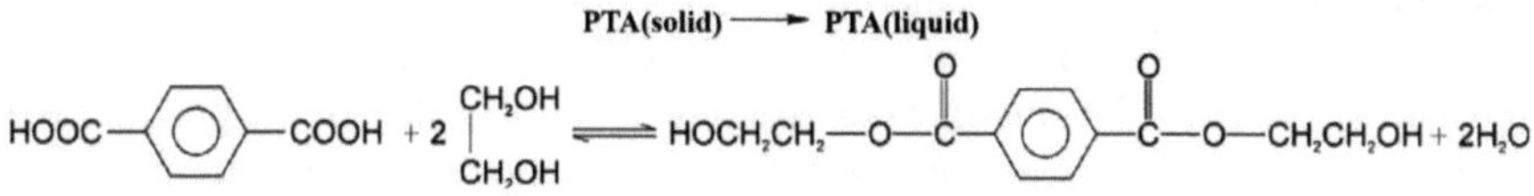

Fig. 5.13 Esterification reaction of PTA and EG to produce BHET. (Deren 1981)

5.1.5 Polyethylene Terephthalate (PET)

PET is a milky white or light yellow, highly crystalline polymer with a smooth and glossy surface. It has excellent physical and mechanical properties over a wide temperature range, excellent electrical insulation, and even at high temperatures and high frequencies, its electrical properties are still well maintained, and its creep resistance, fatigue resistance, friction resistance, dimensional stability and other properties are very outstanding.

1. *PET Chemical Synthesis Route*

The synthesis of PET is achieved by the condensation reaction of monomers terephthalic acid (PTA) and ethylene glycol (EG). In industrial production, it can be divided into three types according to its synthesis route.

(1) Direct Condensation Method

Also known as the direct esterification method, abbreviated as the PTA method. PTA and EG are directly esterified to produce ethylene glycol terephthalate (also known as bis(hydroxyethyl) terephthalate, abbreviated as BHET), and then PET is obtained from BHET through a polycondensation reaction. Because this method involves direct esterification followed by polycondensation, it is called the direct polycondensation method. By reacting high-purity terephthalic acid with ethylene glycol, the production of dimethyl terephthalate and the recovery of methanol can be omitted, thus reducing costs. Due to the advantages of the PTA method's continuous production technology, such as low raw material consumption, short reaction time, high yield, stable quality, and low production cost, it has developed rapidly. Since the 1980s, all newly built PET units have adopted the continuous production technology of the PTA method, which mainly consists of two steps: esterification reaction and polycondensation reaction. The entire process of the esterification reaction is mainly the esterification of dissolved PTA with ethylene glycol (EG), as shown in Fig. 5.13. The esterified product BHET produced by the subsequent reaction undergoes a polycondensation reaction, removing EG and growing the chain to produce PET (Deren 1981).

This method began industrial production in 1963. The mixture of terephthalic acid and ethylene glycol is not the same as the mixture of dimethyl terephthalate and ethylene glycol; the former is a slurry, and the latter is a uniform solution. It is difficult to mix this slurry evenly and heat it under realistic conditions. This is because PTA sublimates easily under high-temperature conditions; not only is the reaction slow, but a series of etherification reactions are also easily produced. To mix the slurry well, an excess of ethylene glycol is often added, but this excessive input will accelerate the etherification reaction, and the quality of the polymer will be even worse. The key to the direct esterification method is to solve the problem of mixing the slurry, increase the reaction speed, meet the requirements of industrial production, and at the same time inhibit the etherification reaction.

(2) Transesterification Method

Also known as the DMT method, or the dimethyl terephthalate method, the purity of the monomer PTA produced in the early days was not high, and there were difficulties in industrial purification. At that time, it was not possible to prepare a qualified PET by the direct method. Therefore, PTA with low purity was first reacted with methanol to produce dimethyl terephthalate (DMT), which is easier to purify. Then, BHET is produced by the transesterification reaction of EG with purified high-purity (≥99.9%) DMT, and then the next step of polycondensation reaction is carried out to produce qualified PET. The transesterification reaction in the synthesis step is the most critical and indispensable, and this method is commonly referred to as the transesterification method in industry. The synthesis of BHET is carried out in the presence of a catalyst, which is often zinc acetate, manganese acetate or cobalt acetate, or mixed with antimony trioxide, and the amount of catalyst used is 0.01–0.05% (DMT weight).

(3) Ethylene Oxide Addition Method

Since ethylene glycol is made from ethylene oxide, if BHET is obtained directly by adding ethylene oxide (EO) and PTA, and then polymerized into PET, this method is called the ethylene oxide method. The advantage of this method is that it can eliminate the step of producing ethylene glycol from EO, significantly reducing costs, and this reaction occurs quickly, which is much better than the direct condensation method from these two aspects. However, this method has the defect that EO is prone to ring-opening to form polyether, and EO is a gas at room temperature, which brings a lot of economic losses in transportation and storage, so it has not been widely used.

In recent years, new progress has been made in the preparation technology of biobased PET: The technology of synthesizing biobased terephthalic acid from cellulose nonfood biomass resources such as corn stalks, sugarcane bagasse, sawdust, straw, wheat straw, and sorghum stalks, and then preparing 100% biobased PET has entered the commercial production stage (Changchun 2017). The Coca-Cola Company in the United States has provided a packaging biobased PET and its preparation method (Coca-Cola Company 2016). Refining sugarcane into molasses and sugar, fermenting molasses to produce ethanol, adding inorganic acid or strong organic acid and catalyst to dehydrate and refine into ethylene, refining ethylene under the action of a catalyst, reacting with oxygen to produce ethylene glycol, adding biobased terephthalic acid to melt and polymerize to form biobased PET, and biobased PET solid polymerizes to form PET.

2. *Side Reactions in the Synthesis of PET*

The synthesis of PET currently basically uses high-purity terephthalic acid (PTA) and ethylene glycol (EG) as raw materials, and PET is generated through the esterification reaction stage and the polycondensation reaction stage. The polycondensation reaction process always needs to be realized through a certain fixed method and precise process, and the methods widely used in the industry currently include melt polycondensation, solution polycondensation, and interfacial polycondensation. The synthesis of PET uses melt polycondensation, that is, no solvent is added in the reaction, and the reaction temperature is above the melting temperature of the raw material monomer and the polycondensation product (generally higher than the melting point temperature by 10–25 °C). The characteristic of the melt polycondensation method is that the reaction temperature is high (generally above 200 °C). High temperature is conducive to increasing the reaction rate and the exclusion of low molecular weight by-products. This method is generally used for reversible polycondensation reactions with very small reaction rates at room temperature.

The melt polycondensation production process is simple, because it does not require solvent, it reduces the loss of solvent evaporation and eliminates the step of returning the solvent, reduces pollution, and is conducive to reducing costs. Since the reaction is a reversible equilibrium, while large molecules are generated, there are several side reactions, and the side reactions of PET synthesis can be divided into the following three aspects: cyclization reactions of monomers or low oligomers, such as the cyclic products shown in Fig. 5.14; side reactions of monomers; side reactions of polymers. And these side reactions can all be regarded as reverse reactions.

Research on the biodegradation of PET has also been in full swing recently. For example, the team led by Gu Lengtao has screened strains capable of degrading PET particles through multiple experiments (Lengtao et al. 2021), and has made

Fig. 5.14 Low oligomer cyclic products in the side reaction of PET melt polycondensation (*n* is generally 2)

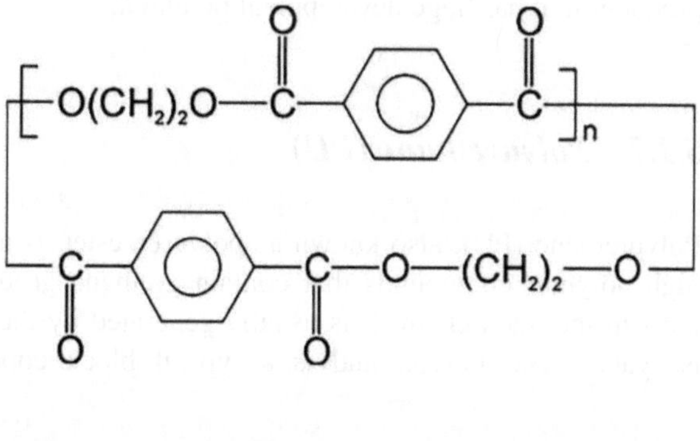

certain progress in the biodegradation process and mechanism of PET. In summary, it can be expected that the research and application development of biobased PET materials will make significant progress in the next few years.

5.1.6 Polycarbonate (PC)

Polycarbonate (PC) is a general term for a class of high polymer compounds that contain carbonate chains in their molecular chains. It can be processed into tasteless, nontoxic, transparent amorphous thermoplastic materials. It is the only product with good transparency among the five major engineering plastics and is also the fastest-growing general engineering plastic in recent years.

In the development history of polycarbonate synthesis technology, there are many synthesis and preparation methods, such as low-temperature solution polycondensation, high-temperature solution polycondensation, pyridine method, and partial pyridine method. New synthesis methods are still being reported, but there are not many industrialized, large-scale production processes, mainly because these methods are costly. At present, most manufacturers in the world generally use the phosgene method to industrialize the production of polycarbonate, but because this process requires a large amount of bisphenol A to be synthesized in an alkaline solution and dichloromethane, and the PC material produced is completely nondegradable, it hinders its sustainable development.

The Gregory team at the University of Bath (Gregory et al. 2017) successfully used CO_2 and sugar to prepare environmentally friendly biobased aliphatic polycarbonate plastic (APC). This method no longer uses bisphenol A, but through a high-yield 3-step process, using CO_2 as a C1 synthesizer, 2-deoxy-D-ribose is converted into a new type of 6-membered carbonate ring for ring-opening polymerization, obtaining a carbohydrate-based APC. Moreover, this environmentally friendly polycarbonate plastic can be biodegraded, and its future application prospects will be very considerable. At the same time, Mitsubishi Chemical Corporation has developed a biobased polycarbonate resin, which uses biobased raw material isosorbide as a comonomer to replace bisphenol A. The finished product produced using isosorbide as a raw material has high transparency and excellent optical properties, and is more wear-resistant than general PC. Once it is put into large-scale industrial production, it has huge development potential.

5.1.7 Polyurethane (PU)

Polyurethane (PU), also known as polyurea ester, is a general term for a class of high polymer compounds that contain urethane groups (–NH–COO–) repeating units in the main chain. It is usually generated by the addition polymerization of isocyanate and polyol, and is a typical block copolymer. PU is one of the

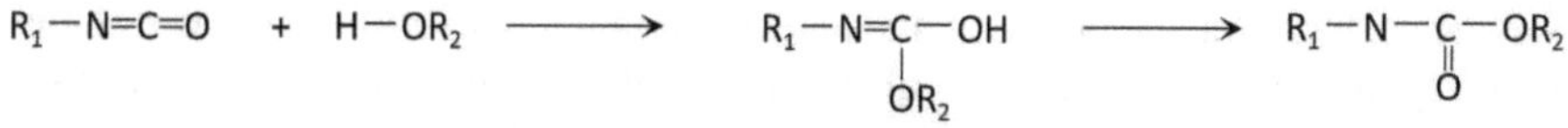

Fig. 5.15 Polyurethane synthesis reaction process

fastest-developing polymer materials, with characteristics such as wear resistance, tear resistance, and good flex resistance, and is therefore widely used.

The most basic reaction of polyurethane synthesis is the reaction between polyisocyanate and polyol to form a urethane group, which is a stepwise addition polymerization reaction involving hydrogen transfer. The synthesis reaction process is shown in Fig. 5.15.

Currently, industrially produced polyurethanes are nonbiodegradable polyester materials that are not derived from biological sources. However, the synthesis of degradable polyurethanes using biobased raw materials has attracted attention. Feng Zhaoxuan and others have synthesized a series of biodegradable self-emulsifying anionic waterborne polyurethanes (WBPU) using aliphatic diisocyanate IPDI, PEG, degradable polyester diol (PBAG), and arginine (Feng et al. 2020). Compared with ordinary polyurethane materials, this material has stronger hydrophilicity, higher mechanical strength, and better flexibility. More importantly, it has controllable degradation properties and the degradation products are noncytotoxic; thus, it has good biocompatibility and promising applications in the field of tissue engineering. In addition, many degradable polyurethanes derived from natural biomaterials have been developed, such as starch-derived polyurethane (Yao et al. 1996; Desai et al. 2000; Dajun and Yaojun 1997; Yong and Dajun 1988), oligosaccharide-derived polyurethane (Zeterlund. 1997), lignin- and tannin-derived polyurethane (Shida 1996; Jinjie and Katsumi 1999), and polyurethane derived from agricultural and forestry waste and plant fibers (Pus 1993). The preparation of biodegradable polyurethane materials from renewable natural plant resources not only solves the problem of environmental pollution caused by discarded polyurethane materials but also reduces the dependence on dwindling petroleum products, especially the comprehensive utilization of agricultural and forestry by-products and waste, which not only reduces the raw material cost of polyurethane, but also enhances the added value of agricultural and forestry by-products and waste. Therefore, such biodegradable polyurethane materials will have a very broad development prospect.

5.1.8 Poly-β-Hydroxybutyrate

(PHB) Poly-β-hydroxybutyrate (PHB) is a linear polymer composed of β-hydroxybutyric acid, existing in many prokaryotic cells in the form of refractive granules. It is a polyester synthesized and stored in cells by microorganisms, and is one of the polyhydroxyalkanoates (PHA). PHB, as an important degradable substance, has attracted attention due to its material properties similar to petroleum-based synthetic plastics (such as polyethylene and polypropylene). The synthesis of PHB is mainly through biological methods at present, but the current level of biological synthesis of PHB has shortcomings such as high production cost, long production cycle, and low synthesis yield, and does not yet have the value of industrial application on a large scale. It needs further research in strain construction and optimization, as well as fermentation technology. The method of synthesizing PHB by chemical method can usually be divided into two categories according to different raw materials: one is to prepare PHB by ring-opening polymerization using β-butyrolactone as the raw material; the other is to produce PHB by condensation using 3-hydroxybutyric acid or its esters as the raw material (Wenchao 2016). Currently, research on the chemical synthesis process of PHB mainly focuses on the β-butyrolactone ring-opening polymerization process. The synthesis of PHB using 3-hydroxybutyric acid is less common, and the industrial synthesis route of the chemical method is still being explored and continuously improved.

5.2 Biobased Polybutyramide

At present, the common polyamide materials (such as nylon 6, and nylon 66) cannot naturally degrade in the natural environment, but polybutyramide (Polyamide 4, PA4) has excellent degradability under natural conditions. It is the only known degradable polyamide. Because it can be prepared or biosynthesized from biomass raw materials, PA4 has a very strong development application prospect in the field of degradable polymer materials.

5.2.1 Chemical Synthesis of PA4

In the 1950s, researchers began to study the synthesis methods of biobased polybutyramide. The most primitive chemical synthesis method is to stir 2-pyrrolidone in a vacuum environment in an alkaline environment and obtain PA4 by controlling the reaction temperature and pressure (Ney et al. 1953). Soon after, researchers successively explored the catalytic conditions of the PA4 synthesis method, such as improving the reaction conditions of PA4 (temperature and dehydration rate), adding alkali metal salts, and using SO_2 as an activator, not only improving the yield but also synthesizing PA4 with suitable melt extrusion thermal characteristics (Barnes and Barnes 1981). Some researchers have obtained alkali metal amides by reacting alkali metals with five to seven ring amides.

Let the alkali metal derivative act as the catalyst of the reaction system in the reaction, and successfully prepare PA4 (Bacskai 1979). In 1972, Jarovitzky improved the previous technology, added a mixture of sulfites and bisulfites,

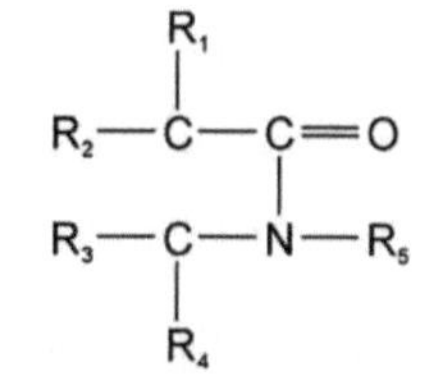

Fig. 5.16 Schematic structure of alkali metal derivative catalyst

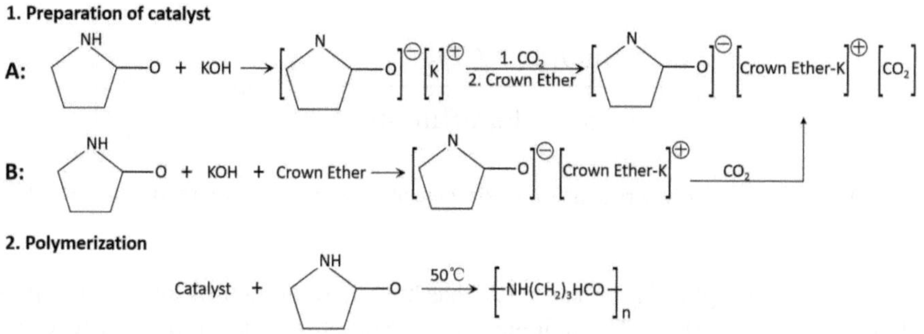

Fig. 5.17 PA4 synthesis route using crown ether as a catalyst. (Bacskai 1984)

controlled the amount of SO_2 input, and controlled the molar ratio of the alkali metal catalyst to 0.20–0.85, obtaining a high conversion rate and high molecular weight white PA4 (Jarovitzky 1972). In the synthesis technology of polybutyramide prepared by alkali metal catalysis, some researchers have proved that the catalyst is the derivative of alkali metal, improved the alkali metal catalytic system, and proposed a C_2 participating in the hybrid nitrogen heterocyclic ketone derivative, its schematic structure is shown in Fig. 5.16 (Jr 1972), where R_1–R_5 can be independent hydrogen atoms or C_1–C_6 alkyl.

At the end of the twentieth century, researchers proposed a synthesis plan for synthesizing ultra-high molecular weight PA4, mainly aimed at accelerating the polymerization reaction while increasing the molecular weight of PA4, using bromide or crown ether as a catalyst for py-K/CO_2 (Bacskai and Fries 1982) (CO_2 and potassium hydroxide catalyzed 2-pyrrolidone system) for synthesis; the synthesis rate is rapidly improved, Fig. 5.17 is a schematic comparison of the technical route for the synthesis of PA4 using crown ether as a catalyst in this work (Bacskai 1984).

Subsequently, researchers have continuously developed more mature and stable PA4 synthesis routes. In a work published by an Italian team, it was proposed to use silicon tetrachloride as a catalyst for nucleophilic addition of easy-to-amide anions (Fig. 5.18), and potassium-substituted 2-pyrrolidone as a polymerization initiator for the suspension polymerization route (Giovanna et al. 1981). Chen et al. also used suspension polymerization to improve the thermal stability of PA4. The synthesis route of the suspension method is shown in Fig. 5.19. Its synthesis route uses CO_2 as a promoter in an alkaline environment, combined with the assistance of organic crown ethers or quaternary ammonium salt compounds to increase yield

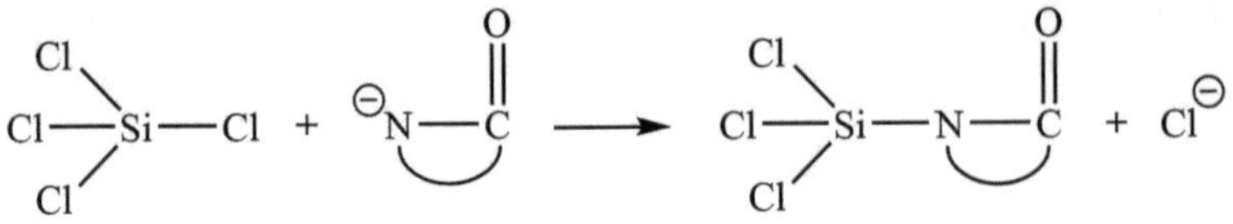

Fig. 5.18 Schematic diagram of the role of silicon tetrachloride in PA4 anionic polymerization reaction. (Giovanna et al. 1981)

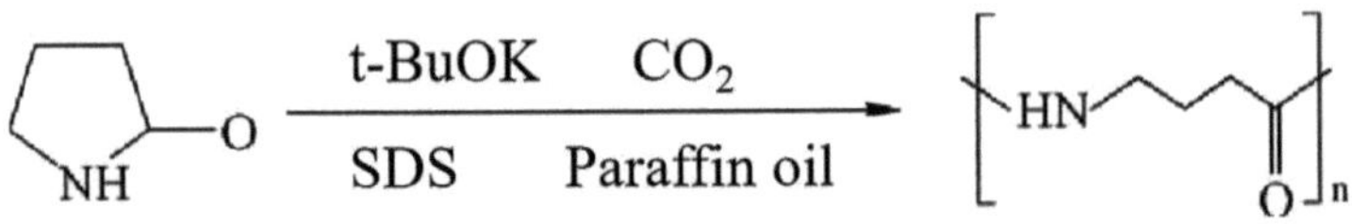

Fig. 5.19 Suspension polymerization synthesis route of 2-pyrrolidone for PA4. (Chen et al. 2020)

and molecular weight. The prepared PA4 has high thermal stability and good dispersion (Chen et al. 2020). The anionic polymerization reaction route has received attention in the later development due to the stability and high purity of the product. In the classic synthesis route, t-BuOK is generally used as the initiator, CO_2 or benzoyl chloride (BZC) as the initiator, and crown ether or TMAC as the catalyst for anionic ring-opening polymerization, which can prepare ultra-high molecular weight PA4 homopolymers and copolymers.

Based on the general route of the above-mentioned anionic ring-opening polymerization, Kim's team has developed a new polymerization method: heterogeneous polymerization (Kim et al. 2013). This method involves the polymerization of 2-pyrrolidone in a nonhomogeneous medium of paraffin oil containing sodium dodecyl sulfate, potassium tert-butanol, and benzoyl chloride, which can produce PA4 microspheres with a yield of up to 76%. Kawasaki and others in Japan have focused on the research of different chain structures of PA4. Their team used polycarboxylic acid chloride as an initiator to synthesize branched PA4 and found that compared with linear PA4, branched PA4 maintains its degradability while its tensile strength increases (Kawasaki et al. 2005).

After several decades of development, the most common method of traditional chemical synthesis of PA4 is still anionic polymerization: using butyramide as a raw material, and sodium hydroxide or potassium hydroxide as an activator, a highly active butyramide anion is obtained through a nucleophilic reaction. Then, acyl chloride reagent is added. During the initiation of the polymerization process, the aminobutyric acid formed by ring-opening under heating conditions has a strong tendency to cyclize, so the reaction can only be carried out at low temperatures, and the polymer obtained at low temperatures has a lower molecular weight. In order to obtain a high molecular weight product, the oligomer obtained in the first step needs to be polymerized at high temperature.

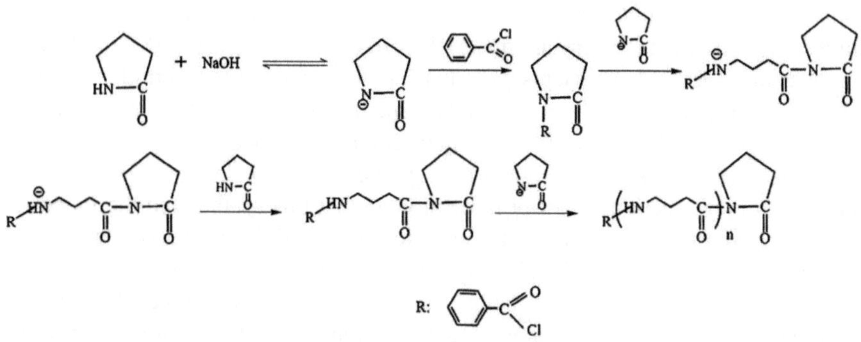

Fig. 5.20 Schematic diagram of PA4 synthesis technology route. (De et al. 2019)

Wu De et al (2019). used the sludge precipitation polymerization method, first pre-treating the monomer, soaking 2-pyrrolidone in a molecular sieve for two days before the reaction to fully remove moisture. After mixing the activator (NaOH, KOH, and BuOK) and 2-pyrrolidone. Nitrogen is introduced to remove air, and stirring is carried out during the vacuum reaction to remove water from the activation process. Then cool down to 45 °C, add the initiator benzoyl chloride to initiate polymerization, and finally obtain pure PA4 powder through the post-treatment process. Figure 5.20 is a schematic diagram of their PA4 synthesis technology route.

In addition to the above chemical synthesis routes and methods, there has been significant progress in the biological synthesis methods of PA4 in recent years, including the synthesis of γ-aminobutyric acid (GABA) using lactic acid lactobacillus and *E. coli* microbial glutamate decarboxylase, and then using chemical synthesis routes to polymerize GABA into PA4 (Saskiawan 2008). Or directly using monosodium glutamate as a raw material, the recombinant *E. coli* expressing glutamate decarboxylase synthesizes GABA into high-concentration GABA, which is then converted into 2-pyrrolidone, and finally used for the chemical synthesis of PA4, with a conversion rate of up to 96% (Si et al. 2013).

5.2.2 Conclusion

The chemical synthesis of biobased degradable materials is still a focus and hotspot of research, with new synthesis methods and technologies emerging and being implemented. For example, the Younes team from McGill University (Younes and Maric 2021) prepared hydroxyl polyurethane by reacting ethyl carbonate with bio-based diamine; then hydroxyl polyurethane initiated ε-caprolactone ring-opening polymerization, forming 5 ε-caprolactone units at both ends of the hydroxyl polyurethane to synthesize biobased, hydrolyzable degradable photocurable thermosetting polyurethane acrylate materials; Wu et al. (2011) using PCL and PLLA diols as biodegradable segments, N,N-bis(2-hydroxyethyl)cinnamamide (BHECA) as a light-responsive monomer, and hexamethylene diisocyanate as a coupling agent, a multi-block polyester polyurethane containing biodegradable segments and cinnamamide groups was designed and synthesized through a two-step addition polymerization reaction. It is worth mentioning that this synthesized thermoplastic elastomer not only degrades but also exhibits superior light-responsive characteristics at room temperature. Hong's research team, after synthesizing vanillin-based dialdehydes and tri-aldehydes with high aromatic content, selected renewable diamines with short fatty chains, and then prepared fully biobased thermosets through the Schiff base reaction of vanillin and diamines. This material not only has good mechanical properties, but can also be completely degraded under mild acidic conditions (Hong et al. 2022).

Of course, many biobased materials can be obtained from biomass, and their new biobased materials synthesized through biosynthesis have been successfully put into the market for use, such as polyhydroxyalkanoates (PHA), polycaprolactone (PCL), and polyglycolic acid (PGA) and its copolymers. These biobased products have been used as biodegradable polymer materials in fields such as drug release and tissue engineering (Fucheng 2008). In addition, some thermoplastic polyesters have also been successfully prepared, only through the simple condensation reaction of biobased diols and diacids. The monomers for preparing biobased polyesters are also prepared from renewable biomass, which makes the sustainable development of the polyester industry possible. At present, biobased polyesters are mainly divided into two types, namely biobased polyesters prepared from biobased monomers entirely from biomass (100% biobased polyesters) and biobased polyesters prepared from biobased monomers from different renewable resources. It can be predicted that in the next few decades, the family of biobased polyesters will greatly expand. However, biobased polyamides in China, especially the research and development and industrialization of fully biobased polyamides, are somewhat behind developed countries due to technical and variety limitations. Although there are still many problems to be solved in the large-scale production of biobased PA, the preparation of high-performance polymer materials using biomass resources is a very critical development hotspot.

References

Abe H, Takahashi N, Kim KJ et al (2004) Thermal degradation processes of end-capped poly (L-lactide)s in the presence and absence of residual zinc catalyst. Biomacromolecules 5(4):1606–1614

Ajioka M, Enomoto K, Suzuki K et al (1995) Basic properties of polylactic acid produced by the direct condensation polymerization of lactic acid. Bull Chem Soc Jpn 68(8):2125–2131

Ajioka M, Suizu H, Higuchi C et al (1998) Aliphatic polyesters and their copolymers synthesized through direct condensation polymerization. Polym Degrad Stab 59(1):137–143

Auras R, Harte B, Selke S (2004) An overview of polylactides as packaging materials. Macromol Biosci 4(9):835–864

Ayyoob M, Lee DH, Kim JH et al (2017) Synthesis of poly(glycolic acids) via solution polycondensation and investigation of their thermal degradation behaviors. Fibers Polym 18(3):407–415

Bacskai R (1979) Catalyst for the polymerization of 2-pyrrolidone from alkali metal: US, US4145519 A

Bacskai R (1984) Synthesis of ultrahigh molecular weight nylon 4 with onium salt and crown ether-containing catalysts. Springer US 279(5):32–39

Bacskai R, Fries BA (1982) The absence of CO_2 in nylon-4 prepared with KOH+ radioactive CO_2 catalyst. J Polym Sci A Polym Chem 20(8):2341–2344

Badens E, Masmoudi Y, Mouahid A et al (2018) Current situation and perspectives in drug formulation by using supercritical fluid technology. J Supercrit Fluids 134:274–283

Bailly M (2002) Production of organic acids by bipolar electrodialysis: realizations and perspectives. Desalination 144(1-3):157–162

Barnes CE, Barnes AC (1981) Polymerization of 2-pyrrolidone with cesium or rubidium catalyst: US, US4247684 A

Basko M, Bednarek M, Kubisa P (2015) Cationic copolymerization of L, L-lactide with hydroxyl substituted cyclic ethers. Polym Adv Technol 26(7):804–813

Berkel JV, Guigo N, Kolstad JJ et al (2015) Isothermal crystallization kinetics of Poly (ethylene 2,5-furandicarboxylate). Macromol Mater Eng 300(4):466–474

Chanfreau S, Mena M, Porras-Domínguez JR et al (2010) Enzymatic synthesis of poly-L-lactide and poly-L-lactide-co-glycolide in an ionic liquid. Bioprocess Biosyst Eng 33(5):629–638

Changchun L (2017) New progress in polyester technology. Synthetic Fiber 6:1–5

Changhui Z, Xia Z (2008) Progress in the synthesis of polybutylene succinate. Polyester Industry:11–14

Changhui Z, Xia Z, Jitao H (2008) Research progress on the synthesis process of PBS-based polyester. Plastics:8–10

Chaorong QI, Huanfeng J (2010) Histidine-catalyzed synthesis of cyclic carbonates in supercritical carbon dioxide. Sci Chin Chem (Engl Ed) 7:5

Chaoyang W, Yaoming Z (2003) Application of chain extension reaction in the synthesis of polymer materials. Chem Propellants Polymeric Mater 19(6):23–26

Chen T, Wang C, Zhao L et al (2020) Suspension polymerization of 2-pyrrolidone in the presence of CO_2 and organic promoters. J Appl Polym Sci

Coca-Cola Company (2016) Bio-based polyethylene terephthalate packaging and method for its preparation: China, 105254858A

Cuiqiong H, Qingzhang Z, Ping W (2004) Suspension polymerization of ethylene glycolate. J Beijing Univ Chem Technol 31(3):62–65

Dajun C, Yaojun L (1997) Biodegradable polyurethane elastomer modified by starch. Synthetic Rubber Industry 20(4):244–248

De W, Liangchen T, Songchao T (2019) Synthesis and properties of bio-based butyramide and polybutyramide. J Funct Polym 32(1):110–116

Deren Z (1981) Polymer synthesis technology. Chemical Industry Press, Beijing

Desai S, Thakore IM, Sarawade BD (2000) Structure-property relationship in polyurethane elastomers containing starch as a crosslinker. Polym Eng Sci 40(5):1200–1210

Discher DE, Ahmed F (2006) Polymersomes. Annu Rev Biomed Eng 8:323–341

Dorgan JR, Lehermeier H, Mang M (2000) Thermal and rheological properties of commercial-grade poly (lactic acid) s. J Polym Environ 8(1):1–9

Eerhart A, Faaij A, Patel MK (2012) Replacing fossil based PET with biobased PEF; process analysis, energy and GHG balance. Energy Environ Sci 5(4):6407–6422

Fan F, Zhang Z, Xing H et al (2017) Progress in synthesis of cyclic carbonates under supercritical carbon dioxide. Chem Eng Prog 36(8):2924–2933

Feng ZX, Wang D, Zheng YD (2020) A novel waterborne polyurethane with biodegradability and high flexibility for 3D printing. Biofabrication 12(3):035015

Ferrari R, Pecoraro CM, Storti G et al (2014) A green route to synthesize poly (lactic acid)-based macromonomers in $ScCO_2$ for biodegradable nanoparticle production. RSC Adv 4(25):12795–12804
Fucheng B (2008) Development of biomass materials and biomass material science. For Products Industry 35(1):3–7
Gadda TM, Pirttimaa MM, Koivistoinen O et al (2014) The industrial potential of bio-based glycolic acid and polyglycolic acid. Appita J 67(1):12
Gandini A, Coelho D, Gomes M et al (2009) Materials from renewable resources based on furan monomers and furan chemistry: work in progress. J Mater Chem 19(45):8656–8664
Giovanna C, Marco N, Saverio R (1981) The anionic polymerization of 2-pyrrolidone in bulk and in suspension. Makromol Chem
Göktürk E, Pemba AG, Miller SA (2015) Polyglycolic acid from the direct polymerization of renewable C1 feedstocks. Polym Chem 6(21):3918–3925
Gregory GL, Kociok-Khn G, Buchard A (2017) Polymers from sugars and CO_2: ring-opening polymerisation and copolymerisation of cyclic carbonates derived from 2-deoxy-D-ribose. Polym Chem 8(1):2093
Griffith LG (2000) Polymeric biomaterials. Acta Mater 48(1):263–277
Guibao W, Aijun C, Qun C et al (2013) Study on the process of preparing polyhydroxyacetic acid by ring-opening polymerization under high pressure. Polym Bull 3:55–60
Hong S, Min KD, Nam BU et al (2016) High molecular weight bio furan-based co-polyesters for food packaging applications: synthesis, characterization and solid-state polymerization. Green Chem 18(19):5142–5150
Hong K, Sun Q, Zhang X (2022) Fully bio-based high-performance thermosets with closed-loop recyclability
Hualin W, Daqing F, Tiejun S et al (2005) Synthesis of D,L-propylene carbonate, an intermediate of poly D,L-lactic acid. Polym Mater Sci Eng 21(5):51–54
Jarovitzky PA (1972) Polymerization of 2-pyrrolidone in the presence of SO_2 with mixture of an alkali metal sulfite and an alkali metal bisulfite: US, US3681294 A
Jiangtao Y, Haihong M (2009) Progress in the synthesis of propylene carbonate. Polyester Industry 22(2):4–7
Jiaping W (2017) Synthesis of biobased poly(2,5-furandicarboxylic acid ethylene glycol ester) and regulation of chain structure and crystallinity. Zhejiang University, Hangzhou
Jinbo L (2013) Synthesis of biodegradable polybutylene succinate. Zhengzhou University, Zhengzhou
Jingchang W, Xuehang S, Weijing W et al (2017) Research progress in enzymatic synthesis of aliphatic polyesters. Chem Industry Eng Progr 36(7):2592–2600
Jinjie G, Katsumi S (1999) Synthesis and property study of biodegradable polyurethane insulation material. Fudan J Natural Sci Ed 38(4):418–421, 427
John A, Katiyar V, Pang K et al (2007) Ni(II) and Cu (II) complexes of phenoxy-ketimine ligands: synthesis, structures and their utility in bulk ring-opening polymerization(ROP) of L-lactide. Polyhedron 26(15):4033–4044
Jr B (1972) Polymerization of 2-pyrrolidone using azetidinone co-activators and carbon dioxide as activator: US, US3681296 A
Juan Y (2006) Synthesis and characterization of polyhydroxyacetic acid (PGA). Wuhan University of Technology, Hubei, pp 1126–1129
Kawasaki N, Nakayama A, Yamano N et al (2005) Synthesis, thermal and mechanical properties and biodegradation of branched polyamide 4. Polymer 46(23):9987–9993
Keyou L, Juhua Z, Furuh X (1999) Principles and technology of polymer synthesis. Science Press, Beijing
Kim NC, Kim JH, Kim JH et al (2013) Preparation of Nylon 4 microspheres via heterogeneous polymerization of 2-pyrrolidone in a paraffin oil continuous phase. Polymer Korea 37(2):211–217
Knoop RJI, Vogelzang W, Van Have J et al (2013) High molecular weight poly(ethylene-2,5-furanoate); critical aspects in synthesis and mechanical property determination. J Polym Sci Part A 51(19):4191–4199

Köhn RD, Pan Z, Sun J et al (2003) Ring-opening polymerization of D, L-lactide with bis (trimethyl triazacyclohexane) praseodymium triflate. Catal Commun 4(1):33–37
Lasprilla A, Martinez G, Lunelli BH et al (2012) Poly-lactic acid synthesis for application in biomedical devices—a review. Biotechnol Adv 30(1):321–328
Lengtao G, Zhengfei Y, Jing W et al (2021) Screening, identification and degradation characteristics of a PET-degrading bacterial strain. Genom Appl Biol 40(3):1179–1186
Li G, Zhao M, Xu F et al (2020) Synthesis and biological application of polylactic acid. Molecules 25(21)
Liu S, Qin S, He M et al (2020) Current applications of poly (lactic acid) composites in tissue engineering and drug delivery. Compos Part B: Eng:108238
Lowe CE (1954) Preparation of high molecular weight polyhydroxyacetic ester: U.S. Patent 2668162
Lu D, Zhang X, Zhou T et al (2008) Biodegradable poly (lactic acid) copolymers. Prog Chem 20(203):339–350
Min Z, Xiaomei T, Xiaoxia W (2006) Study on the factors affecting the relative molecular weight of biodegradable polyester PBS. J Shaanxi Univ Sci Technol 24:8–11
Moon SI, Lee CW, Miyamoto M et al (2000) Melt polycondensation of L-lactic acid with Sn(II) catalysts activated by various proton acids: a direct manufacturing route to high molecular weight Poly (L-lactic acid). J Polym Sci Part A Polym Chem 38(9):1673–1679
Moon SI, Lee CW, Taniguchi I et al (2001) Melt/solid polycondensation of L-lactic acid: an alternative route to poly (L-lactic acid) with high molecular weight. Polymer 42(11):5059–5062
Morales-Huerta JC, Antxon M, MuOz-Guerra S (2016) Poly(alkylene 2,5-furandicarboxylate)s (PEF and PBF) by ring opening polymerization. Polymer:148–158
Nan L, Wenfang J, Jingbo Z et al (2003) Synthesis and purification of D, L-propylene carbonate. Petrochemicals 32(12):1073–1077
Nan L, Wenfang J, Jingbo Z et al (2005) Preparation of D,L-propylene carbonate and its ring-opening polymerization in xylene solution. Polym Mater Sci Eng 21(2):73–76
Nef JU (1914) Dissoziationsvorgänge in der Zuckergruppe. Justus Liebigs Ann Chem 403(2-3):204–383
Nejib K, George P, Dimitris A et al (2018) Solid-state polymerization of poly (ethylene furanoate) biobased polyester, II: an efficient and facile method to synthesize high molecular weight polyester appropriate for food packaging applications. Polymers 10(18):471
Ney WO, Nummy WR, Barnes CE (1953) Polymers from pyrrolidone: US2638463
Nofar M, Sacligil D, Carreau PJ et al (2019) Poly(lactic acid) blends: processing, properties and applications. Int J Biol Macromol 125:307–360
Papageorgiou GZ, Bikiaris DN (2005) Crystallization and melting behavior of three biodegradable poly(alkylene succinates). A comparative study. Polymer 26:12081–12092
Pingkai O (2012) Biobased polymer materials. Chemical Industry Press, Beijing
Pus SRS (1993) Liquefaction of wood without a catalyst. Mokukzai Gakkaishi 39(4):446–458
Qimei S, Chonghui W (2015) Current status and progress of L-propylene carbonate synthesis technology. Chem Industry Progr 34(3):802–809
Rosenboom JG, Hohl DK, Fleckenstein P et al (2018) Bottle-grade polyethylene furanoate from ring-opening polymerisation of cyclic oligomers. Nat Commun 9(24):2701
Saeidlou S, Huneault MA, Li H et al (2012) Poly(lactic acid) crystallization. Prog Polym Sci 37(12):1657–1677
Saskiawan I (2008) Biosynthesis of polyamide 4, a biobased and biodegradable polymer. Microbiology Indonesia 2(3)
Shaohui C, Tao L (2020) Current status and future prospects of biodegradable plastics industry. Modern Plast Process Appl 2:50–54
Shida H (1996) Studies on liquefaction of wood meals. J Appl Polym Sci 60(7):1187–1198
Si JP, Kim EY, Noh W et al (2013) Synthesis of nylon 4 from gamma-aminobutyrate (GABA) produced by recombinant Escherichia coli. Bioprocess Biosyst Eng 36(7):885–892
Sinclair RG (1977) Copolymers of D, L-lactide and epsilon caprolactone: U.S. Patent 4045418
Singhvi MS, Zinjarde SS, Gokhale DV (2019) Polylactic acid: synthesis and biomedical applications. J Appl Microbiol 127(6):1612–1626

Södergård A, Stolt M (2010) Industrial production of high molecular weight poly(actic acid). Poly (Lactic Acid) synthesis, structures, properties, processing, and applications. Wiley, New York, pp 27–41

Takahashi K, Kimural Y (2000) Melt/solid polondesycation of glycolic acid to obtain high-molecular-weight poly (glycolic acid). Polymer 41:8725–8728

Tang SY, Zhang R, Liu F et al (2015) Hansen solubility parameters of polyglycolic acid and interaction parameters between polyglycolic acid and solvents. Eur Polym J 72:83–88

Tao S, Chaowen L (1998) Synthesis of D,L-propylene carbonate without the need for high vacuum. Chem Industry Times 12(6):14–17

Wang J, You G, Wang F et al (2004) Synthesis and characterization of polyglycollic acid via direct melt polymerization. Chin Synthetic Fiber Industry 19:44–49

Wei S, You X, Lv L et al (2012) Discussion on the preparation of PLA from the mixture of L-lactide and D, L-lactide. New Chem Mater 40(5):45–47

Weiliang L, Aijun C, Zeyun W et al (2013) Process for preparing polyglycolic acid by suspension polymerization. Chem Industry Progr 32:652–656

Weiwei J (2009) Further optimization study on the synthesis process of high molecular weight polylactic acid. Nanjing University of Science and Technology, Nanjing

Wenchao Y (2016) Preparation of biodegradable material poly-β-hydroxybutyrate. Guangzhou University, Guangzhou

Whulanza Y, Rahman SF, Suyono EA et al (2018) Use of Candida rugosa lipase as a biocatalyst for L-lactide ring-opening polymerization and polylactic acid production. Biocatal Agric Biotechnol 16:683–691

Woo SI, Kim BO, Jun HS et al (1995) Polymerization of aqueous lactic acid to prepare high molecular weight poly (lactic acid) by chain-extending with hexamethylene diisocyanate. Polym Bull 35(4):415–421

Wu L, Jin C, Sun X (2011) Synthesis, properties, and light-induced shape memory effect of multiblock polyesterurethanes containing biodegradable segments and pendant cinnamamide groups. Biomacromolecules 12(1):235

Xia L, Chenguang L, Aihua H (2011) Research on the purification of L-propylene carbonate. J Qingdao Univ Sci Technol Natural Sci Ed 32(5):509–513

Xiabin J, Xuesi C, Xinzao Z et al (2001) Synthesis method of biodegradable aliphatic polyester. CN 1306019A

Xie Hongzhou W, Linbo LB (2019) Research progress on the synthesis and modification of biobased polyester—poly(2,5-furandicarboxylic acid ethylene glycol ester). Bioprocessing 5:23–26

Yamane K, Sato H, Ichikawa Y et al (2014) Development of an industrial production technology for high-molecular-weight polyglycolic acid. Polym J 46(11):769–775

Yao G, Yoshio K, Nobuos. (1996) Water absorbing polyurethane foams from liquefied starch. J Appl Polym Sci 60(11):1939–1949

Yaoming Z, Chaoyang W, Hangzhen M et al (2002) Research on the direct synthesis of polylactic acid by melt-solid polymerization. J South Chin Univ Technol Natural Sci Ed 30(11):155–159

Yibin W, Chaoyang S, Wei L et al (2017) Method for preparing high molecular weight polyglycolic acid from glycolic acid or glycolic acid methyl ester. CN 107177032A

Yong L, Dajun C (1988) A new method for preparing biodegradable polyester. Synthetic Rubber Industry 21(6):359

Yoo YD, Km SC (1988) Crystallization behavior of semi-flexible liquid crystalline polyesters and their blends. Polym J 20:1117–1124

Younes GR, Maric M (2021) Bio-based and hydrolytically degradable hydroxyurethane acrylates as photocurable thermosets. J Appl Polym Sci 3:41

Yuxiang S, Jin Z, Lübing Y (2000) One-step synthesis of polyhydroxyacetic acid from chloroacetic acid. Chem Industry Times 12:50–51

Zeterlund. (1997) Stereochemical studies of nitrosamines: the induced circular dichroism of achiral nitrosasmines. Polym Int 42(12):1–8

Zhao D, Zhu T, Li J et al (2021) Poly(lactic-co-glycolic acid)-based composite bone-substitute materials. Bioact Mater 6(2):346–360

Chapter 6
Modification of Biobased Materials

Yajing Zhang, Tao Chen, and Liming Zhao

Due to their green, environmentally friendly, and resource-saving characteristics, biobased materials are gradually becoming another new leading industry that guides contemporary scientific and technological innovation and economic development. However, because of some defects such as their hard nature, poor toughness, and poor heat resistance, the appropriate modification of these emerging biobased materials is required to meet the needs of practical applications (Weng and Fu 2010). The general modification methods for polymer materials can be used to modify biobased materials, mainly include physical and chemical modification. Physical modification changes the processability of biobased materials by adding modifiers or blending with other materials, while chemical modification improves the performance of biobased materials by changing the main chain chemical or surface structure using branching, copolymerization, end group modification, surface modification, etc.

Y. Zhang
State Key Laboratory of Bioreactor Engineering, East China University of Science and Technology, Shanghai, China

T. Chen (✉)
School of Materials Science and Engineering, East China University of Science and Technology, Shanghai, China

Key Laboratory of Biobased Material Engineering of China Light Industry, East China University of Science and Technology, Shanghai, China
e-mail: tchen@ecust.edu.cn

L. Zhao
State Key Laboratory of Bioreactor Engineering, East China University of Science and Technology, Shanghai, China

Key Laboratory of Biobased Material Engineering of China Light Industry, East China University of Science and Technology, Shanghai, China
e-mail: zhaoliming@ecust.edu.cn

L. Zhao (ed.), *Research and Applications of Bio-based Degradable Materials*,
https://doi.org/10.1007/978-981-95-1188-4_6

6.1 Physical Modification

6.1.1 Toughening Modification

1. *Polyamide*

Similar to petroleum-based ones, biobased polyamides have excellent bending strength, good wear resistance, and creep resistance, but poor impact strength and high water absorption rate at low temperature and dry state. Therefore, toughening is needed. Toughening methods commonly applied to petroleum-based polyamides can be used for reference, which includes mixing with rubber or thermoplastic elastomer, other organic polymers, inorganic rigid particle fillers, or core–shell copolymers. Among them, rubber and thermoplastic elastomer are the most effective for toughening polyamides. Adding a small amount of rubber or elastomer to the polyamide can effectively improve the impact strength and low-temperature performance.

(1) Rubber

Rubber has high elasticity and a very low glass transition temperature, and a good toughening effect. Commonly used rubbers are ethylene propylene diene monomer (EPDM) rubber, ethylene propylene rubber, nitrile rubber, styrene-butadiene rubber, polybutadiene rubber, and so on. Fletes mixed styrene-butadiene rubber of recycled tires with PA6 to produce thermoplastic elastomer blends (Fletes et al. 2021). The elongation at break and impact strength significantly increased to 167% and 131% compared with the neat PA6 matrix, which greatly improved the mechanical properties and realized the added-value application of recycled tires.

(2) Thermoplastic Elastomers

Elastomer-modified polyamides have the advantages of low cost, excellent processing performance, and comprehensive mechanical properties. The elastomer generally acts as a soft segment and polyamide chain segment as a hard segment, which can form crystalline microregions and also hydrogen bonds with the soft segment to play the role of physical crosslinking points. It effectively improves the toughness of the modified polyamide, in which the elastomer content affects the hardness, tensile strength, and other physical and mechanical properties of the blend, and the lower the melt flow rate of the elastomer, the better the toughening effect (Ma et al. 2017). Since elastomers are mostly nonpolar materials such as polyolefin whereas polyamides are polar polymers, the compatibility problem that prevents effective blending is generated in research and development of products. The toughening of polyamides with elastomers primarily relies on the reaction compatibility technology of elastomer functionalization, and the graft or block polymers produced by chemical reactions during the processing serve as blend compatibilizers. For example, during melt blending, the grafted maleic anhydride in the compatibilizer undergoes in situ compatibilization reaction with the amine and carboxyl groups at the end of the polyamide chain, which increases the interfacial compatibility, and the dispersion effect of the toughener is significantly improved. When the material is subjected to impact stress, the matrix material can effectively transfer the stress to the dispersed phase of the toughener, thereby significantly improving the toughness of the material. Polymers commonly used for grafting include ethylene-octene copolymer (POE), styrene-butadiene-styrene copolymer (SBS), hydrogenated SBS (SEBS), and ethylene-vinyl acetate copolymer (EVA). As a thermoplastic elastomer, POE has not only the thermoplasticity of plastic but also the crosslinking property of rubber, and it has attracted much attention in toughening polyamide (Zhu 2011).

Yu et al. (2011) melt-blended PA1010 with three types of toughening agents, SEBS, POE, and EVM, in a twin-screw extruder. Using the maleic anhydride grafts of the three toughening agents as compatibilizers. All three kinds of blends showed high impact strength of more than 50 kJ/m^2 when the mass fraction of toughening agent was greater than 20%. As the mass fraction of the compatibilizer was increased, the dispersed particle size decreased from 1 μm to 0.1 μm in the system using POE and EVA as tougheners. The highest low-temperature impact strength (80 kJ/m^2) was obtained when the mass ratio of PA to POE grafted toughener was 80:20.

(3) Other Organic Polymers

Other commonly used organic polymers include polypropylene (PP), polyethylene (PE), acrylonitrile-butadiene-styrene plastic (ABS), liquid crystal polymers, etc. From the perspective of toughening theory, these polymers with lower glass transition temperatures than polyamide will have a certain toughening effect on polyamide, but the toughening range is limited. Most polymers are quite different from polyamide in polarity and are thermodynamically incompatible with polyamide. A "sea-island" structure with high interfacial tension and weak interfacial interaction usually forms when they blend directly to prepare an alloy, for which the mechanical properties are poor, and the introduction of a compatibilizer is required to reduce the interfacial tension. The commonly used compatibilizers are usually maleic anhydride grafted compatibilizers, such as maleic anhydride grafted PP (PP-*g*-MAH), butyl acrylate grafted PP (PP-*g*-BA), dibutyl maleate grafted PP (PP-*g*-DBM), and glycidyl methacrylate grafted PP (PP-*g*-GMA), and so on (Liu et al. 2020).

Kawada prepared bio-alloys based on PA11, PP, and maleic anhydride grafted ethylene-butene rubber copolymer (m-EBR) by a twin-screw extruder and found that the bio-alloy had a flexural modulus of the blend of 1090 ± 20 MPa and the Charpy notched impact strength was 98 ± 5 kJ/m^2 when the mass ratio of PA11:PP:m-EBR was 60:30:10 (Kawada et al. 2017). The field emission-scanning and transmission electron microscopy observations showed that the bio-alloy has a unique "salami-like structure in a co-continuous phase," and some m-EBR formed 10–20 nm wide continuous interphases between the PP and PA11 matrices. Due to the presence of this nanostructured elastomer, the Charpy notched impact strength of the bio-alloy was nearly 10 times and 40 times higher than that of pure PA and

PP, respectively, along with good flexural modulus, making the bio-alloy feasible for applications in the automobile field.

The biopolymer blends of PA1010, biobased high-density polyethylene (bio-HDPE), gallic acid (GA), and maleinized linseed oil (MLO) using a twin-screw extruder were reported by Quiles-Carrillo (Quiles-Carrillo et al. 2019). It was found that the addition of GA significantly improved the thermal stability of bio-HDPE, and MLO could significantly improve the plasticity and impact toughness of the blend. When the mass ratio of PA1010:bio-HDPE:GA:MLO was 70:30:0.8:5, the impact strength of the blend increased from 2.8 ± 0.2 kJ/m^2 without a modifier to 4.3 ± 0.5 kJ/m^2, while the tensile strength dropped from 26.9 ± 1.9 MPa to 23.3 ± 0.6 MPa. It could be attributed to the addition of MLO, which decreased the particle size of bio-HDPE embedded in the PA1010 matrix from 15 μm to 5 μm. The significantly enhanced compatibilization of the blend is beneficial in improving impact strength.

(4) Inorganic Rigid Particles

If the particle size of the inorganic filler reaches the nanoscale, it will have a certain toughening effect. Common inorganic rigid particles include calcium carbonate, talcum powder, montmorillonite, silica, wollastonite, and so on. Li et al. modified hyperbranched PA6 with montmorillonite (MMT) by the in situ polymerization (Li et al. 2019). After adding MMT, the tensile strength, elongation at break, and notch impact strength of the material were significantly increased. The tensile strength, elongation at break, and notch impact strength of the composite material increased, respectively, from 52.67 MPa, 12.65%, 4.75 kJ/m^2 of the pure hyperbranched PA6s to 81.78 MPa, 26.31%, 6.98 kJ/m^2. The nanoscale lamellar structure of MMT greatly enhanced the interaction with PA6. When the composite material was subjected to external force, MMT particles adhered to the matrix and were difficult to detach, which effectively transferred and withstood external stress, causing the surrounding matrix to yield, and hence significantly improved the tensile strength and impact strength. On the other hand, the viscous effect of the large specific surface area of MMT on the relative molecular mass of the polymer acted as crosslinking points. When the material was stretched by external force, the polymer chains oriented along the stretching direction, and the slip between molecular chains increased the tensile strength and elongation at break of the composites. The blend of PA11 and polyphenylene ether (PPO), PA11/PPO 80/20, was modified by amino-functionalized silica (A-SiO_2) (Jeziórska et al. 2021). The amino groups on A-SiO_2 could react with the carboxyl group of PA11 to form a graft copolymer, which significantly improved the hardness of the blend. When the addition amount of A-SiO_2 was 3 wt%, the impact toughness of the blend reached the highest of 19 kJ/m^2.

(5) Core–shell Copolymers

The so-called core–shell copolymer is a copolymer with a flexible molecule as the core and a rigid molecule as the shell. Such a kind of copolymer has a strong toughening effect on polyamides. Liu prepared a unique "core–shell" structure based on high-density polyethylene (PE), high-density polyethylene grafted with

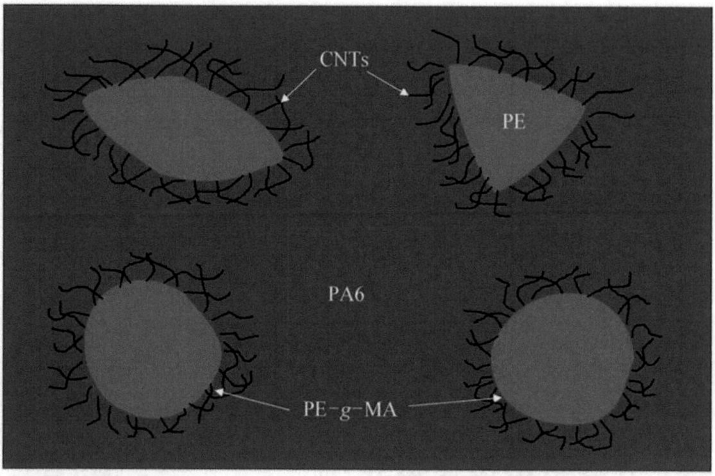

Fig. 6.1 Schematic of "core–shell" structure of PE/PE-*g*-MA/CNT composite (see color insert). (Liu et al. 2021)

maleic anhydride (PE-*g*-MA), and multi-walled carbon nanotubes (CNTs) (Fig. 6.1), and then introduced it into boron nitride (BN) filled PA6 (Liu et al. 2021). In the "core–shell" structure, PE domains acted as the "core," while CNTs and their localizing PE-*g*-MA constituted the "shell." Benefiting from the optimized morphology of the composite introduced with "core–shell" structure, the tensile strength, elongation at break, and impact strength of the composite were improved by 7.3%, 86.7%, and 57.9%, respectively (Liu et al. 2021).

2. *Polyester*

PLA has high tensile strength, but the defects, such as hard nature, poor toughness, and poor heat resistance, limit the application. Toughening modification of PLA can greatly expand the application. Common toughening agents include rubber, elastomers, natural polymers, and their derivatives.

(1) Rubber

PLA was modified by poly (methyl methacrylate)-grafted natural rubber (NR-PMMA) and natural rubber (NR), from which a ternary thermoplastic vulcanizate (TPV) with balanced stiffness and toughness was successfully prepared by peroxide (dicumyl peroxide, DCP) induced dynamic vulcanization (Chen et al. 2018). The preparation process and toughening mechanism are shown in Fig. 6.2.

With 10% of NR and NR-PMMA, the yield stress of the PLA/NR-PMMA/NR ternary TPV was 41.7 MPa, only 38% loss compared with neat PLA, and the impact strength was as high as 91.30 kJ/m^2, about 32 times that of neat PLA. The flexibility of the "soft" NR core and outer "hard" NR-PMMA shell with excellent PLA/rubber interfacial adhesion is responsible for the supertoughness and considerable tensile strength of the ternary TPVs (Chen et al. 2018).

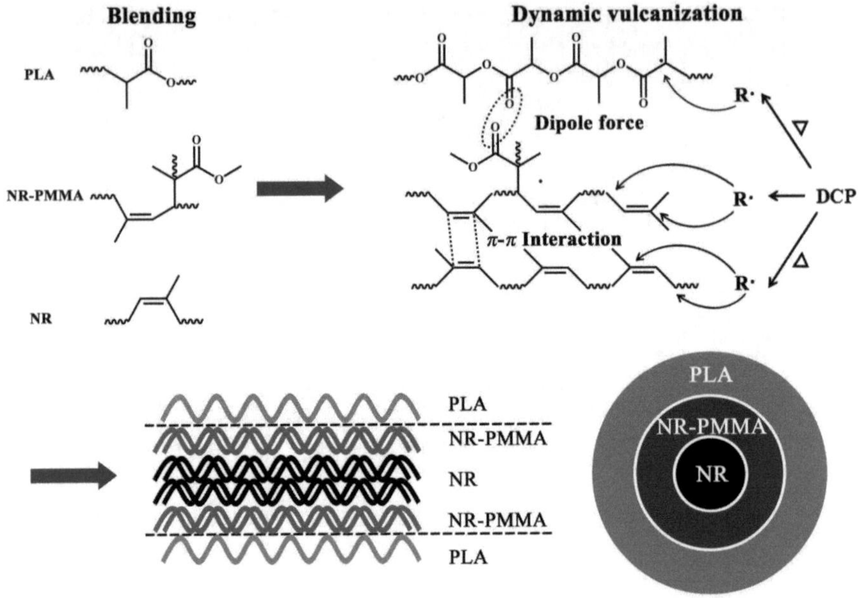

Fig. 6.2 Schemes of in situ compatibilization at the interface between PLA and rubber phases and the final core–shell rubber phase structure (see color insert). (Reprinted with permission from *ACS Sustain Chem Eng*, 2018, 6(5): 6488. © 2018, American Chemical Society (Chen et al. 2018))

Huang used a metal–ligand coordination-induced dynamic vulcanization method to prepare a PLA/butadiene acrylonitrile rubber (NBR)/cupric sulfate ($CuSO_4$) TPV with balanced stiffness-toughness and shape memory performance (Huang et al. 2020a). During dynamic vulcanization, micron-sized $CuSO_4$ particles were converted into nanoparticles by the process of "dissolution-reaction-redissolution-reaction" and dispersed in the continuous NBR phase. $CuSO_4$ nanoparticles cross-linked NBR by the coordination interaction between nitrile groups and copper ions, and simultaneously reinforced the NBR phase, which could withstand greater impact stress and transmit stress more effectively. As the impact stress transfers within the NBR phase, the regular NBR network could absorb impact energy effectively, and then stretched to a larger strain along the impact direction, triggering the formation of obvious plastic deformation to absorb more impact energy, giving the TPV balanced stiffness-toughness properties (notch Izod impact strength of 152.88 kJ/m^2 and tensile strength of 22.30 MPa) (Huang et al. 2020a). Huang modified PLA with a series of bromobutyl rubber (BIIR) based imidazole ionomers (*i*-BIIRs) and found that, compared with pure BIIR and *i*-BIIRs without functional polar groups, introducing polar hydroxyl groups with different alkyl lengths or amide groups into the imidazolium cations of *i*-BIIRs significantly improved the compatibility and impact toughness of PLA/*i*-BIIR blends (Huang et al. 2020b). The cavitation of *i*-BIIR ionomer particles and interfacial debonding, which triggered the yielding of the surrounding matrix, was responsible for the remarkable increase in the compatibility and impact toughness of the PLA/i-BIIR blends (Huang et al. 2020b).

Eucommia ulmoides gum (EUG), an isomer of natural rubber, is a natural polymer derived from Eucommia ulmoides Oliv with both rubber and plastic characteristics. EUG has advantages of high hardness, low melting point, and good processability, and is widely used in plastic products, submarine cables, rubber blends, medical materials, and so on. A fully biobased PLA/EUG blend with high-toughness was prepared by melt blending (Kang et al. 2017). The blends exhibited a maximum elongation at break of 81% and an impact strength of 21.1 kJ/m^2 compare with values of 5% and 2.4 kJ/m^2, respectively, for the neat PLA. Thermal analysis indicated that PLA and EUG were thermodynamically immiscible, so the tensile strength decreased with increasing EUG content, but the elongation at break and the impact strength were enhanced (Kang et al. 2017). Wang modified EUG by bulk-free radical polymerization of glycidyl methacrylate (GMA) to improve its compatibility with PLA (Wang et al. 2019a). The obtained EUG-*g*-GMA was then melt-blended with PLA by in situ dynamic vulcanization to prepare a fully biobased shape memory thermoplastic vulcanizate (TPVs). The TPV with 40% EUG-*g*-GMA displayed the highest toughness with elongation at break of 285% and impact strength of 54.8 kJ/m^2. Compared with the unmodified EUG, the effect of EUG-*g*-GMA greatly improved the toughness of PLA.

(2) Elastomers

Because of their low modulus, high elongation at break, and good resilience, elastomers can effectively toughen PLA. Using both biomass-derived hydrogenated dimer acid (HDA) and excess L-lysine ethyl ester diisocyanate (LDI) as toughening monomers, Liu accomplished the in situ formation and self-cross-linking of flexible biopolyamide (HDAPA) toughener, as well as its reactive compatibilization with PLA in a single melt-blending step, and obtained fully biobased PLA blends with excellent impact toughness and melt-flowability (Liu et al. 2019). When the amount of HDAPA was higher than 10 wt%, a sharp and persistent brittle–ductile transition occurred. When the HDAPA content was 40 wt%, the notched impact strength of the composite was as high as 1347 J/m^2, and the elongation at break was 539%. The matrix yielding triggered by internal cavitation of percolated HDAPA domains, together with the pull-out of many in situ formed block copolymers located at the interfaces, is believed to be the major impact-toughening mechanism (Liu et al. 2019). Ethylene-methyl acrylate-glycidyl methacrylate ternary copolymer (EMA–GMA) elastomer and nanofiller zirconium phosphate (ZrP) were introduced into the PLA matrix by reactive melt blending method to improve its toughness (Wu et al. 2019). The impact strength of PLA/EMA–GMA/ZrP (82/15/3) nanocomposites was improved about 22 times that of pure PLA to 65.5 kJ/m^2. SEM observation of the fracture surface showed a typical core–shell morphology in the nanocomposites, i.e., ZrP was wrapped by the EMA–GMA phase, leading to plastic deformations. Therefore, the supertough effect of PLA was mainly confirmed by effective interfacial compatibilization and massive shear-yielding deformation achieved by the synergy of EMA–GMA with ZrP in the PLA matrix (Wu et al. 2019). Poly (L-lactide)

(PLLA) was modified by melt blending using a thermoplastic elastomer poly (ester-amide) (PEA), synthesized from fatty acid-based precursors, as an impact modifier (Lebarbé et al. 2015). The analysis showed that PEA could effectively toughen PLLA, and beneficial effects of PEA on the crystallization rate of PLLA were observed.

Thermoplastic polyurethane (PU) elastomers have excellent mechanical properties and biocompatibility. Using a continuously and intensively elongational flow field created by an original and revolutionary eccentric rotor extruder, He prepared super tough PLA/thermoplastic polyurethane (TPU) nanofiber composites with in situ formed aligned TPU nanofibers (TNFs), which acted as effective heterogeneous nucleating sites and elevated extraordinarily notch impact strength (He et al. 2019). When the TPU fraction reached 25 wt%, the notch Izod impact strength reached 73.5 and 58.3 kJ/m^2 for the impact direction parallel and perpendicular to the TNFs, respectively, which were 30 and 20 times that of neat PLA, and the elongation at break reached 113%, approximately 12 times (He et al. 2019). Especially, when TPU content was 25 wt% and the impact direction was parallel to TNFs, the composite materials were not completely fractured, and the unfractured part of the nanofiber composites would absorb more impact energy, and the impact strength reached 97.7 kJ/m^2 when it was completely fractured. Super toughened PLA-based blends were prepared by compatibilizing PLA/thermoplastic polyurethane blends with polyurethane elastomer prepolymer (PUEP) as an active compatibilizer (Zhang et al. 2020). The introduction of a compatibilizer improved the interfacial compatibility of the PLA/TPU blend and provided an effective network for impact energy transfer and dissipation. When TPU content was 26 wt% and PUEP was 4 wt%, the tensile strength, elongation at break, and impact strength were 44 MPa, 92.6%, and 81.3 kJ/m^2, respectively. Super toughened polylactic acid/crosslinked polyurethane (PLA/CPU) binary blends with CPU dispersed in the PLA matrix were prepared by reactive blending of PLA with poly (ethylene glycol) (PEG) and polymeric methylene diphenylene diisocyanate (PMDI) (Liu et al. 2014a). The elongation at break and notched Izod impact strength were increased to more than 20 times and 30 times those of neat PLA, respectively. The well dispersion of CPU in PLA matrix and toughness improvement were attributed to the fact that the dispersed CPU was formed by the in situ polymerization of PEG and PMDI in PLA matrix, and the interfacial compatibility between CPU and PLA improved by the reaction of NCO of PMDI with terminal hydroxyl groups of PLA. In addition, the soft PEG segment of the CPU was also a plasticizer for PLA.

Blending PLA with rubber or elastomers has become one of the effective methods for toughening PLA. However, although the addition of elastomers improves the toughness of PLA, it also significantly reduces its strength. Therefore, it is particularly important to find an elastomer that can toughen PLA without reducing its strength.

(3) Natural Polymers and their Derivatives

Natural polymers are macromolecules found in animals, plants, and organisms, such as starch, cellulose, and vegetable oils. Because the raw materials are cheap

Table 6.1 Properties of PLA toughened by natural polymers and their derivatives

Toughener	Amount (wt%)	Tensile strength (MPa)	Tensile modulus (GPa)	Elongation at break (%)	Impact strength (kJ/m^2)	References
–	–	72.0	3.69	2.57	15.3	Thakur et al. (2019)
MIECO	15	37.1	2.7	47.14	35.48	
MHO	10	64	2.7	11.65		Quiles-Carrillo and Montanes (2018)
MLO	2.5		1.0		14.9	Agüero et al. (2020)
–	–	60.2		6.7	3.5	Wang et al. (2018)
Starch	20	52.8		412	31.4	
CCS-D	15	45		361		Dong et al. (2021)
CNCaq-*r*D/CNCaq-*r*D–PDLA	10/10			798	7.9	Muiruri et al. (2019)

and easy to obtain, they are favored by researchers and widely used in toughening PLA. The studies on toughening PLA using vegetable oils, cellulose, and starch are listed in Table 6.1.

Vegetable oil is a renewable resource with biodegradability, environmental friendliness, and easy availability. Thakur synthesized monomethyl itaconated epoxidized downstream corn oil (MIECO) from downstream corn oil derivatives, a coproduct of the bioethanol industry, and itaconic acid, as a green multifunctional toughness modifier for PLA (Thakur et al. 2019). During melt blending, MIECO self-polymerized and reacted with functional groups of PLA to produce a rubbery part within the PLA matrix and compatibilized the blend. When 15 wt% MIECO was incorporated, the elongation at break of PLA-MIECO was 47.14%, the notched Izod impact strength was enhanced to 35.48 kJ/m^2, but the tensile strength was deteriorated to 37.1 MPa.

It was confirmed that, when the material was fractured, the cavities were generated by debonding of polymerized MIECO from PLA matrix, which provided plastic deformation and excellent toughness in the PLA–MIECO blends. Quiles-Carrillo modified PLA with maleinized hemp seed oil (MHO) (Quiles-Carrillo and Montanes 2018). When 10 wt% MHO was incorporated, the tensile modulus of PLA/MHO blend was 2.7 GPa, and the tensile strength and elongation at break were 64 MPa and 11.6%, respectively. Unlike typical plasticizers, MHO not only improved the elongation at break and impact resistance, but also enhanced the elastic modulus and tensile strength. Agüero melt-blended PLA with epoxidized linseed oil (ELO) and maleinized linseed oil (MLO) at 2.5 phr (per hundred resin) and 5 phr, respectively, in the presence of 30 wt% flaxseed flour (FSF) (Agüero et al. 2020). Both multifunctional linseed oils effectively improved ductility, toughness, and thermal stability of the composites, while water diffusion was reduced. Ascribed to the higher solubility of MLO with the PLA matrix, the most optimal performance was obtained for the MLO-containing green composites, which demonstrate the potential utilization of flax by-products or waste from flax.

Starch is a cheap, renewable polymer material that has been used as an environmentally friendly plastic filler. Blending starch with PLA can reduce the cost of PLA and expand its application. Core–shell starch-based nanoparticles (CSSNPs) were prepared via gelatinization of the esterified starch to form a hard core, followed by soap-free emulsion copolymerization of ethyl acrylate to form poly(ethyl acrylate) (PEA) as the outer shell (Wang et al. 2018). The synthetic scheme is illustrated in Fig. 6.3. PLA-based nanocomposites with enhanced toughness were fabricated by blending PLA with CSSNPs. The synergistic association of hard starch and soft PEA domains in the core–shell structure led to unique toughening effects on PLA-based composites (Wang et al. 2018). The "hard" starch core could maintain the strength of PLA, while PEA improved the toughness/ductility. With the addition of 20 wt% CSSNPs, the elongation at break and notched impact strength of PLA composites reached 412% and 31.4 kJ/m^2, respectively, which were 61 times and 9 times higher than those of pure PLA, and the tensile strength was 52.8 MPa. Core–shell starch nanoparticles (CSS-D) with double polymer shells were prepared via a soap-free emulsion polymerization method and then melt-blended with PLA (Dong et al. 2021). With the addition of 15 wt% CSS-D, the elongation at break of PLA/CSS-D reached 361%, which was 46 times that of PLA. The toughness of PLA/CSS-D blends was as high as 130.71 MJ/m^3, which was 49 times that of pure PLA.

By ring-opening polymerization of ε-caprolactone (ε-CL) and D-lactide (D-LA) using cellulose nanocrystal (CNC) as initiator, two fillers with CNC core and a rubber layer (*r*D) composed of ε-CL and D-LA copolymers, and an outer layer of PDLA blocks, named CNC_{aq}-*r*D and CNC_{aq}-*r*D–PDLA, respectively, were obtained (synthesis route is shown in Fig. 6.4), and were then applied to modify PLLA (Muiruri et al. 2019). When compared with the neat PLLA, incorporating 10% CNC_{aq}-*r*D and 10% CNC_{aq}-*r*D–PDLA into the PLLA matrix, the strain at break and

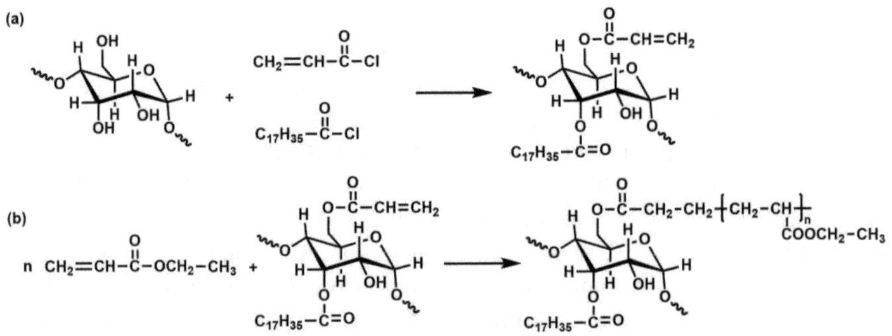

Fig. 6.3 Synthetic scheme of (**a**) esterified starch particles and (**b**) core–shell starch nanoparticles. (Reprinted with permission from *Ind Eng Chem Res*, 2018, 57(39): 13048. © 2018, American Chemical Society (Wang et al. 2018))

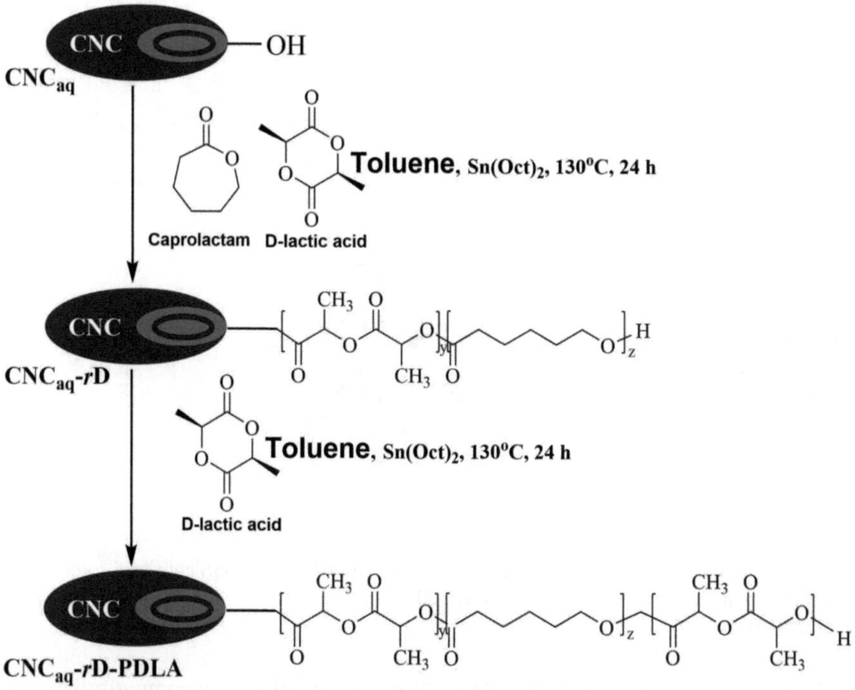

Fig. 6.4 Synthesis route for CNC_{aq}-*r*D and CNC_{aq}-*r*D–PDLA fillers. (Reprinted with permission from *ACS Appl Polym Mater*, 2019, 1(3): 509–518. © 2019, American Chemical Society (Muiruri et al. 2019))

impact strength increased by 100-fold and 3-fold, respectively. Dualism in deformation mechanisms was revealed. The CNC_{aq}-*r*D filler facilitated cavitation under stress and created voids, thus absorbing a lot of fracture energy. The PDLA outer layer of CNC_{aq}-*r*D–PDLA could form strong interactions with the PLLA matrix via stereocomplexation, and CNC_{aq}-*r*D–PDLA filler acted as a craze nucleation site, thereby inducing crazing in the PLLA matrix. The cooperative effect of cavitation of the CNC_{aq}-*r*D and matrix crazing induced by CNC_{aq}-*r*D–PDLA led to pervasive plastic deformation and resulted in huge improvements in both tensile toughness and impact strength (Table 6.1) (Muiruri et al. 2019).

(4) Other Toughening Modifications

Eight-armed polyhedral oligomeric silsesquioxane (POSS) cored star-shaped polycaprolactones (PCL) (SP) were prepared via combining “click” chemistry and ring-opening polymerization (Doganci et al. 2019). The synthetic procedure is shown in Fig. 6.5. SPs with different arm lengths were melt-blended with PLA to improve the toughness of PLA.

When the PLA:SP ratio was 90:10, SP with an arm length of 20 repeating units had the maximum effect on toughening PLA, and the impact strength was twice that

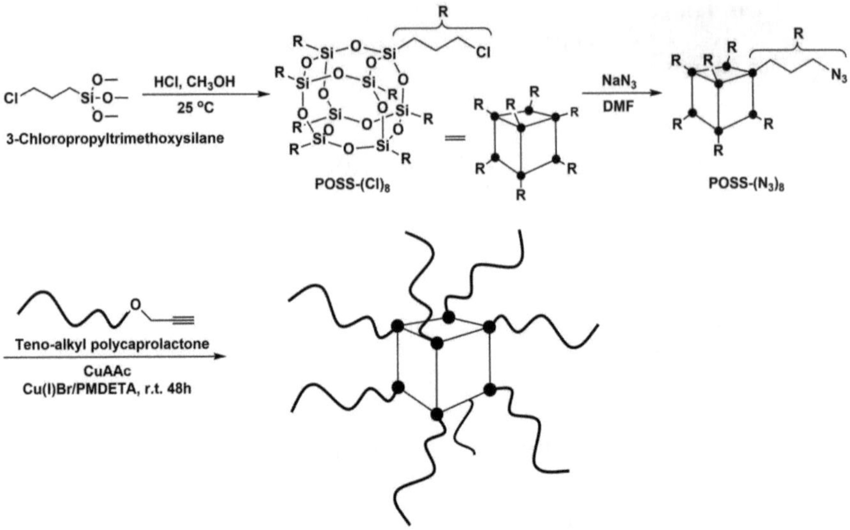

Fig. 6.5 Synthetic procedure of 8-homoarm star-shaped PCL with POSS core (see color insert). (Reprinted with permission from *Eur Polym J*, 2019, 121: 109316. © 2019 Elsevier Ltd (Doganci et al. 2019))

of PLA. The addition of 1,4-phenylene diisocyanate (PDI) as a compatibilizer into PLA/SP blends also significantly enhanced the toughness of the material, and the impact strength reached two to three times that of PLA.

6.1.2 Reinforcement Modification

Fiber reinforcement is an important method to enhance the mechanical properties of polymers. There are many types of fibers used for fiber-reinforced composites, among which the natural fibers mainly include hemp fiber, bamboo fiber, wood fiber, cotton fiber, and so on, whereas the nonnatural fibers mainly include glass fiber, carbon fiber, aramid fiber, polyester fiber, boron fiber, etc. Whiskers can also be used to prepare reinforced composites (Wang and Wang 2008).

1. *Natural Fibers*

In the research on reinforcing biobased polyamides, the most widely investigated reinforcing fillers are lignocellulose, lignin, and mixed fibers. Such a type of natural fiber exhibits good effects on reinforcing biobased polyamides. There are a large number of hydroxyl and oxygen-containing functional groups in natural fibers such as cellulose or lignin. During melt blending, these groups can chemically react or hydrogen-bond with the amide groups in biobased polyamides, or with the terminal amine and carboxyl groups, thereby significantly improving the dispersion effect of natural fibers in nylon matrix. When the composite is subjected to external impact, good interfacial adhesion can effectively transmit the impact stress from the matrix to fibers, thereby improving the mechanical properties of the material.

Newspaper is a very important recycled material, mainly composed of cellulose fibers and some inorganic fillers. Application of recycled newspaper fiber (RNF) can make the material reusable, and is beneficial to protect the environment. Recycled newspaper fiber-reinforced PLA composite fabricated had similar mechanical properties to talc-filled ones (Huda et al. 2005). Compared with neat PLA, the tensile and flexural moduli of PLA composites were significantly improved, and the tensile modulus reached up to 6.3 GPa.

Compared with natural fibers from plants, the benefit of artificial cellulosic fibers is their high cellulose content and the constant quality in regard to their shape and mechanical properties. Feldmann reinforced biobased PA610 and PA1010 with artificial cellulose fibers and found that when fiber content was 30%, tensile modulus and tensile strength of the composite reached 5000 MPa and 120 MPa, respectively, and about two times higher than the tensile strength of neat polyamide (Feldmann and Bledzki 2014).

Flax fiber (FF) is a commonly used reinforcement material that has the benefit of being rich in cellulose, high strength, thin fiber, and durability. Oksman prepared PLA/FF composites with a twin-screw extruder (Oksman et al. 2003). The mechanical properties of the composites were promising, and the strength was about 50% better than that of PP/FF composite used in many automotive panels. Moreover, PLA/FF composites showed no difficulty in extrusion and compression molding processes and can be processed in a similar way as PP-based composites. Lebaupin prepared a flax fiber-reinforced PA11 composite by hot press molding and found that the composite, using type B flax fiber and film-like PA11 as raw materials, manufactured with a temperature value of 210 °C and using gradual levels of pressure had the best mechanical properties with Young's modulus of 36 GPa and tensile strength of 174 MP (Lebaupin et al. 2016).

Oliver-Ortega reinforced PA11 with stone ground wood fibers (SGW) and found that the incorporation of SGW promoted the α' crystalline form of PA11 (Oliver-Ortega et al. 2017). The presence of the fibers formed an amorphous phase with more restricted mobility, which contributed to thermally stabilizing the composites. Sallem-Idrissi prepared fully bio-composites based on PA11 and lignin and found that the incorporation of lignin significantly delayed the kinetics of PA11 crystallization and led to an increase in both the yield stress and the Young's modulus (Sallem-Idrissi et al. 2018).

2. *Nonnatural Fibers*

(1) Glass Fiber-Reinforced Polyamide

Glass fiber (GF) is obtained by drawing molten glass, which is formed in a refractory furnace at 1260–1500 °C using dry mixtures of quartz sand, feldspar, limestone, boric acid, $MgCO_3{\cdot}Al_2O_3$, and so on, into continuous long fibers at a very high speed through a spinneret at the refractory furnace bottom. According to the diameter, GF is divided into primary, intermediate, advanced, and super, with diameters of more than 20 μm, 10–20 μm, 3–9 μm, and less than 3 μm, respectively. The smaller the diameter of GF, the better the flexural performance of reinforced plastic. Glass fiber has the advantage of high tensile strength, excellent chemical resistance, no moisture absorption, low thermal expansion coefficient, high thermal conductivity, good thermal stability, etc. Moreover, it is nonconducting and an ideal electrical insulator, heat and fire resistant, and is important for reinforcing PA. Compared with the neat PA, the tensile strength, flexural strength, hardness, creep resistance, heat distortion temperature, and fatigue resistance of GF-reinforced PA are all greatly improved.

The interface combination between GF and plastic matrix significantly affects the mechanical properties of the composite, which generally needs to be treated with additives. Xie used maleated styrene-ethylene/butylene-styrene triblock copolymer (SEBS-*g*-MA) and epoxy monomer, individually or in combination, to toughen PA6/GF composites (Xie et al. 2010). They found that most PA6/GF binary composites were quite brittle. The addition of epoxy monomer alone did not change the brittle features of PA6/GF composites, because the enhanced interface did not help improve the ductility of the composites. Similarly, without epoxy monomers, adding SEBS-*g*-MA to PA6/GF composites did not increase the ductility much, even when 20 wt% of SEBS-*g*-MA was added. The composites showed prominent ductile characteristics only when both SEBS-*g*-MA and epoxy monomers were added in a certain combination. Especially, when epoxy was 1.0 phr and SEBS-*g*-MA was 10 wt% or epoxy was ≥0.67 phr when SEBS-*g*-MA was 20 wt%, PA6/GF composites showed very prominent ductile characteristics, such as stress-whitening, necking, and large elongation at break exceeding 250%.

(2) Carbon Fiber-Reinforced Polyamide

Carbon fiber (CF) is made by carbonizing polyacrylonitrile fiber, viscose, or asphalt filament with characteristics such as high-temperature resistance and conductivity. The relative density of CF ranges from 1.3 to 1.8, while that of glass fiber is about 2.5. CF reinforced composites are lightweight and high-strength materials not only widely used in the aviation and aerospace industry but also applied in sports and daily necessities, for which the modulus is significantly higher than that of GF reinforced composites.

Kuciel melt-blended short CFs and two types of commercially available bio-based PA1010 with different tensile strengths in a twin-screw extruder (Kuciel et al. 2016). For PA1010 with higher strength and stiffness, the tensile strength of the composites increased from 51.4 MPa of neat PA1010 to 158.0 MPa and 184.9 MPa, respectively, when filled with 20 wt% and 40 wt% of CFs, and the strain at break decreased from an initial 89% to 4.5% and 3.8%. As for PA1010 with lower strength, the tensile strength increased from 26.7 MPa of neat matrix to 71.1 MPa and 102.8 MPa, respectively, when 10 wt% and 30 wt% CFs were introduced, and strain at break decreased from 277% to 10% and 6.2%. In addition, the crystallization temperature of the composites gradually increased while the crystallinity decreased with the increase in CF fraction.

However, the interfacial bonding of CF/PA composites is weak due to the chemical inertness, hydrophobicity, and smooth graphite surface of CF. Moreover, the enormous modulus difference between CF and PA makes it difficult to construct an interface layer with a gentle modulus transition, which may result in stress concentration, resin peeling, and final interface failure. A promising way to improve interface performance is to introduce a rigid or flexible transition layer on the fiber surface to balance the significant modulus difference between the fiber and the matrix. The "flexible" aramid nanofibers (ANFs) were electrophoretically deposited on the CF surface, and then grafted with the "rigid" MXene sheets by ultrasonic impregnation (Zhang et al. 2024). This two-step treatment introduced a hierarchical "rigid-flexible" structure at the CF/PA interface. Results demonstrated that this "rigid-flexible" multilayer structure improved the roughness, chemical bonding, mechanical interlocking, and wettability of CF/PA composites. Meanwhile, the modulus variation between the fiber and the matrix was significantly smoothed due to the increased thickness of the interfacial layer, increasing the payload transfer from the PA matrix to the fiber and decreasing the stress concentration. Compared with the desized CF, the interlaminar shear strength and tensile strength of the modified composite increased by 50.02% and 36.11%, respectively (Zhang et al. 2024).

(3) Other Fiber-Reinforced Polyamides

Whiskers are small single crystals that exist in the form of monofilaments. There are many types of whiskers and the representative varieties are silicon carbide whiskers and calcium sulfate whiskers. Whiskers have very high strength and modulus. Silicon carbide whiskers are about four times and three times those of steel wire in modulus and tensile strength, respectively. In addition, whiskers have finer dimensions and a larger aspect ratio than other fibers used for reinforcement. The length of the calcium sulfate whisker is 100–200 μm, whereas the diameter is only 1–4 μm (Ge et al. 1997). Therefore, introducing whiskers to the polymer matrix can not only increase the melt viscosity but also improve the processing fluidity. Whiskers are also excellent in heat resistance and lightweight. Calcium sulfate whisker has high cost performance due to its high strength but lower price than other whiskers. For maximum reinforcement, whiskers must have the right length, shape, and appropriate cross-section. If the whiskers can be well wetted and oriented in the polymer melt, the tensile strength of the material can be increased by 10 to 20 times. Whiskers used to reinforce polyamides mainly include potassium titanate, aluminum borate, magnesium borate, zinc oxide, calcium sulfate, and calcium carbide whiskers, and so on.

PA1012/calcium silicate whisker nanocomposites were obtained using a twin-screw extruder without any addition of coupling agent (Wang et al. 2016). Attributed to the intercalation of the needle-like shape whisker in polyamide matrix and mutual affinity caused by hydrogen bonding interaction between carbonyl group in polyamide chain and hydroxyl group in whisker surface, even without surface treatment, calcium silicate whisker homogeneously dispersed in polyamide, directly leading to the enhanced mechanical properties and thermal stability of PA1012 (Wang et al. 2016).

Halloysite nanotubes (HNTs) were functionalized by grafting styrene-ethylene-butylene-styrene (SEBS) copolymer chains and then incorporated HNTs into the blends of PA11 and SEBS grafted maleic anhydride (SEBS-*g*-MA) to improve their compatibility (Sahnoune et al. 2017). The thermal performance and toughness of PA11 were highly improved without affecting the PA11 crystallization. By blending spinifex cellulose nanofibers derived from *Triodia pungens* with biobased PA11, bio-nanocomposites, in which the cellulose nanofibers act as a new nano-reinforcement agent, significantly enhanced the tensile strength and toughness of the composites (Rohner et al. 2018).

6.1.3 Flame Retardant Modification

As an engineering plastic, polyamide has poor ablation resistance and cannot maintain its mechanical properties and stability under flame ablation. The vertical combustion grade of PA is UL94V-2, and the limiting oxygen index (LOI) is around 24%. It can self-extinguish at room temperature, showing a certain flame-retardant property. However, with the continuous expansion of applications, higher flame retardants, such as the vertical combustion grade UL94V-0 and LOI more than 28%, is required when polyamides are used in the industries dealing with construction, electrical, and electronic components (Li et al. 2020). The flame-retardant performance and thermal stability of polyamide can be significantly improved by compounding with flame-retardant materials.

The flame-retardant grade of PLA is only UL94HB, with an LOI of only 21%. When it is burned, only a visible carbonized layer is formed, and then liquefied, dripped, and burned quickly.

The flame retardance of polymers is generally improved by the following methods: ① adding flame retardant additives during the compounding process; ② grafting or bonding flame retardant groups on the polymer chains or surface; and ③ copolymerizing with flame retardant monomers (lactams, diamines, or dicarboxylic acids). Adding flame retardants is the commonly used method currently. Although this method can easily affect the mechanical properties of materials, it involves simple steps, has low equipment investment, and has wide applications.

1. *Halogen Flame Retardants*

Halogen flame retardants mainly include brominated flame retardants and chlorinated flame retardants. Halogen flame retardants are well compatible with PA and less affect the mechanical properties of materials, along with high flame retardant efficiency and low cost, and are widely used in the industrial preparation of flame-retardant PA. Traditional halogen flame retardants include dechlorane Plus (DCRP), decabromodiphenyl ether (DBDPO), tetrabromobisphenol A (TBBPA), and so on (Wu et al. 2010). The mechanism mainly relies on the termination of chain reactions during combustion by halogen hydrides in the vapor phase generated by thermal decomposition, which slows down the combustion rate of the material to a stop.

Meanwhile, the generated halogen hydrides lower the surface temperature of the material, dilute the concentration of combustible gases, and achieve the goal of suppressing combustion. Although halogen flame retardants have many advantages, the generated toxic gases, such as CO, HCN, halogenated hydrogen, and dioxins during the flame retardant process not only pollute the environment but inevitably damage the eyes and respiratory system of humans (Lu and Hamerton 2002). Therefore, many Western countries have adopted a comprehensive review and assessment and have strictly restricted and even forbidden the sale of halogen flame retardants. In order to comply with the concepts of environmental protection and safety, new high-efficiency flame retardants represented by brominated polystyrene (BPS) and decabromodiphenylethane have been developed, which are more environmentally friendly and safe while maintaining flame retardant performance, indicating an important development direction for halogen flame retardants. Antimony trioxide, zinc stannate, and zinc borate exhibit a good synergistic effect with halogen flame retardants. When combined with BPS or decabromodiphenylethane, the halogen flame retardant PA with excellent flame retardant performance and smoke suppression can be achieved (Zhang 2014; Horrocks et al. 2012). It is also an important direction to develop compound synergistic flame retardants composed of halogen flame retardants and nitrogen or phosphorus flame retardants. Sulfamic acid-intercalated MgAl-layered double hydroxide was prepared and then added with aluminum phosphate into PA11 (Zhang et al. 2016). The flame retardant performance, thermal performance, and mechanical properties of PA11 were all improved.

2. *Nitrogen Containing Flame Retardants*

Nitrogen-containing flame retardants mainly include melamine (MA), melamine polyphosphate (MPP), and melamine cyanurate (MCA), which are environmentally friendly, UV-resistant, and highly efficient in flame retardancy (Jiang and Yu 2014), and are becoming a hot research and development topic. Nitrogen-containing flame retardants act either in the vapor phase or the condensed phase. Nitrogen-containing flame retardants release nonflammable gases such as N_2, CO_2, and H_2O in the vapor phase by decomposition reaction under the heat, which dilute the combustion supporting gas, lower the surface temperature of the material, and inhibit the proceed of the chain reaction. In the condensed phase, they promote the formation of a charred layer on the surface, which shields the polymer surface from heat and air and suppresses smoke (Ma 2014). In order to fulfill the actual flame retardant demands, nitrogen-containing flame retardants are often combined with antimony trioxide (Li et al. 2008), halogen flame retardants, and phosphorus flame retardants.

MCA is currently the most researched and widely used nitrogen-containing flame retardant (Yang et al. 2016). MCA not only has the flame-retardant ability of general nitrogen-containing flame retardant but also can decompose cyanuric acid under heat to accelerate the material degradation and form droplets to remove heat. Generally, the flame retardant performance can be significantly improved by introducing only 10% MCA. However, MCA has poor compatibility with PA and is prone to moisture absorption, which can affect the mechanical and electrical properties of the material. For better actual utilization, cooperation with other additives is required. Liu synthesized MCA/phosphoric acid melamine (MP) composite flame retardants in phosphoric acid/PA6 solution (Liu et al. 2006). Phosphoric acid acted as the solvent of PA6, the catalyst of the melamine-cyanurate self-assembly reaction, and the reactant of the melamine-phosphoric acid reaction. With the consumption of the acid, the pH value of the system increased, and the dissolved PA6 precipitated on the surface of flame retardant particles to form a polymer encapsulation (Liu et al. 2006). It realized the synthesis and surface modification of the flame retardants in one process. The catalyst and solvent, phosphoric acid, were finally converted into MP product, without an additional removing process. The encapsulated MCA/MP (EMCMP) composite flame retardants were successfully applied in the fire-resistance to glass fiber (GF)-reinforced PA6 and enhanced the flame retardancy of PA6. Since the encapsulation layer of EMCMP was also PA6, good interfacial compatibility and effective dispersion of EMCMP in PA6 resin were obtained, and the corresponding flame retardant materials showed excellent flame retardancy (1.6 mmV-0 level) and mechanical properties (tensile strength of 101 MPa, impact strength of 81 J/m^2, and flexural strength of 116 MPa).

3. *Phosphorus-Containing Flame Retardants*

Phosphorus-containing flame retardants are the most important type, with environmentally friendly properties, smoke suppression, and good stability, and are one of the most widely researched and used halogen-free flame retardants with the best market application prospects. Phosphorus-containing flame retardants are divided into inorganic phosphorus and organic phosphorus categories. Inorganic phosphorus-containing flame retardants mainly include red phosphorus and ammonium polyphosphate (APP), phosphates and their derivatives, etc. Organic phosphorus-containing flame retardants include phosphite esters, organophosphate, and organophosphite, and so on (Zhao et al. 2018). The flame retardant mechanism of phosphorus-containing flame retardants depends on the type of phosphide, the chemical structure of polymers, and the combustion conditions: ① Decomposes into phosphoric acid or polyphosphoric acid during combustion and forms a high-viscosity molten glassiness or dense carbon, which shields the substrate from heat and oxygen; ② captures free radicals. PO·or HPO· and other radicals are decomposed during combustion, which capture active H or·OH radicals in the vapor phase; and ③promotes the formation of a foamed char layer with highly porous, acting as a barrier to heat, air, and pyrolysis products.

Based on the excellent comprehensive performance of phosphorus-containing flame retardants, they have been introduced into the PLA by many researchers. Triphenyl phosphate (TPP) is the first generation of phosphate flame retardants with poor hydrolysis stability and thermal stability. However, due to the low processing temperature of PLA, TPP can be applied. For example, the results of using TPP from Supresta LLC (Phosflex TPP, phosphorus mass fraction is 9.5%) to flame-retardant PLA indicated that incorporating 10% TPP increased the LOI from 21.4% of neat PLA to 25.4% and flame-retardant performance reached the UL94V-0 level. PLA products maintained good thermal stability at 180 °C and no precipitation phenomenon on the surface (Ren and Li 2014). Under different chemical environments, a novel flame retardant (FR 1-4) with two phosphorus groups based on polymeric acrylate was synthesized in three steps (Sag et al. 2020). The result of the UL94 vertical burning test showed that 10% FR 1-4 is sufficient to get a V-0 rating for PLA. By bonding phosphate with silicon-based polysilsesquioxane (PSQ), phosphorus–silicon flame retardant (P_5PSQ) was prepared and used as a flame retardant for PLA (Zhu et al. 2019). The test results in cone calorimetry showed that the LOI value of PLA containing 10 wt% P_5PSQ was 24.1%, and the peak heat release rate (PHRR) and total heat release rate (THR) were 21.8% and 25.2% lower than that of pure PLA, respectively, indicating that the flame retardancy of P_5PSQ was better than PSQ. In addition, the gas release during combustion and the morphology and composition of carbon residue after the combustion of PLA were studied. The results proved that P_5PSQ had flame retardancy in both condensed and gas phases. In the condensed phase, phosphorus in phosphate promoted the formation of a better and more stable carbon layer containing P and Si, inhibiting heat and O_2 transfer during combustion (Zhu et al. 2019). Whereas in the gas phase, the phosphorus-containing compound released from the phosphate in P_5PSQ could inhibit the release of C-O containing products, and had a certain flame-retardant effect on PLA. Therefore, P_5PSQ was a good phosphorus–silicon synergistic flame retardant.

A novel, eco-friendly, phosphorus- and nitrogen-containing flame retardant, ammonium phytate (AP) was developed to improve the flame retardant properties of PA66 textiles (Rahman et al. 2023). The results showed that all flame-retardant-coated fabrics could completely reduce the tendency for melt-dripping and achieve B-3 to B-2 ratings in the UL94 vertical burning test. In addition, the LOI values of treated textiles range from 18.5% to 23.5%, greater than the LOI values of pure fabrics.

4. *Other Flame Retardants*

Magnesium hydroxide (MH) and aluminum hydroxide (ATH) are the most widely used metal oxide flame retardants. The demand for ATH is the largest among all flame retardants in China, up to 31% of the total. The mechanism of metal oxide flame retardants is that, when they are subjected to heat, metal oxide absorb heat and decompose into H_2O and metal oxides, among which H_2O reduces the concentration of O_2 on the material surface and also absorbs some heat, while metal oxides act as a barrier to O_2 and heat and protect the residual material (Guo et al. 2010). Metal oxide flame retardants are halogen-free, low-toxic, thermally stable, and low-cost, making them a hot research topic of environmentally friendly flame retardants. However, in order to obtain the flame retardant performance that satisfies actual requirements, a large amount of metal oxide (>50%) needs to be added. However, due to their hydrophilic and oleophobic surface and poor compatibility with PA, metal oxide flame retardants affect the mechanical properties of PAs (Zhou 2014). Surface modification, ultra-fine processing, microencapsulation, and combined with other flame retardants may resolve these problems (Li and Li 2017). PA6 composites with red phosphorus masterbatch/magnesium hydroxide synergistic flame retardant were prepared using flaky graphite as thermally conductive fillers, which passed UL94-1.6 mm V-0 grade (Li et al. 2016). Woo modified PLA with ATH (0 wt%, 10 wt%, 20 wt%, 30 wt%, and 50 wt%) using chopped kenaf fibers (40 wt% content) as the primary reinforcement (Woo and Cho 2013). Compared with the kenaf/PLA blend without ATH, the flame retardant efficiency of kenaf/PLA/ATH composites improved by 66%, and the storage modulus and tensile modulus were significantly enhanced by 136% and 59%, respectively.

Inorganic nano flame retardants have the advantages of high flame retardant efficiency, being environmentally friendly, and capable of improving compatibility, which currently coincide with discovering new halogen-free environmentally friendly flame retardants. Halloysite nanotubes (HNTs) are nanomaterials with hollow tubular nanostructures and high aspect ratio and have hydroxyl groups on the surface. HNTs provide exceptional mechanical, thermal, and biological properties. When incorporated with polymer, HNTs can simultaneously improve flame retardancy and mechanical properties of the composites, making them excellent flame retardants. The flame-retardant mechanism of HNTs can be ascribed to two aspects. On one hand, HNTs can promote the char formation, which forms a protective "barrier" against heat and oxygen when burning the nanocomposites. On the other hand, due to their tubular structure, HNTs can also form a spatial network that entraps decomposition products in the lumen and impedes the diffusion toward flame and oxygen, delaying mass transport and increasing flame resistance significantly (Liu et al. 2014b). 25% Metal phosphate and 2.5% HNTs were melt-compounded with PA11 via twin-screw extrusion (Hao et al. 2015). The flame-retardant performance of the prepared flame-retardant PA11 composite reached UL94V-0 level, and the toughness of the material was also improved. HNTs and bisphenol-A bis(diphenyl phosphate) (BDP) were introduced into PA6 to prepare PA6/HNT/BDP flame retardant nanocomposites (Boonkongkaew and Sirisinha 2018). The nanocomposites achieved UL94V-0 level, and the tensile strength and thermal stability of the PA6 were retained without loss of stiffness.

The cylindrical nanostructure gives carbon nanotubes (CNTs) excellent electrical and thermal conductivity and mechanical properties, and is also suitable for preparing flame-retardant nanomaterials. Incorporation of CNTs into PA6 significantly improved UL94 level and increased LOI value from 22.7% to 26.5%, and the heat release rate of the composite was also significantly reduced (Sun et al. 2014).

With the rapid development of nanocomposites, the characteristic that various layered silicate particles (kaolin, montmorillonite, and other clay substances) can be homogeneously distributed in the PA matrix has facilitated a new flame retardant process method, i.e., melt intercalation (Fang et al. 2017). By this method, PA6/organically modified montmorillonite (OMMT) composites were prepared (Liu and Zhang 2012). OMMT was found to accelerate the formation of char layers, which enhanced the flame retardancy of the material and significantly reduced the heat release rate. The mechanical properties of PA6 incorporated with 5% OMMT were the best.

Although the modified inorganic nano-scale flame retardants can improve flame retardancy and some mechanical properties when used alone, most of them still cannot satisfy the requirements of actual flame retardancy. Therefore, inorganic nano-scale flame retardants need to broaden their applications through synergies. The synergistic effect between HNTs and MCA was discovered in PA6/MCA/HNTs composites (Li et al. 2015). The incorporation of HNTs increased the LOI value of flame-retardant PA6 composites to a maximum of 31.7%, along with the improvement in mechanical properties. The synergistic effect between CNT and OMMT was found not only improved the flame retardancy and thermal stability of PA6 but also to enhance PA6 (Li et al. 2018). Moreover, in situ polymerization and sol-gel methods are also important methods for developing flame-retardant nano PA.

6.1.4 Alloying

Polymer alloys refer to composites composed of two or more polymer materials. The purpose of developing polymer alloys is to combine and enhance the beneficial properties of different polymers, overcome the original performance defects, facilitate new useful properties and functions, and obtain materials that satisfy actual demand. Generally, polymer alloys should have higher mechanical properties and can be used as engineering plastics. Therefore, they are often termed plastic alloys in industry.

1. *Polyamide*

The dimensional stability, heat resistance, and low-temperature impact strength of polyamides need to be improved, and alloying is an important way to improve the performance of polyamides. There are strong polar amide groups in the polyamide main chain and highly reactive amino and carboxyl functional groups at the chain ends; therefore, polyamide can be easily mixed with and modified by other polymers. Alloying can complement each other's strengths, improve polyamide performance, and meet the special requirements of users in variety and performance with flexible and changeable formulas.

(1) Alloying Between Polyamides

Alloying between polyamides is to combine the advantages and disadvantages of different polyamides through mixing, and obtain alloys with excellent comprehensive properties, and meanwhile, reduce costs to a certain extent. Since the amide bond in polyamide is easy to open and repolymerize at high temperature, the alloy preparation process is often accompanied by the amide exchange reaction. By opening the amide bonds of different polyamides, the exchange between alkyl chains is realized, resulting in various polyamide copolymers. The copolymers obtained through the amide exchange reaction can enhance the interfacial bonding force and interfacial compatibility between different polyamides, and form polyamide alloys with good compatibility. The transesterification reactions of PA1012 and PA612 were investigated (Fig. 6.6) (Wang et al. 2015). Based on the increase in storage modulus with scanning time, the existence of reactive chain ends was confirmed, which promoted chain growth in polyamides and exchange reactions in a binary

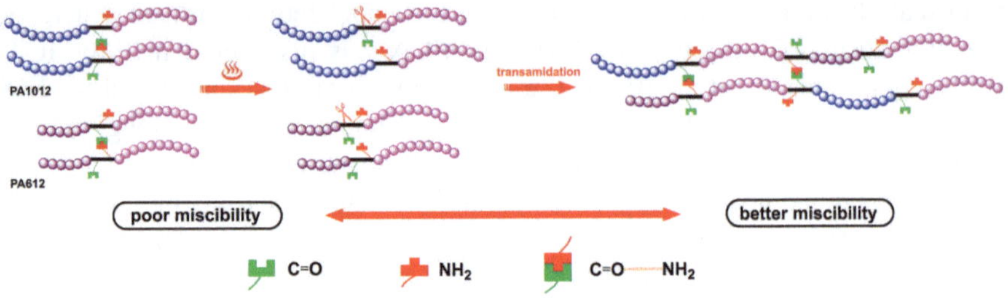

Fig. 6.6 Schematic of the mechanism of improved miscibility of PA1012/PA612 blends (see color insert). (Reprinted with permission from *Polymer*, 2015, 59: 16. © 2014 Elsevier Ltd. (Wang et al. 2015))

blend at high temperatures. NMR and VT-FTIR test signals fully confirmed the expected exchange reactions. Quantitative data analysis of DSC and NMR showed that increasing the reaction temperature or prolonging the reaction time would accelerate the speed and degree of transesterification. The formation of copolyamides could improve the interfacial miscibility between the two long-chain polyamides. The crystallization behavior of PA4 blended with PA56 and PA11 was examined using molecular dynamics simulations and non-isothermal crystallization kinetics (Zhang et al. 2023a). Increasing PA56 and PA11 content reduced crystallinity. PA4/PA56 blends showed phase integration and easier crystallization at higher PA4 content. PA4/PA11 blends exhibited less PA11 participation in crystallization and higher activation energy, indicating poorer crystallization ability.

(2) Alloying of Polyamides with Other Polymers (Wang et al. 2015)

PLA and PA11 were blended by continuous reactive extrusion processing in the presence of *p*-toluenesulfonic acid catalyst (Gug and Sobkowicz 2016). The copolymerization between two immiscible biobased polymers took place to a certain degree through ester-amide exchange reaction, which reduced interfacial tension, improved compatibility, and enhanced mechanical properties. The investigation on interfacial tension, morphology, PLA phase continuity, and thermal behavior of PLA/biobased PA11 blends discovered significant differences in the static and dynamic properties of the blends (Heshmati et al. 2017). Under static conditions, the poor interfacial interaction indicated a highly immiscible system between two polymers, while under dynamic conditions, a highly interactive interfacial compatibility system with significantly lower interfacial tension was formed. The difference was related to the limited mobility of PLA chains, indicating that the static interfacial tension could be significantly reduced by plasticizing PLA.

A chain extender, JoncrylADR®-6368 (Fig. 6.7), was applied to compatibilize PLA and PA11 blends (Walha et al. 2016). The results showed that premixing PLA with the compatibilizer before adding PA11 was responsible for good adhesion between the matrix and the dispersed phase. This result is highlighted by the complete disappearance of the gaps at the interface and the empty cavities corresponding to the extracted particles, indicating the decrease of the interfacial tension. This

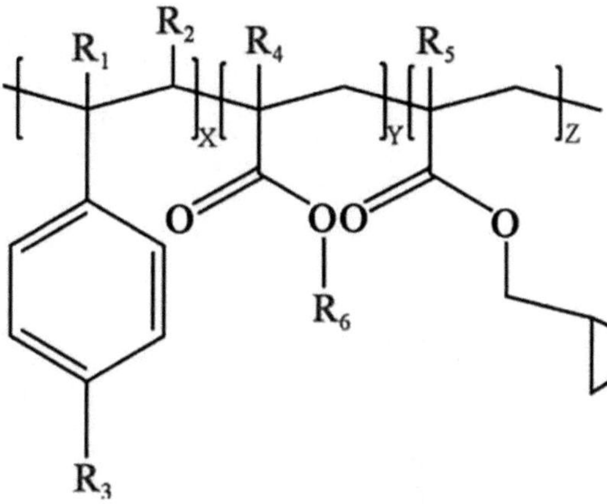

Fig. 6.7 Chemical structure of JoncrylADR®-6368 and the general structure of styrene–acrylic multifunctional oligomeric chain extenders. (Walha et al. 2016)

suggests that by modifying the macromolecular PLA chains, the blends could become more compatible.

R_1-R_5 are H, CH_3, a higher alkyl group, or their combination; R_6 is alkyl group; X, Y, and Z are between 1 and 20 (Walha et al. 2016).

The shape memory behavior, interfacial interaction, morphology, dynamic mechanical, rheological, and mechanical behavior of biobased blends of TPU and PA11 were investigated (Rashmi et al. 2017). A strong interface between two polymeric phases due to hydrogen bonding observed through morphology indicated that TPU and PA11 are well compatible. Strong hydrogen bonding resulted in a strong interface between the two polymeric phases, ensuring good compatibility. Increasing TPU content lowered the melting point of PA11, and enhanced ductility and impact strength with the small reduction in tensile strength. The blends exhibited 95% shape memory recovery, making them suitable for ergonomic and sports applications.

2. *Polyesters*

Since they both have ester groups, biodegradable polyester and PLA are well compatible, and polyester can be used to toughen PLA. Ideal polyester polymers can be obtained by selecting different monomer structures and properties and adjusting the monomer ratio. Table 6.2 lists the common biodegradable polyesters, including poly(adipic acid/butylene terephthalate) (PBAT), poly(butylene succinate) (PBS), and polyhydroxyalkanoates (PHA) for toughening PLA.

(1) PBAT

PBAT is a polyester synthesized through a condensation polymerization of 1, 4-butanediol, butanedioic acid, and terephthalic acid. It has excellent properties such as high toughness, high elongation at break, and biodegradability, making it one of the ideal materials for modifying PLA. PLA was melt-blended with high molecular weight PBAT using a twin-screw extruder (Gigante et al. 2019). When PBAT content was 25 wt%, the elongation at break and notch impact strength increased to 332% and 9.1 kJ/m^2, respectively, indicating that the addition of PBAT

Table 6.2 Performance of PLA blended with biodegradable polyesters

Polyester	Content (wt%)	Compatibilizer and content	Tensile strength (MPa)	Tensile modulus *E* (GPa)	Elongation at break (%)	Impact strength (kJ/ m^2)	Notched Izod impact strength (J/m)	References
–	–			3.04	7.1	3.2		Gigante et al. (2019)
PBAT	25			2.28	332.3	9.1		
PBAT	40		31.87		483.7	5.11		Wang et al. (2019b)
PBAT	40	Joncryl ADR 4370S-0.75 phr	40.88		579.91	29.62		
PBAT	30	PLA–PEG–PLA 3%	54.5	1.14	173.4	–		Ding et al. (2018)
–	–		70.4	1.82	7.4		23.6	He et al. (2020)
PBS	10	MFC-EPI 2%	71.4	1.67	273.6		116.8	He et al. (2020)
PBS	10	EGMA 12%	37		460	98		Xue et al. (2019)
PBSA	40	Joncryl 0.6%			179	38.4		Ojijo and Ray (2015)
PHA	20		40	3.3	4	113		Burzic et al. (2019)

greatly improved the toughness of PLA. To further improve the compatibility between PLA and PBAT, a multifunctional epoxy oligomer, Joncryl ADR-4370S (ADR), was used as a reactive compatibilizer to reactive blend PLA/PBAT with different ratios (Wang et al. 2019b). The elongation at break and notch impact strength of the PLA/PBAT/ADR (60/40/0.75) blend reached 579.9% and 29.6 kJ/m^2, respectively, which were 75.3 times and 12.3 times that of neat PLA. The shear yield deformation results of the impact fracture surface indicate that the high toughness could be attributed to the large number of branched copolymers formed by the reaction of the epoxy group of the reactive compatibilizer with the terminal carboxyl and hydroxyl groups of PLA and PBAT (Wang et al. 2019b). PLA–PEG–PLA triblock copolymers with different chain lengths were synthesized and used to improve the interfacial bonding strength of PLA/PBAT blend using the emulsion interfacial layer method (Ding et al. 2018). When the segment ratio of ethylene glycol/lactic acid was 2:1, the triblock copolymer exhibited attractive compatibilizing efficiency for PLA/PBAT blends and the elongation at break reached 173.4%, which was eight times more than the pristine blends. Therefore, PLA–PEG–PLA triblock copolymers could be used as an effective compatibilizer for PLA/PBAT blends to improve the interfacial interactions and enhance the toughness.

(2) PBS and Copolymers

PBS is synthesized through the condensation reaction using 1,4-butanediol and butanedioic acid as raw materials. PBS is a semi-crystalline polymer with a glass transition temperature of about −35 °C, a melting temperature of 114 °C, excellent mechanical properties, and elongation at break exceeding 300%. Epoxy microfibrillated cellulose (MFC-EPI) was employed as an interfacial compatibilizer as well as a reinforcement filler to prepare PLA/PBS composites (He et al. 2020). When 2 wt% MFC-EPI was added, the elongation at break and tensile strength of PLA/PBS/MFC-EPI (90/10/2) composite increased to 273.6% and 71.4 MPa, respectively, and the notch impact strength was 116.8 J/m^2. The presence of MFC-EPI contributed to the obviously increased tensile and impact toughness of the composites due to energy transfer and dissipation during fracture as a result of the bridge effect. Super-tough ternary blends were prepared by melting ternary blends of PLA, PBS, and ethylene-methyl acrylate-glycidyl methacrylate (EGMA) (Fig. 6.8) (Xue et al. 2019). EGMA was selectively localized at the PLA–PBS interface and resulted in improved interfacial adhesion. Compared with virgin PLA, the ternary blend with 10 wt% PBS and 12 wt% EGMA exhibited a 32 times higher impact strength (98 kJ/m^2), and superior stretchability (elongation at break 460%), with only 44% loss of tensile strength (38 MPa) (Xue et al. 2019). PLA/PBS-*g*-EGMA copolymers were formed at the PLA–PBS interface, which acted as a "bridge" connecting these phases and significantly improved interfacial adhesion between immiscible PLA and PBS phases. The number of the "bridges" increased with EGMA content and led to improved compatibility. When subjected to external forces, these "bridges" transfer the force from the brittle PLA matrix to the ductile PBS phase, causing an elongational deformation in the PBS phase and consequently dissipating the fracture energy. PLA was melt-blended with poly [(butylene succinate)-*co*-adipate

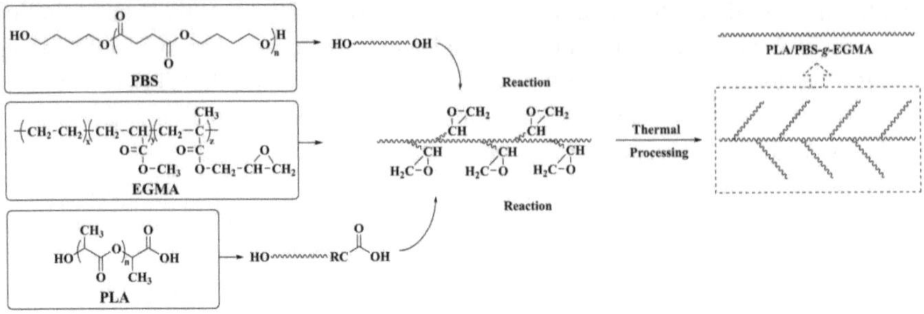

Fig. 6.8 Possible in situ reactions between PLA, PBS, and EGMA occurring during melt compounding (see color insert). (Reprinted with permission from *Compos Part B-Eng*, 2019, 172: 743. © 2019 Elsevier Ltd (Xue et al. 2019))

(PBSA) using Joncryl as the compatibilizer (Ojijo and Ray 2015). Due to the presence of multifunctional reactive Joncryl, a nonlinear copolymer architecture was formed at the interface. Adding 40% PBSA and 0.6% compatibilizer, the impact strength of PLA increased from 4.6 kJ/m^2 to 38.4 kJ/m^2, and the elongation at break increased from 6% to 179%.

(3) PHA

PHA is an intracellular polyester synthesized by microorganisms. It is a natural bio-macromolecule with good biocompatibility, biodegradability, and plastic processability that can be used as biomedical materials and biodegradable packing materials. Two different amorphous PHA copolymers were used to improve the impact toughness of PLA without compromising the biobased carbon content and compostability (Burzic et al. 2019). When PHA-2 with a higher weight average molecular weight was 20% and after being annealed, the tensile modulus of PLA/PHA-2 blend was 3.3 GPa, and tensile strength, fracture elongation rate, and unnotched impact strength were 40 MPa, 3%, and 113 kJ/m^2, respectively. The properties of PLA/PHA hybrid materials for 3D printing were investigated (Torabi et al. 2023). Adding PHA reduced shear and extensional viscosities, improving elongation at break and adhesion strength while maintaining comparable tensile strength with PLA. Nozzle temperature had the most significant impact on adhesion strength and tensile characteristics. Samples printed at the nozzle temperature of 215 °C showed an order of magnitude higher adhesion strength and up to 34% enhancement in elongation at break compared with the printed samples at 185 °C.

6.1.5 Other Functional Modifications

The conductivity mechanism of biobased PA can be described by percolation theory. When the mass fraction of conductive filler is small, the composite exhibits insulator characteristics. When the mass fraction of the filler reaches a specific value, the composite conductivity will change abruptly, indicating that the dispersion state of the filler in the matrix has changed, and a percolation network is formed. As the mass fraction of the filler continuously increases, the composite conductivity will not increase significantly.

The modification of the electrical properties of biobased PA mainly focuses on improving the conductivity and dielectric properties. Biobased PA11 and graphene nanoplatelets (GNP) were blended by melt extrusion using a masterbatch dilution process (Rashmi et al. 2018). It was found that GNP was uniformly distributed in PA matrix, which significantly improved the electrical conductivity and dielectric constant of PA11. With an increase in GNP loading, crystallization temperature, and degree of crystallinity of the nanocomposites increased. When 5% GNP was incorporated, the dielectric constant reached 9.2 at 1000 Hz, which was three times higher than that of neat PA11, and the conductivity reached 5.2×10^{-6} S/m. Compared with neat PA11, tensile strength and modulus of nanocomposites increased by 25% and 56%, respectively, while elongation at break decreased by 80%. Nanocomposites with PA11 and layered clays were prepared by an extrusion process (Leveque et al. 2017). The effects of filler type (Cloisite 20A, 10A, and Na^+) and content on the piezoelectric and dielectric properties of the blend film were examined. The results showed that the piezoelectric constant of nanocomposites at room temperature was related to the crystalline phase and the nature of the filler. The nanocomposites with Cloisite Na^+ loaded showed the best piezoelectric constant, the highest polarization performance, and the largest polarization response. It was also found that by adding 5% silicate filler, the mechanical properties of nanocomposites were improved, and the tensile modulus increased from 840.5 MPa of pure PA11 to 1107.5 MPa.

6.2 Chemical Modification

6.2.1 Branching Modification

1. *Polyamide*

A series of linear PA4 and branched PA4 were prepared to study the effect of the molecular topological structure on the properties (Kawasaki et al. 2005). The results showed that polyacyl chlorides were effective initiators for synthesizing branched PA4. The branched structure had less effect on the melting point of PA4. Compared with linear PA4 with a similar molecular weight, the tensile strength of branched PA4 was significantly higher, and the biodegradability was better. PA4/6 branched copolymers were prepared by using polyacyl chloride as an initiator (Fig. 6.9) (Kawasaki et al. 2020). It was found that as the proportion of caprolactam monomer increased, the melting point of the copolymer decreased, while the thermal decomposition temperature increased, and the elongation at break of the copolymers with an intermediate monomer ratio also significantly increased.

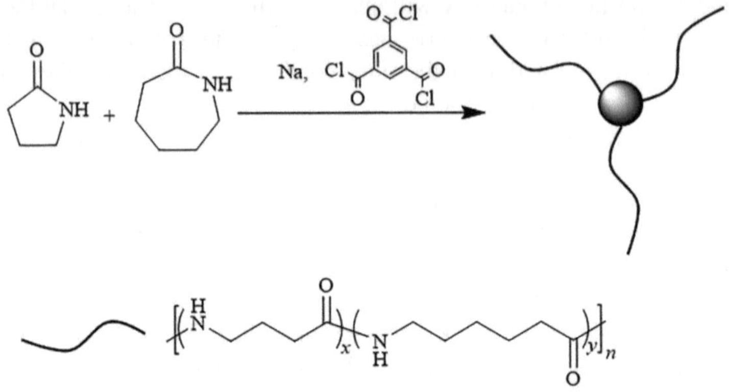

Fig. 6.9 Synthesis and structure of branched polyamide 4/6. (Kawasaki et al. 2020)

2. *Polyester*

To improve the performance of PLA, four-arm star-shaped PLA was synthesized and functionalized with glycidyl methacrylate (GMA) (Pasanphan et al. 2021). The results showed that PLA–GMA exhibited good compatibility with PLA and promoted PLA crystallization. Further structural modification was induced by electron beam (EB) irradiation, as shown in Fig. 6.10, and it was found that the crystallization performance, thermal performance, and mechanical properties of the modified material were all improved. Change of the crystallization led to an increase in the thermal stability of the modified PLA by about 14 °C and to increase the flexibility due to increasing the elongation at break by twofold The EB irradiated star–PLLA–GMA/PLA exhibited evidence of microwave-resistant property and was expected to be used in the field of microwave bioplastics for food packaging. Highly branched PCL was synthesized via bulk ring-opening polymerization catalyzed by $Sn(Oct)_2$ and employed hyperbranched polyamide (HPPA) as a macro-initiator (Ziolo et al. 2024). The obtained products showed core–shell structure with HPPA core and linear PCL chains as shell; PCL-HPPA displayed a similar degree of crystallinity to linear neat PCL, but branched products possessed smaller and less regular crystallites.

6.2.2 Copolymerization Modification

1. *Copolymerization from Monomers*

The properties of biobased polymers can also be changed by copolymerization. By adding crown ether, the reactivity of the activated monomer anion was increased, and PA4/6 copolymers with higher yield and higher molecular weight were obtained (Kim et al. 2013). Although the molecular weight less affected the melting point and

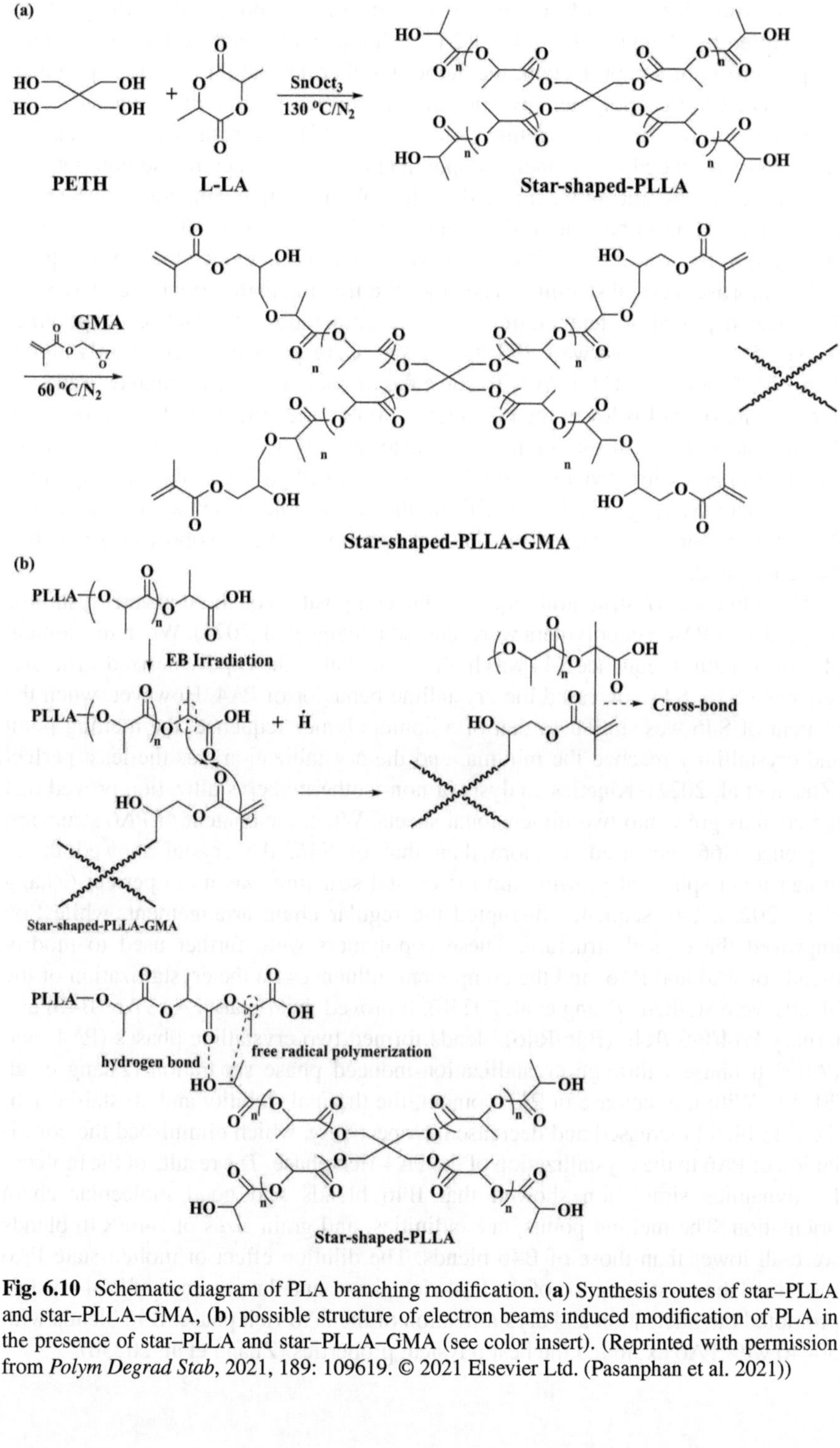

Fig. 6.10 Schematic diagram of PLA branching modification. (**a**) Synthesis routes of star–PLLA and star–PLLA–GMA, (**b**) possible structure of electron beams induced modification of PLA in the presence of star–PLLA and star–PLLA–GMA (see color insert). (Reprinted with permission from *Polym Degrad Stab*, 2021, 189: 109619. © 2021 Elsevier Ltd. (Pasanphan et al. 2021))

thermal properties of PA4/6 copolymers less, the thermal properties of the products were varied by changing the ratio of 2-pyrrolidone and ε-caprolactam. The higher the proportion of ε-caprolactam, the higher the thermal stability of the copolymer. A series of PA4/6 copolymers was prepared to further disclose the influence of the feed ratio and reaction temperature (Kim et al. 2014). The results showed that the final products tended to be random copolymers, and the structure and composition of the products significantly depended on the polymerization temperature. The thermal stability of the copolymer is determined by chemical composition and sequence distribution. Incorporation of PA6 units was beneficial to reduce the melting point and to improve thermal stability. When the PA6 fraction in the copolymer was 35%, the onset degradation temperature and the temperature at which each polymer showed 50% weight loss were 292 °C and 359 °C, respectively, significantly higher than 270 °C and 288 °C for PA4. To solve the problem that the thermal degradation temperature of PA4 is lower than the melting point, the influence of the initiator on the thermal stability of the products was also investigated. Using an initiator containing aromatic nitro groups ensured PA4 maintained good thermal stability in the low-temperature range (210–230 °C), and the above conclusion that the increase in PA6 unit proportion increased the thermal stability of PA4/6 copolymers was further confirmed.

The influences of structural sequence and composition on thermal and crystalline properties of PA4/6 copolymers were studied (Zhang et al. 2022). When the content of PA4 structural sequence S44 was higher than that of the copolymerized structural sequence S46, S44 possessed the crystalline behavior of PA4. However, when the content of S46 was similar to that of a homopolymer sequence, the melting point and crystallinity reached the minima, and the crystallization was the least perfect (Zhang et al. 2022). Kinetics analysis of non-isothermal crystallization proved that the crystals grew into two-dimensional sheets. When the content of PA6 structural sequence S66 increased to more than that of S46, the crystal showed three-dimensional spherical growth, and the crystal structure was more perfect (Zhang et al. 2022). S46 sequence disrupted the regular chain arrangement, while S66 improved the crystal structure. These copolymers were further used to modify blends of PA4 and PA6, and the component influences on the crystallization of the blends were studied (Zhang et al. 2023b). It proved that binary PA4/PA6 (B46) and ternary PA4/PA6/R46 (B46/R46) blends formed two crystalline phases (PA4- and PA6-rich phases) through crystallization-induced phase separation (Zhang et al. 2023b). With the increase in PA6 content, the thermal stability and crystallinity of the B46 blend increased and decreased, respectively, which diminished the contribution of PA6 to the crystallization of the PA4-rich phase. The results of the molecular dynamics simulation showed that B46 blends had good molecular chain orientation. The melting points, crystallinities, and grain sizes of B46/R46 blends were all lower than those of B46 blends. The dilution effect of molten-state PA6 restrained the crystallization of the PA4-rich phase, and the presence of PA4 crystals promoted the nucleation and crystallization of the PA6-rich phase. B46 blends with 30–40 wt% PA6 exhibited the best physical properties (Zhang et al. 2023b).

By using caprolactam, PA66 salt, and PA1010 salt as raw materials, a low-melting-point thermally bonded ternary copolyamide was prepared (Gong et al. 2003). Compared with homopolymer PA, the copolymerization destroyed the chain regularity and the amide bonds deviated from the best position to form a hydrogen bond, which decreased hydrogen bond density, weakened the intermolecular forces, and resulted in a reduction in melting point of copolyamides and more obvious shear-thinning.

Dimer acid (DA)-modified polyamide 11 copolymers with a certain hydrophobicity were synthesized via melt polycondensation using the hydrolysis product of waste vegetable oil as raw material (Park et al. 2019). The thermal and mechanical properties were modified by increasing the composition of DA in copolymers, increasing the chain irregularity, thereby reducing the hydrogen bonding strength of the amide groups (Park et al. 2019).

2. *Block Copolymers*

Block copolymerization can also significantly improve the performance of PA4. By using 4,4′-azobiscyanopentanoyl chloride as an initiator, PA4 with azo group end (azo-PA4) was synthesized through ring-opening polymerization of 2-pyrrolidone (Kawasaki et al. 2012). It proved that azo-PA4 had initiating activity for vinyl monomers and could be applied to prepare block copolymers by free radical polymerization. PA4-polyvinyl acetate block copolymers were then synthesized using azo-PA4 as an initiator (Fig. 6.11), and the properties of which combine amorphous polyvinyl acetate and crystalline PA4 were examined (Kawasaki et al. 2015). It was found that the heat of fusion was linearly dependent on PA4 content. The mechanical properties of the block copolymers were monotonically dependent on the composition, i.e., increasing PA4 content increased tensile strength, whereas increasing poly(vinyl acetate) content increased elongation at break. Quite different from that of both the PA4 and PVAc, the morphology of the block copolymers suggested the appearance of microphase separation.

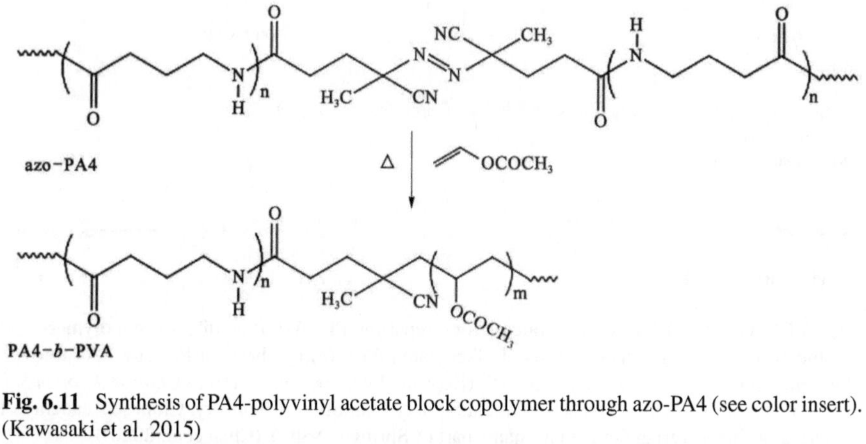

Fig. 6.11 Synthesis of PA4-polyvinyl acetate block copolymer through azo-PA4 (see color insert). (Kawasaki et al. 2015)

PA4-polyurethane (PU)-PA4 triblock copolymers were synthesized and applied to investigate the effect of the block length of hard PA4 and soft PU on the properties of the triblock copolymer as a thermoplastic elastomer (Park et al. 2014). The results showed that the mechanical strength and melting point of the copolymer increased with the increase in molecular weight of the PA4 block, while elongation at break increased with the increase in that of the PU block.

In order to develop a simple and efficient method for synthesizing PA4-based biodegradable block copolymers, "click" reaction was applied to prepare PLLA-*b*-PA4 block copolymers using PLLA and PA4 as raw materials (Fig. 6.12), which were then electrospun into nanofibers (Chen et al. 2020). The results showed that PLLA and PA4 were incompatible and formed individual crystalline regions, respectively, and faster-formed PA4 crystalline regions inhibited the crystallization of PLLA. The crystallinity of either PLLA or PA4 decreased after electrospinning, and PLLA-rich fibers were almost amorphous. Electrospun fibers were highly hydrophobic, even if hydrophilic PA4 was rich. The fibers electrospun from block copolymers were relatively homogeneous due to micro-phase separation between PLLA and PA4, while those electrospun from PLLA and PA4 blends exhibited a sheath-core structure with a PLLA outer layer (Fig. 6.13), attributed to macro-phase separation between PLLA and PA4 homopolymers. Since both components are completely biobased and biodegradable, this new type of fiber has potential applications in biomedicine, packaging materials, and other fields.

Due to PA6's insolubility and high melting point, transesterification is easy to take place. Therefore, the synthesis of di- or triblock copolymers with individual blocks of PA6 and any other polyester becomes a challenge. Dou et al. proposed a new mild reaction condition for the synthesis of PCL-*b*-PA6-*b*-PCL triblock copolymer by melt polycondensation and ROP in solvents (Fig. 6.14) (Dou et al. 2022).

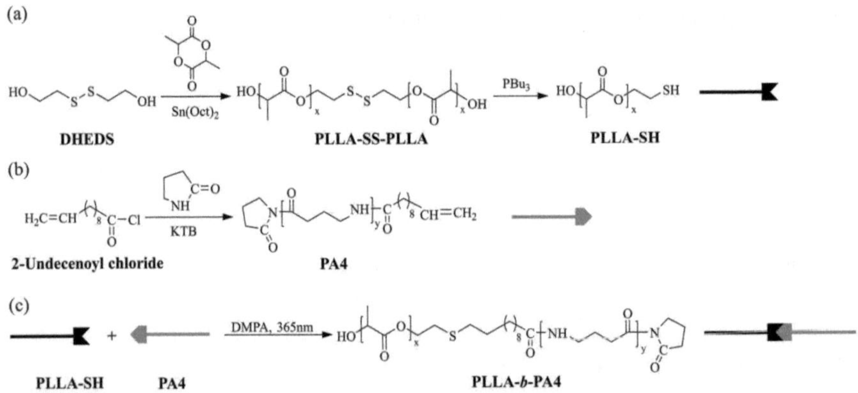

Fig. 6.12 The "click" reaction strategy for preparing PLLA-*b*-PA4 diblock copolymers. (**a**) Synthesis of PLLA-SH, (**b**) synthesis of alkenylated PA4, (**c**) synthesis of PLLA-*b*-PA4 through thiol-ene "click" reaction (see color insert). (Reprinted with permission from *Chinese J Polym Sci*, 2019, 38(1): 53. © 2019, Chinese Chemical Society Institute of Chemistry, Chinese Academy of Sciences Springer-Verlag GmbH Germany, part of Springer Nature (Chen et al. 2020))

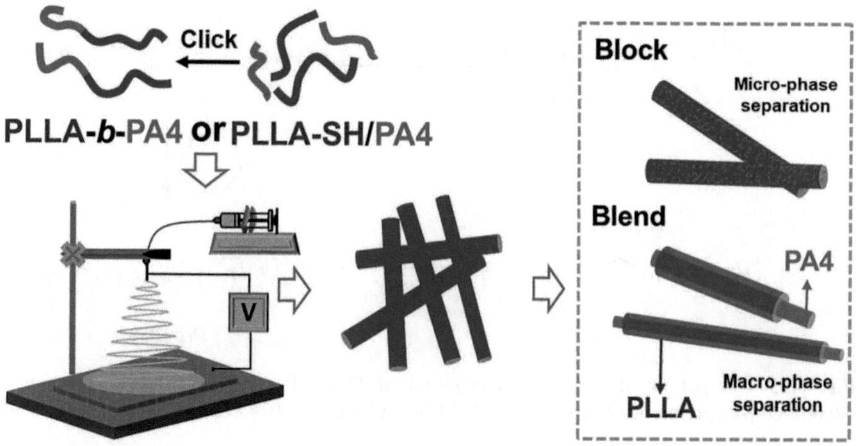

Fig. 6.13 Schematic diagram of the fibrous structure of block copolymers and blends (see color insert). (Reprinted with permission from *Chinese J Polym Sci*, 2019, 38(1): 53. © 2019, Chinese Chemical Society Institute of Chemistry, Chinese Academy of Sciences Springer-Verlag GmbH Germany, part of Springer Nature (Chen et al. 2020))

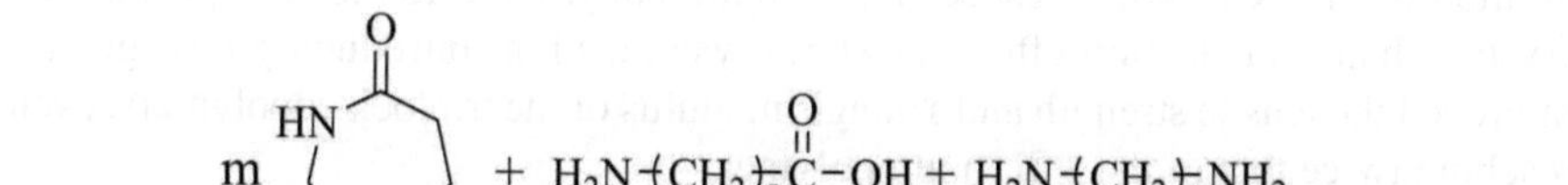

Fig. 6.14 Schematic diagram of the synthesis of PCL-*b*-PA6-*b*-PCL. (Reprinted with permission from *Macromolecules*, 2022, 55(22): 10090. © 2022 American Chemical Society (Dou et al. 2022))

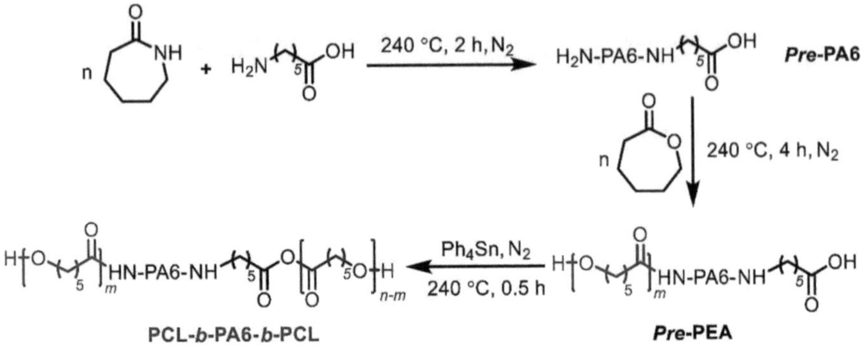

Fig. 6.15 One-pot in situ bulk polymerization of PCL-*b*-PA6-*b*-PCL triblock copolymers. (Reprinted with permission from *Polymer*, 2024, 300:126968. © 2024 Elsevier Ltd (Gu et al. 2024))

The lower reaction temperature and solution polymerization inhibited the ester-amide exchange, resulting in well-defined block copolymers. The thermal property and crystallization behavior of triblock copolymers depended highly on their compositions. By increasing the length of the PCL block, the cooling crystalline temperature of PA6 first increased and then decreased due to the plasticization and dilution of the PCL block, respectively. Additionally, the PA6 block, especially its crystals, had a nucleation effect on PCL crystallization. Introducing PA6 greatly improved the tensile strength and Young's modulus of the triblock copolymers, even reaching twice that of the PCL homopolymer.

The high quality image has been uploaded.More recently, Gu et al. proposed a simple and green method for solvent-free preparation of PCL-*b*-PA6-*b*-PCL triblock copolymer using tetraphenyl tin as a catalyst (Fig. 6.15) (Gu et al. 2024). The results showed that for ROP of both CPL and CL, the catalyst Ph_4Sn greatly limited the formation of cyclic oligomers. The obtained copolymers had a higher number-averaged molecular weight of more than 25,000 and the polydispersity index was narrower than 2. The crystallization behavior was greatly affected by PCL block length. Few CL units existed in the PA6 crystalline, acting as a defect to induce a large proportion of amorphous area during crystallization (Gu et al. 2024). When the PCL segment was long enough, microphase separation between PA6 and PCL occurred during cooling, resulting in independent crystallization of the PA6 and PCL segments. The copolymer consisting of 80% PCL had a minimum water absorption of 2.62%, which facilitates the process of PA6 into precision electronic devices.

Block copolyamide 611 used as an engineering plastic was prepared using PA6 prepolymer and PA11 prepolymer as raw materials (Gao 2006). The incorporation of PA6 prepolymer improved the tensile strength of the block copolyamide 611, and the maximum tensile strength was almost a onefold increase in comparison with pure nylon 11 when the content of PA6 prepolymer was 10 wt%.

A series of segmented copolymers consisting of PA1012 crystalline blocks and polytetramethylene oxyether (PTMO) amorphous soft blocks was synthesized (Zhu et al. 2017), which greatly improved the toughness of PA1012. Due to the strain-induced crystallization of the polyether soft segment under large strain, the true stress-strain curve of the prepared copolymer elastomer deviated from the classical Gaussian model of plastic copolymers, making PA1012 a good elastic engineering plastic.

Fully biobased polyamide 11 (PA11)/polylactide (PLA) diblock copolymers were prepared adopting a two-step synthesis strategy involving first preparation of an amino-terminated PA11 prepolymer (i.e., with an amino-end group and a carboxyl group protected), which was subsequently employed as a macro-initiator in the ring-opening polymerization of D-lactide (or L-lactide) (Fig. 6.16) (Gardella et al. 2018). By adjusting lactide conversion, diblock copolymers with different PA11/PLA ratios were obtained. Due to that PLA segments were quickly (and exclusively) involved in the formation of the stereocomplex even in the case of fast cooling (quenching) from the melt, fast stereocomplexation was demonstrated when the enantiomeric diblock pairs with similar composition were mixed, and hence the phase separation, happened in the case of incompatible PLA/PA11 blends, was not observed. In addition, compared with PLA homopolymers, the stereocomplex has better chemical/thermal resistance, which can improve the quality of the material.

3. *Graft Copolymers*

Graft copolymerization changes the molecular structure of polymers, thereby affecting the crystallization and mechanical properties of the materials. By adopting the "one-pot" method, long-chain PLA was grafted onto SiO_2 nanoparticles and a series of PLA/SiO_2-*g*-PLA nanocomposites with different nanofiller loadings were

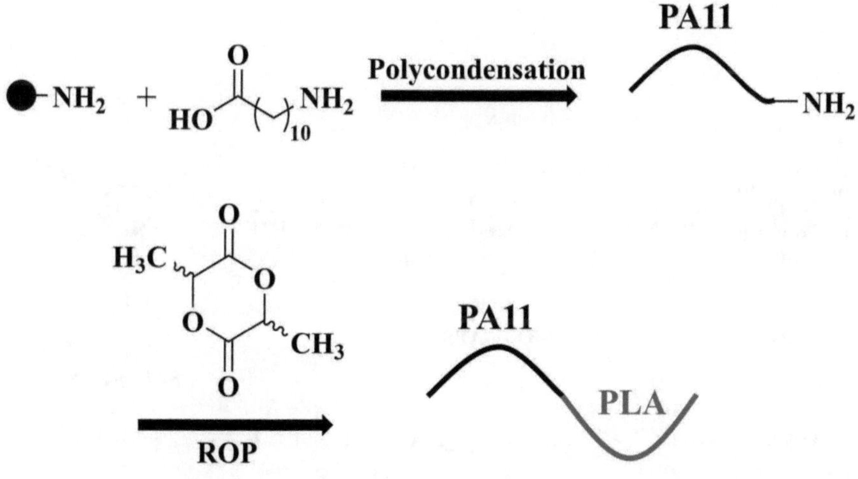

Fig. 6.16 Scheme of two-step synthetic strategy of PA11/PLA block copolymers (see color insert). (Reprinted with permission from *Eur Polym J*, 2018, 98: 83. © 2017 Elsevier Ltd (Gardella et al. 2018))

prepared by melt compounding (Fig. 6.17) (Wen 2019). The results showed that when the grafting rate of PLA on the surface of SiO_2 was 19.5%, the nanoparticles exhibited a core–shell structure. The grafted silica acted as an effective nucleating agent and positively improved the crystallization rate and increased the crystallinity. Meanwhile, the PLA nanocomposites presented outstanding mechanical properties, including excellent toughness and high stiffness. With 3 wt% SiO_2-*g*-PLA, the PLA nanocomposites showed the optimal comprehensive performance, with the crystallinity as high as 56.3%, and elongation at break and impact strength of 118.6% and 14.5 kJ/m^2, respectively (Wen 2019).

Maleic Anhydride (MAH) was grafted onto PLLA using dicumyl peroxide (DCP) as a radical initiator (Hwang et al. 2012). With the addition of MAH, the glass transition temperature and crystallinity decreased significantly. The thermal decomposition of PLLA film was affected by MAH content, whereas the mechanical properties were almost unchanged. Due to the branching reaction of MAH or a possible cross-linking reaction between PLLA chains, the molecular weight slightly increased.

Dendritic PA6/MMT hybrid materials were prepared via in situ graft polymerization using a dendritic initiator and amino-functionalized MMT (AMT) (Fig. 6.18) (Lai et al. 2020). The results showed that the amino functional groups of MMT were the initiators of PA6 polymerization, which function as branching points and inhibit spherulite growth. When adding AMT to PA6, the α-form of PA6 degraded, and the crystallinity of PA6 decreased. With the increase of AMT content, the strength and toughness of composites gradually increased due to the refining effect of AMT on the crystal particle (Lai et al. 2020).

Using ethylene maleic anhydride syndiotactic copolymer resin (ZeMac) and triglycidyl isocyanurate (TGIC) as modifiers, a grafting structure and a long-chain

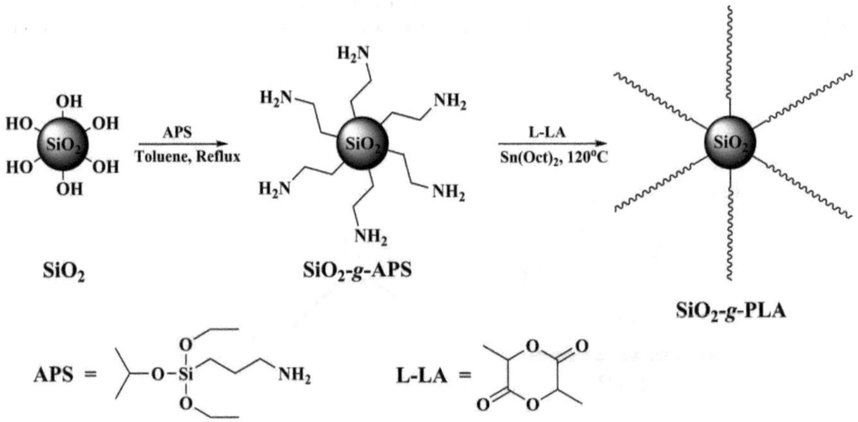

Fig. 6.17 The scheme of surface modification of silica nanoparticles (see color insert). (Wen 2019)

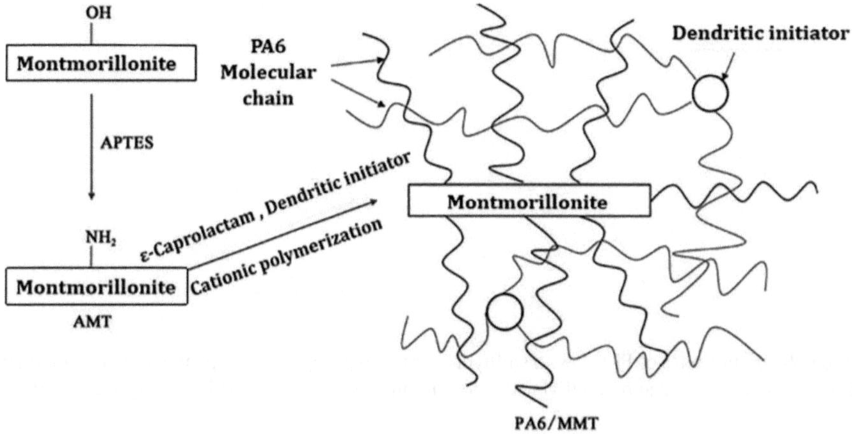

Fig. 6.18 Cationic polymerization of PA6 initiated by amino-functionalized MMT and dendritic initiator. (Reprinted with permission from *Mater Today Commun*, 2020, 25(1): 101578. © 2020 Elsevier Ltd (Lai et al. 2020))

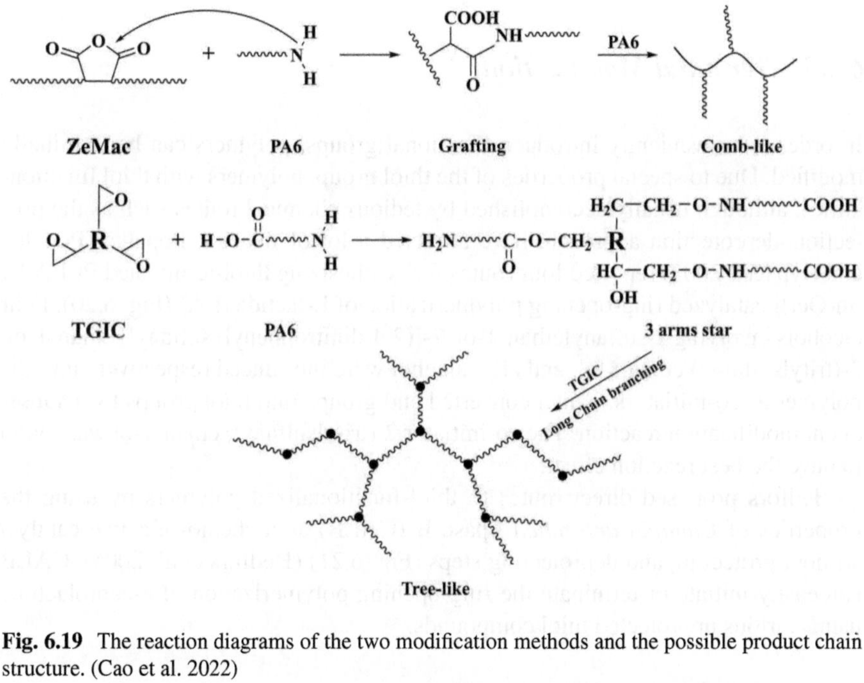

Fig. 6.19 The reaction diagrams of the two modification methods and the possible product chain structure. (Cao et al. 2022)

branching structure were modified from PA6, respectively (Fig. 6.19) (Cao et al. 2022). The results showed the two modification methods had different crystallization properties, but both significantly improved the rheology, foaming performance, and mechanical properties of PA6. Particularly, TGIC-modified PA6 with

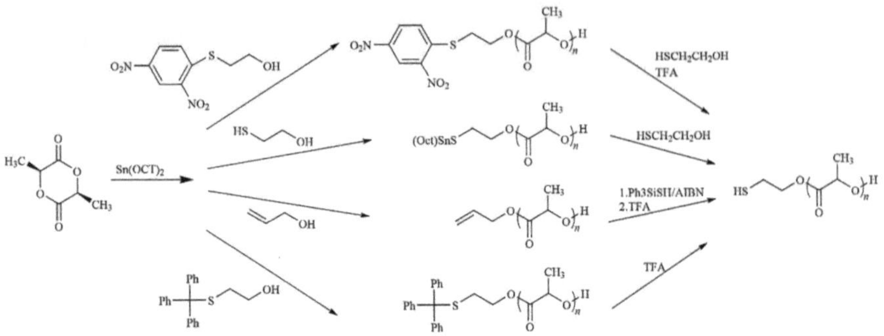

Fig. 6.20 Four routes of PLLA with sulfhydryl terminal group by ring-opening polymerization of L-lactide (LA) catalyzed by 2-ethylhexanoic acid tin [$Sn(OCT)_2$]. (Popelka and Rypáček 2003)

long-chain branching structure showed better performance in various physicochemical characterizations (Cao et al. 2022). The mechanical properties of unfoamed or foamed samples were improved after modification.

6.2.3 *Terminal Modification*

In order to conveniently introduce functional groups, polymers can be terminally modified. Due to special properties of the thiol group, polymers with thiol functionalities, although usually accomplished by tedious chemical routes such as the protection–deprotection approach, have attracted a lot of interest. Popelka (Popelka and Rypáček 2003) reported four routes for synthesizing thiol-terminated PLLA by $Sn(Oct)_2$ catalyzed ring-opening polymerization of L-lactide (LA) (Fig. 6.20). Four alcohols involving 2-sulfanylethan-1-ol, 2-[(2,4-dinitrophenyl) sulfanyl] ethan-1-ol, 2-(tritylsulfanyl) ethan-1-ol, and allyl alcohol were introduced respectively into the polymer as co-initiators, which converted end groups into thiol groups by a subsequent modification reaction. The co-initiator 2-(tritylsulfanyl) ethan-1-ol was found to have the best reaction effect.

Hedfors proposed direct routes to thiol-functionalized polymers by using the properties of *Candida antarctica* lipase B (CALB) as a chemo-selective catalyst without protecting and deprotecting steps (Fig. 6.21) (Hedfors et al. 2005). CALB can easily initiate or terminate the ring-opening polymerization of ε-caprolactone using various unprotected thiol compounds.

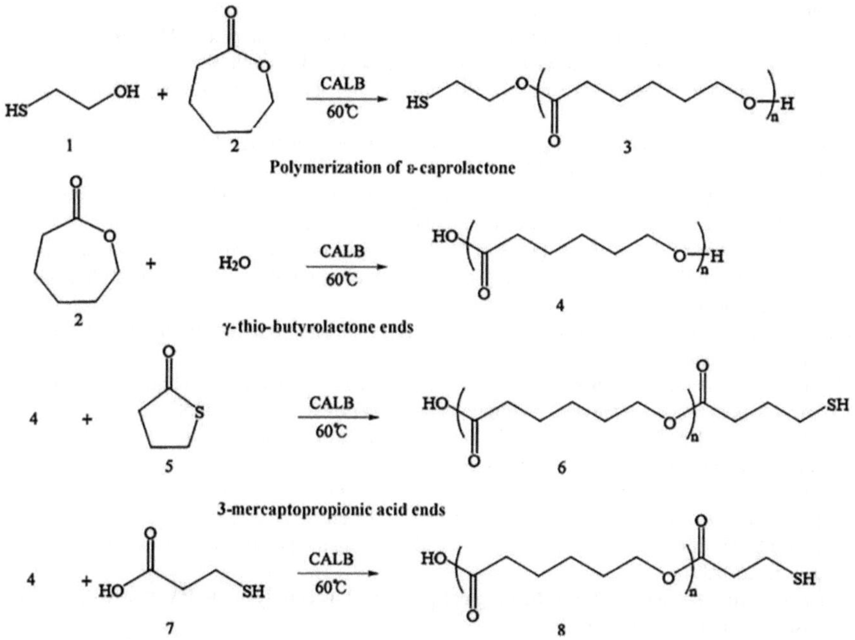

Fig. 6.21 Terminal thiol functionalization of polycaprolactone. (Reprinted with permission from *Macromolecules*, 2005, 38(3): 647. © 2018, American Chemical Society (Hedfors et al. 2005))

References

Agüero A, Lascano D, Garcia-Sanoguera D et al (2020) Valorization of linen processing by-products for the development of injection-molded green composite pieces of polylactide with improved performance. Sustainability 12(2):652

Boonkongkaew M, Sirisinha K (2018) Halloysite nanotubes loaded with liquid organophosphate for enhanced flame retardancy and mechanical properties of polyamide 6. J Mater Sci 53(14):10181–10193

Burzic I, Pretschuh C, Kaineder D et al (2019) Impact modification of PLA using biobased biodegradable PHA biopolymers. Eur Polym J 114:32–38

Cao L, Jiang T, Liu B et al (2022) Effects of grafting and long-chain branching structures on rheological behavior, crystallization properties, foaming performance, and mechanical properties of polyamide 6. E-Polymers 22(1):249–263

Chen YK, Wang WT, Yuan DS et al (2018) Bio-based PLA/NR-PMMA/NR ternary thermoplastic vulcanizates with balanced stiffness and toughness: "soft-hard" core-shell continuous rubber phase, in situ compatibilization, and properties. ACS Sustain Chem Eng 6(5):6488–6496

Chen T, Zhong GC, Zhang YT et al (2020) Bio-based and biodegradable electrospun fibers composed of poly(L-lactide) and polyamide 4. Chin J Polym Sci 38(1):53–62

Ding Y, Feng WT, Lu B et al (2018) PLA-PEG-PLA tri-block copolymers: effective compatibilizers for promotion of the interfacial structure and mechanical properties of PLA/PBAT blends. Polymer 146:179–187

Doganci MD, Aynali F, Doganci E et al (2019) Mechanical, thermal and morphological properties of poly(lactic acid) by using star-shaped poly(epsilon-caprolactone) with POSS core. Eur Polym J 121:109316

Dong XY, Wu ZG, Wang Y et al (2021) Strikingly toughening polylactide by using novel core-shell starch-based nanoparticles with double polymer shells. Mater Lett 289:129400

Dou YY, Tian N, Ning ZB et al (2022) Facile Method for the synthesis of PCL-*b*-PA6-*b*-PCL using amino-terminated PA6 as a macroinitiator and its characterization. Macromolecules 55(22):10090–10099

Fang HY, Li HL, Li L (2017) Property study of PA6 nano-composits. J Heilongjiang Univ Tech (Comprehensive Ed) 17(7):35–38

Feldmann M, Bledzki AK (2014) Bio-based polyamides reinforced with cellulosic fibres-processing and properties. Compos Sci Technol 100:113–120

Fletes RCV, López EOC, Gudiño PO et al (2021) Ground tire rubber/polyamide 6 thermoplastic elastomers produced by dry blending and compression molding. Prog Rubber Plast Recycl Technol 38(1):38–55

Gao FZ (2006) Investigation on synthesis, characterization and properties of block copolyamide 6 11. North University China, Taiyuan

Gardella L, Mincheva R, Winter JD et al (2018) Synthesis, characterization and stereocomplexation of polyamide 11/polylactide diblock copolymers. Eur Polym J 98:83–93

Ge TJ, Yang HY, Han YX (1997) Study on the performance of the calcium sulfate matrix compositing reinforced polypropylene. Plastics Sci Technol 1:16–19

Gigante V, Canesi I, Cinelli P et al (2019) Rubber toughening of polylactic acid (PLA) with poly(butylene adipate-co-terephthalate) (PBAT): mechanical properties, fracture mechanics and analysis of ductile-to-brittle behavior while varying temperature and test speed. Eur Polym J 115:125–137

Gong XY, Xu DZ, Guo J (2003) Rheological properties of copolyamide PA6/66/1010. J Dalian Inst Light Ind 22(3):194–196

Gu Y, Liu JH, Gao W et al (2024) One-pot in situ bulk polymerization of triblock copolymer PCL-*b*-PA6-*b*-PCL based on PA6 by using tetraphenyl tin. Polymer 300:126968

Gug J, Sobkowicz MJ (2016) Improvement of the mechanical behavior of bioplastic poly(lactic acid)/polyamide blends by reactive compatibilization. J Appl Polym Sci 133(45):43350

Guo BH, Zhang ZM, Xu J (2010) Polyamide alloy technology and applications. China Mechine Press, Beijing

Hao A, Wong I, Wu H et al (2015) Mechanical, thermal, and flame-retardant performance of polyamide 11-halloysite nanotube nanocomposites. J Mater Sci 50(1):157–167

He Y, Yang ZT, Qu JP (2019) Super-toughened poly(lactic acid)/thermoplastic poly(ether)urethane nanofiber composites with in-situ formation of aligned nanofibers prepared by an innovative eccentric rotor extruder. Compos Sci Technol 169:135–141

He L, Song F, Li DF et al (2020) Strong and tough polylactic acid based composites enabled by simultaneous reinforcement and interfacial compatibilization of microfibrillated cellulose. ACS Sustain Chem Eng 8(3):1573–1582

Hedfors C, Östmark E, Malmström E et al (2005) Thiol end-functionalization of poly(ε-caprolactone), catalyzed by Candida antarctica lipase B. Macromolecules 38(3):647–649

Heshmati V, Zolali AM, Favis BD (2017) Morphology development in poly (lactic acid)/polyamide 11 biobased blends: chain mobility and interfacial interactions. Polymer 120:197–208

Horrocks AR, Smart G, Kandola B et al (2012) Zinc stannate interactions with flame retardants in polyamides; Part 1: synergies with organobromine-containing flame retardants in polyamides 6 (PA6) and 6.6 (PA6.6). Polym Degrad Stab 97(12):2503–2510

Huang JR, Fan JF, Cao LM et al (2020a) A novel strategy to construct co-continuous PLA/NBR thermoplastic vulcanizates: metal-ligand coordination-induced dynamic vulcanization, balanced stiffness-toughness and shape memory effect. Chem Eng J 385:123828

Huang D, Ding YL, Jiang H et al (2020b) Functionalized elastomeric ionomers used as effective toughening agents for poly(lactic acid): enhancement in interfacial adhesion and mechanical performance. ACS Sustain Chem Eng 8(1):573–585

Huda MS, Drzal LT, Misra M et al (2005) A study on biocomposites from recycled newspaper fiber and poly(lactic acid). Ind Eng Chem Res 44(15):5593–5601

Hwang SW, Lee SB, Lee CK et al (2012) Grafting of maleic anhydride on poly(L-lactic acid). Effects on physical and mechanical properties. Polym Test 31(2):333–344

Jeziórska R, Szadkowska A, Studziński M et al (2021) The use of modified silica to control the morphology of polyamide 11 and poly(phenylene oxide) blends. Polimery 66(7-8):399–410

Jiang JZ, Yu XH (2014) Research progress of nitrogen-containing flame retardants applied in PA6 engineering plastics. Synth Tech Appl 29(3):9–12, 17

Kang HL, Yao L, Li YS et al (2017) Highly toughened polylactide by renewable Eucommia ulmoides gum. J Appl Polym Sci 135(12):46017

Kawada J, Kitou M, Mouri M et al (2017) Super impact absorbing bio-alloys from inedible plants. Green Chem 19(19):4503–4508

Kawasaki N, Nakayama A, Yamano N et al (2005) Synthesis, thermal and mechanical properties and biodegradation of branched polyamide 4. Polymer 46(23):9987–9993

Kawasaki N, Yamano N, Takeda S et al (2012) Synthesis of an azo macromolecular initiator composed of polyamide 4 and its initiation activity for the radical polymerization of vinyl monomers. J Appl Polym Sci 126(S2):E425–E432

Kawasaki N, Yamano N, Nakayama A (2015) Polyamide 4-block-poly(vinyl acetate) via a polyamide 4 azo macromolecular initiator: thermal and mechanical behavior, biodegradation, and morphology. J Appl Polym Sci 132(37):42466

Kawasaki N, Yamano N, Nakayama A (2020) Synthesis, properties, and biodegradability of three-branched copolyamide (4/6). J Appl Polym Sci 137(39):e49165

Kim NC, Kim JH, Nam SW et al (2013) Synthesis and characterization of very high molecular weight nylon 4 and nylon 4/6 copolymers. Polym-Korea 37(2):211–217

Kim NC, Kamal T, Park SY et al (2014) Preparation, chemical, and thermal characterization of nylon 4/6 copolymers by anionic ring opening polymerization of 2-pyrrolidone and epsilon-caprolactam. Fiber Polym 15(5):899–907

Kuciel S, Kuznia P, Jakubowska P (2016) Properties of composites based on polyamide 10.10 reinforced with carbon fibers. Polimery 61(2):106–112

Lai DW, Li YH, Wang CH et al (2020) Inhibition effect of aminated montmorillonite on crystallization of dendritic polyamide 6. Mater Today Commun 25(1):101578

Lebarbé T, Grau E, Alfos C et al (2015) Fatty acid-based thermoplastic poly(ester-amide) as toughening and crystallization improver of poly(L-lactide). Eur Polym J 65:276–285

Lebaupin Y, Chauvin M, Hoang T et al (2016) Influence of constituents and process parameters on mechanical properties of flax fibre-reinforced polyamide 11 composite. J Thermoplast Compos Mater 30(11):1503–1521

Leveque M, Douchain C, Rguiti M et al (2017) Vibrational energy-harvesting performance of bio-sourced flexible polyamide 11/layered silicate nanocomposite films. Int J Polym Anal Charact 22(1):72–82

Li M, Li YF (2017) Application research progress on aluminum hydroxide and its composite flame retardant system. Fine Spec Chem 25(3):46–49

Li MY, Fan ZF, Zhang Y et al (2008) Influence of synergistic flame retardance of nano-Sb_2O_3 with melamine cranurate acid on Nylon6. Chin Plast Ind 36(12):42–45

Li LL, Wu ZH, Jiang SS et al (2015) Effect of halloysite nanotubes on thermal and flame retardant properties of polyamide 6/Melamine cyanurate composites. Polym Compos 36(5):892–896

Li M, Cui H, Li Q et al (2016) Thermally conductive and flame-retardant polyamide 6 composites. J Reinf Plast Compos 35(5):435–444

Li JH, Huang YW, Sun J et al (2018) The synergistic effects between carbon nanotubes and montmorillonite in enhancing the flame retardancy of polyamide 6. Plastics 47(4):55–58, 81

Li SJ, Zhang YW, Wang WZ et al (2019) Crystallization and rheological properties of montmorillonite modified hyperbranched polyamide 6 nanocomposites. Polym Mater Sci Eng 35(6):59–65

Li SQ, Tang YR, Yang YB (2020) Research progress on fire retarded nylon. Eng Plast Appl 48(2):136–138

Liu Y, Zhang J (2012) Effect of residue structure on flame retardant properties of PA6/montmorillonite composits. Modern Plast Process Appl 24(2):12–15

Liu Y, Wang Q, Fei GX et al (2006) Preparation of polyamide resin-encapsulated melamine cyanurate/melamine phosphate composite flame retardants and the fire-resistance to glass fiber-reinforced polyamide 6. J Appl Polym Sci 102(2):1773–1779

Liu GC, He YS, Zeng JB et al (2014a) In situ formed crosslinked polyurethane toughened polylactide. Polym Chem-UK 5(7):2530–2539

Liu MX, Jia ZX, Jia DM et al (2014b) Recent advance in research on halloysite nanotubes-polymer nanocomposite. Prog Polym Sci 39(8):1498–1525

Liu HZ, Chen N, Shan PJ et al (2019) Toward fully bio-based and supertough PLA blends via in situ formation of cross-linked biopolyamide continuity network. Macromolecules 52(21):8415–8429

Liu LW, Yin CQ, Ding MD (2020) Study on glass fiber reinforced polypropylene (PP)/polyamide 6 (PA6) alloy. Chin Plastics Industry 48(10):65–69

Liu R, Han H, Wu X et al (2021) Construction of "core-shell" structure for improved thermal conductivity and mechanical properties of polyamide 6 composites. Polym Bull 78(5):2791–2803

Lu SY, Hamerton I (2002) Recent developments in the chemistry of halogen-free flame retardant polymers. Prog Polym Sci 27(8):1661–1712

Ma Y (2014) Reaserch progress in inorganic flame retardant. Chem Ind Times 28(4):33–34

Ma Y Y, Yan C, Zhu Y D, et al. Research progress in toughening modification of anionic polyamide 6. Plastics Sci Technology, 2017, 45(2): 91-97.

Muiruri JK, Liu SL, Yeo JCC et al (2019) Synergistic toughening of poly(lactic acid)-cellulose nanocrystal composites through cooperative effect of cavitation and crazing deformation mechanisms. ACS Appl Polym Mater 1(3):509–518

Ojijo V, Ray SS (2015) Super toughened biodegradable polylactide blends with non-linear copolymer interfacial architecture obtained via facile in-situ reactive compatibilization. Polymer 80:1–17

Oksman K, Skrifvars M, Selin JF (2003) Natural fibres as reinforcement in polylactic acid (PLA) composites. Compos Sci Technol 63(9):1317–1324

Oliver-Ortega H, Méndez J, Mutjé P et al (2017) Evaluation of thermal and thermomechanical behaviour of bio-based polyamide 11 based composites reinforced with lignocellulosic fibres. Polymers 9(10):522

Park KW, Kim DH, Kim HJ (2014) Preparation and characterization of polyamide4(PA4)-polyurethane (PU)-PA4 triblock copolymers. Polym-Korea 38(1):9–15

Park MS, Lee S, Kim AR et al (2019) Toughened and hydrophobically modified polyamide 11 copolymers with dimer acids derived from waste vegetable oil. J Appl Polym Sci 136(10):47174

Pasanphan W, Haema K, Kongkaoroptham P et al (2021) Glycidyl methacrylate functionalized star-shaped polylactide for electron beam modification of polylactic acid: synthesis, irradiation effects and microwave-resistant studies. Polym Degrad Stab 189:109619

Popelka S, Rypáček F (2003) Synthesis of polylactide with thiol end groups. Collect Czechoslov Chem Commun 68(6):1131–1140

Quiles-Carrillo L, Montanes N (2018) Reactive toughening of injection-molded polylactide pieces using maleinized hemp seed oil. Eur Polym J 98:402–410

Quiles-Carrillo L, Montanes N, Fombuena V et al (2019) Enhancement of the processing window and performance of polyamide 1010/bio-based high-density polyethylene blends by melt mixing with natural additives. Polym Int 69(1):61–71

Rahman MZ, Wang X, Song L et al (2023) A novel sustainable phosphorus-containing heteroatom-based coating for polyamide 6.6 textiles: enhanced flame retardancy and hydrophilicity. Ind Crop Prod 205:117425

Rashmi BJ, Loux C, Prashantha K (2017) Bio-based thermoplastic polyurethane and polyamide 11 bioalloys with excellent shape memory behavior. J Appl Polym Sci 134(20):44794

Rashmi BJ, Prashantha K, Lacrampe MF et al (2018) Scalable production of multifunctional bio-based polyamide 11/graphene nanocomposites by melt extrusion processes via masterbatch approach. Adv Polym Technol 37(4):1067–1075

Ren J, Li JB (2014) Polylactic acid. Chemical Industry Press, Beijing
Rohner S, Humphry J, Chaléat CM et al (2018) Mechanical properties of polyamide 11 reinforced with cellulose nanofibres from Triodia pungens. Cellulose 25(4):2367–2380
Sag J, Kukla P, Goedderz D et al (2020) Synthesis of novel polymeric acrylate-based flame retardants containing two phosphorus groups in different chemical environments and their influence on the flammability of poly (Lactic Acid). Polymers 12(4):778
Sahnoune M, Taguet A, Otazaghine B et al (2017) Effects of functionalized halloysite on morphology and properties of polyamide-11/SEBS-g-MA blends. Eur Polym J 90:418–430
Sallem-Idrissi N, Velthem PV, Sclavons M (2018) Fully bio-sourced nylon 11/raw lignin composites: thermal and mechanical performances. J Polym Environ 26(12):4405–4414
Sun J, Gu X, Zhang S et al (2014) Improving the flame retardancy of polyamide 6 by incorporating hexachlorocyclotriphosphazene modified MWNT. Polym Adv Technol 25(10):1099–1107
Thakur S, Cisneros-Lopez EO, Pin JM et al (2019) Green toughness modifier from downstream corn oil in improving poly(lactic acid) performance. ACS Appl Polym Mater 1(12):3396–3406
Torabi H, McGreal H, Zarrin H et al (2023) Effects of rheological properties on 3d printing of poly (lactic acid) (PLA) and poly (hydroxy alkenoate) (PHA) hybrid materials. ACS Appl Polym Mater 5(6):4034–4044
Walha F, Lamnawar K, Maazouz A et al (2016) Rheological, morphological and mechanical studies of sustainably sourced polymer blends based on poly(lactic acid) and polyamide 11. Polymers 8(3):61
Wang GQ, Wang XF (2008) Polymer modification. China Light Industry Press, Beijing
Wang LL, Dong X, Gao YY et al (2015) Transamidation determination and mechanism of long chain-based aliphatic polyamide alloys with excellent interface miscibility. Polymer 59:16–25
Wang LL, Dong X, Wang XR et al (2016) High performance long chain polyamide/calcium silicate whisker nanocomposites and the effective reinforcement mechanism. Chin J Polym Sci 34(8):991–1000
Wang Y, Hu QE, Li T et al (2018) Core-shell starch nanoparticles and their toughening of polylactide. Ind Eng Chem Res 57(39):13048–13054
Wang Y, Liu JH, Xia L et al (2019a) Fully biobased shape memory thermoplastic vulcanizates from poly(lactic acid) and modified natural eucommia ulmoides gum with co-continuous structure and super toughness. Polymers 11(12):2040
Wang X, Peng SX, Chen H et al (2019b) Mechanical properties, rheological behaviors, and phase morphologies of high-toughness PLA/PBAT blends by in-situ reactive compatibilization. Compos Part B-Eng 173:107028
Wen X (2019) One-pot route to graft long-chain polymer onto silica nanoparticles and its application for high-performance poly(L-lactide) nanocomposites. RSC Adv 9(24):13908–13915
Weng YX, Fu Y (2010) Biodegradable plastics and bio-based plastics. Chemical Industry Press, Beijing
Woo Y, Cho D (2013) Effect of aluminum trihydroxide on flame retardancy and dynamic mechanical and tensile properties of kenaf/poly(lactic acid) green composites. Adv Compos Mater 22(6):451–464
Wu C, Liu H, Dai W et al (2010) Research progress on fire retarded nylon. Guangdong Chem Ind 37(3):24–25
Wu H, Hou AL, Qu JP (2019) Phase morphology and performance of supertough PLA/EMA-GMA/ZrP nanocomposites prepared through reactive melt-blending. ACS Omega 4(21):19046–19053
Xie XL, Yu ZZ, Zhang QX et al (2010) Synergistic effect of SEBS-g-MA and epoxy on toughening of polyamide 6/glass fiber composites. J Polym Sci Polym Phys 45(12):1448–1458
Xue B, He HZ, Huang ZX et al (2019) Fabrication of super-tough ternary blends by melt compounding of poly(lactic acid) with poly(butylene succinate) and ethylene-methyl acrylate-glycidyl methacrylate. Compos Part B-Eng 172:743–749
Yang YX, Li YC, Wang P et al (2016) Effect of talc on flame retardant property of PA6/MCA composites. Plastics Sci Technol 44(1):64–68
Yu H, Yong Z, Ren W et al (2011) Comparison of the toughening effects of different elastomers on nylon 1010. J Appl Polym Sci 121(6):3340–3346

Zhang H (2014) Study on the microstructure of brominated polystyrene in coordination with antimony trioxide on the flame retarding mechanism of nylon 6. Fire Tech Prod Inf 5:12–14
Zhang S, Tang W, Gu X et al (2016) Flame retardancy and thermal and mechanical performance of intercalated, layered double hydroxide composites of polyamide 11, aluminum phosphinate, and sulfamic acid. J Appl Polym Sci 133(20):43370
Zhang HC, Kang BH, Chen LS et al (2020) Enhancing toughness of poly (lactic acid)/Thermoplastic polyurethane blends via increasing interface compatibility by polyurethane elastomer prepolymer and its toughening mechanism. Polym Test 87:106521
Zhang YJ, Cai ZL, Wang YB et al (2022) Effect of sequence distribution on the non-isothermal crystallization of copolyamide 4/6. J Mater Sci 57(37):17883–17901
Zhang Y, Wang M, Wang L et al (2023a) Molecular dynamics simulation and non-isothermal crystallization kinetics of polyamide 4 and different bio-based polyamide blends. Phys Chem Chem Phys 25(37):25309–25321
Zhang YJ, Wang MD, Zhang D et al (2023b) Crystallization and performance of polyamide blends comprising polyamide 4, polyamide 6, and their copolymers. Polymers 15(16):3399
Zhang Z, Cao W, Yuan X et al (2024) Development of a "rigid-flexible" structure at the interface through aramid nanofibers@mxene to enhance mechanical properties of carbon fiber/polyamide composites. ACS Appl Mater Interfaces 16(12):15514–15524
Zhao ZW, Xiao SJ, Yu SW (2018) Research progress in nonhalogen flame retardant of nylon. Plastics Sci Technol 46(12):128–132
Zhou YX (2014) Research progress in halogen-free flame retardant of polyamide 6. J Dongguan Univ Tech 21(3):59–64
Zhu JM (2011) Polyamide resin and its application. Chemical Industry Press, Beijing
Zhu P, Dong X, Wang DJ (2017) Strain-induced crystallization of segmented copolymers: deviation from the classic deformation mechanism. Macromolecules 50(10):3912–3922
Zhu S, Gong W, Luo J et al (2019) Flame retardancy and mechanism of novel phosphorus-silicon flame retardant based on polysilsesquioxane. Polymers 11(8):1304
Zioło A, Mossety-Leszczak B, Walczak M et al (2024) Synthesis and morphology characteristics of new highly branched polycaprolactone PCL. Molecules 29(5):991

Chapter 7
Development of Biobased Materials Applications

Yulin Li, Zhili Cai, and Fangrui Liu

Biobased degradable materials, which use abundant natural resources such as agriculture, forestry, animal husbandry, marine products, or waste as raw materials, reduce the dependence on nonrenewable resources such as petroleum to a certain extent. Biobased degradable polymers have become alternative materials for many industrial applications to control the risk of environmental damage caused by non-biodegradable plastics, with the advantages of environmental protection and sustainable development. As representatives of biodegradable polymers, poly(lactic acid) (PLA), polyamides, and polyurethanes all contain heteroatoms. They may degrade easily due to the hydrolysis of ester (–COO–), amide (–CONH–), or ether bonds (–O–). In addition, they have excellent mechanical properties, thermal properties, and barrier properties, and therefore can compete with traditional materials on the market, with a wide range of application prospects. According to statistics, from 2015 to 2023, the scale of China's biodegradable plastics market grew steadily, reaching 18.9 billion yuan in 2023 (Fig. 7.1) (Industrial Information Network 2020; Ganyi et al. 2022). With the enhancement of people's environmental awareness and the implementation of the "plastic restriction order", biodegradable plastics will have a good development prospect. In this chapter, we mainly introduce biobased degradable materials whose monomers come from biomass or are produced by microorganisms, such as PLA, polyhydroxyalkanoates (PHA), polybutylene

Y. Li (✉) · F. Liu
The Key Laboratory for Ultrafine Materials of Ministry of Education, Frontiers Science Center for Materiobiology and Dynamic Chemistry, Engineering Research Center for Biomedical Materials of Ministry of Education, School of Materials Science and Engineering, East China University of Science and Technology, Shanghai, China
e-mail: yulinli@ecust.edu.cn; Y82210340@mail.ecust.edu.cn

Z. Cai
State Key Laboratory of Bioreactor Engineering, East China University of Science and Technology, Shanghai, China

L. Zhao (ed.), *Research and Applications of Bio-based Degradable Materials*,
https://doi.org/10.1007/978-981-95-1188-4_7

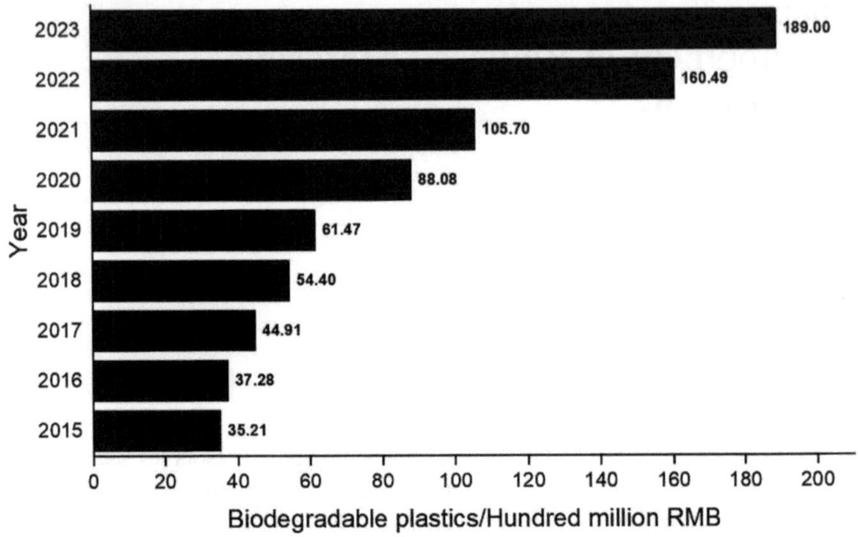

Fig. 7.1 Scale of biodegradable plastics from 2015 to 2023. (Industrial Information Network 2020; Ganyi et al. 2022)

succinate (PBS), polyamide 4 (PA4) and the second largest natural polymer chitosan (CS), and their applications in packaging, agriculture, and medical and fiber fields.

7.1 Application of Biobased Materials in the Packaging Field

Biobased plastics are used in various industries. Nonbiodegradable biobased plastics, composed of polyethylene (PE), polyethylene terephthalate (PET), and nondegradable polyamide (PA), account for about 41.9% of the global bioplastic production. Biodegradable plastics include PLA, PHA, starch blends, PBS, polybutylene adipate terephthalate (PBAT), etc., accounting for 58.1% of the global bioplastic production. Packaging is their largest application field, with biobased plastics for food packaging accounting for over 47% (990,000 tons) of the total bioplastic production in 2020. Due to environmental pressures and various measures taken by governments to reduce plastic waste, food packaging and disposable tableware are the main drivers for the rapid growth of biobased degradable plastics (European Plastics).

Packaging is a component of production, storage, distribution, preservation, and other unit operations. It can protect goods from external biological, physical, and chemical damage, thus having certain economic value in the composition of goods. The life cycle of packaging is short, often discarded after use, with low reuse and recycling rates. Therefore, packaging waste is an important part of urban solid waste. It is mainly made of petroleum-based polymers, which are usually nonbiodegradable. On one hand, landfill measures will leave waste in landfills for a long time, occupying a large amount of land resources; on the other hand, waste incineration will produce carcinogens like dioxins and acidic gases, causing significant environmental damage. In recent years, the development of biodegradable packaging materials from renewable natural resources has received increasing attention, especially in EU countries. Compared with traditional materials made from nonrenewable raw materials, the use of biodegradable materials is expected to have a smaller impact on the environment. This section mainly introduces the application of biobased degradable materials in food packaging and daily necessities packaging, as well as their shortcomings.

7.1.1 Food Packaging

Food packaging is one of the important factors to ensure food safety. It can not only prevent food spoilage and contamination but also provide great convenience for production, storage, transportation, and sales, and has the function of maintaining the stable quality of the food itself. It needs to facilitate food consumption and first display the appearance of the food to attract consumption. It has a value beyond material cost, becomes a part of food commodities, and can bring certain economic effects to food manufacturers, distribution companies, and consumers.

Plastic resins and films, as the largest industry for food packaging, have many advantages such as large usage, light weight, and easy carrying. Currently, nondegradable materials such as polyvinyl chloride (PVC), polystyrene (PS), polyamide (PA), polyvinylidene chloride (PVDC), polyethylene (PE), and polypropylene (PP) are widely used. Data shows that food packaging plastics already account for over 60% of China's total plastic production, and flexible packaging films account for 55% of food packaging plastics (Industrial Information Network 2020), which shows that food packaging plastics have a broad market in China.

However, these nondegradable plastic packaging materials cannot be reused and are often disposable, causing serious environmental pollution after being discarded. Using biobased degradable materials for food packaging, they can be recycled for composting after use, will not produce harmful substances, and have the same printing function as traditional nondegradable plastics. Currently, the biobased degradable plastics that can be used for food packaging are mainly polyester materials represented by PLA, PBS, PHA, etc., which have many advantages: ① They have a wide range of raw material sources and can be obtained from natural substances such as corn, beets, sweet potatoes, cellulose, and other starch materials through microbial fermentation; ② they have excellent gas barrier properties; ③ they have good oil resistance and printing performance; ④ they have high transparency; ⑤ some biobased degradable materials such as PLA and CS have excellent antibacterial properties; and ⑥ under suitable temperature, humidity, and O_2 atmosphere in the natural environment, these biobased degradable materials can be degraded into

CO_2 and H_2O by microorganisms through metabolism in a short time, which helps to reduce the problem of urban solid waste.

1. *Research on the Application of Polylactic Acid in Food Packaging*

On January 19, 2020, the National Development and Reform Commission and the Ministry of Ecology and Environment announced the "Opinions on Further Strengthening Plastic Pollution Control." The opinion clearly states that in 2020, some areas and fields will take the lead in banning and restricting the production, sale, and use of some plastic products. By 2022, the consumption of disposable plastic products will be significantly reduced, alternative products will be promoted, and the proportion of resource and energy utilization of plastic waste will be greatly increased. Starting from January 2021, degradable plastic bags have gradually appeared in the public eye, among which PLA degradable plastic bags are the most used.

The characteristics of PLA include high transparency, gloss, and breathability, high modulus, complete foldability and entanglement retention, low-temperature heat sealing and easy opening, excellent flexibility, high flatness and good barrier properties, etc., which can replace PS, PP, acrylonitrile (ABS) and other petrochemical plastics (Shuhan 2016). Through different processing methods such as melt extrusion, injection molding, blow molding, foaming, and vacuum forming, PLA can be made into products of various shapes, such as food containers (bottles, trays, etc.), films, heat-shrink packaging, breathable packaging, fragrance-retaining packaging, shopping bags, and garbage bags, mainly used in the field of food packaging (Yi 2019; Tripathi et al. 2017). For example, many candy packages on the market use PLA packaging film. This packaging film has a similar appearance and performance to traditional candy packaging film, has high transparency and excellent barrier properties, and can better retain the aroma of candy. PLA is currently used for packaging foods with a short shelf life, such as organic salads, lettuce, yogurt cups, and ice cream, and its composites are also used in the packaging of bottled water and yogurt. These containers have reached the relevant standards of Germany and other EU countries (Zhe 2018).

In addition to excellent mechanical properties, PLA also has excellent antibacterial properties and can be used for soft packaging of fruits and vegetables. Niu et al. (Niu et al. 2018) modified cellulose (R-CNF) was added to PLA to make a PLA film, which was then coated with chitosan (CHT) to prepare a double-layer composite film (R-CNF/PLA/CHT composite film) for antimicrobial food packaging. Experimental results show that the R-CNF/PLA/CHT composite film has good antimicrobial properties. Fathima et al. (Fathima et al. 2018) found that PLA film has good antimicrobial effects and can extend the shelf life of Indian white shrimp in refrigerated environments.

PLA has similar physical and mechanical properties to polyethylene terephthalate (PET) and oriented polystyrene (PS), and can also be made into packaging bottles or beverage bottles by blow molding. At present, countries and regions such as Europe and the United States have developed and applied packaging materials made from PLA. In 2000, the German Danone Group first used PLA materials in the packaging of commercial organic yogurt cups, and McDonald's in Switzerland has also used PLA materials to package fresh salads. BASF has produced a biodegradable product—Ecovio, which contains 45 wt% PLA, for the manufacture of transport bags and compostable can liners. In addition, since 2005, Walmart in the United States has been using PLA materials on the packaging boxes of strawberries and bean sprouts. In 2010, PolenghiLAS in Italy was the first to use PLA blow-molded bottles for bottled lemon juice in the EU region (Castro-Aguirre et al. 2016).

In the field of food packaging, polylactic acid has advantages such as safety, non-toxicity, degradability, and good transparency, and can be used for food packaging bags, fresh-keeping films, lunch boxes, and tableware. However, due to the poor heat resistance and wear resistance of polylactic acid, it is only suitable for use at room temperature or for disposable products. At the same time, the slow degradation rate is also a problem in the disposal of packaging films and other daily necessities (Castro-Aguirre et al. 2016; Rydz et al. 2014), which pose great challenges to the widespread application of polylactic acid in the packaging field.

2. *Research on the Application of Polyhydroxyalkanoates in Food Packaging*

The PHA family is a type of biodegradable thermoplastic polyester, including common ones such as poly-3-hydroxybutyrate (PHB), polyhydroxyvalerate (PHV), and copolymers of 3-hydroxybutyrate and 3-hydroxyvalerate (PHBV). PHA can be synthesized by nearly 300 kinds of microorganisms, and *Ralstonia eutropha* is one of the representative microorganisms for industrial production of PHA, which can accumulate more than 90% of PHB in its own cells. PHA has high CO_2 and O_2 barrier properties, so it is more suitable for fresh food packaging, such as fresh meat, fruits, clean vegetables, and other fresh products.

PHA has excellent film-forming and printing properties. It alone or in combination with synthetic plastics or starch can produce high-performance food packaging films, making it more suitable for commercial use. The main global PHA-based commodity manufacturers include Ningbo Tianan Biomedical Materials Co., Ltd., Meridian (Bainbridge, Georgia, USA), and Kaneka Corporation (Japan).

Currently, many PHAs are used for packaging high-oil foods, frozen foods, and organic foods. Bucci and others (Bucci et al. 2005) have tested the possibility of using PHB plastic for food packaging. They used injection molding to manufacture packaging with PHB and evaluated it through dimensional tests (size, volume capacity, weight, and thickness) and mechanical tests (dynamic compression and impact resistance), comparing it with PP (polypropylene) packaging of the same format. The results showed that in strength tests, the deformation value of PHB was 50% lower than that of PP, showing a higher brittleness. In terms of high and low temperature resistance, PHB material is better than PP material at high temperatures, while PP material is better at low temperatures. Additionally, the sensory evaluation values of products such as margarine, mayonnaise, and cream cheese packaged with PHB material showed no significant difference at the 5% level. Levkane and others (Levkane et al. 2008) studied the effect of pasteurization on traditional packaging (PE, PP) and biobased packaging (PLA, PHB) for meat salads, and found that PHB film can be successfully used for packaging such foods.

Haugaard and others (Haugaard et al. 2003) found that compared with high-density polyethylene (HDPE), orange juice simulants and condiments packaged with PHB undergo the same quality changes, indicating that PHB has great potential for application in the packaging of commercial fruit juices, other acidic beverages, condiments, and other fatty foods. According to www.ptonline.com, all biopolymers of Danimer Scientific, including Nodax PHA, have received FDA food contact approval. Nodax PHA can biodegrade in anaerobic soil, freshwater, and marine environments, and is 100% biobased with a short life cycle (Fig. 7.2). Additionally, bottles made from PHA also have good biodegradability (Fig. 7.3). It also has seven TUV Austria certifications and statements of industrial and home compostability. The company announced in November 2020 that biodegradable straws produced by Danimer Scientific and Eagle Beverages can be used in fast food restaurants. These straws were available for purchase by Eagle Beverage's QSR customers in early 2021, indicating that PHA has potential applications in the packaging industry, especially in food packaging.

However, the practical application of PHA is limited due to its high cost, unstable quality of biologically synthesized products, high brittleness, and poor mechanical properties, especially PHB, which can have a crystallinity of up to 80%. Therefore, solving the brittleness problem of PHB is key to its application in packaging materials: on the one hand, nucleating agents can be added to accelerate crystallization; on the other hand, plasticizers can be added to prevent crystallization,

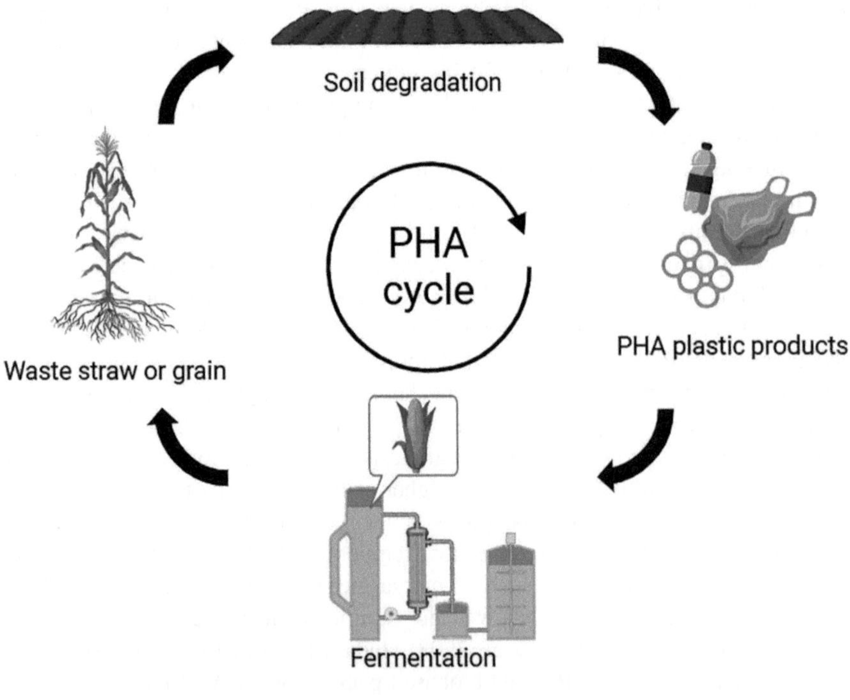

Fig. 7.2 PHA full life cycle degradation schematic

Fig. 7.3 Cups and packaging bottles made from PHA. (Mehrpouya et al. 2021)

thereby improving product flexibility and elongation at break. In order for PHA to be applied to the plastics industry, ICI (Imperial Chemical Industries) synthesized Biopol, a high polymer PHBV copolymerized with 3-hydroxybutyric acid and 3-hydroxyvaleric acid in any ratio, by hindering the crystallization of the main chain. The decrease in crystallinity is due to the decrease in the regularity of its molecular chain; therefore, PHBV is an ideal choice for thermoforming and manufacturing flexible bioplastics.

3. *Research on the Application of Polybutyramide in Food Packaging*

Polyamide is nontoxic and tasteless. Due to its high hydrogen bond density and good barrier properties, it can prevent O_2 from entering the packaging, prevent moisture and aroma loss, and can be used for food and beverage packaging. Compared to other common plastic films (such as PE and PP), polyamide film has many advantages, such as good O_2 barrier properties, puncture strength and tear strength, high-temperature resistance, and printing performance. It also has environmental advantages compared to barrier materials such as ethylene-vinyl alcohol copolymer (EVOH) and polyvinylidene chloride (PVDC). In addition, polyamide film also has high transparency and safety. However, in practical applications, single-layer polyamide film is rarely used directly due to its poor heat sealing and relatively high cost. Furthermore, polyamide has good mechanical properties and extrusion processing performance, and can also be used as a barrier layer in single-layer or multi-layer films. As a polyamide core film in multi-layer films, it can improve the mechanical properties and barrier properties of the multi-layer films, for example, high barrier laminated linear low-density polyethylene (LLDPE)-polyamide-ethylene-vinyl alcohol-polyamide-LLDPE and medium barrier laminated LLDPE-polyamide-LLDPE combinations, to meet various packaging requirements.

It is expected that in the next few years, the global demand for polyamide in the field of packaging materials will grow at a rate of 4%. It is mainly used for soft packaging of food, especially suitable for greasy food and high-temperature steamed food packaging. As polybutyramide (PA4) in the polyamide family, it has characteristics such as high strength, wear resistance, and biocompatibility, and can degrade in activated sludge, seawater, and the human body, showing great application potential in food packaging. Zhang Tiantian et al. (2021) used biobased PA4 as the main body, blended with chitosan with antibacterial and antioxidant functions, and prepared a degradable PA4 sausage casing film by the casting method. The results showed that the antibacterial performance of the PA4 sausage casing film was 3.13 [lg(CFU/cm^2)], with good antibacterial performance. Under 4 °C storage conditions, the shelf life of pork sausage packaged with PA4 sausage casing film was 28 days, and its comprehensive performance was significantly better than that of commercially available collagen sausage casing.

4. *Research on the Application of Polybutylene Succinate in Food Packaging*

The glass transition temperature of polybutylene succinate (PBS) is far below room temperature, which gives it excellent melt processing capabilities and outstanding mechanical properties, very similar to widely used polyethylene (PE) and polypropylene (PP). In the 1990s, the Showa Corporation in Japan produced a product named Bionole using PBS as a raw material, which can be used to make various packaging materials. Compared with degradable plastics such as polycaprolactone (PCL) and PHA, PBS has the advantages of low cost and excellent comprehensive performance; compared with PLA, which is close in price, PBS has the advantages of easy processing, can be directly processed on ordinary plastic processing equipment, has a larger processing window, and good heat resistance, with its heat deformation temperature exceeding 100 °C. However, high molecular weight PBS is quite brittle, with a very low elongation at break, but the use of chain extenders like isocyanate can significantly improve its elongation (Fabbri et al. 2014). Liu Mengyu and others (2019a) blended PBAT and PBS and prepared a blend film using a twin-screw extruder casting machine. The results showed that the addition of PBAT reduced the crystallinity of PBS, enhanced the flexibility of the PBS film, and the maximum elongation at break of the blend reached 832.1%. Currently, there are reports of PBAT and PBS being blended to prepare modified atmosphere packaging films. The good gas permeability of the PBS/PBAT film keeps the concentration of CO_2 and O_2 inside the packaging bag always stable within the range suitable for preserving cherry tomatoes, effectively inhibiting the spoilage of cherry tomatoes, maintaining their good sensory quality and nutritional value (Mengyu et al. 2019b). Among these fully degradable aliphatic polyesters, PBS, due to its relatively low cost and excellent processing performance, has become one of the most promising general-purpose degradable plastics for industrialization.

7.1.2 Daily Goods Packaging

Daily goods packaging materials are a major application in the plastics industry, including products such as packaging boxes, bottles, and plastic bags. In recent years, the problem of "white pollution" caused by nondegradable plastics has become increasingly severe, so the development and application of degradable packaging materials have gradually become a research focus. With the development of a sustainable economy, green packaging is gradually replacing traditional packaging. By blending PLA with other polymers, the performance of packaging is continuously improved. Due to its good biodegradability, nontoxic safety, high breathability, high tensile strength, and excellent transparency, PLA can be processed into various film products through vacuum forming and pressure forming, applied in the fields of fresh-keeping film, shopping bags, and vacuum packaging, and currently, the packaging field is the largest application market for PLA. PLA can be used to prepare various types of packaging materials through injection, extrusion, and blow molding. Compared with other general plastics, PLA not only has biodegradability, but also has high transparency, good barrier properties, and excellent mechanical strength and processing performance (Zhe 2018; Ruixia et al. 2019).

PHA not only has thermoplasticity but also has good biodegradability. It can be completely degraded by microorganisms into hydroxy fatty acids, causing almost no environmental pollution. It is a new type of excellent low-carbon environmental protection material, expected to replace petrochemical plastics. It is widely used in the manufacture of shopping bags, daily chemical product packaging, utensils, feminine hygiene products, packaging bags, and many other disposable items. It can reduce the dependence on nonrenewable resources such as petrochemicals and reduce the pollution and destruction of solid waste to the environment. The first to use PHA material as a biodegradable plastic product was the British ICI company. The product named Biopol, made from PHBV, was successfully launched in the German market and was later acquired by the Metabolix company for its patent rights and production technology. PHBV has been developed as a packaging material by many companies, including Biomers, ADM, P&G, Metabolix, ICI, Biocycle, etc. In 2009, Shiseido China first applied the composite material of PHA and high-density polyethylene (HDPE) to its bottled shampoo. However, due to the low yield of PHA, expensive purification procedures, and limited commercial applications, it has a smaller market share in the field of low-added-value daily product packaging.

7.1.3 Automotive Field with the Development of Raw Materials and Processing Technology

Durable biobased materials have been developed, and more and more biobased degradable materials show application potential in the field of automotive interiors. PLA is brittle and has weak impact resistance, so using natural fibers to enhance the mechanical properties of PLA is a good improvement method. Fibers such as hemp, ramie, jute, straw, wood, coconut fiber, sisal, sword hemp, bamboo, rice husk, oil palm, and flax are often used to enhance PLA. Researchers have used alkali and silane-treated ramie fiber to enhance PLA and found that ramie/PLA composites have a higher storage modulus than PLA, and the tensile, impact, and bending strengths of the composites are all higher than unmodified PLA. In addition, the thermal decomposition temperature of the ramie/PLA composite material treated with ramie fiber is also increased (Tao et al. 2010). In addition, GMA (Khan et al. 2016) and others have prepared plain woven jute fabric/PLA composites by hot pressing, and the results show that the average tensile strength, tensile modulus, bending strength, bending modulus, and impact strength of the composite material have increased by about 103%, 211%, 95.2%, 42.4%, and 85.9%, respectively. This shows that the method of adding natural fibers to PLA can greatly improve the overall performance of PLA. Ejaz et al. (2020) and others have made PLA/flax, PLA/jute, and PLA/flax/jute (mixed composite material) by hot pressing molding. The research results show that when the single fiber addition amount is 40%, the tensile performance of the PLA composite material is the best, and the impact resistance is also improved. Basalt fiber (BF) is used as a reinforcing material to enhance the tensile mechanical properties of PLA to meet the mechanical performance requirements of nonmetallic products in automobiles. In order to increase the interaction between the two, some researchers have used the silane coupling agent KH550. One end of the KH550 molecule reacts with the hydroxyl group on the BF surface to form a chemical bond, and the other end of the molecule reacts with the terminal hydroxyl or terminal carboxyl group of the PLA molecule to form a chemical bond, thereby improving the tensile strength of the BF/PLA composite material (Liu et al. 2019). In 2003, Shah Jiangzi's reports (Jiangzi 2003) indicate that Japan's Toray Company has turned PLA fiber (ECODEAR) into automotive interior parts and has officially started production. The company is developing car doors, seats, wheel rims, ceilings, and other automotive interior parts made from PLA fiber and is attempting to add natural jute fibers to spare tire covers. They have already supplied Toyota with spare tire covers and carpets. These results prove that polylactic acid composites can be used as alternatives to petroleum-based materials in automotive product development.

PLA is a rather brittle material, with a small elongation at break, poor melt strength, low heat bending temperature, and poor thermal stability (Auras et al. 2011). These drawbacks make PLA unsuitable for processing and injection molding, and its industrial application is somewhat limited. In particular, its low toughness restricts its use in many industrial applications that require plastic deformation under high stress levels. Therefore, addressing the shortcomings of PLA and expanding its range and scale of use is still the research direction of current researchers.

7.2 Applications of Biobased Materials in Agriculture

The consequence of consuming large amounts of nondegradable plastics in agriculture is the continuous increase in non-environmental pollution related to poor waste management. Many discarded polymer materials are mostly incinerated, a large part of which is left in the fields or buried underground, leading to the degradation of the natural environment, and the occupied land cannot be restored for a long time, affecting the sustainable use of the land. Currently, only a small portion of polymer waste is recycled, and due to the high cost and labor-intensive nature of this process, collecting polymer waste from the fields requires a lot of work, thus requiring considerable financial expenditure. Biobased degradable materials have special advantages in the field of agriculture, especially in a large agricultural country like ours. Agricultural film is a notable example. After use, the biobased degradable film can be plowed directly into the ground after agricultural production, instead of being cleaned and then recycled, which practically improves the economy of operation.

It takes decades or even hundreds of years for general plastics to degrade themselves in nature and return to the natural cycle. However, degradable biobased materials can degrade in a year or even a few months, and these plastics can be composted with the help of microorganisms and enzymes. The compost can be used for crop growth, thus achieving the maximum recycling and utilization of resources. Nonrecyclable biobased degradable materials can degrade under the action of solar radiation, O_2, H_2O, and microorganisms, and the products are nontoxic and harmless small molecules, leaving no harmful residues. This section mainly introduces the specific applications of biobased degradable materials in agriculture, forestry, and fisheries.

7.2.1 Applications in the Planting Industry

Mulch covering technology can increase crop yields, reduce the use of herbicides and pesticides, and also has important functions such as moisture retention and fertilization. It has become an important means in agricultural production and is therefore widely promoted. The widespread use of plastic mulch has brought many environmental problems, because ordinary plastics take hundreds of years to degrade, and are accompanied by the leaching of a large amount of harmful substances. Residual plastic mulch hinders the flow of water, gas, and fertilizer in the soil, causing soil structure to compact, hindering root absorption of water and fertilizer, reducing soil fertility, and causing secondary salinization, seriously affecting the sustainable development of agricultural production and the ecological environment. The soil damage caused by agricultural mulch has become a global problem. In agricultural production, a large amount of discarded mulch and other plastic products destroys the soil structure, endangers crop growth and development, and seriously affects the balance of the soil micro-ecosystem. In recent years, biobased degradable plastics have become a research hotspot to solve this problem.

In order to improve the growth environment of crops and increase crop yields, mulching technology is widely used in the field of agricultural production. In addition, most of the mulch is currently made of polyethylene material. The process of removing polyethylene mulch is very time-consuming (about 16 hours per hectare), and the process of recycling used polyethylene mulch is very complicated. Even with the help of modern agricultural machinery, the entire recycling process still requires a lot of manual labor, making the entire recycling process face huge costs. Despite the huge cost, it is still impossible to completely recycle all discarded mulch. The biobased degradable material mulch not only retains the functions of traditional mulch to suppress weeds, increase temperature, and retain moisture, but also completely solves the problem of recycling and treatment after the use of mulch (Ding Qian et al. 2019; Wang et al. 2019).

PLA can be applied to mulch, drip irrigation pipes, greenhouse film, pesticide and fertilizer slow-release materials, etc. Ren Xiang et al. (2014) studied the effect of PLA mulch on oat yield and water use efficiency in the arid area of Dingxi, Gansu Province, and found that in terms of soil water storage, PLA mulch is better than ridge tillage, and ridge tillage is better than flat cultivation. Covering PLA mulch, the water usage efficiency and grain yield of oats increased by 14% and 4%, respectively, compared to flat cultivation. Yang Lin et al. (2018) studied the effect of PLA mulch on the weed suppression effect and yield of chrysanthemum, and found that the relative weed suppression rate of PLA mulch during the seedling stage was 52.2%, and the yield could be increased by 16.9%. Zhang et al. (2018) studied the mechanical properties and degradation behavior of PE, PLA/PHA, starch–polyester mixtures, and cellulose-based mulch in soil, and found that under soil weathering, the mechanical properties of starch-polyester mixtures and cellulose-based mulch significantly decreased, while PLA/PHA mulch still has mechanical properties similar to PE mulch. PLA agricultural mulch has entered the production stage, such as BASF expanding the PLA production line and developing EcovioF mulch (a mixture of PLA and Ecoflex degradable polyester) (Chiyu 2012). In addition, related research on degradable mulch found that PLA/PBAT coverings can be degraded within 3 months, and the loss of crop biomass is only about 5% (Touchaleaume et al. 2016). In agricultural production, the use of agricultural film can increase soil temperature and inhibit weed growth, but large-scale discarded film can cause serious soil pollution problems. The usage of biodegradable agricultural film made from PLA as raw material can reduce environmental and soil pollution.

Materials such as PHA and PLA, due to their biodegradability and controllable degradation time, have been used as slow-release carriers and environmental remediation materials in agricultural production. PHA can also be made into slow-release carriers for biochemical fertilizers, herbicides, and pesticides, which can slowly release the drug effect, maintain the long-term effectiveness of the drug, and are more conducive to the growth of crops (Kadouri et al. 2005). The removal of pollutants from the environment is called environmental remediation or bioremediation. During the remediation process, adsorbents such as activated carbon, zeolite, and polymers are often added. Biodegradable polymers, such as PCL, PBS, PHBV, and PLA, have the potential for environmental remediation (Perego and Cella 2010). These polymers work by absorbing pollutants from any contaminated system (absorption mechanism) or by supplying carbon and energy to microorganisms to promote denitrification mechanisms. Because they can be used as raw materials and their prices are relatively low, many studies are currently focused on the possible uses of PLA and other biodegradable polymers in environmental remediation. However, compared with other degradable polymers (such as PCL), PLA has higher resistance to microbial activity (Caruso 2015). In terms of slow-release carrier applications, You Shengyong and others used PLA-PEG copolymers as wall materials and prepared slow-release fertilizers through the complex coacervation method (Shengyong et al. 2015). Some researchers have prepared metformin slow-release microspheres with PLA as the wall material, and the results show that its slow-release performance is significant (Xinyan et al. 2013). In addition, PLA and PHB, due to their good mechanical properties, can be used to make degradable seedling pots, seedling trays, etc.

7.2.2 Applications in the Forestry Field

In forestry, biobased degradable materials can be used for fruit protection sleeves, compost bags, bundling ropes, etc. Bilck and others used thermoplastic starch and PBAT as raw materials to make starch-based sleeves for the growth of pomegranates. Compared with PP nonwoven sleeves, it was found that the use of this degradable plastic sleeve does not affect the quality of the pomegranate (Bilck et al. 2011). The Dunhuang Gobi Desert Ecology and Environment Research Station/Key Laboratory of Desert and Desertification of the Northwest Institute of Eco-Environment and Resources of the Chinese Academy of Sciences used PLA as the main raw material for the first time to weave into a grid structure product for sand fixation. By fixing the column on the sand surface, a PLA grid sand barrier similar to a semi-hidden wheat grass sand barrier was formed to explore the service life and sand prevention effect of the PLA grid sand barrier under the extreme environment of the plateau (wind permeability rate 20–49%). The results show that the PLA grid sand barrier is not only radiation-resistant and aging-resistant but also has good sand prevention effects. The growth of vegetation in the early, middle, and late stages is shown in Fig. 7.4. After more than 90 days, the vegetation inside the sand barrier grows well. Therefore, the PLA grid sand barrier can become a new type of sand barrier to replace the grass square grid under harsh climatic conditions (Suchang 2022).

7.2.3 Applications in the Fisheries Sector

China is a major country in aquaculture, especially after the reform and opening up. To encourage the development of aquaculture, the state has vigorously adjusted the development policy of the fisheries, making aquaculture the focus of national fisheries development. In 2019, the national output of aquatic products was 64.8036 million tons, of which the aquaculture output was 50.7907 million tons, accounting for 78.38% of the total output (Journal News 2020).

In the fisheries sector, biobased degradable materials can be used for fishing lines, fishing nets, etc. Fishing net materials mainly include polyethylene and nylon,

Fig. 7.4 Sand fixation effect of polylactic acid sand barrier (see color insert). (**a**, **b**) Sand dune topography before sand fixation; (**c**) initial growth of vegetation; (**d**, **e**) mid-term growth of vegetation; (**f**) late-stage growth of vegetation. (Reprinted with permission from *Journal of arid land*, 2022, 14(9): 978–992. © 2022 Springer Nature Ltd (Suchang 2022))

and more than 90% of the sturdy and durable fishing nets are made of synthetic fibers. Nylon is a corrosion-resistant, difficult-to-degrade polymer material. These fishing nets will cause a great environmental problem after their service life. At the same time, fishing gear made for fishing or farming not only becomes garbage polluting the marine environment after being discarded, but also becomes a "killer" threatening marine life. Organized recycling of discarded fishing nets from thc sea is a solution, but it involves a wide range. The labor cost of fishing net recovery is high, and the difficulty is high. The degree of recycling is low, the recycling process is complex, and it cannot solve the problem from the source. PA4 has high strength, good wear resistance, impact resistance, flexibility, and appropriate elongation at break. As a member of the nylon family, it is degradable in activated sludge and the sea, so it will have a good application prospect in the fisheries sector.

7.3 Applications of Biobased Materials in the Medical Field

In recent years, clinical practice has shifted from replacing lost or damaged tissues with permanent implant materials to using regenerative scaffold materials that can help the body self-repair. It is this current situation that has stimulated the interest of medical workers and related researchers in developing biodegradable materials. Currently, about 70–80% of biomedical implants are made of metal materials. Some problems associated with these materials are potential stress shielding and biological complications related to their corrosion. Ceramic-based materials, especially those based on calcium phosphates, are generally the materials of choice for applications in musculoskeletal, oral, and maxillofacial areas. However, their use is mainly limited by their inherent brittleness, low fracture strength, and manufacturing difficulties.

Using biodegradable internal fixation materials instead of metal materials can avoid the disadvantages of stress shielding and secondary extraction. Degradable internal fixation materials can stably fix fractures and are ideal orthopedic internal fixation materials. Natural biodegradable polymers have excellent biocompatibility, but poor mechanical strength. Synthetic biodegradable materials can artificially adjust the degradation rate, have better mechanical strength, but lack biocompatibility. Composite materials can complement each other; therefore, compared to the first two materials, they have obvious advantages. Polymers used for biomedical applications can be biodegradable or nonbiodegradable, depending on their ability to degrade when implanted in biological systems. Polymers with a carbon–carbon backbone have strong resistance to degradation, while the presence of heteroatoms in the form of esters, anhydrides, amides, etc., makes them easy to degrade, can avoid pain and infection when removing implants after tissue healing, and reduce carcinogenic or other biological effects caused by corrosion. These factors will reduce the cost of treatment and the incidence of patients. This section mainly introduces the application of biobased degradable materials in the medical field, such as artificial stents, absorbable surgical sutures, bone nails, drug-controlled slow-release carriers, and medical dressings.

7.3.1 Tissue Engineering Scaffold

Tissue engineering is a developing science and technology that can be used to improve a variety of clinical conditions, including spinal fusion, joint replacement, nonhealing fractures, and pathological bone loss etc. Tissue engineering is a science that combines tissue cells with specific biological activities with biomaterials to construct tissues and organs in vitro or in vivo to maintain, repair, regenerate, or improve the function of damaged tissues and organs. The four elements of tissue engineering mainly include seed cells, biomaterials, cell proliferation, and in vivo transplantation. Among them, the biomaterial scaffold needs to provide a growth and differentiation environment with sufficient nutrition and suitable conditions for cells. Biobased materials represented by PLA, PHA, etc., have excellent biocompatibility, good mechanical properties, strong plasticity, and adjustable and controllable degradation performance, which meet the requirements of tissue engineering scaffold materials.

PLA is widely used in tissue engineering scaffold materials due to its bioabsorbability and biocompatibility (Lin et al. 2019). A composite scaffold for an artificial intervertebral disc (IVD) can be constructed by combining 3D printing and electrospinning. The scaffold is constructed using PLA as the framework structure and is made of porous poly (L-lactide)/octahedral oligomeric silsesquioxane (PLLA/POSS—(PLLA) 8) fiber bundles to create fiber rings (AF). The test results show that the structure and mechanical properties of this composite scaffold can be comparable to natural in vitro diagnostic scaffolds. In addition, the porosity and mechanical properties of this scaffold can be adjusted by 3D model design, indicating that the composite biomimetic IVD scaffold based on PLA through 3D printing can be personalized for repair and regeneration (Zhu et al. 2021). PLLA/3D microfiber scaffolds prepared using electrospinning technology have been applied to bone repair, and PLLA/hydroxyapatite and PLLA/hydroxyapatite/silk composite materials prepared using 3D printing show mechanical properties close to original bone tissue and good biological activity (Zhang et al. 2013). A polylactic acid bone regeneration scaffold (SSL scaffold) was prepared using two-photon polymerization technology (Timashev et al. 2016). This scaffold can provide the microenvironment needed for bone marrow mesenchymal stem cells to differentiate and form new bones in vitro and in vivo for patients with bone defects.

PHA, due to its good biodegradability, tissue compatibility, strong plasticity, and surface modifiability, is an ideal cell scaffold material in the field of tissue engineering. Currently, tissue-engineered joints and tracheal cartilage have been successfully cultivated in animals, which can replace missing tissues or organs in the human body. Boeree et al. (1993) studied the characteristics of this composite material and found that PHB can stimulate the formation of new bones. Hydroxyapatite and bone tissue have a good affinity. After the two are combined, different strengths of bone repair materials can be made by adjusting the proportion of components. Yue Pengju et al. (2009) adhered chondrocytes to the PHBV scaffold, cultured them in vitro for 28 days, and then implanted them into the damaged rabbit knee cartilage. After 30 days, a smooth knee cartilage-like structure that was well integrated with the rabbit knee joint was formed, and no rejection occurred. Using PHA to process into a tubular support to culture endothelial cells, after the formation of new blood vessels, the scaffold is completely degraded, which can achieve the purpose of repairing damaged tissues (Jin et al. 2016). Shi Peiliang et al. (2001) injected the bone marrow stromal cells of New Zealand rabbits into PHB foam material, and after being implanted subcutaneously in the back of nude mice for 8 weeks, a large amount of new bone tissue was formed.

7.3.2 Absorbable Surgical Sutures

Among the biomaterials used for implantation in the human body, sutures are the most used material, with a huge market of over 1.3 billion US dollars each year. With the development of the absorbable polymer polyglycolic acid (PGA) in the early 1970s, a new chapter for absorbable polymer sutures began. Absorbable sutures refer to a new type of suture material that can be degraded and absorbed by the human body after being implanted into human tissues during surgical suturing and do not need to be removed, thus eliminating the pain of removing sutures. They can be widely used in gynecology, obstetrics, surgery, plastic surgery, urology, pediatrics, dentistry, otolaryngology, ophthalmology, and other surgeries and suturing of subcutaneous soft tissues. According to the absorbability of the suture material, absorbable sutures can be divided into catgut, synthetic polymer sutures, and pure natural collagen sutures. Compared with the other two types of natural absorbable surgical sutures, biodegradable polymer sutures have the advantages of a wide source of raw materials, suitable for large-scale production, low cost, and long-term maintenance of high strength.

1. *Research on the Application of PLA in the Field of Absorbable Surgical Sutures*

PLA is a polymer with excellent biocompatibility and biodegradability, and has been approved by the FDA for use as a surgical suture. According to reports by Li Xiaohong et al. (1999), the final metabolites of PLA in the body are CO_2 and H_2O; the intermediate product, lactic acid, is also a normal product of sugar metabolism in the body, so it will not accumulate in important organs. Polylactic acid and its copolymers are used as surgical sutures, which degrade and absorb automatically after wound healing, eliminating the need for suture removal. Lactide has three optical isomers, namely L-lactide, D-lactide, and DL-lactide. PLA can be divided into three different types: polylactic acid (PLLA), poly-D-lactic acid (PDLA), and poly-DL-lactic acid (PDLLA). PLLA and PDLA have higher crystallinity due to their more regular molecular chain arrangement, while PDLLA is an amorphous material with poor mechanical properties, so the latter degrades faster and PDLA is rarely used alone (Middleton and Tipton 2000). PLLA is more suitable for surgical sutures due to its good mechanical properties and higher tensile strength. There have been reports of PLLA being used as surgical sutures as early as the 1960s.

Since 1962, when the American Cyanamid company launched the first polylactic acid surgical suture, research on polylactic acid-based absorbable sutures has developed rapidly, but there are still many problems limiting the performance of this system of products (Liang and Guoming 2001). In 1975, polylactic acid–hydroxyethyl acid copolymer (PLGA) (LA:GA = 90:10) began to be used as surgical sutures on the market. Currently, by synthesizing PLA with higher molecular weight and improving processing technology to enhance the mechanical strength of the suture, and further regulating the mechanical properties and degradation rate of PLLA fibers through copolymerization, blending, or grafting. Compared with PHBV, PLA is easier to degrade and crystallize. Therefore, Yu et al. (2014) blended PHBV and

Fig. 7.5 PHBV/PLA surgical sutures. (He et al. 2014)

PLA to prepare multifilament fibers (Fig. 7.5), and PHBV/PLA multifilament fibers were implanted in rats as medical sutures. Compared with catgut, PHBV/PLA fibers have less inflammatory reaction, better cell affinity, and lower degradation rate, so PHBV/PLA multifilament fibers have the potential as surgical sutures. In addition, in order to improve the biological activity and anti-bacterial and anti-inflammatory functions, blending or doping drugs is also one of the few ways that researchers focus on (Tiejun and Zhixian 2001).

2. *Research on the Application of Polyhydroxyalkanoates in the Field of Absorbable Surgical Sutures*

Polyhydroxyalkanoates have good thermoplasticity and elastic strength. Because they come from microbial fermentation products, they can reduce the occurrence of chronic immune reactions and cytotoxicity during implantation. PHB and PHBV have good biocompatibility. When surgical sutures made of PHB and PHB/PHV (PHBV) are implanted into muscle tissue, histological observation results show that, except for short-term postoperative reactions, the quality, strength, and inflammation time of polyhydroxyalkanoate sutures are similar to those of implanted silk, and the inflammatory response is significantly reduced compared to when catgut is implanted, indicating that PHB and PHBV will significantly reduce the incidence of inflammation after implantation compared to catgut. At the same time, the fibers of PHBV can provide sufficient mechanical strength to meet the requirements of muscle tissue (Volova et al. 2003; Shishatskaya et al. 2004). Tepha Inc. has produced surgical sutures made from polyhydroxyalkanoates and has received approval from the US FDA. Among them, TephaFLEX sutures made from P4HB are the most widely used. These sutures do not need to be removed after surgery and can naturally degrade in the human body, providing great convenience for patients and doctors.

3. *Research on the Application of Chitin in Absorbable Surgical Sutures*

Chitin has many advantages such as good biocompatibility, a wide source of raw materials, simple and low-cost manufacturing, moderate mechanical strength and stability in an vivo environment, and antibacterial and anti-inflammatory functions, can promote wound healing, and simple disinfection, dyeing, and preservation treatment (Qingji et al. 1998). Since the 1970s, many companies, represented by Unigeca, have focused on Chitin, a resource-rich material, for absorbable surgical sutures (Qiaoling et al. 2003). They dissolve high-purity Chitin in organic solvents such as trichloroacetic acid, use wet spinning to form, and adjust the solvent and coagulant to improve spinning and posttreatment processes, producing a variety of types and performances of absorbable Chitin sutures.

Research by Goosen et al. (1997) shows that chitin sutures have good tolerance in the body. However, compared with sutures like PGA, their tensile strength and acid corrosion resistance are somewhat lacking, unable to meet the requirements of high-strength suturing in harsh environments. To solve practical problems, Su Xiurong et al. (1996) used chitin derivative chitosan to prepare a high-performance surgical suture. In vivo experiments show that this material is nontoxic and harmless, can maintain good mechanical strength in the body, and its degradation rate is much higher than that of the control sheep gut line. Due to the lack of breakthrough progress in the research on the preparation process, in vivo degradation, and immune mechanism of C=chitin and chitosan, the clinical application of chitin implant materials is still very limited. In addition, some scholars have studied the application of alginate fiber in surgical sutures, which is a long-chain copolymer composed of rhamnose aldehyde acid and mannose aldehyde acid. The special molecular structure provides alginate fiber with high absorbency, hemostasis, and tissue affinity (Yonglian and Bing 2001).

7.3.3 Bone Pins

From minor fracture treatments to major craniotomy surgeries, bone screws that serve a fixation function are used. After 20 years of in-depth research and development on biodegradable metal materials, magnesium-based alloys remain the most popular research subjects in metallurgical laboratories, especially in the field of orthopedic applications. In 2019, the biodegradable pure magnesium bone screw product developed in cooperation between the Institute of Metal Research of the Chinese Academy of Sciences and Dongguan Yi'an Technology Co., Ltd. received clinical approval from the National Medical Products Administration, becoming the first biodegradable magnesium-based metal Class III (implantable) medical device product in China to receive clinical approval. This signifies a breakthrough in the clinical transformation of products in this field in China. In June 2021, Xijing Hospital of Air Force Medical University successfully performed an internal fixation surgery on a patient with a distal femoral fracture using a self-made biodegradable silk screw, which is the world's first biodegradable medical silk bone screw used for human surgical fixation.

1. *Application of PLA in the Field of Bone Nails*

In the field of biomedicine, people have been studying the use of PLA materials to replace metal medical devices. Compared with traditional materials, PLA is non-toxic, nonirritating, has low immunogenicity and biodegradability, and its degradation product, lactic acid, has no known toxic effects on the human body (Singhvi et al. 2019). PLA does not have a negative impact on biological tissues it comes into contact with, has excellent compatibility, and its hydrolysis product hydroxy acid does not interfere with tissue healing, does not produce toxicity or carcinogenic effects in local tissues, and can be introduced into the human body's tricarboxylic acid cycle and excreted from the body (Tokiwa and Calabia 2006). PLA materials can be used in biomedicine for bone nails, oral materials, absorbable bone fixation implants, etc. (Sakaguchi and Kobayashi 2016; Kanno et al. 2018). In the development of tissue engineering, the initial scaffold materials used were all metal materials. Although metals have good mechanical properties, their biological activity is poor. Polymer materials can be flexibly designed with different components and structures to achieve the desired performance. For example, PLLA/octadecylamine-functionalized nanodiamond composite materials have good biocompatibility and can resist thrombosis formation, with broad application prospects in vascular transplantation. Ritz et al. (2017) tested the biocompatibility of PLA 3D printed parts coated with collagen and found that their endotoxin contamination levels were significantly lower than the FDA's limits, and various cells grew well on PLA printed parts. Semi-crystalline PLA with high mechanical strength, combined with other polymers or proteins such as glass fiber, polyglycolic acid, carbon fiber, collagen, and hydroxyapatite can be used in orthopedic surgery, such as plates, pins, and screws (Hamad 2015; Serra et al. 2013; Liu et al. 2016).

2. *Application of Polyhydroxyalkanoates in the Field of Bone Nails*

In addition to being biodegradable, polyhydroxyalkanoates also possess excellent biocompatibility, providing an environment for the growth of various tissue and organ cells, and are noncarcinogenic. Therefore, they are widely used in tissue engineering and medical health fields. For example, PHB has good biocompatibility, and its monomer 3-hydroxybutyric acid is endogenously present in animals and can be degraded into CO_2 and H_2O in the body. Compared with other polyester polymers, the acidity and inflammation caused by the released hydroxy acid monomers are smaller. Studies have shown that there is a concentration of 0.3–1.3 mmol/L of 3-hydroxybutyric acid in the blood, and low oligomers of 3-hydroxybutyric acid exist on the cell membrane of prokaryotes. This phenomenon has been found in almost all true bacteria, archaea, plants, and mammals (Guoqiang and Daixu 2014). Therefore, based on the characteristics and advantages of polyhydroxyalkanoate materials, polyhydroxyalkanoate materials can be made into bone repair materials, bone nails, and other internal fixation devices (Xu et al. 2010).

7.3.4 *Drug-Controlled Slow-Release Carrier*

Nanoparticles prepared by oil–water amphiphilic polymer polymers usually have biodegradability and good biocompatibility, which has attracted the attention of the academic community in the field of drug-controlled release, and more and more research uses them as drug carriers for tumor treatment. Degradable polymers represented by PLA, due to their adjustable degradation rate in the body, have received widespread attention in the field of drug slow release. The construction of drug-controlled release systems ensures that drugs maintain an effective concentration in the body for a long time, while eliminating or reducing drug side effects, and have excellent performance in the field of in-body implantation or subcutaneous injection.

1. *Research on the Application of PLA in the Field of Drug-Controlled Slow-Release Carrier*

Due to the presence of a large number of ester bonds on the main chain, PLA is a relatively hydrophobic polymer, which results in a low affinity between the material and cells when used in the body environment (Burg et al. 1999). In addition, due to its lack of sufficient reactive side chain groups, pure PLA is a relatively chemically inert polymer, which makes it difficult to modify the surface and body. To solve the application limitations of polylactic acid polymer in drug delivery due to its low cell affinity and difficulty in modification, researchers introduced hydrophilic material chain segments such as CS, polyethylene glycol, polypeptide, polyacrylic acid, and polyurethane into the hydrophobic main chain to synthesize PLA-based amphiphilic copolymers, achieving improvements in polymer biological properties.

Polyethylene glycol is a common substance used to modify PLA as a hydrophilic chain segment. Ren et al. (2010) copolymerized lactide with PEG esterified by biotinyl chloride, prepared the copolymer into biotinized PEG–PLA nanoparticles by nanoprecipitation technology, and conjugated transferrin (Tf) and nanoparticles by using the biotin groups on the surface. These protein-loaded nanoparticles were injected into a rat model with C6 glioma. Fluorescent microscopy of brain sections intuitively demonstrated that Tf-functionalized PEG–PLA nanoparticles can penetrate into tumors in the body. Flow cytometry results showed that these nanoparticles have the ability to target tumor cells in vitro. These biotinylated PEG–PLA nanoparticles are expected to be used as carriers for targeted drug delivery in gliomas.

In the treatment of corneal diseases, the drug cyclosporin A (CycA) can treat chronic dry eye disease through immunosuppression and increasing mucin secretion levels. It is safe for long-term use in clinics and has few side effects. However, as a lipophilic large peptide drug, CycA is usually made into oily eye drops because it is not easily soluble in water, which can be irritating to patients due to the large dosage. Using nanoparticles (NPs) as drug carriers to control the release of encapsulated substances can target the eye with a smaller dosage, thus alleviating this irritation. Because dextran has abundant hydroxyl groups on its skeleton, which has a higher hydroxyl group functional group density than PEG, researchers have prepared poly(D,L-lactide)-b-dextran (PLA-Dex) core-shell structure nanoparticles. The surface of Dex was modified with phenylboronic acid (PBA), allowing it to form a complex with the diol groups of ocular mucosal sialic acid at physiological pH (Fig. 7.6). Then, CycA was loaded into the hydrophobic region of the PLA inside the nanoparticles through hydrophobic interactions. In this way, the carrier with CycA can further enhance mucosal adhesion through the hydrophilic outer layer of PBA, allowing the drug carrier to concentrate in the cornea, thereby prolonging the action time of the drug on the corneal surface and effectively improving dry eye symptoms (Liu et al. 2012).

Gamma-polyglutamic acid is nontoxic and harmless to the human body and has strong hydrophilicity. It can also replace PEG and PLA to prepare amphiphilic block copolymers. Because the amino and carboxyl groups on the surface of polyglutamic acid have strong reactivity, Liang et al. attached galactose to gamma-polyglutamic acid, using galactose as the targeting molecule for liver cancer cells, and finally obtained the liver cancer-targeting nanoparticle carrier Gal-P/NPs. Compared with the commercially available paclitaxel formulation Phyxol®, the paclitaxel loaded by Gal-P/NPs has more significant activity in inhibiting the growth of HepG2 cells. The experimental group injected with Gal-P/NPs can more effectively reduce tumor size, indicating that the Gal-P/NPs drug carrier prepared based

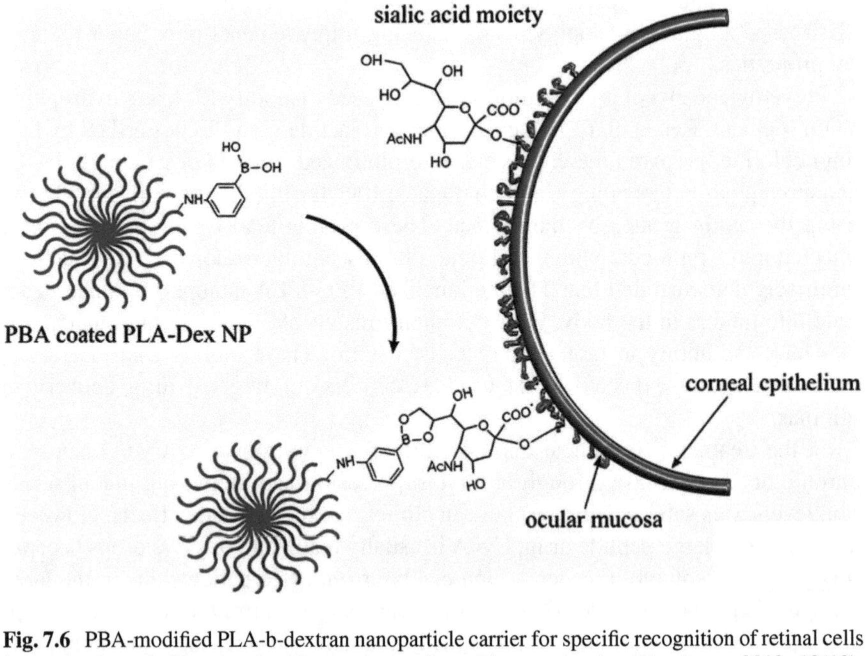

Fig. 7.6 PBA-modified PLA-b-dextran nanoparticle carrier for specific recognition of retinal cells (see color insert). (Reprinted with permission from *Macromolecular Bioscience*, 2012, 12(12): 1622–1626. © 2012 Wiley Online Library Ltd. (Liu et al. 2012))

on PLA with specific recognition of liver cancer has great application potential in targeted treatment of liver cancer (Liang et al. 2006).

The application of self-assembled polyester micelles in drug slow-release has received much attention, such as the rapidly degrading amorphous PLA often used as a drug slow-release excipient (Peres et al. 2017). By controlling the degradation rate of the PLA carrier, the controlled release of the drug can be achieved, ensuring that the drug maintains an effective concentration for a long time, eliminating or reducing side effects, especially in the field of in vivo implantation or subcutaneous injection. Feng et al. (2017) prepared an efficient anti-tumor drug carrier, using 4-arm ethylene glycol-(L-lactide/polyD-lactide)-cholesterol block copolymer (CSCM) and poly-cyclodextrin (PCD) as raw materials, and prepared a series of polylactic acid stereocomplex micelles by self-assembly in an aqueous phase containing doxorubicin (DOX). This drug delivery system can enter tumor cells and continuously release DOX and is expected to be used for the treatment of malignant tumors. Chen Xin et al. (2019) prepared paclitaxel and sorafenib drug-loaded poly(lactic acid-hydroxyethyl acid) microspheres; the obtained microspheres are round in shape, paclitaxel and sorafenib slowly release in vitro environment at 37 °C, and the drug release rates after 41 days are 71.83% and 81.44%, respectively.

2. *Application of PHA in the Field of Drug-Controlled Release Carriers*

The design of drug release systems is a promising and rapidly developing field. In addition to PLA-based biodegradable polymers, a large number of clinical studies have been carried out on the use of PHA series polyesters in the field of drug coating. In the early 1990s, PHA was already a candidate material for drug carriers due to its biodegradability and biocompatibility. After using PHA to coat drugs, they can be injected intravenously or taken orally, and the drug is gradually released in the body, controlling the drug concentration in the patient's blood or tissues within an appropriate range, prolonging the drug effect, and making the drug effect reach difficult-to-administer parts. In the body, it can ultimately be degraded into CO_2 and H_2O, and will not produce toxicity to the human body.

PHA can be loaded in various forms such as gels, microclusters, microspheres, nanoparticles, and porous scaffolds. Currently, PHB, PHBV, and copolymers of 3-hydroxybutyrate and 4-hydroxybutyrate (P3HB4HB) and other PHA copolymers have all been studied to prepare micronanoparticles for drug loading (Jin et al. 2016). However, due to the high crystallinity of PHB, up to 60–90%, it is not easily degraded by enzymes in the body; therefore, the copolymer PHBV has a wider application in the medical field due to its lower crystallinity. Some researchers have integrated the antibiotic sulbactam-cefoperazone into a carrier made of poly(3-hydroxybutyrate-co-22 mol%-3-hydroxyvalerate) ($1 \times 0.3 \times 0.3$ cm^3, 100 mg) and implanted it into the tibia of a rabbit artificially infected with Staphylococcus aureus. Half a month later, the infection symptoms subsided, and it was almost completely healed after 1 month (Yagmurlu et al. 1999). Masood and others have prepared PVA/PHBV drug-loaded nanoparticles with an average diameter of 200 nm through emulsification and solvent evaporation methods. The drug contains rosetree alkaloid, which can be used for tumor treatment. Compared with nondrug-loaded particles, the average diameter of drug-loaded particles is slightly larger. In vitro cytotoxicity experiments show that nondrug-loaded nanoparticles only act as a placebo; not only do they not affect the normal growth of tumor cells, but they also have good biocompatibility; thus, they will not interfere with the comparison between drug-loaded nanoparticles and unencapsulated drug results. The results show that the PHBV nanoparticles loaded with rosetree alkaloid have an even better inhibitory effect on tumor cells than the unencapsulated drug, which also indicates that the drug-controlled slow-release carrier can improve the bioavailability of the drug (Masood et al. 2013).

7.3.5 Medical Dressings

Medical dressings are medical materials used to cover wounds or other injuries. Natural gauze is the earliest and most widely used type of dressing, but its nondegradable and easy adhesion to the wound surface limit its further development in dressing applications. However, biobased polymers such as PLA and chitin, due to their good physical and chemical properties, biological activity, excellent biocompatibility, and biodegradability, have received much attention in the field of medical auxiliary material application research.

1. *Application of PLA in the Field of Medical Dressings*

The modulus of polylactic acid fiber is between polyester and polyamide; it has good acid and weak-alkali resistance. When used as a dressing, it comes into contact with a wide range of drugs and has a certain degree of extensibility and comfort. At present, PLA-based polymers, including pure PLA, polylactic acid–polyethylene glycol, and poly(lactic acid–ethanol acid), are receiving much attention in the research of wound dressings.

PLA has adjustable and controllable properties and can be made into various dressing products, such as nonwoven fabrics, electrospun fiber films, gel particles, and nanoparticles. Yagi et al. from Japan (Wakabayashi et al. 2019) compared the nonwoven PLA cloth product of Toray Industries in Japan with traditional gauze, studying their performance when used as wound dressings. When the two types of dressing samples were laid on the cut surface of a mouse's liver, the results showed that due to the finer fiber diameter of PLA (≤ 1 μm), the PLA nonwoven fabric dressing was more likely to adhere to the wound surface. Under an electron microscope, it was also observed that red blood cells showed better adhesion on PLA fibers. In addition, the average hemostasis time and bleeding volume of the PLA dressing were only 270 s and 0.7 g, respectively, far superior to the 495 s and 2.1 g of ordinary gauze, indicating that this PLA nonwoven fabric product can be used as a hemostatic material in situations such as minimally invasive liver surgery. Schneider et al. (Wyrwa et al. 2018) mixed L-lactic acid-D-lactic acid random copolymer with adrenaline (AD) and tranexamic acid (TA) hemostatic agents and made a thin and soft fiber film dressing through electrospinning. In vitro drug loading and release experiment results show that the loading amounts of AD and TA on this dressing can reach up to 20 wt% and 50 wt%, respectively, and the release rates are 50% and 85%, respectively. At the same time, mechanical testing found that the tensile strength of this dressing can reach 14MPa. More experimental tests show that this dressing has the advantages of low cytotoxicity, fast hemostasis, etc., and is suitable for hemostasis in minor wound surgery and dressing use in difficult-to-fix scenarios.

The use of various bioactive materials combined or coated with PLA for wound repair has also been clinically proven to be feasible and effective. Lee et al. (Birajdar et al. 2018) coated a hyaluronic acid–iron chloride mixture on electrospun PLA fibers to produce a type of gel particle. These gel particles can, like platelets, target and aggregate at the bleeding site under the action of an external magnetic field, forming a stable artificial thrombus, which can increase the modulus of the thrombus by 300% and the hardness by 50%, playing a role similar to fibrin in increasing the strength of the thrombus, thus helping the hemostasis process. Lavik et al. from the United States (Lashof-Sullivan et al. 2016) synthesized a type of hemostatic nanoparticle by coating pure PLA polymer with adhesive peptides. Animal experiment results showed that this particle can quickly stop bleeding, reduce the amount of bleeding in mouse wounds, and increase the survival rate of mice. In addition, this nano-hemostatic material can be stably stored under high-temperature conditions. These particles can still maintain their original shape after being stored at a temperature of 50 °C for 7 days and can still effectively stop bleeding in mouse wound repair experiments. This hemostatic material is expected to be used in harsh storage and use environments such as deserts and battlefields.

In addition to pure PLA polymers, PLA copolymers are also widely used in the manufacture of wound dressings. The introduction of polyethylene glycol (PEG) or polyglycolic acid (PGA) can effectively improve the hydrophilicity and tissue adhesiveness of polylactic acid. Chitrattha and Phaechamud (2016) prepared PLLA/PEG drug-loaded porous films by dissolving gentamicin sulfate (GS) or metronidazole (MZ) with the polymer in dichloromethane (DCM) and applying them to a glass plate. The pore size of this film is about 20μm, and the porosity is about 50%. The GS-loaded porous film has a significant inhibitory effect on *Staphylococcus aureus*, *Bacteroides fragilis*, and *Pseudomonas aeruginosa*, while the MZ-loaded porous film has an inhibitory effect on *Bacteroides fragilis*. Both have antibacterial activity for more than 7 days. This drug-loaded porous film is expected to be used for wound antibacterial. In addition, compared with PLLA, the crystallinity, water oxygen permeability, degradation rate, and drug release capacity of the PLLA/PEG drug-loaded porous film have been significantly improved. To prevent postoperative intestinal adhesion, Xia et al. prepared a three-layer barrier wound dressing, with a carboxymethyl chitosan (CMCS) sponge layer sandwiched between a PLGA/PLA-b-PEG electrospun layer. The former plays a role in rapid hemostasis, while the latter mainly serves as an anti-adhesion material. In vivo test results show that this three-layer composite material can effectively reduce the level and incidence of postoperative intra-abdominal intestinal adhesion (Xia et al. 2015).

2. *Research on the Application of Chitosan in Medical Dressings*

Chitosan (CS) achieves hemostatic effects by stimulating the adhesion and aggregation of platelets, a function derived from its high positive charge density. Wound dressings made from chitosan as the main raw material have many advantages, such as antibacterial, hemostatic, good biocompatibility, promoting wound healing, and tissue growth. In recent years, more and more chitosan-based medical dressings have been developed.

Chitosan dressings are generally composed of chitosan, other excipients or drugs. In a rat liver bleeding model, the chitosan sponge has a better hemostatic effect and faster degradation than gel sponge (GS), indicating that chitosan has excellent hemostatic performance. Research results show that chitosan has a strong hemostatic effect when the degree of deacetylation is 40% (Huang et al. 2015). Although chitosan has certain antimicrobial properties, it cannot effectively inhibit the growth of bacteria. Therefore, many studies are dedicated to enhancing the antimicrobial properties of chitosan. Among them, a good choice is to use antibiotics that can inhibit bacterial growth together with chitosan to prepare medical dressings. Some researchers have prepared ampicillin-grafted chitosan (CSAP) sponges. These sponges, after cytotoxicity tests, not only have good biocompatibility but also can accelerate wound healing, enhance the antimicrobial activity of chitosan sponges, and significantly inhibit the growth of *Staphylococcus aureus*, *Candida albicans*, and *Escherichia coli*. In addition, they can also reduce the use of antibiotics (Wu et al. 2018). Similarly, Ren et al. (2019) successfully prepared medical dressings by using chitosan, silk fibroin (SF), and halloysite nanotubes (HNTs) loaded with the broad-spectrum antimicrobial drug chlorhexidine digluconate (CHD) through electrospinning technology. The in vitro coagulation test results show that the addition of HNTs can accelerate coagulation, and the drug-loaded HNTs can extend the drug release for about 8 days. In addition to antibiotics, metal nanoparticles (Chakraborty et al. 2019) are often used as additives. The addition of metal nanoparticles will reduce the antimicrobial properties of chitosan itself, but due to the metal chelation effect, they improve the overall antimicrobial properties of the chitosan-based dressings.

The mechanism of chitosan dressings mainly lies in promoting wound healing, hemostasis, moisture retention, analgesia, and inhibiting scar hyperplasia, etc. Chitosan dressings are mainly used on skin-surface cuts, burns, and inflammations. The related manufacturers include Henan Chengdong Biotechnology Co., Ltd., Hubei Puai Pharmaceutical Co., Ltd., and Shanghai Changjia Medical Technology Co., Ltd.

7.4 Biobased Degradable Fiber Materials

In 2018, the production of chemical fibers in China reached 50.11 million tons, of which the production of synthetic fibers was 45.626 million tons, accounting for more than 70% of the world's total production of chemical fibers (China Chemical

Fiber Industry Association 2018), which is enough to show that China is a major producer of chemical fibers. Polyester, nylon, and acrylic fibers are not easily degradable and have problems with recycling. Whether it is landfill or incineration, it will cause harm to the environment. Biobased degradable chemical fibers, as a new type of fiber material that is expected to alleviate resource crises and environmental pollution, are made from biodegradable polymer materials through a series of processes such as preparing spinning solutions, spinning molding, and post-processing. The products not only have advantages such as affinity and comfort, but also have ecological and environmental protection advantages, which are of great significance for achieving sustainable development in China.

In the production process, chemical fibers mainly use two types of spinning methods: melt spinning (Fig. 7.7) and solution spinning (Fig. 7.8). Solution spinning can be divided into wet spinning and dry spinning according to different coagulation methods. Manufacturers often combine the advantages of wet spinning and dry spinning to use dry-wet spinning methods for spinning. This spinning method is especially suitable for the molding and processing of liquid crystal polymers, so it is also called liquid crystal spinning. At present, fibers produced by this method include polylactic acid fibers, chitosan fibers, aromatic polyamide fibers, etc.

The development of PLA fiber can be traced back to the late 1980s, which was jointly completed by Japan's Kanebo and Shimadzu Corporation, with raw materials from Shimadzu Corporation and CDP Company, and was launched in 1995 under the brand name "Lactron". China did not realize the production of direct-spun PLA filament until 2006, and the development and production company was Changshu Yangtze Chemical Fiber Co., Ltd. In 2019, Hengtian Yangtze Bio Co., Ltd. built a continuous polymerization melt direct spinning PLA filament production line with a scale of 2000 t/a. Because the melt direct spinning eliminates the steps of pelletizing and drying, the production cost of PLA fiber is greatly reduced, which will be the main trend of its future development.

At present, the spinning processes for producing PLA fibers mainly include melt spinning, wet spinning, dry spinning, and dry-wet spinning. Among biobased degradable materials, only PLA can be melt-spun on a large scale into textile fibers with sufficient strength. Almost all commercially available PLA is made by melt spinning because this process can not only produce at high speed but also is environmentally friendly and economical. Melt spinning is also the most researched and applied method for manufacturing PLA fibers. Ghosh et al. explored the effects of different spinning speeds and stretching and heat-setting posttreatments on the morphology of PLA filaments. The results showed that PLA produced at a spinning

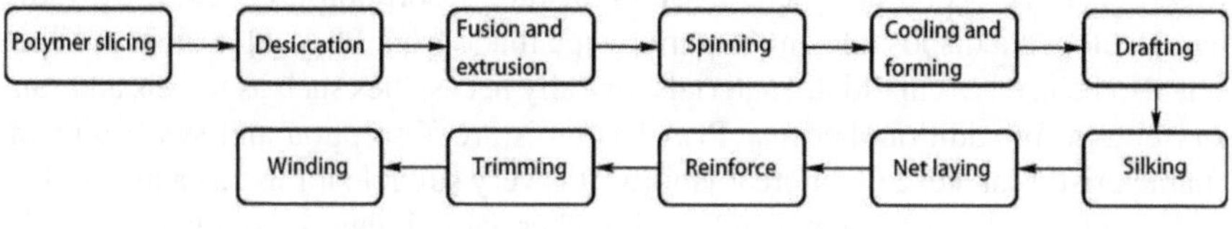

Fig. 7.7 Melt spinning process flow

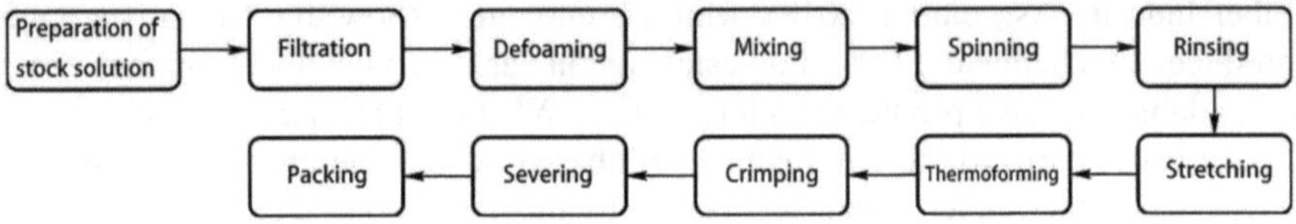

Fig. 7.8 Solution spinning process flow

Table 7.1 Comparison of the performance of polylactic acid fiber and common fibers (Hong 2005)

Fiber type	Nylon 6	Polyester	Acrylic	Polylactic acid	Viscose fiber	Cotton	Silk	Wool
Melting point (°C)	215	255	320 *	130–170	– *	– +	– +	– +
Strength (g/d)	5.5	6	44	6	2.5	44	44	1.6
Moisture absorption (%)	4.1	0.2–0.4	1.0–2.0	0.4–0.6	11	7.5	10	14–18
Specific gravity	1.14	1.39	1.18	1.25	1.52	1.52	1.34	1.31
Limiting oxygen index (%)	20–24	20–22	18	26	17–19	17	–	24–25
Heat of combustion (MJ/kg)	31	25~30	31	19	17	17	–	21
Refractive index	1.52	1.54	1.5	1.35~1.45	1.52	1.53	1.54	1.54

*Acrylic and viscose fibers decompose when melted; +: Cotton, silk, and wool decompose without melting; – No relevant data

speed of 500 mm^{-1} is almost amorphous, while the crystallinity of the filament spun at a speed of 1850 mm^{-1} is 6%. After stretching and heat setting, the crystallinity of PLA filament is about 60% higher than that of the initial spinning (Ghosh and Vasanthan 2010). Therefore, the main factors affecting the mechanical properties of PLA fibers are melt draw ratio, solid-state draw ratio, and draw temperature. In addition, one of the key issues of PLA melt spinning is degradation, which may occur during crushing, spinning, and post-spinning operations. The types of degradation involved include hydrolysis, depolymerization, cyclic oligomerization, and intermolecular and intramolecular ester exchange. Therefore, in order to prevent the ester bond on the main chain from hydrolysis and depolymerization due to H_2O molecules, PLA must be thoroughly dried before melt spinning.

PLA fiber products made through melt spinning have a wide range of applications in the textile field. The physical and mechanical properties of PLA fibers are also excellent, and the performance comparison of their fibers is shown in Table 7.1. Based on PLA's good antibacterial and moisture absorption properties, Biovation has developed a disposable antibacterial wipe made from PLA. PLA fiber products can also be used as candidate materials for daily necessities such as towels and sanitary cotton. In addition, because PLA has moisture absorption and sweat-wicking characteristics and excellent breathability, it is very suitable for use as a material for making clothes, so it is often blended with other natural fibers or synthetic fibers for use in underwear and sportswear fabrics. When made into a jacket, it has high resilience, and PLA's washability has met the standards of the American Association of Textile Chemists and Colorists (AATCC). PLA fiber has good UV resistance, flame retardancy, retention, and wrinkle resistance, and can also be used for household textiles, such as curtains, carpets, and bedding (Zhen 2008).

Currently, the main methods for preparing PHA fibers are electrospinning, melt spinning, and dry spinning. PHA is sensitive to temperature and easily decomposes during heat processing, so the main technical means of its fiber production is electrospinning, which is mainly used in the medical field, such as stents and medical sutures. The difficulty in the production of PHA fiber is (Jing et al. 2010): pure PHA has high crystallinity, poor flexibility, and mechanical properties; high cost, small dye uptake; low spinning speed, high draw ratio is not easy to achieve. Extruded PHA fibers are generally used in cigarette filter tow, nonwoven fabrics, and the surface yarn of car carpets. The biobased polyester PHA mainly used in medical textiles is P4HB, P3HB4HB, PHA4400, etc. The types of medical fibers include PHA non-crimped short fibers and medical monofilaments, and the products have been used in plastic repair, regenerative medicine, surgical mesh, and tissue engineering (Changchun 2013).

Chitosan fiber is primarily made from chitosan derived from the shells of shrimp, crabs, and other arthropods. It is produced through a spinning process that involves dissolution, coagulation, and stretching, resulting in a tightly structured and high-performing regenerated biomass fiber. Its strength generally ranges from 0.97 to 2.73 cN/dtex. Common chitosan production processes include wet spinning, dry-wet spinning, electrospinning, and liquid crystal spinning, with wet spinning being the most common method for industrial production of chitosan fiber. The representative company in China is Hesmore Biotechnology Co., Ltd., a global leader in the research, production, processing, and sales of chitosan fiber. In 2012, it was the first to achieve industrial production of pure chitosan fiber at the kiloton level. Fabrics made from chitosan fiber are soft, comfortable, and elastic after finishing, and also have antibacterial properties. The company has launched antibacterial socks, underwear, bras, towels, bed sheets, sanitary pads, and other textile products on the market, which are exported to more than a dozen countries and regions and are well-received by consumers.

Compared with other biodegradable polyester materials such as PLA and PHA, PBS is cheaper and has good physical and mechanical properties, heat resistance, and ease of processing and molding. However, due to its wide molecular weight distribution, poor thermal stability, low melt strength, and poor spinnability, its application and development in the fiber field are limited. Currently, Shaoxing Jiuzhou Chemical Fiber Co., Ltd. has a patent for PBS melt spinning, but the development of PBS fiber is still in its infancy and has not yet been industrialized. Gu Jingjun (2007) prepared PBS primary fibers in the laboratory, but they could not be further drawn. Zhou Dengfei (2016) used melt polycondensation to synthesize PBS and then melt-spin it. The results showed that when the temperature was controlled at 240–260 °C and the speed was 2000–2400 m/min, the spinning effect was better. In addition, when the drawing temperature was below 90 °C, the drawing speed was within 50 m/min, and the drawing ratio was within 1.7 times, the thermal stress of the fiber was smaller, and the drawing was smooth.

7.5 Conclusion

Biobased biodegradable materials, distinguished by their exceptional biodegradability and biocompatibility, present an environmentally benign alternative. Consequently, their primary applications reside within domains characterized by abbreviated life cycles and elevated value, encompassing packaging, fibers, and the medical field. In contrast to their ubiquitous petroleum-derived counterparts, biobased biodegradable materials represent a nascent field, characterized by evolving developmental processes. Presently situated at the nascent stages of industrialization, their effective production capacity remains circumscribed, culminating in elevated costs. Moreover, inherent limitations persist, such as the pronounced brittleness of PLA, the suboptimal barrier properties of PHA, its compromised mechanical and thermal attributes, the modest mechanical performance of PBS, and the heightened cost and dimensional instability of PA4. These constraints necessitate refinement to facilitate the widespread adoption and dissemination of these biobased biodegradable materials.

Biobased biodegradable materials derive their constituent elements from renewable biomass sources, mitigating reliance on finite petroleum reserves. Furthermore, throughout their entire life cycle, encompassing production and utilization, these materials exert a considerably diminished environmental burden in comparison to their nondegradable counterparts. The enactment of policies prohibiting the sale of single-use, nondegradable plastics, coupled with initiatives promoting degradable alternatives, portends a sustained escalation in the demand for biodegradable plastics within China's future. As burgeoning population pressures coincide with the depletion of nonrenewable resources, ecological consciousness has ascended to the forefront of societal concerns. Concurrently, the refinement and development of biobased biodegradable materials have attained a heightened level of sophistication, accompanied by the implementation of increasingly economical, efficient, and scientifically grounded technological methodologies. Certain biobased biodegradable materials, exemplified by polylactic acid, have demonstrably fulfilled the requisites of both quotidian and medical applications. Consequently, a significant expansion and utilization of these materials is anticipated across diverse domains, encompassing packaging, agriculture, medicine, and fibers.

References

Auras R, Lim LT, Selke S et al (2011) Poly(lactic acid): synthesis, structures, properties, processing, and applications. Wiley

Bilck AP, Roberto SR, Grossmann MVE et al (2011) Efficacy of some biodegradable films as pre-harvest covering material for guava. Sci Hortic 130(1):341–343

Birajdar MS, Halake KS, Lee J (2018) Blood-clotting mimetic behavior of biocompatible microgels. J Ind Eng Chem 63:117–123

Boeree NR, Dove J, Cooper JJ et al (1993) Development of a degradable composite for orthopaedic use: mechanical evaluation of an hydroxyapatite-polyhydroxybutyrate composite material. Biomaterials 14(10):793–796

Bucci DZ, Tavares L, Sell I (2005) PHB packaging for the storage of food products. Polym Test 24(5):564–571

Burg K, Holder WD, Culberson CR et al (1999) Parameters affecting cellular adhesion to polylactide films. J Biomater Sci Polym Ed 10(2):147–161

Caruso G (2015) Plastic degrading microorganisms as a tool for bioremediation of plastic contamination in aquatic environments. J Pollut Eff Cont 3(3):1–2

Castro-Aguirre E, Iniguez-Franco F, Samsudin H et al (2016) Poly(lactic acid)-Mass production, processing, industrial applications, and end of life. Adv Drug Deliv Rev 107:333–366

Chakraborty P, Ghosh M, Schnaider L et al (2019) Composite of peptide-supramolecular polymer and covalent polymer comprises a new multifunctional, bio-inspired soft material. Macromol Rapid Commun 40(18):1900175

Changchun L (2013) Current status of technology development of biobased polyester and its fiber. Textile Guide 2:35–40

China Chemical Fiber Industry Association (2018) Analysis and forecast of China's chemical fiber economic situation in 2018. China Textile University Press, Beijing

Chitrattha S, Phaechamud T (2016) Porous poly (DL-lactic acid) matrix film with antimicrobial activities for wound dressing application. Mater Sci Eng C 58:1122–1130

Chiyu L (2012) Biodegradable polylactic acid blend for agricultural mulch. Foreign Plastics 30(5):64

Dengfei Z (2016) Synthesis and fiber preparation of biodegradable PBS. Zhejiang Sci-Tech University, Hangzhou

Ding Qian Y, Jia JX et al (2019) Research progress on biodegradable mulch materials. Eng Plastics Appl 47(12):150–153

Ejaz M, Azad MM, Ur A et al (2020) Mechanical and biodegradable properties of jute/flax reinforced PLA composites. Fiber Polym 21(11):2635–2641

European Plastics. https://www.european-bioplastics.org/market/

Fabbri M, Gigli M, Gamberini R et al (2014) Hydrolysable PBS-based poly(ester urethane)s thermoplastic elastomers. Polym Degrad Stab 108(oct.):223–231

Fathima PE, Panda SK, Ashraf PM et al (2018) Polylactic acid/chitosan films for packaging of Indian white prawn (Fenneropenaeus indicus). Int J Biol Macromol 117:1002–1010

Feng X-R, Ding J-X, Gref R et al (2017) Poly(beta-cyclodextrin)-mediated polylactide-cholesterol stereocomplex micelles for controlled drug delivery. Chin J Polym Sci 35(6):693–699

Ganyi H, Yunxuan W, Xiaoqian D et al (2022) The current development situation and future development of biodegradable plastic industry. Mater Chin 41(01):52–67

Ghosh S, Vasanthan N (2010) Structure development of poly(L-lactic acid) fibers processed at various spinning conditions. J Appl Polym Sci 101(2):1210–1216

Goosen MFA (1997) Applications of chitin and chitosans. Lancaster, p 297

Guoqiang C, Daixu W (2014) Microbial polyhydroxyalkanoates. Chemical Industry Press, Beijing

Hamad K (2015) Properties and medical applications of polylactic acid: a review. Express Polym Lett 9(5):435–455

Haugaard V, Danielsen B, Bertelsen G (2003) Impact of polylactate and poly (hydroxybutyrate) on food quality. Eur Food Res Technol 216(3):233–240

He Y, Hu Z, Ren M et al (2014) Evaluation of PHBHHx and PHBV/PLA fibers used as medical structures. J Mater Sci Mater Med 25(2):561–571

Hong H (2005) Environmentally friendly biodegradable polylactic acid (PLA) fiber. Tianjin Textile Sci Technol 43(2):11–17

Huang X, Sun Y, Nie J et al (2015) Using absorbable chitosan hemostatic sponges as a promising surgical dressing. Int J Biol Macromol 75:322–329

Industrial Information Network (2020). https://www.chyxx.com/industry/202010/901791.html

Jiangzi S (2003) Corn can be used for car interior parts. World Invent 12:20

Jin Y, Xuemei C, Guoqiang C (2016) Research progress on polyhydroxyalkanoates. J Biotechnol 6:726–737

Jing G, Hengxue X, Qianqian W (2010) Research progress on spinning technology of polyhydroxyalkanoates. Synthetic Fiber Indus 33(4):46–49

Jingjun G (2007) Study on the properties and fiber preparation of biodegradable polyester polybutylene succinate (PBS) and polyterephthalate-co-butylene succinate (PBST). Donghua University, Shanghai

Journal News (2020) 2019 National fisheries economic statistics bulletin. Chin Aquat Products 7:2–3

Kadouri D, Jurkevitch E, Okon Y et al (2005) Ecological and agricultural significance of bacterial polyhydroxyalkanoates. Crit Rev Microbiol 31(2):55–67

Kanno T, Sukegawa S, Furuki Y et al (2018) Overview of innovative advances in bioresorbable plate systems for oral and maxillofacial surgery. Jpn Dent Sci Rev 54(3):127–138

Khan G, Terano M, Gafur MA et al (2016) Studies on the mechanical properties of woven jute fabric reinforced poly(l-lactic acid) composites. J King Saud Univ Eng Sci 28(1):69–74

Lashof-Sullivan M, Holland M, Groynom R et al (2016) Hemostatic nanoparticles improve survival following blunt trauma even after 1 week incubation at 50 degrees C. ACS Biomater Sci Eng 2(3):385

Levkane V, Muizniece-Brasava S, Dukalska L (2008) Pasteurization effect to quality of salad with meat in mayonnaise. Foodbalt 1:69–73

Liang L, Guoming L (2001) Current status of polylactic acid synthesis and its applications in biomedicine. Anhui Chem Industry 6:15–18

Liang HF, Chen CT, Chen SC et al (2006) Paclitaxel-loaded poly(gamma-glutamic acid)-poly(lactide) nanoparticles as a targeted drug delivery system for the treatment of liver cancer. Biomaterials 27(9):2051–2059

Lin Y, Li Z, Lin L et al (2018) The effect of covering biodegradable mulch on the weed suppression and growth of chrysanthemum. Shanxi Agricultural. Science 46(4):623–626, 633

Lin M, Firoozi N, Tsai C-T et al (2019) 3D-printed flexible polymer stents for potential applications in inoperable esophageal malignancies. Acta Biomater 83:119–129

Liu S, Jones L, Gu FX (2012) Development of mucoadhesive drug delivery system using phenylboronic acid functionalized poly(D,L-lactide)-b-dextran nanoparticles. Macromol Biosci 12(12):1622–1626

Liu A, Xue GH, Sun M et al (2016) 3D printing surgical implants at the clinic: an experimental study on anterior cruciate ligament reconstruction. Sci Rep 6(1):1–13

Liu SQ, Wu GH, Yu JJ et al (2019) Surface modification of basalt fiber (BF) for improving compatibilities between BF and poly lactic acid (PLA) matrix. Compos Interf 26(4):275–290

Masood F, Chen P, Yasin T et al (2013) Encapsulation of Ellipticine in poly-(3-hydroxybutyrate-co-3-hydroxyvalerate) based nanoparticles and its in vitro application. Mater Sci Eng C 33(3):1054–1060

Mehrpouya M, Vahabi H, Barletla M et al (2021) Additive manufacturing of polyhydroxyalkanoates (PHAs) biopolymers: materials, printing techniques, and applications. Mater Sci Eng C 127:112216

Mengyu L, Limei W, Zhixin S et al (2019a) Study on the thermal, mechanical and barrier properties of PBS/PBAT blend film. Plast Technol 47(4):41–47

Mengyu L, Yujiao Q, Minhuan Z et al (2019b) The spontaneous atmosphere adjustment preservation effect of modified PBS film on cherry tomatoes. Food Industry 40(8):169–174

Middleton JC, Tipton AJ (2000) Synthetic biodegradable polymers as orthopedic devices. Biomaterials 21(23):2335–2346

Niu X, Liu Y, Song Y et al (2018) Rosin modified cellulose nanofiber as a reinforcing and co-antimicrobial agents in polylactic acid/chitosan composite film for food packaging. Carbohydr Polym 183:102–109

Pengju Y, Jianning Z, Zhiwei H et al (2009) Experimental study on repairing articular cartilage defects with polyhydroxybutyrate-hydroxyvalerate scaffold combined with allogeneic chondrocytes. J Med Postgrad 22(1):20–23

Perego G, Cella GD (2010) Poly(lactic acid): synthesis, structures, properties, processing, and applications. Poly(Lactic Acid)
Peres C, Matos AI, Conniot J et al (2017) Poly(lactic acid)-based particulate systems are promising tools for immune modulation. Acta Biomater 48:41–57
Qiaoling H, Zhongming Z, Xiaoli W et al (2003) Research progress on absorbable chitosan and chitosan biomedical implant materials. J Funct Polym 16(2):293–298
Qingji W, Shiying L, Min Z (1998) Preparation and research of chitosan sutures. J Chin Textile Univ 5:18–22
Ren W, Chang J, Yan C et al (2010) Development of transferrin functionalized poly(ethylene glycol)/poly(lactic acid) amphiphilic block copolymeric micelles as a potential delivery system targeting brain glioma. J Mater Sci 21(9):2673–2681
Ren X, Xu Z, Wang L et al (2019) Silk fibroin/chitosan/halloysite composite medical dressing with antibacterial and rapid haemostatic properties. Mater Res Express 6(12):125409
Ritz U, Gerke R, Goetz H et al (2017) A new bone substitute developed from 3D-prints of polylactide (PLA) loaded with collagen I: an in vitro study. Int J Mol Sci 18(12):2569–2569
Ruixia D, Wentao L, Jinzhou C et al (2019) Progress in modification research of polylactic acid for packaging. Packag Eng 40(5):109–116
Rydz J, Sikorska W, Kyulavska M et al (2014) Polyester-based (bio)degradable polymers as environmentally friendly materials for sustainable development. Int J Mol Sci 16(1):564–596
Sakaguchi M, Kobayashi S (2016) Effect of extrusion drawing and twist-orientation on mechanical properties of self-reinforced poly(lactic acid) screws. Adv Compos Mater 25(5):443–456
Serra T, Mateostimoneda MA, Planell JA et al (2013) 3D printed PLA-based scaffolds: a versatile tool in regenerative medicine. Organogenesis 9(4):239
Shengyong Y, Runying D, Yanhua C et al (2015) Preparation of degradable slow-release microcapsule fertilizers by complex phase emulsion method. Jiangsu Agric Sci 43(11):418–421
Shi Peiliang H, Ping GX et al (2001) Experimental study on the construction of tissue engineered bone with polyhydroxybutyrate and bone marrow stromal cells. J PLA Med Sch 26(4):244–245
Shishatskaya EI, Volova TG, Puzyr AP et al (2004) Tissue response to the implantation of biodegradable polyhydroxyalkanoate sutures. J Mater Sci Mater Med 15(6):719–728
Shuhan H (2016) Corbion: biobased PLA resin for extrusion, thermoforming, injection molding and fiber spinning. Int Textile Guide 44(7):14
Singhvi MS, Zinjarde SS, Gokhale DV (2019) Polymmactic acid (PLA): synthesis and biomedical applications. J Appl Microbiol 127(6):1612–1626
Suchang Y, Zhun Q (2022) Cost analysis of sand barriers in desertified regions based on the land grid division model. J Arid Land 14(9):978–992
Tao Y, Jie R, Li S et al (2010) Effect of fiber surface-treatments on the properties of poly(lactic acid)/ramie composites. Compos Part A Appl Sci Manuf 41(4):499–505
Tiantian Z, Qiming C, Liming Z et al (2021) Performance analysis of bio-based polybutyramide enteric film. Food Sci 42(3):236–242
Tiejun S, Zhixian D (2001) Properties, synthesis methods, and applications of polylactic acid. New Chem Mater 29(5):13–16
Timashev P, Kuznetsova D, Koroleva A et al (2016) Novel biodegradable star-shaped polylactide scaffolds for bone regeneration fabricated by two-photon polymerization. Nanomedicine 11(9):1041–1053
Tokiwa Y, Calabia BP (2006) Biodegradability and biodegradation of poly(lactide). Appl Microbiol Biotechnol 72(2):244–251
Touchaleaume F, Martin-Closas L, Angellier-Coussy H et al (2016) Performance and environmental impact of biodegradable polymers as agricultural mulching films. Chemosphere 144:433–439
Tripathi N, Monika, Katiyar V (2017) Poly(lactic acid)/modified gum arabic based bionanocomposite films: thermal degradation kinetics. Polym Eng Sci 57(11):1193–1206
Volova T, Shishatskaya E, Sevastianov V et al (2003) Results of biomedical investigations of PHB and PHB/PHV fibers. Biochem Eng J 16(2):125–133

Wakabayashi T, Yagi H, Tajima K et al (2019) Efficacy of new polylactic acid nonwoven fabric as a hemostatic agent in a rat liver resection model. Surg Innov 26(3):312–320

Wang Y, Chaoyun W, Yongjian Y et al (2019) Research progress on domestic biodegradable mulch. Plast Technol 47(12):156–165

Wu J, Su C, Lei J et al (2018) Green and facile preparation of chitosan sponges as potential wound dressings. ACS Sustain Chem Eng 6(7):9145–9152

Wyrwa R, Otto K, Voigt S et al (2018) Electrospun mucosal wound dressings containing styptics for bleeding control. Mater Sci Eng C 93:419–428

Xia Q, Liu Z, Wang C et al (2015) A biodegradable trilayered barrier membrane composed of sponge and electrospun layers: hemostasis and antiadhesion. Biomacromolecules 16(9):3083–3092

Xiang R, Qi W, Enhe Z et al (2014) The effect of covering material and ridge ratio on oat yield and water use efficiency. Chin J Eco-Agric 22(8):945–954

Xiaohong L, Minglong Y, Chengdong X et al (1999) Synthesis and biomedical applications of polylactic acid and its copolymers. Polym Bull 1:24–32

Xin C, Xiang L, Xiaojian L et al (2019) Preparation and in vitro drug release characteristics of paclitaxel-sorafenib-poly(lactic acid)-hydroxyethyl acid drug-loaded embolic microspheres. Chin Pharm 30(10):1327–1333

Xinyan Z, Mingming Y, Fuliang C (2013) Preliminary determination of the slow-release properties and indoor toxicity of maverick salt polylactic acid microspheres. Pesticides 52(9):653–655

Xiurong S, Taiwu L, Mingyong Z (1996) Research on the preparation of absorbable surgical sutures with chitosan. J Liaoning Normal Univ (Natural Sci Ed) 19(4):321–329

Xu XY, Li XT, Peng SW et al (2010) The behaviour of neural stem cells on polyhydroxyalkanoate nanofiber scaffolds. Biomaterials 31(14):3967–3975

Yagmurlu MF, Korkusuz F, Gürsel I et al (1999) Sulbactam-cefoperazone polyhydroxybutyrate-co- hydroxyvalerate (PHBV) local antibiotic delivery system: In vivo effectiveness and biocompatibility in the treatment of implant-related experimental osteomyelitis. J Biomed Mater Res 46(4):494–503

Yi T (2019) Development status and trend of new biobased material polylactic acid industry. Chin Grain Econ 8:49–53

Yonglian L, Bing C (2001) Suture materials and their clinical applications. Chin J Med Dev 25(2):99

Zhang H, Fang J, Ge H et al (2013) Thermal, mechanical, and rheological properties of polylactide/poly(1,2-propylene glycol adipate). Polym Eng Sci 53(1):112–118

Zhang L, Sintim HY, Bary AI et al (2018) Interaction of Lumbricus terrestris with macroscopic polyethylene and biodegradable plastic mulch. Sci Total Environ 635:1600–1608

Zhe H (2018) Research progress and design application of polylactic acid degradable plastic food packaging. Plastic Sci Technol 46(6):131–134

Zhen T (2008) Study on the mechanical properties of polylactic acid fiber blended yarn and its fabric. Qingdao University, Qingdao

Zhu M, Tan J, Liu L et al (2021) Construction of biomimetic artificial intervertebral disc scaffold via 3D printing and electrospinning. Mater Sci Eng C 128:112310

Chapter 8
Intelligent Biobased Biodegradable Materials

Fangke Zhan and Liming Zhao

Intelligent materials, as the fourth generation of materials following natural materials, synthetic polymer materials, and artificially designed materials, possess intelligent characteristics. And they are one of the important directions in the development of modern high-tech new materials, thereby supporting the development of future high technologies, gradually blurring the boundary between traditional functional materials and structural materials, and realizing structural functionalization and functional diversification. Moreover, some scientists predict that the development and large-scale application of intelligent materials will lead to a major revolution in the development of material science. Biobased biodegradable materials have a significant impact on the composition of the future world of new environmental protection materials. Additionally, the combination of these two types of materials, each with its unique skills, will be favored by future scientific researchers. This chapter summarizes the current research progress of biobased intelligent materials.

F. Zhan
State Key Laboratory of Bioreactor Engineering, East China University of Science and Technology, Shanghai, China

L. Zhao (✉)
State Key Laboratory of Bioreactor Engineering, East China University of Science and Technology, Shanghai, China

Key Laboratory of Biobased Material Engineering of China Light Industry, East China University of Science and Technology, Shanghai, China
e-mail: zhaoliming@ecust.edu.cn

L. Zhao (ed.), *Research and Applications of Bio-based Degradable Materials*,
https://doi.org/10.1007/978-981-95-1188-4_8

8.1 Shape Memory Materials

The definition of shape memory materials is: materials that can remember their original shape, obtain one or more temporary shapes through programming design, and spontaneously recover from temporary deformation to the original shape when stimulated by the external environment. In contrast, shape memory polymers are generally defined as polymer materials with a certain initial shape under certain conditions, which can recover to their initial shape through the stimulation of external conditions after changing their initial shape and fixing the shape (Fig. 8.1). In this section, we focus on the relatively novel shape memory biobased degradable polymer materials in recent studies.

As shown in Fig. 8.1, the principle of shape memory mainly consists of four steps: ① Shape memory after polymer material molding and cooling; ② Under the action of external force or other factors such as light and heat, the amorphous part undergoes free deformation due to rubber elasticity; ③ Molding and cooling under Tg (glass transition temperature); ④ Recover shape by heating above Tg in free form.

As a type of biobased degradable material, polylactic acid has excellent mechanical properties, good biocompatibility, and degradability. How to combine the excellent characteristics of polylactic acid with the advanced nature of intelligent materials is the current focus of researchers. Lu et al. (2006) introduced poly ε-caprolactone into PLA to synthesize a two-component copolymer, thereby obtaining an intelligent, advanced copolymer composite material with shape memory function. The increase in poly ε-caprolactone makes the shape recovery rate larger and the fixation rate smaller. In the polymer with the same composition, the increase of its relative molecular weight improves the shape memory function. After adjusting the composition and structure of the material, a copolymer composite material with a high shape recovery rate and shape fixation rate can be obtained, which will have an excellent application blueprint in future advanced medical aspects. The research topic of Zhang et al. (2012; Tang and Cai 2007) is to obtain shape memory properties by mixing polylactic acid and polyamide elastomer at low temperatures,

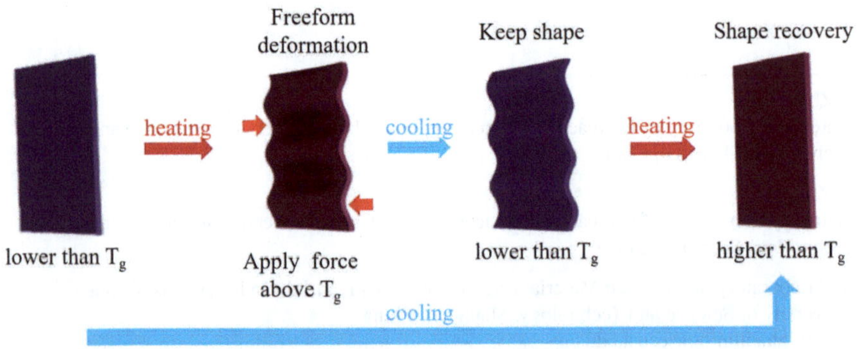

Fig. 8.1 Schematic diagram of the molecular model of the shape memory mechanism of shape memory polymers (see color insert)

and to perform high elastic deformation on the previous materials, and then study their shape memory mechanism and their performance under cold deformation conditions. The research results of the team show that the increase of the elastomer will reduce the recovery rate of the composite material, and after the strain is reduced, the recovery rate and tensile strength recovery rate of the material will relatively increase. In addition, they finally developed a product with a high shape recovery force and relatively high recovery speed, which can shorten the recovery time when the recovery rate does not change. Jiang et al. (Dong et al. 2010) prepared PLLA/polyglycolic acid copolymer, and deeply explored the relationship between its composition and memory performance and mechanical properties. This work shows that the polymer material with a composition ratio of 4:1 has the best memory performance. Ma et al. (Yan et al. 2011) used the triblock low unsaturated polypropylene oxide/polylactic acid diblock copolymer as the raw material, and obtained the degradable polypropylene oxide/polylactic acid-based polyurethane through toluene diisocyanate crosslinking. After adjusting the structure, a PLLA–PU material with low modulus and high fracture elongation was obtained, and the deformation temperature was 96–153 °C. In addition, this advanced material with a fixation rate of 65–100% and a recovery rate of 100% has a recovery time far less than 20 s at a temperature of 140 °C, and it has good toughness and is degradable. Bertmer et al. (2005) prepared polylactic acid–ethylene glycol copolymer material by ultraviolet crosslinking. The specific method is: in the molten state at 90 °C, the lactic acid-ethylene glycol prepolymer is irradiated under ultraviolet light with an intensity of 120 W/cm^2 for 25 min to prepare a polylactic acid–ethylene glycol copolymer material with a covalent crosslinked network structure, which has a very prominent shape memory function.

Min et al. (Min et al. 2005) first copolymerized polycaprolactone (PCL) and ethyl ester together, and then used 1,6-hexanediisocyanate (HDI) to couple the material to prepare a thermoplastic copolymer. After that, the material with excellent memory performance was obtained by precisely adjusting the composition and chain segment length of the material, and its shape recovery rate and fixation rate were both above 90%. While changing the time required for degradation, the physical and mechanical properties can be precisely controlled. Zheng and Zhou (Zheng et al. 2006, 2008; Zhou et al. 2007) explored the effect of degradable poly-D-lactic acid and poly-L-lactic acid and inorganic material composites on shape memory behavior. Using PLA and β-tricalcium phosphate and other materials to prepare a composite, a shape memory performance material with a shape recovery rate of 95% was obtained. Nagate et al. (Huang et al. 2012) noted that the mixed material containing cinnamic acid and hexadiacyl chloride groups is added to the PCL/PLA copolymer to obtain a triblock copolymer. This material has a good memory effect and a recovery rate of up to 100%. Notably, in subsequent research, when they put this polymer into a phosphate buffer solution for degradation performance verification experiments, they found that after 27 days, the remaining material mass was 10%, indicating that the material has good degradability. In addition, this triblock copolymer has good biocompatibility, good degradability, and a low glass transition temperature. Zheng et al. (2017) used polylactic acid as the raw material, and through physical blending, used PBS, PCL, PBS+PEG as the modified materials, and Fe_3O_4 as the doping material, successfully prepared shape memory polylactic acid with a good driving effect and shape memory polylactic acid with magnetic driving function.

8.2 Stimulus-Responsive Biobased Degradable Materials

Nowadays, more and more research focuses on stimulus-responsive polymers or smart polymers, which are gradually replacing traditional polymer materials and are widely used in various fields (Gao et al. 2019). When exposed to external or internal stimuli, including light, temperature, pH, ion strength, mechanical force (such as compression, tension, and shear), biomolecules, and magnetic or electric fields, stimulus-responsive polymers will change the structure and other physicochemical properties, and their various properties, such as wettability, absorbance, color, and magnetism, will also change (Fig. 8.2) (Zhang et al. 2008; Kuckling and Wycisk 2013; Xia et al. 2006; Jochum and Theato 2013; Liang et al. 2019; Wiggins et al. 2013; Sharifzadeh and Hosseinkhani 2017; Huang 2011). Stimulus-responsive materials are also widely used in some biobased degradable materials, and the following will introduce the application of these biobased materials according to the response classification.

In recent years, people are becoming more and more interested in the chemical modification of biobased or natural polymers. Many types of monomers can be obtained, and the modification methods are changing with each passing day. This

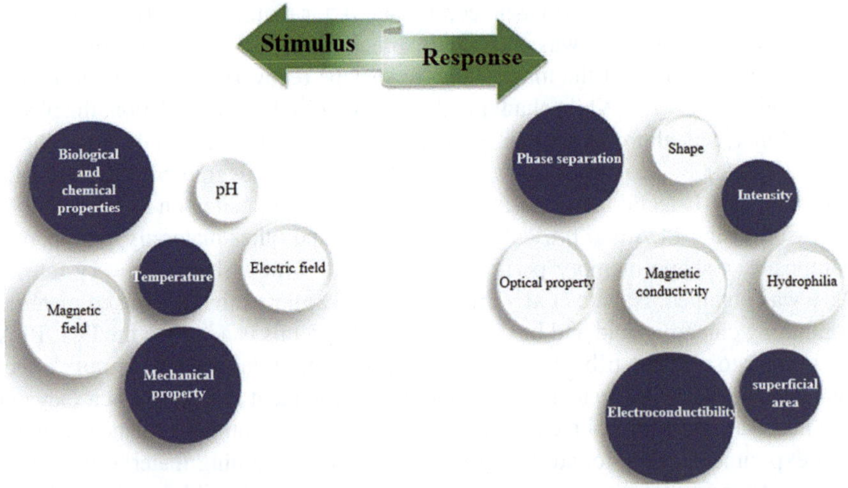

Fig. 8.2 Schematic diagram of the response changes corresponding to the stimulation methods of potential polymers (see color insert)

provides more possibilities for the development of biobased polymer materials in the direction of intelligence. At present, there are quite a few biobased degradable polymers that can respond to external stimuli. By combining biobased polymers with other materials, transformational composite materials with excellent properties such as biocompatibility, degradability, and non-toxicity can be obtained. The application of biobased degradable polymers in the direction of stimulus response is mainly divided into the following aspects.

8.2.1 Smart Drug Delivery Systems

Drug delivery refers to the method of accurately delivering drugs to the target site within the human body for the treatment of diseases. A drug delivery system refers to a technical system that comprehensively regulates the distribution of drugs in space and time, as well as the dosage of drugs in the body, to achieve the required concentration, maintain the effect level of the drug for the necessary time, and improve the efficiency of drug treatment, reducing costs and side effects (Allen 2004). The current drug delivery system has entered a new era, with various excellent drug release systems emerging, and intelligent drug delivery systems (also known as on-demand or intelligent drug delivery systems) are receiving increasing attention (Xiao and Brazel 2001). Intelligent drug release systems can release active molecules or drugs at appropriate sites, and the release amount can be adjusted according to the progress of the disease or certain biochemical characteristics, or biological rhythms of the organism. Intelligent drug release systems aim to adjust the release site and/or release rate according to the patient's physiological condition, the progress of the disease, or the circadian rhythm. Unlike traditional preprogrammed controlled release formulations, new devices aim to provide the most suitable drug release pattern for each patient's needs (Oh et al. 2009). Intelligent drug delivery systems are mainly based on stimulus-responsive polymers, sensing changes in specific variables and activating delivery; this phenomenon is reversible. Biodegradable polymers are widely used in the field of drug delivery due to their unique advantages. Because of their biodegradability, biocompatibility, natural abundance, and unique chemical structure and physical, chemical, and biological characteristics, they have considerable application value in the field of biomedical engineering (Concheiro 2008).

8.2.2 Intelligent Hydrogels

Hydrogels are ideal materials that simulate extracellular matrix conditions, due to their soft tissue biocompatibility, and drugs are easy to disperse in the matrix; they are often combined with medical drugs for application. A particularly noteworthy point in the medical community is that hydrogels can regulate the physicochemical characteristics of the polymer network by selecting various raw materials, including synthetic and natural polymers, thereby achieving highly controlled and flexible use in controlled-release drugs. Biodegradable polymers, due to their high water solubility, good biocompatibility, and biodegradability, are good candidate materials for forming hydrogels as drug carriers (He 2001). Therefore, a large number of biopolymer hydrogels have been prepared and used in intelligent responsive hydrogel materials, including light-responsive, ultrasonic, magnetic field-responsive, pH-responsive, electric field-responsive, and mechanical-responsive hydrogels.

8.2.3 *Electrospun Nanofibers*

Many biopolymers are excellent electrospun materials, including polylactic acid, polycaprolactone, and natural polymers, such as gelatin (Wang et al. 2013), chitosan (Chen et al. 2008) , and cellulose (Frey 2008). These biopolymers have good spinnability and can significantly improve the biocompatibility and degradability of nanofibers. Therefore, biopolymers used for electrospinning are receiving increasing attention in the field of biomedical engineering.

Electrospun nanofiber membranes are structurally similar to the extracellular matrix, and the use of biobased polymers can significantly improve the biocompatibility of electrospun scaffolds. The controllable preparation process of electrospinning and the effective combination of the unique properties of biobased polymers have more attractive functions and advantages (Fig. 8.3). Therefore, combining

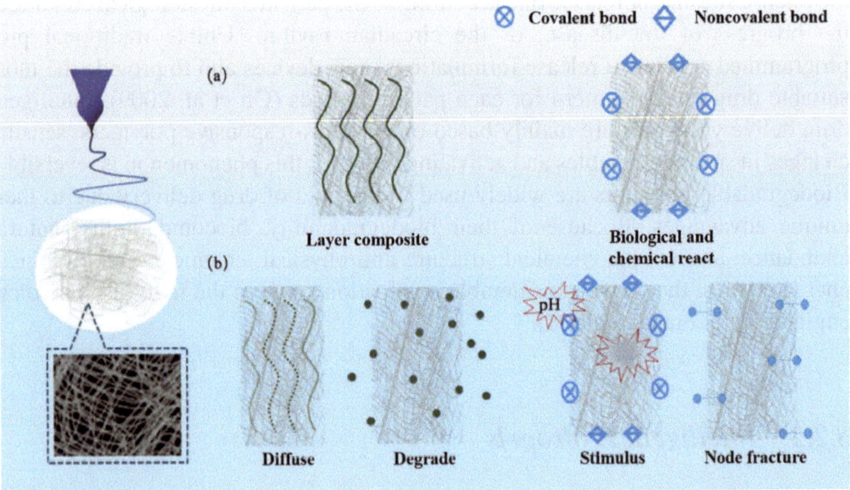

Fig. 8.3 Schematic diagram of the preparation of polymer electrospinning materials and controlled release of drugs (see color insert). (**a**) Common methods for preparing composite spun fiber mats, (**b**) mechanism of stimulus-responsive controlled release of drugs from composite spun fiber mats

electrospun technology with stimulus-responsive materials can effectively control the drug delivery process of nanofibers not only in vitro but also in complex in vivo environments, thereby enabling electrospun materials to be more widely used.

8.2.4 Responsive Intelligent Food Packaging

In recent years, intelligent food packaging has received increasing attention. Some experts define "intelligent packaging" as any type of packaging that provides specific functions in addition to the physical barrier function between food and the surrounding environment. The so-called intelligent packaging is based on the interaction between the packaging and the food and/or its direct environment and can be simply regarded as an upgraded enhanced system in addition to the main basic functions of traditional food. Intelligent packaging can detect, perceive, and record the status of food throughout the food chain, track or convey information about the quality and/or status of food during its production, transportation, shelf life, and storage period. It can also respond to various stimuli generated by food or external environmental stimuli (including the effects of light, oxygen, water vapor, microorganisms, mechanical stress, etc.) (Liao et al. 2016).

Biobased degradable materials are very suitable as candidate materials for food packaging due to their nontoxic, degradable, selectively permeable, safe, and strong sterilization capabilities. Polybutylene succinate, polylactic acid, etc., are especially suitable for food packaging materials in biobased degradable materials and have been put into production. The food packaging materials made from these materials have excellent mechanical properties and barrier properties, and can be biodegraded at the end of their service life. Biobased polymer packaging materials can also serve as gas and solute barriers and extend the shelf life of food by improving quality. In addition, biobased polymer packaging materials are excellent carriers for various additives, such as antioxidants, antifungal agents, antibacterial agents, dyes, and other nutrients (Han 2003; Imran et al. 2010; Clarinval and Halleux 2005). At the same time, the development of packaging materials with antioxidant and anti-corrosion antibacterial functions by adding active substances to biobased degradable carriers, or the preparation of effective food detection biosensors using biobased degradable materials as sensor carriers, is also a topic of interest to researchers (Halonen et al. 2020). Despite the increasingly broad field of scientific research related to biobased degradable materials and renewable materials and the increasingly mature and stable production technology, more work is needed to evaluate the real environmental and socioeconomic impacts of each potential viable solution before they can be added to the queue for smart responsive food packaging.

The concept of biobased degradable materials is gradually becoming familiar to people, and its international influence is also growing. However, the intelligent research on biobased degradable materials is still in its infancy, with more mature and prominent results revolving around polylactic acid and its derivative polymers. The intelligent research work on other types of biobased degradable materials is not much. However, the research significance of using them to prepare high-performance intelligent materials is significant and will be an extremely important research direction in the future.

References

Allen TM (2004) Drug delivery systems: entering the mainstream. Science 303:1818–1822

Bertmer M, Buda A, Blomenkamp-Hofges I et al (2005) Shape- memory polymer networks: characterization with solid-state NMR. Macromolecules 38(25):3793–3799

Chen Z, Mo X, He C et al (2008) Intermolecular interactions in electrospun collagen-chitosan complex nanofibers. Carbohydr Polym 72:410–418

Clarinval AM, Halleux J (2005) Classification of biodegradable polymers, vol 14. Woodhead Publishing Ltd, Cambridge, pp 3–31

Concheiro AL (2008) Intelligent drug delivery systems: polymeric micelles and hydrogels. Mini-Rev Med Chem 8:1065–1074

Dong W, Jiang J, Xie M et al (2010) Preparation and performance study of shape memory poly(lactic acid-ethanol acid) (PLLGA). Acta Chim Sin 68(21):2243–2249

Frey MW (2008) Electrospinning cellulose and cellulose derivatives. Polym Rev 48(6):378–391

Gao S, Tang G, Hua D et al (2019) Stimuli-responsive bio-based polymeric systems and their applications. J Mater Chem B

Halonen N, Palvolgyi PS et al (2020) Bio-based smart materials for food packaging and sensors—a review. Front Mater 7(5):6671–6683

Han JH (2003) Antimicrobial food packaging. Novel Food Packag Techniq 2001(2):50–70

He W (2001) Intelligent hydrogels and their application. Inner Mongolin Petrochem Ind:45–47

Huang T (2011) Stimuli-responsive electrospun fibers and their applications. Chem Soc Rev:2898–2899

Huang X, Xu Z, Wang S et al (2012) J Polym Sci, Part A: Polym Chem

Imran H, Revol-Junelles AM, Martyn A et al (2010) Active food packaging evolution: transformation from micro- to nanotechnology. Crit Rev Food Sci Nutr 50(3):799–821

Jochum FD, Theato P (2013) Temperature- and light-responsive smart polymer materials. Chem Soc Rev 42:7468–7483

Kuckling D, Wycisk A (2013) Stimuli-responsive star polymers. J Polym Sci A Polym Chem 51:2980–2994

Liang S, Guan Y, Zhang Y (2019) Layer-by-layer assembly of microgel colloidal crystals via photoinitiated alkyne-azide click reaction. ACS Omega 4:5650–5660

Liao Y, Chen D, Li W et al (2016) Research and application progress of intelligent packaging. Green Packag:39–46

Lu X, Cai W, Gao Z (2006) Shape memory effect of poly L-lactic acid/poly ε-caprolactone block copolymer. Funct Mater 37(2):1795–1797

Min C, Cui W, Bei J et al (2005) Biodegradable shape memory polylactide-co-poly (glycodide-co-caprolactone) multiblock copolymer. Polym Adv Technol 16(8):608–615

Oh YK, Senter PD, Song SC (2009) Intelligent drug delivery systems. Bioconjug Chem 20:1813–1815

Sharifzadeh G, Hosseinkhani H (2017) Biomolecule-responsive hydrogels in medicine. Adv Healthc Mater 6:1700801

Tang W, Cai W (2007) Cold deformation shape memory effect of PLLA-PCL random copolymer. Funct Mater 69:89–93

Wang HY, Feng YK, Fang ZC et al (2013) Fabrication and characterization of electrospun gelatin-heparin; nanofibers as vascular tissue engineering. Macromol Res 21(7):860–869

Wiggins KM, Brantley JN, Bielawski CW (2013) Methods for activating and characterizing mechanically responsive polymers. Chem Soc Rev 42:7130–7147

Xia F, Feng L, Wang S (2006) Dual-responsive surfaces that switch between superhydrophilicity and superhydrophobicity. Adv Mater 18(4):432–436

Xiao H, Brazel CS (2001) On the importance and mechanisms of burst release in matrix-controlled drug delivery systems. J Control Release 73:121–136

Yan M, Wenpeng S, Chenyang Z (2011) Preparation, structure and properties of degradable shape memory polymer based on polylactic acid. 6:719–724

Zhang H, Meng X, Li P (2008) Light and temperature responsive materials. Progr Chem 20:657–672

Zhang W, Wei F, Zhang Y (2012) Shape memory effect and mechanism of PLA/PAE composite materials. J Nantong Univ 11(2):52–56

Zheng Z (2017) Performance study of polylactic acid-based shape memory polymer and its 4D printing. Harbin Institute of Technology, Harbin

Zheng X, Zhou S, Li X (2006) Shape memory properties of poly(D, L-lactide)/hydroxyapatite composites. Biomaterials 27(25):4288–4295

Zheng X, Zhou S, Yu X (2008) Effect of in vitro degradation of poly(D, L-lactide)/β-tricalcium composite on its shape-memory properties. J Biomed Mater Res B 86:170–180

Zhou S, Zheng X, Yu X (2007) Hydrogen bonding interaction of poly(D, L-lactide)/hydroxyapatite nanocomposites. Chem Mater 19(13):247–253

Chapter 9
Degradation of Biobased Materials

Di Zhang, Ziwei Zhang, Tao Chen, and Liming Zhao

Over the past few decades, plastic products have brought great convenience to people's lives, but due to improper disposal, a large amount of plastic waste has caused many environmental problems. Currently, the global annual production of disposable plastic products reaches 120 million tons, of which only 10% is recycled, another 12% is incinerated, and more than 70% is discarded into the soil, air, and oceans (Yujun and Dawei 2018). Therefore, it is important to increase consumer awareness of biodegradable polymer materials and introduce them to the market. Presently, biodegradable plastics in China are still in the early stages of industrialization, and the issuance of plastic restriction and ban orders has provided a huge market opportunity for degradable plastics.

D. Zhang · Z. Zhang
State Key Laboratory of Bioreactor Engineering, East China University of Science and Technology, Shanghai, China
e-mail: zhangdi3@jahwa.com.cn

T. Chen
School of Materials Science and Engineering, East China University of Science and Technology, Shanghai, China
e-mail: tchen@ecust.edu.cn

L. Zhao (✉)
State Key Laboratory of Bioreactor Engineering, East China University of Science and Technology, Shanghai, China

Key Laboratory of Biobased Material Engineering of China Light Industry, East China University of Science and Technology, Shanghai, China
e-mail: zhaoliming@ecust.edu.cn

L. Zhao (ed.), *Research and Applications of Bio-based Degradable Materials*,
https://doi.org/10.1007/978-981-95-1188-4_9

9.1 Basic Concepts

The definitions related to material degradation are gradually refined. Material degradation refers to the process by which the physical properties, chemical properties, and appearance of the material are destroyed. Material degradation generally includes thermal degradation, oxidative degradation, photodegradation, biodegradation, hydrolytic degradation, and radiation degradation, etc. The following will elaborate on degradable plastics and related important concepts.

9.1.1 Degradable Plastics

According to the guidelines issued by the China National Light Industry Council (CNLIC) in September 2020, degradable plastics refer to plastics that degrade under the action of microorganisms existing in nature, such as soil, sand, freshwater environments, seawater environments, and specific conditions (composting conditions or anaerobic digestion conditions, etc.), and eventually completely degrade into CO_2 and/or CH_4, H_2O and the mineralized inorganic salts of its contained elements, as well as new biomass (such as microbial corpses, etc.).

9.1.2 Photodegradation

Photodegradation refers to the process in which the material's molecular chain breaks down under the action of light (Linyu et al. 2020). Under certain temperature, humidity, and O_2 conditions, the material's molecular chain undergoes photooxidation reactions, and the material's molecular chain degrades and breaks into soluble small molecule substances, thereby achieving material degradation. Due to the limitations of light exposure, the use of photodegradable materials has significant limitations (Yangyang et al. 2019). Ultraviolet light in natural light plays a role in the degradation process of photodegradable materials. After absorbing ultraviolet light, some segments and groups of photodegradable materials are excited and then undergo degradation reactions, causing their chemical bonds and chains to break. Ordinary plastics have a low absorption capacity for light and a limited absorption rate. The addition of colored groups will activate the plastic, thereby accelerating the absorption of light (Wei and Jun 2006; Maraveas et al. 2024), ultimately leading to an accelerated degradation process of the plastic in natural environments such as oxygen, heat, and water (Yanxia et al. 2012). Adding photosensitive groups and substances to polyethylene can accelerate the photodegradation reaction of PE molecules under light. Photodegradation mainly includes photochemical degradation and photooxidation degradation. After absorbing ultraviolet light, high molecular weight polymers undergo photochemical degradation reactions, high molecular

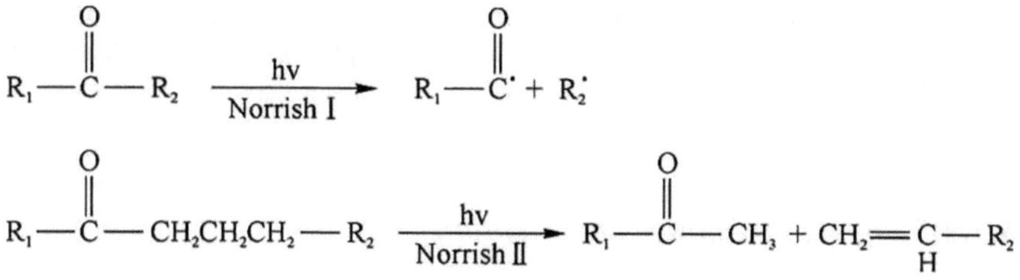

Fig. 9.1 Photodegradation Norrish reaction equation (Grause et al. 2020). (Reprinted with permission from Polym. Degrad. Stab, 2020, 181: 109364.@ 2020, Elsevier Ltd)

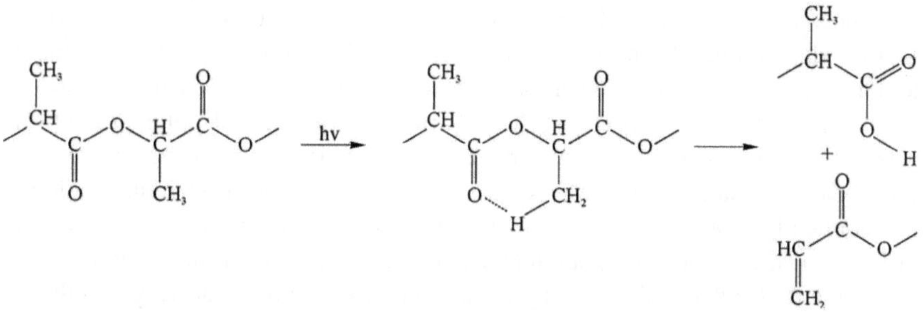

Fig. 9.2 Photodegradation process of PLA (Grause et al. 2020). (Reprinted with permission from Polym Eng Sci, 2020, 60: 2061–2075.@ 2020, Wiley Online Library)

long chains decompose into low molecular weight short chains, and then undergo photooxidation degradation reactions in the air, degrading into low molecular compounds that can be biodegraded, and finally completely decompose into CO_2 and H_2O; the whole process is called the Norrish reaction. The Norrish reaction is divided into Norrish I and Norrish II reactions (Grause et al. 2020), and the reaction mechanism is shown in Fig. 9.1. The photodegradation process is affected by many factors, mainly the type of photosensitizer, wavelength, atmospheric conditions, and the molecular structure of the material. When the molecule contains C=O, –N=N–NH–, –NH–NH–, –S–, –O–, $–CH_2N=N–$, and $–CH_2–CH_2–$groups, it is prone to photodegradation.

Taking polylactic acid (PLA) as an example (Fig. 9.2), PLA undergoes main chain scission during the photodegradation process, forming C=C double bonds and carboxyl end groups, and the reaction is triggered by the electron transition of the C=O bond. When ultraviolet radiation penetrates the polymer, degradation will continue through overall corrosion, where light penetrates the polymer, regardless of the chemical structure and crystallinity of the polymer; its strength will not significantly decrease, and photodegradation will also change the physical and optical properties of the plastic. The most noticeable changes are visual effects (yellowing), loss of polymer mechanical properties, changes in molecular weight under the same molecular weight, and changes in molecular weight distribution (Zaaba and Jaafar 2020).

9.1.3 Thermo-oxidative Degradation

The thermo-oxidative degradation of polymer materials is essentially the result of complex changes in chemical structure due to the oxidation reaction, causing molecular chain breakage and cross-linking (Maoying 2010). The main manifestations of plastic thermo-oxidative degradation are discoloration, yellowing, weight loss, decrease in transparency, surface cracking, and powdering. PLA is prone to thermal degradation during processing, which leads to a decrease in mass fraction and a decrease in the rheological properties of the melt and mechanical properties of the product during processing. The thermal degradation of polymers not only relates to the treatment of polymers but also has important significance in analyzing their heat resistance, polymer processes (such as extrusion or injection molding), and application pathways. The thermal degradation of polymers includes two different reactions, which occur simultaneously in the reactor. One is the irregular breakage of the molecular chain, leading to a decrease in the molecular weight of the raw material polymer; the other is the end breakage of the C-C bond, producing volatile products. The thermal degradation of polymers follows the chain end degradation (also known as the unzipping pathway) (Eqs. 9.1 and 9.2) or the random degradation pathway (Eq. 9.3).

$$M_n^* \rightarrow M_{n-1}^* + M \tag{9.1}$$

$$M_{n-1}^* \rightarrow M_{n-2}^* + M \tag{9.2}$$

$$M_n \rightarrow M_X + M_Y \tag{9.3}$$

Chain degradation starts from the chain end and releases monomer units in sequence. This degradation pathway is also known as depolymerization. This reaction is the opposite of the propagation step in addition to polymerization and occurs through a free radical mechanism. In this degradation, the molecular weight of the polymer decreases slowly, while a large amount of monomer is released. Random degradation occurs at any random point along the polymer chain, and the polymer chain does not need to have any active sites to undergo random degradation. This is contrary to the condensation process. Condensation reaction refers to a reaction in which one or more monomers condense to form a polymer, and the main product is called a condensate. The monomer of the condensation reaction is a compound with two (or more) reactive functional groups that forms a polymer by eliminating small molecules when polymerizing, so the molecular weight of the repeating unit of the polymer is smaller than the monomer.

9.1.4 Hydrolysis

Research shows that in a natural environment, the degradation of biodegradable materials can be roughly divided into two processes: simple hydrolysis and enzyme-catalyzed degradation. Simple hydrolysis is the reaction of H_2O attacking the ester bonds in polyester molecules, causing them to decompose into carboxylic acids and alcohols. This is mainly affected by factors such as the temperature, humidity, acidity of the hydrolysis environment, and the properties of the polymer itself.

Taking the simple hydrolysis of polylactic acid (PLA) as an example, the simple hydrolysis of PLA is due to the presence of ester bonds in the molecular chain. Ester bonds easily break under the action of hydrogen ions to form carboxylic acids and alcohols, and the carboxyl groups at the free ends of the main chain will catalyze the degradation of PLA. As the degradation progresses, the amount of carboxyl groups increases, and the degradation rate accelerates, forming a self-catalytic effect.

9.1.5 Biodegradation

Albertsson and Karlsson define biodegradation as reactions that occur through the action of enzymes and/or chemical decomposition related to organisms and their secretions (Luckachan and Pillai 2011). In this process, non-biological actions such as photodegradation, oxidation, and hydrolysis must also be considered. Due to environmental factors, these non-biological actions may replace biodegradation before or during the process and alter the polymer. Therefore, strictly speaking, the definition of "biodegradation of polymers" refers to the deterioration of the physicochemical properties of polymers and the reduction of molecular weight under the assistance of non-biological chemical reactions such as photodegradation, oxidation, and hydrolysis, under aerobic and anaerobic conditions and the influence of microorganisms, until the formation of CO_2, H_2O, CH_4 and other low molecular weight products.

Biodegradation generally consists of two stages (Fig. 9.3): The first stage is the depolymerization of large molecules into shorter chains. Due to the length of the polymer chain and the insolubility of many polymers, this step usually occurs outside the organism. Extracellular enzymes (endo- or exo-enzymes) and non-biological reactions are the cause of polymer chain breakage. In this stage, the contact area between the polymer and the microorganisms increases. The second stage corresponds to mineralization. Once sufficiently small oligomer fragments are formed, they are transported into the cell and assimilated by microorganisms, then mineralized. Depending on the presence of O_2, biodegradation can be divided into two types—aerobic biodegradation (under aerobic conditions) and anaerobic biodegradation (under anaerobic conditions). Complete biodegradation or mineralization occurs when there are no residues, i.e., when the original product is completely converted into gaseous products and salts.

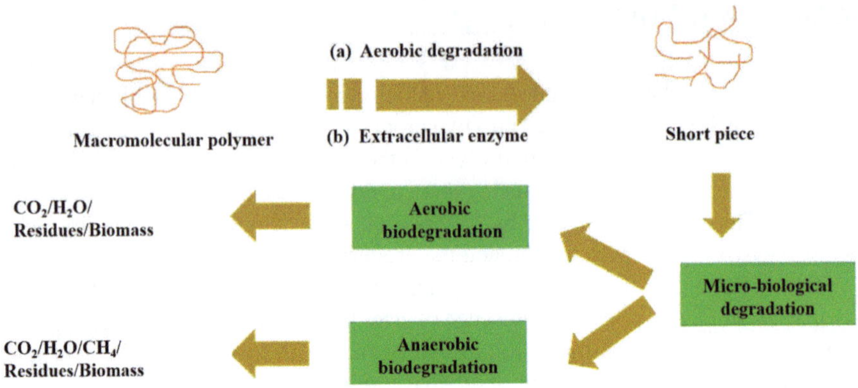

Fig. 9.3 Biodegradation process (see color insert)

9.2 Evaluation Methods of Degradation Performance

Currently, many materials have been proven to degrade naturally in a short period of time, but the types of materials and degradation environments are different, and the degradation rate also has certain differences. In order to standardize the comparison of the degradation performance of different materials, specific evaluation methods (Table 9.1) have been given internationally.

9.2.1 Freshwater Environment Degradation

1. GB/T 19276.1 (GB/T 19276.1-2003 2003)

Determination of the ultimate aerobic biodegradability of materials in an aqueous medium by measuring the oxygen demand in a closed respirometer.

(1) Test principle: The biodegradation rate of materials is determined in an aqueous system using aerobic microorganisms. The test mixture includes an inorganic culture medium, a test material with an organic carbon concentration between 100 and 2000 mg/L (the only source of carbon and energy), and a culture solution made from a suspension of activated sludge, compost, or active soil. This mixture is stirred and cultured in a sealed flask in a respirometer for a certain period of time, and the test period cannot exceed 6 months. The CO released in the upper part of the flask is absorbed by an appropriate absorber. Measurement of biochemical oxygen demand (BOD). The level of biodegradation is determined by the ratio of BOD to theoretical oxygen demand (ThOD), expressed as a percentage.

Table 9.1 Standards for degradable plastics

Types of degradable plastics		According to standard number	Standard name	References
Freshwater environment degradation		GB/T 19276.1	Determination of the ultimate aerobic biodegradability of materials in an aqueous medium by measuring the oxygen demand in a closed respirometer	GB/T 19276.1-2003 (2003)
		GB/T 19276.2	Determination of the ultimate aerobic biodegradability of materials in an aqueous medium by measuring the amount of CO_2 released	GB/T 19276.2-2003 (2003)
		GB/T 32106	Determination of the ultimate anaerobic biodegradability of plastics in an aqueous medium by measuring the biogas production	GB/T 32106-2015 (2015)
Degradable	Industrial compostable	GB/T 19277.1	Determination of the ultimate aerobic biodegradability of materials under controlled composting conditions by measuring the amount of CO_2 released—Part 1: General method	GB/T 19277.1 2011 (2011)
		GB/T 19277.2	Determination of the ultimate aerobic biodegradability of materials under controlled composting conditions by measuring the amount of CO_2 released—Part 2: Method by analysis of evolved carbon dioxide under laboratory conditions	GB/T 19277.2 2013 (2013)
		GB/T 19811	Determination of the disintegration degree of plastic materials under defined composting conditions in a pilot-scale test	GB/T 19811-2005 (2005)
		GB/T 28206	Technical requirements for compostable plastics	GB/T 28206 (2011)
Compostable in backyard		AS5810-2010	Technical specifications for compostable plastics in backyard	AS 5810-2010 (2010)
Soil degradable		GB/T 22047	Determination of the ultimate aerobic biodegradability of plastic materials in soil by measuring the oxygen demand in a closed respirometer or the amount of carbon dioxide evolved	GB/T 22047 (2009)

(continued)

Table 9.1 (continued)

Types of degradable plastics	According to standard number	Standard name	References
Marine environment degradation	ISO 18830	Determination of the ultimate aerobic biodegradability of non-floating plastic materials in a seawater/sediment interface by measuring the oxygen demand in a closed respirometer	ISO 18830:2016 (2016)
	ISO 19679	Determination of the ultimate aerobic biodegradability of non-floating plastic materials in a seawater/sediment interface by measuring the amount of carbon dioxide evolved	ISO 19679 (2016)
	ISO 22404	Determination of the ultimate aerobic biodegradability of plastic materials buried in marine sediment by measuring the amount of carbon dioxide evolved	ISO 22404:2019 (2019)
Sludge anaerobic digestion degradation	GB/T 38737	Determination of the ultimate anaerobic biodegradation of plastic materials in controlled sludge digestion systems by measuring the biogas evolved	GB/T 38737-2020 (2020)
High solid anaerobic digestion degradation	GB/T 33797	Determination of the ultimate anaerobic biodegradability of plastic materials under high solid composting conditions by measuring the biogas evolved	GB/T 33797-2017 (2017)

(2) Degradation period: 180 days.
(3) Scope of application: Natural and/or synthetic polymers, copolymers, or their mixtures; plastic materials with additives such as plasticizers, pigments, or other compounds; water-soluble polymers.

2. GB/T 19276.2 (GB/T 19276.2-2003 2003)

Determination of the ultimate aerobic biodegradability of materials in aqueous medium by measuring the amount of CO_2 released.

(1) Test principle: The biodegradation rate of the test material in an aqueous system is determined using aerobic microorganisms. The test mixture includes an inorganic culture medium, test material with an organic carbon concentration between 100 and 2000 mg/L (the only source of carbon and energy), and a culture medium made from a suspension of activated sludge or compost, or

active soil. The mixture is stirred in a test flask and aerated to remove CO_2; the test period depends on the biodegradability of the test material, but cannot exceed 6 months. The CO_2 released when the microorganisms decompose the material can be measured by a suitable method. The degree of biodegradation is determined by the ratio of the amount of CO_2 released to the theoretical amount of CO_2 released ($ThCO_2$), expressed as a percentage.

(2) Degradation period: 180 days.

(3) Scope of application: Natural and/or synthetic polymers, copolymers or their mixtures; plastic materials with additives such as plasticizers, pigments or other compounds; water-soluble polymers.

3. GB/T 32106 (GB/T 32106-2015 2015)

Determination of the ultimate anaerobic biodegradability of materials in an aqueous medium by measuring the biogas produced.

(1) Test principle: The biodegradability of plastics in an aqueous medium under anaerobic conditions is determined. The digested sludge is first washed before use to contain a very small amount of inorganic carbon (IC), and diluted to a total dry solid concentration of 1–3 g/L. The test material with an organic carbon (OC) concentration of 20–200 mg/L is cultured with the digested sludge in a sealed container at a temperature of 35 ± 2 °C under anaerobic conditions for a period of time (usually not exceeding 60 days). Under these conditions, the test material will biodegrade into CO_2 and CH_4; the production of CO_2 and CH_4 will cause an increase in pressure or volume at the top of the test container, so the amount of biogas released can be obtained by measuring the increase in pressure or volume. The percentage of the total carbon converted into biogas and inorganic carbon to the total carbon contained in the test material itself (which can be obtained by measurement or calculation from the molecular formula) is the biodegradation percentage of the test material.

(2) Degradation period: 60 days.

(3) Scope of application: Natural and/or synthetic polymers, copolymers, or their mixtures; plastic materials with additives such as plasticizers, pigments, or other compounds; water-soluble polymers.

9.2.2 Compostable Degradation

1. *GB/T 19277.1* (GB/T 19277.1-2011 2011)

Determination of the ultimate aerobic biodegradability of materials under controlled composting conditions, using the method of measuring the released CO_2—Part 1: General method.

(1) Test principle: This test method determines the ultimate aerobic biodegradability and disintegration degree of the test material under simulated intense aerobic composting conditions. The inoculum used comes from stable, mature compost, and if possible, can also be obtained from the compost of organic matter in municipal solid waste. After mixing the test material with the inoculum, it is introduced into a static composting container. In this container, the mixture undergoes intense aerobic composting at specified temperature, oxygen concentration, and humidity. The test period does not exceed 6 months. During the aerobic biodegradation process of the test material, CO_2, H_2O, mineralized inorganic salts, and new biomass are all products of the ultimate biodegradation. The cumulative amount of CO_2 produced by the test container and the blank container is continuously monitored and measured regularly during the test. The ratio of the actual amount of CO_2 produced by the test material in the test to the theoretical amount of CO_2 that the material can produce is the biodegradation percentage.
(2) Degradation period: 180 days.
(3) Applicable scope: Plastics and other organic polymer materials.

2. *GB/T 19277.2* (GB/T 19277.2-2013 2013)

Determination of the ultimate aerobic biodegradability of materials under controlled composting conditions, using the method of measuring the released CO_2—Part 2: Determination of the release of CO_2 under laboratory conditions by weight analysis.

(1) Test principle: The aim is to determine the ultimate biodegradability of the test material using a small reactor. By controlling the humidity, aeration rate, and temperature of the composting container, the biodegradation rate of the test material under mature composting conditions is determined. The test material is made up of inoculum from mature compost and inert materials, such as sea sand. The biodegradation rate (expressed as a percentage) of the material is obtained by comparing the amount of CO_2 released with the theoretical CO_2 release ($ThCO_2$). The test ends when biodegradation reaches a steady state.
(2) Degradation period: 45 days, up to a maximum of 180 days.
(3) Applicable scope: Natural and/or synthetic polymers, copolymers, and their mixtures; plastics containing additives such as plasticizers, pigments; water-soluble polymers; materials that will not inhibit the microbial activity in the inoculum under test conditions.

3. *GB/T 19811* (GB/T 19811-2005 2005)

Determination of the disintegration degree of plastic materials under defined composting test conditions.

(1) Test principle: Used to determine the degree of disintegration of plastic materials in aerobic composting tests under defined test conditions. The test method specified in this standard can be used to determine the impact on the test material during the composting process and the quality of the compost obtained, but

it cannot be used to determine the aerobic biodegradation ability of the test material. After the test material is mixed with fresh biomass waste in an exact ratio, it is placed in a defined composting environment. The composting process is naturally triggered by the microbial populations that are ubiquitous in nature, usually after about 12 weeks. The disintegration of the test material is evaluated by the ratio of the amount of test material fragments on the 2 mm test sieve to the total dry solids.

(2) Degradation period: 12 weeks or actual composting cycle.

(3) Scope of application: Plastics and other organic polymer materials.

4. *OECD 208* (OECD 1994)

Requirements for plant toxicity tests.

(1) Test principle: Aimed at assessing the potential impact of chemicals on seed germination and growth. In the evaluation of compostable plastics, it mainly evaluates the ecological toxicity of compost after the biodegradation material is composted. During the test, the average survival rate of germinated seedlings must reach at least 90%.

(2) Degradation period: Plant growth time.

(3) Scope of application: Compost after the biodegradation material is composted.

5. *AS5810-2010* (AS 5810-2010 2010)

Technical specifications for compostable plastics in the courtyard.

(1) Test principle: The standard specifies the requirements and evaluation procedures for plastic products suitable for home composting. When all components in the plastic product meet all requirements, it is considered home compostable. Composting plastic products that meet the requirements of this standard does not necessarily produce high-quality compost.

(2) Degradation period: For biodegradability (i.e., the original biodegradability of the material), the test method is GB/T 19277.1, 19277.2, and the degradation period is 365 days; for the disintegration performance during the composting process, the test method is GB/T 19811 or ISO 20200, and the degradation period is 180 days; for the adverse effects produced during the biodegradation process, the test method is OECD 208, and the degradation period is the plant growth time.

(3) Scope of application: Plastic materials.

9.2.3 Soil Degradable

GB/T 22047 (GB/T 22047 2009): Determination of the ultimate aerobic biodegradability of plastic materials in soil by measuring the oxygen demand in a closed respirometer or by measuring the amount of CO_2 released.

(1) Test principle: This standard specifies the method of determining the ultimate aerobic biodegradability of plastic materials in soil by measuring the oxygen demand in a closed respirometer or by measuring the amount of CO_2 released. The biodegradation rate is obtained by the ratio of BOD to ThOD or by the ratio of the amount of CO_2 released to the $ThCO_2$, and the results are expressed as a percentage.
(2) Degradation period: 180 days.
(3) Scope of application: Natural and/or synthetic polymers, copolymers, and their mixtures; plastics containing additives such as plasticizers, pigments; water-soluble polymers; materials that do not inhibit the microbial activity in the inoculum under test conditions.

9.2.4 Marine Environment Degradation

1. *ISO 18830* (ISO 18830:2016 2016)

The ultimate aerobic biodegradation capacity of non-floating plastic materials at the plastic-seawater sediment interface can be determined by measuring the oxygen consumption in a closed respirometer.

(1) Test principle: Plastic products are directly discarded or flow into the open ocean area (free water area) with freshwater, and then may sink to the subcoast and reach the seabed surface due to the influence of material density, tides, ocean currents, and ocean folds. Many biodegradable plastics tend to sink to the seabed because their density is greater than 1. From the surface (interface with seawater) to the deep layer, the sediment state changes dramatically from aerobic to anaerobic to anoxic, presenting a sharp oxygen gradient. This standard specifies a test method for determining the degree and rate of aerobic biodegradation of plastic materials at the seawater-sediment interface by measuring the oxygen demand in a closed respirometer. The determination of aerobic biodegradation can also be achieved by measuring the release of CO_2. This method is a simulation of the habitat environment of different seawater sediment areas in the ocean under laboratory conditions, such as the benthic zone (phototrophic zone) where sunlight can shine, which is called the sublittoral zone in marine science. The level of biodegradation is obtained by the ratio of BOD to ThOD, expressed as a percentage.
(2) Degradation period: 2 years.
(3) Applicable scope: Non-floating plastic materials.

2. *ISO 19679* (ISO 19679 2019)

The ultimate aerobic biodegradation capacity of non-floating plastic materials at the plastic-seawater sediment interface can be determined by measuring the release of CO_2 .

(1) Test principle: This standard specifies a test method for determining the degree and rate of aerobic biodegradation of plastic materials at the marine sandy sediment–seawater interface by measuring the release of CO_2. This method is also applicable to other solid materials. The release of CO_2 during microbial degradation can be determined using appropriate analytical methods. The biodegradation rate of the material (expressed as a percentage) is obtained by the ratio of the amount of CO_2 released to the $ThCO_2$.
(2) Degradation period: 2 years.
(3) Applicable scope: Non-floating plastic materials.

3. *ISO 22404* (ISO 22404:2019 2019)

The ultimate aerobic biodegradation capacity of non-floating materials exposed to marine sediments can be determined by analyzing the released CO_2 .

(1) Test principle: A laboratory test method for determining the degree and rate of aerobic biodegradation of plastic materials. The CO_2 emission amount determines the biodegradation rate. Use appropriate analytical methods to measure the CO_2 emitted during the microbial degradation process. The biodegradation rate is determined by the ratio of CO_2 emission amount to the $ThCO_2$, expressed as a percentage.
(2) Degradation period: 2 years.
(3) Applicable scope: Non-floating plastic materials.

9.2.5 Anaerobic Digestion Degradation of Sludge

GB/T 38737 (GB/T 38737-2020 2020): The final anaerobic biodegradation rate of materials in controlled sludge digestion systems can be determined by measuring the release of biogas.

(1) Test principle: Evaluate the anaerobic biodegradation ability of plastics in controlled sludge digestion systems, where the solid content of the system does not exceed 15%. This system is common in the treatment of sludge, sewage, livestock manure, or garbage. This method aims to measure the conversion rate of organic carbon in the material to biogas, such as CO_2 and CH_4 .
(2) Degradation period: General 60 days. The test period can be shortened or extended until a stable decomposition stage is reached, but generally not exceeding 90 days.
(3) Applicable scope: Natural and/or synthetic polymers, copolymers, and their mixtures; plastics containing additives such as plasticizers, pigments; water-soluble polymers; materials that will not inhibit the microbial activity in the inoculum under experimental conditions.

9.2.6 *High Solid Anaerobic Digestion Degradation*

GB/T 33797 (GB/T 33797-2017 2017): The final anaerobic biodegradation ability of plastics under high solid composting conditions can be determined by analyzing and measuring the release of biogas.

(1) Test principle: A method to evaluate the anaerobic biodegradation ability of plastics under high solid anaerobic digestion conditions by measuring the release of biogas. This method simulates typical anaerobic digestion conditions with urban organic solid waste. The test material is exposed to the inoculum of household garbage that has undergone anaerobic digestion treatment in the laboratory. Anaerobic decomposition occurs in a high solid content environment (total dry solid content greater than 20%) and is left undisturbed under unmixed conditions. This test method is used to determine the carbon content in the test material and its conversion to CO_2 and CH_4 percentages.
(2) Degradation period: 15 days, if the biodegradation phenomenon is still obvious after 15 days, the cultivation period can be extended until the biodegradation of the test material reaches a stable period.
(3) Applicable scope: Plastics and other organic polymer materials.

9.3 Degradation Mechanism Research Methods

With the increasing prominence of white pollution in recent years, research on the degradation process of biodegradable materials has received more and more attention. Hypotheses on degradation mechanisms have been proposed for different biodegradable materials, but these all need to be proven through systematic research. In natural environments such as soil, seawater, and compost, materials may undergo different degradation behaviors due to environmental temperature, humidity, microbial distribution, light, etc. To clarify the degradation mechanism of materials, it is often necessary to combine multiple methods for research. Generally speaking, the weighing method and apparent morphology observation method can intuitively reflect the degradation process of materials. If further exploration of hydrolysis, enzymolysis, oxidation, etc. that occur during the degradation process of materials is needed, it is necessary to conduct a comprehensive analysis of the material's structure and environmental factors and conduct in-depth research from different aspects such as the relative molecular weight, functional groups, crystallinity, and physical and chemical properties of the material. The following will introduce the main research methods for analyzing the degradation process and degradation mechanism of materials.

9.3.1 Degradation Degree Analysis Method

1. *Disintegration Degree (Mass Loss)*

Testing the weight loss rate during the degradation process of the material is the simplest and most direct method to judge whether the material has undergone degradation. The calculation method for evaluating the disintegration degree of the material is stipulated in GB/T 19811-2005 (GB/T 19811-2005 2005) "Determination of the Disintegration Degree of Plastic Materials under Defined Composting Test Conditions":

$$D_i = \frac{m_1 - m_2}{m_1} \times 100\% \tag{9.4}$$

In the formula, D_i represents the disintegration degree of the test material, expressed in %; m_1 is the total dry solid amount of the test material put in at the beginning of the test, in grams; m_2 is the total dry solid amount of the test material collected after the test, in grams.

2. *Oxygen Demand/Carbon Dioxide Release*

Evaluating the degree of degradation by measuring the mass loss of the material often has certain limitations. For example, after being buried in the soil for a period of time, the sample surface may have already seeped into the soil components and is difficult to clean, or the material degrades into small particles that cannot be sampled, which has a great impact on the calculation of the material's degradation rate. At present, the standard method for evaluating the biodegradation rate of materials internationally is by measuring the oxygen demand in a sealed respirometer or measuring the released CO_2 amount to evaluate the ultimate aerobic biodegradation ability of plastic materials in soil.

GB/T 22047-2008 (GB/T 22047 2009) "Determination of the ultimate aerobic biodegradability of plastic materials in soil by measuring the oxygen demand in a closed respirometer or the amount of carbon dioxide evolved" stipulates the method of determining the ultimate aerobic biodegradation ability of plastic materials in soil by measuring the oxygen demand in a closed respirometer or the amount of CO_2 evolved. The plastic material is mixed with soil as the only source of carbon and energy, and the mixture is placed in a narrow-necked bottle. The biochemical oxygen demand can be determined by maintaining a constant volume of gas in the flask within the respirometer and automatically or manually measuring the volume or pressure changes (or both), and the amount of CO_2 evolved can be determined by passing air without CO_2 through the soil, and then measuring the amount of CO_2 evolved during the biodegradation of the test material. The biodegradation rate is obtained by the ratio of the biochemical oxygen demand to the theoretical oxygen demand or the ratio of the amount of CO_2 evolved to the theoretical amount of CO_2 evolved, and the result is expressed as a percentage. The following number of test bottles are generally prepared during the measurement process: ① Two flasks containing the test material (F_T); ② Two flasks for blank tests (F_B); ③ Two flasks using reference materials to test soil activity (F_C); ④ One flask to check for possible non-biological decomposition or nonmicrobial changes such as hydrolysis (F_S); ⑤ One flask to check for possible inhibitory effects of the test material on microbial activity (F_I).

Taking the measurement of CO_2 evolved as an example, the formula for calculating the biodegradation percentage is as follows:

A. Theoretical CO_2 evolved from the test material

$$ThCO_2 = m \times \omega_C \times \frac{44}{12} \tag{9.5}$$

where $Th\mathrm{CO_2}$ is the theoretical CO_2 evolved, in milligrams (mg); m released amount of CO_2 , in milligrams; ω_C is the carbon content in the test material, determined by the chemical molecular formula, calculated by elemental analysis, and expressed in %; 44 and 12, respectively, represent the molecular mass of CO_2 and the atomic mass of carbon.

The theoretical release amount of CO_2 in the reference material and the mixture of test material and reference material in the test bottle is calculated in the same way.

B. Calculation of Biodegradation Percentage

$$D_t = \frac{\Sigma(m)_T - \Sigma(m)_B}{ThCO_2} \times 100 \tag{9.6}$$

where D_t is the biodegradation percentage at each measurement interval of test bottle F_T , in %; $\sum(m)_T$ is the amount of CO_2 released from bottle F_T from the start of the test to time t, in milligrams (mg); $\sum(m)_B$ is the amount of CO_2 released from the blank bottle F_B from the start of the test to time t , in milligrams. amount of CO_2 released from the bottle, in milligrams (mg); Th CO_2 is the theoretical release amount of CO_2 from the test material, in milligrams (mg).

The same method is used to calculate the biodegradation percentage of the reference material in the soil activity detection bottle F_C. The BOD values or CO_2 amounts and biodegradation percentages for each flask at each measurement period are compiled into a table. For each flask, a curve of BOD or CO_2 release amount and biodegradation percentage is plotted with time as the horizontal axis. The maximum biodegradation rate is determined from the average or maximum value of the stable stage of the biodegradation curve to characterize the degree of biodegradation of the test material.

3. *CH_4 Generation Amount*

Under anaerobic conditions, organic compounds are decomposed into CO_2 , CH_4 , H_2O, and mineralized inorganic salts of their contained elements, as well as new biomass. GB/T 32106-2015 (GB/T 32106-2015 2015) “Determination of the ultimate anaerobic biodegradation capability of plastics in aqueous culture medium: by measuring the production of biogas” provides a test method for evaluating the ultimate anaerobic biodegradation capability of materials by measuring the gas release amount of CO_2 and CH_4.

Under anaerobic conditions, the test material decomposes into CO_2 and CH_4 under the action of microorganisms. The production of these two gases will cause an increase in pressure or volume at the top of the test container, so the amount of biogas released can generally be obtained by measuring the increase in pressure or volume. It should also be considered that under test conditions, a portion of CO_2 will dissolve in the aqueous culture medium or convert into bicarbonate or carbonate, and the organic carbon contained in this part of CO_2 is also a part of the biodegradation of the organic carbon of the test material.

9.3.2 Molecular Weight Analysis Method

The molecular weight of a polymer is one of the most basic structural parameters of polymeric materials. By measuring the change in molecular weight during the degradation process of the material, we can understand the chain-breaking situation of the molecule, which is of great significance for analyzing the degradation mechanism of the material.

1. *Viscosity Method*

There are many methods to determine the molecular weight of polymers, such as end-group titration, permeation, light scattering, ultracentrifugation, and viscosity, and the average relative molecular weight measured by different methods is also different. Due to the simplicity of the equipment and the convenience of operation, the viscosity method is the most widely used. The viscosity method measures the molecular weight based on the friction generated by large molecules moving in solution. The viscosity generally represents the resistance of the polymer solution during flow, and the average relative molecular weight measured is called the viscosity-average relative molecular weight.

At a certain temperature and solvent condition, the intrinsic viscosity η is related to the relative molecular weight M of the polymer. The relationship between them is usually expressed by the following empirical equation:

$$\eta = KM^{\alpha} \tag{9.7}$$

In the formula, K and α are constants related to temperature, solvent, and the nature of the polymer. Usually, for each polymer solution, the values of K and α are determined by polymers with known average relative molecular weights. Then, the viscosity-average relative molecular weight of the polymer to be tested can be obtained using these K and α values and the experimental value of the intrinsic viscosity of the same polymer solution. For many polymer solutions, their K and α values can be found in relevant manuals or books.

2. *Gel Permeation Chromatography*

The molecules of a polymer are not composed of molecules with the same molecular weight, but are composed of components with different molecular weights distributed in a certain way. The most commonly used, fast, and effective method for determining the relative molecular weight and its distribution of a polymer is gel permeation chromatography (GPC), which is a special type of liquid chromatography based on the size of polymer molecules, separating different components of the polymer. The polymer can be separated on the separation column according to the size of the molecular hydrodynamic volume. A known amount of polymer dilute solution is injected into the solvent flow, which can carry it through the gel permeation chromatography column at a steady rate. The concentration of separated molecular components in the solvent flow is measured using an appropriate detector. By using a calibration curve, the number-average molecular weight ($\bar{M}_n$) and weight-average molecular weight ($\bar{M}_w$) of the analyzed sample can be determined from the retention time and corresponding concentration. When measuring polymer properties, the molecular weight distribution is also an important parameter, which can be represented by the polymer dispersion D , given by the following formula:

$$D = \bar{M}_w / \bar{M}_n \tag{9.8}$$

9.3.3 Apparent Morphology Analysis Method

Among the methods of analyzing material degradation, observing the changes in the apparent morphology of the material is one of the most important steps. The erosion traces on the material surface, the degree of cracking, etc., are all reactions to the material degradation behavior, which are of great significance for analyzing the mechanism of material degradation.

1. *Scanning Electron Microscope*

The scanning electron microscope (SEM) is a large precision instrument used for high-resolution micro-area morphology analysis. By scanning the sample with a high-energy electron beam, various physical information of the material can be obtained. By receiving, amplifying, and displaying the information in images, the surface morphology of the test sample can finally be obtained. Compared with optical microscopes and transmission electron microscopes, the scanning electron microscope has simple sample preparation, can directly observe the structure of the sample surface, has a large depth of field, and the image is rich in three-dimensionality, and can intuitively observe the surface microscopic morphology changes of polymer materials before and after degradation.

2. *Transmission Electron Microscope (TEM)*

TEM is an electron optical microscope that uses a very short wavelength electron beam as a light source and focuses on imaging with an electromagnetic lens. The transmission electron microscope has two major functions: phase analysis and tissue analysis. Phase analysis uses the characteristic that electrons can diffract when interacting with crystalline substances to obtain phase diffraction patterns; tissue analysis uses the principle of Abbe imaging, followed by electron waves to obtain various contrast images through interference imaging. The transmission electron microscope has the advantages of high resolution, high magnification, and rich stereo information. It is widely used in the field of material science for composition and structure analysis. However, due to its imaging principle, its sample preparation is destructive, and the sampling rate is relatively low.

3. *Atomic Force Microscope (AFM)*

AFM uses a microcantilever to sense and amplify the interaction force between the fine probe on the cantilever and the atoms of the tested sample, thereby achieving the purpose of detection, with atomic-level resolution. Compared with the scanning electron microscope, the most significant advantage of the atomic force microscope is that it can provide a true three-dimensional surface map and calculate the surface roughness, while the electron microscope can only provide a two-dimensional image.

9.3.4 Structural Analysis Method

The process of material degradation often involves changes in structure. By analyzing the breaking of molecular chains during the degradation process, changes in groups in the structure, changes in crystallinity, and the components produced, we can help explain the degradation mechanism of the material.

1. *Fourier Transform Infrared Spectroscopy (FTIR)*

The vibrational energy of a molecule is greater than its rotational energy. When a vibrational energy level transition occurs, it inevitably accompanies a rotational energy level transition. Therefore, it is impossible to measure a pure vibrational spectrum, and only the vibrational–rotational spectrum of the molecule can be obtained. This spectrum is called an infrared absorption spectrum. FTIR is an analytical identification method that combines computer technology with infrared spectroscopy. It mainly consists of an optical detection part and a computer part. When the sample is placed in the interferometer light path, the resulting interference pattern intensity curve changes accordingly due to the absorption of energy at certain frequencies. Through the mathematical Fourier transform technique, each frequency on the interference pattern can be converted into the corresponding light intensity, thus obtaining the entire infrared spectrum. Based on the different features of the spectrum, it is possible to identify the functional groups of unknown substances, measure chemical structures, observe the course of chemical reactions, distinguish isomers, analyze the purity of substances, etc. To analyze the spectrum, the infrared spectrum is usually divided into two areas, namely the functional group area and the fingerprint area. The frequency range of 4000–1400 cm^{-1} is the functional group area, where absorption is mainly caused by the stretching vibration of the molecule. Common functional groups generally have specific absorption peaks in this area; the area below 1400 cm^{-1} is called the fingerprint area, which has a relatively large number of absorption peaks. This is caused by the bending vibration of chemical bonds and the stretching vibration of some single bonds. The position and intensity of the absorption band vary with the compound. The fingerprint absorption of different compounds is different, so it is important to identify the same compound with the peak shape and peak intensity of known substances.

2. *Raman Spectroscopy*

Raman spectroscopy is a type of scattering spectroscopy. When light is irradiated onto a substance, elastic scattering and inelastic scattering occur. The scattered light of elastic scattering is the component with the same wavelength as the excitation light, and the scattered light of inelastic scattering has components longer and shorter than the wavelength of the excitation light, collectively referred to as the Raman effect. Raman spectroscopy is an analytical method based on the Raman scattering effect, which analyzes the scattered light spectrum different from the incident light frequency to obtain information about molecular vibration and rotation, and is used for molecular structure research. Through the analysis of Raman spectroscopy, the vibrational and rotational energy levels of the substance can be determined, thereby identifying the substance and analyzing the properties of the substance. In essence, both Raman spectroscopy and infrared spectroscopy are vibrational spectroscopies, measuring the excitation or absorption of the ground state, and their energy ranges are the same. However, the difference is that infrared spectroscopy detects the dipole moment changes produced by molecular vibration, which is more sensitive to polar groups, while Raman spectroscopy is more sensitive to the shape and polarization of molecules, and is sensitive to nonpolar groups. Some weak bands in the infrared spectrum may be strong in the Raman spectrum, and the two complement each other in structure identification.

3. *Nuclear Magnetic Resonance Spectroscopy (NMR)*

NMR spectroscopy studies the absorption of radio frequency radiation by atomic nuclei. It is one of the most powerful tools for qualitative analysis of the composition and structure of various organic and inorganic substances, and sometimes it can also be used for quantitative analysis. In a strong magnetic field, atomic nuclei undergo energy level splitting. When they absorb external electromagnetic radiation, nuclear energy level transitions occur, resulting in the so-called NMR phenomenon. Therefore, a specific atomic nucleus in a given external magnetic field only absorbs energy provided by a specific frequency radio field, forming a nuclear magnetic resonance signal. So far, only atomic nuclei with spin quantum numbers equal to 1/2 can be utilized, and the atomic nuclei often used are ^{1}H, ^{11}B, ^{13}C, ^{17}O, ^{19}F, and ^{31}P. Among them, the most commonly used in structural analysis are proton NMR (^{1}H NMR) and carbon NMR (^{13}C NMR). Nuclear magnetic resonance is an important means of identifying the structure of organic compounds. It generally identifies groups based on chemical shifts, determines group linkage relationships from coupling splitting peaks and coupling constants, and determines the proton ratio of each group based on the integral area of each H peak. NMR spectra can be used for studies in chemical kinetics, such as intramolecular rotation, chemical exchange, etc., as they all affect the condition of the external chemical environment of the nucleus, and therefore are reflected in the spectrum.

4. *X-Ray Diffraction (XRD)*

The basic principle of XRD is: when a monochromatic X-ray beam is incident on a crystal, because the crystal is composed of atoms arranged in a regular lattice, the distances between these regularly arranged atoms are of the same order of magnitude as the wavelength of the incident X-ray. Therefore, X-rays scattered by different atoms interfere with each other, producing strong X-ray diffraction in certain special directions. The distribution and intensity of the diffracted rays in space are closely related to the crystal structure. The composition of the substance, crystal form, bonding method within the molecule, molecular configuration, conformation, etc., will all affect the diffraction spectrum produced by the substance. In the analysis of polymer materials, XRD is the main method used to analyze the crystal form and crystallinity of materials.

9.3.5 Mechanical Property Analysis Method

Changes in material properties are important reflections of changes in material molecular chain segments, crystallinity, and molecular weight. By analyzing the changes in properties during the degradation process of materials, such as the change pattern and rate, on the one hand, the changes in the degradation process of materials can be intuitively obtained, and on the other hand, combined with the relevant content of structural characterization, it can help analyze the degradation mechanism of materials. Among them, the most common performance test is the mechanical property test during the degradation process. The mechanical properties of a material refer to the mechanical characteristics exhibited by the material under various external loads (tension, compression, bending, torsion, impact, alternating stress, etc.) in different environments (temperature, medium, humidity). The structure of polymer materials is a key factor determining their mechanical properties, which, in addition to chemical composition, also includes structural factors such as relative molecular mass and its distribution, cross-linking and branching, crystallinity, and crystal morphology. For different materials and different forms of materials, many types of mechanical properties can be tested, generally including elasticity, hardness, strength, plasticity, toughness, fatigue performance, fracture toughness, etc. In current research on material degradation, materials are mostly in the form of thin films, and their mechanical properties are generally evaluated by testing their tensile properties.

The main test indicators are tensile strength, which represents the strength of the material, and elongation at break, which represents the toughness of the material. Tensile strength (σ_t, N/m^2), also known as tensile strength, is the maximum load per unit area that a standard specimen can withstand at a specified test temperature, humidity, and tensile rate until the specimen breaks, i.e.:

$$\sigma_t = \frac{p}{bd} \tag{9.9}$$

where, p is the maximum load the specimen can withstand before breaking, N; b is the width of the specimen, m; d is the thickness of the specimen, m. The ratio of the elongation length before and after stretching to the length before stretching is called the elongation at break, expressed as a percentage. The elongation at break can represent the ability of the material to deform under the maximum load, and its calculation formula is as follows:

$$\varepsilon = \frac{(L - L_0)}{L_0} \times 100\% \tag{9.10}$$

In the formula, L is the distance of the marking line when the sample breaks; L_0 is the original marking distance.

Polymer material samples are usually placed on a tensile machine and stretched at a uniform speed until they break. If the stress and deformation experienced by the sample during the stretching process are recorded, a stress-strain curve can be obtained. The specific operation standard can refer to GB/T 1040-2006 "Determination of Tensile Properties of Plastics" (GB/T 1040-2006 2006).

9.4 Research Progress of Biobased Degradable Materials

9.4.1 Polylactic Acid

In the natural environment, the degradation of polylactic acid (PLA) can be achieved through various forms, such as biodegradation, hydrolysis, photolysis, radiation degradation, and thermal degradation, among which biodegradation plays a major role. However, due to the important influence of environmental factors on microbial populations and the activity of different microorganisms themselves, the degradation of PLA is related to environmental factors, such as temperature, humidity, pH, the presence of O_2, and the supply of different nutrients, etc.; all have an important impact on PLA degradation. In addition, the degradation effect of PLA is also related to its own chemical and physical properties (morphology, porosity, mechanical strength, purity, heat resistance, and electromagnetic radiation resistance) (Shi Ke and Tingting 2019). Generally speaking, when exposed to suitable conditions composed of H_2O, O_2, and natural microorganisms, PLA will decompose into H_2O, CO_2, and a small number of other oligomers. The biodegradation of PLA in the natural environment mainly includes four degradation mechanisms: hydrolysis degradation, photodegradation, microbial degradation, and enzymatic degradation, among which enzymatic and microbial degradation mechanisms are the most important (Zaaba and Jaafar 2020).

1. *Hydrolysis*

Polylactic acid is prone to hydrolytic degradation under humid conditions, mainly including four processes: water absorption, ester bond breakage, diffusion of soluble oligomers, and fragment decomposition. After polylactic acid absorbs moisture, H_2. The O molecule starts from the amorphous region, diffuses into the vicinity of the ester bond or other hydrophilic groups, and under the action of acidic or alkaline media, the ester bond undergoes free hydrolysis and breaks, that is, the reverse reaction of esterification. Subsequently, the molecular weight of the material slowly decreases. When the molecular weight is small enough to be soluble in water, the material structure deforms, the sample begins to dissolve, and soluble degradation products are generated. The lactic acid oligomers generated after the ester bond breaks increase the concentration of carboxylic acid groups in the medium. These carboxylic acid functional groups can catalyze the hydrolytic degradation reaction in reverse (Fig. 9.4). The internal degradation of polylactic acid is faster than surface degradation, because the degradation products with terminal carboxylic groups remain in the sample, producing a self-accelerating effect, so the

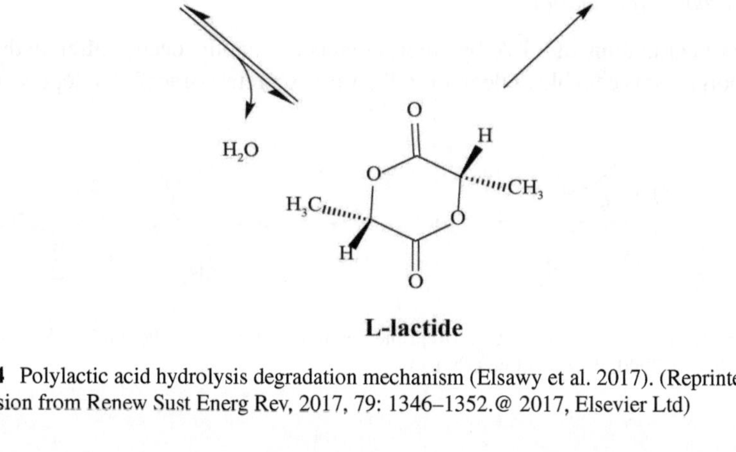

Fig. 9.4 Polylactic acid hydrolysis degradation mechanism (Elsawy et al. 2017). (Reprinted with permission from Renew Sust Energ Rev, 2017, 79: 1346–1352.@ 2017, Elsevier Ltd)

hydrolysis degradation of polylactic acid can be regarded as a self-catalyzing and self-sustaining process.

The hydrolysis degradation process of polylactic acid is mainly controlled by four basic parameters: water absorption, diffusion rate of segments in the polymer, diffusion coefficient, and solubility of degradation products. This is related to factors such as chemical structure, molar mass and distribution, material crystallinity, material morphology, and hydrolysis conditions. Studies have shown (Elsawy et al. 2017) that when polylactic acid is immersed in 50% ethanol, the degradation rate is faster because the diffusion rate of ethanol molecules in the polymer matrix is faster than H_2O molecules. Since the hydrolysis process preferentially occurs in the amorphous region, the hydrolysis rate of amorphous polylactic acid is much higher than that of semi-crystalline polylactic acid. In addition, filling nanoparticles in polylactic acid will also affect its hydrolysis rate. The stronger the hydrophilicity of the filler, the more obvious the degradation effect (Iñiguez-Franco et al. 2017).

2. *Photodegradation*

Light plays an important role in the degradation of polylactic acid. Photodegradation promotes changes in materials, thereby affecting their characteristics during their life cycle. Current research believes that the photodegradation mechanism of polylactic acid is initiated by photoionization (Norrish I), followed by polymer chain scission (Norrish II). The Norrish II reaction mechanism of polylactic acid is shown in Fig. 9.5 (Zaaba and Jaafar 2020).

In recent years, scholars have found that during the UV irradiation process of PLA, the band at 1845 cm^{-1} gradually becomes significant, which means the formation of anhydride during the irradiation process. Therefore, they proposed a photo-oxidation free mechanism of polylactic acid with hydrogen peroxide as an intermediate: under ultraviolet light, the tertiary hydrogen in the PLA chain is extracted to form a free radical, which then reacts with oxygen to form a peroxide free radical. This free radical can easily extract another hydrogen from the tertiary carbon to form a hydrogen peroxide free radical. The photolysis of hydrogen peroxide triggers the breakage of the PLA chain, as shown in Fig. 9.6 (Bocchini et al. 2010).

3. *Microbial degradation*

The degradation of PLA by microorganisms mainly occurs after hydrolysis. Microorganisms capable of degrading PLA first secrete extracellular depolymerases

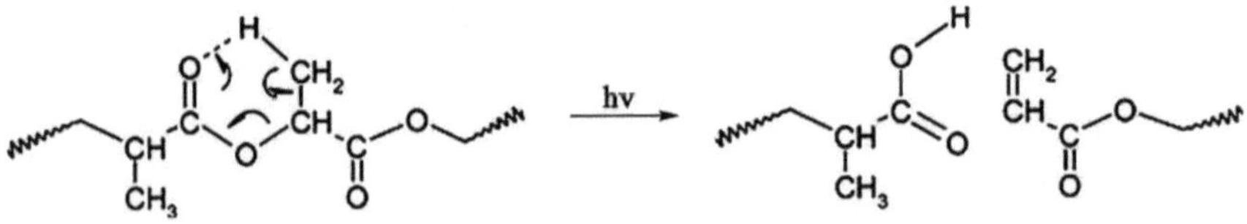

Fig. 9.5 Norrish II reaction of PLA. (Reprinted with permission from Polym Eng Sci, 2020, 60: 2061–2075.@ 2020, Wiley Online Library)

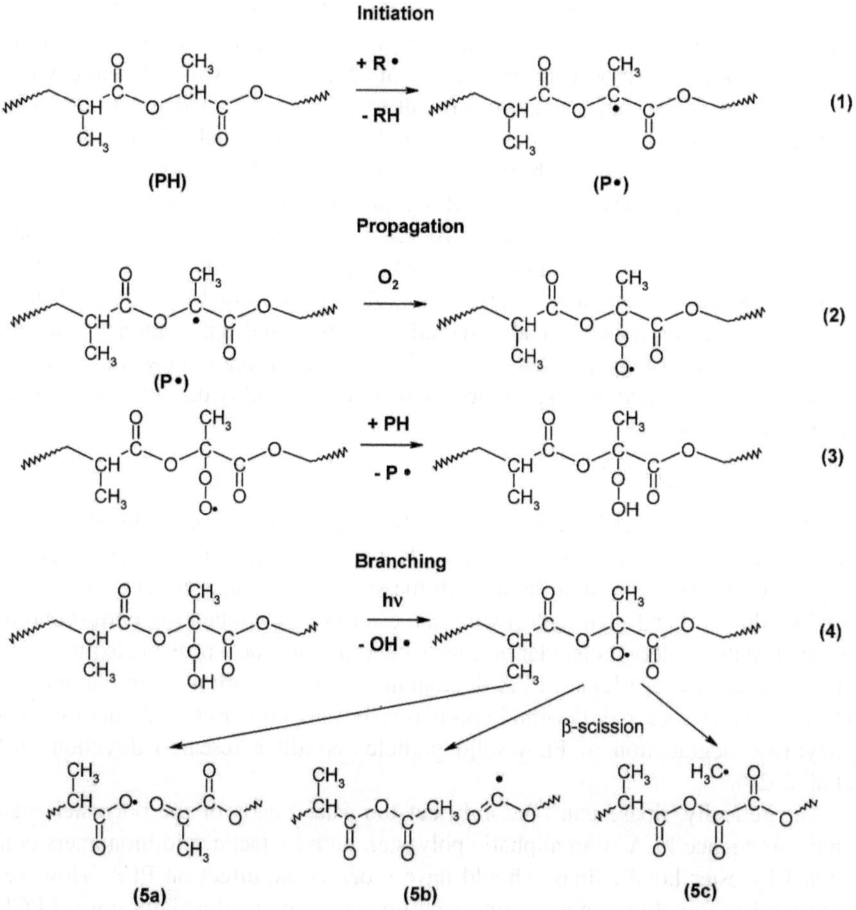

Fig. 9.6 Photo-oxidation mechanism of polylactic acid (Bocchini et al. 2010). (Reprinted (adapted) with permission from Biomacromolecules, 2010, 11(11): 2919–26.© 2010 American Chemical Society)

to target and break down the ester bonds within the molecule. The depolymerase is stimulated by inducers such as silk protein, elastin, gelatin, and some peptides and amino acids, which accelerate degradation, producing PLA oligomers, dimers, or monomers. These low molecular weight compounds can penetrate the microbial membrane and further decompose into CO_2, H_2O, or CH_4 (Qi et al. 2017) under the action of intracellular enzymes. The absorption of PLA degradation products by microorganisms is an important step in the PLA degradation process, and the small molecules produced after degradation can also provide nutrients for microbial growth.

Compared with other biodegradable materials, the degradation rate of PLA in soil is relatively slow. Studies have shown that PLA begins to degrade only after being buried in the soil for 6 weeks, and complete degradation takes several years.

However, it can be completely degraded within 90 days in a composting environment. In this process, microorganisms play an irreplaceable role. After analyzing the distribution of microorganisms capable of degrading PLA in different environments, it was found that it is more difficult to screen for PLA-degrading bacteria compared to those that can degrade other polyester compounds. So far, dozens of PLA-degrading bacteria have been discovered, most of which belong to the actinomycetes, with only a few being fungi (Rongqiu 2016). Microorganisms that degrade PLA also include some anaerobic microorganisms, which have a slower degradation rate. The effects of crystallinity, thermal properties, thickness, and additives on degradation were investigated (Mario et al. 2025). In addition to the study of PLA-degrading microorganisms, it also includes the potential application of enzyme engineering technology in PLA degradation. In recent years, new universal test methods have emerged, through which more biodegradability data can be collected (Lors et al. 2025).

4. *Enzymatic Degradation*

The essence of PLA biodegradation is the result of the action of depolymerases secreted by microorganisms. Enzymatic degradation of aliphatic polymers generally includes two steps: first, binding to the substrate through the enzyme surface binding domain, and then hydrolyzing the ester bond to obtain the corresponding product. With modern molecular biology techniques, it is now possible to purify and characterize extracellular PLA-degrading enzymes from microorganisms. Degrading enzymes mainly contain proteases, lipases, cutinases, and esterases. The enzymatic degradation of PLA solid particles is still a research direction (Adi et al. 2024).

Theoretically, lipase can randomly cut the ester bonds of the polyester main chain, and since PLA is an aliphatic polyester, with its lactic acid monomers connected by ester bonds, lipase should have a degrading effect on PLA. However, research has found that most commercial lipases show good activity toward PCL, PBS, etc., but can hardly degrade PLA, with only one lipase from *Alcaligenes sp.* being able to degrade PLA at 55 °C, under the conditions of $pH<8.5$. Complete degradation of PLA was achieved after 20 days. It is speculated that the degradation of this enzyme is based on hydrolysis at high temperature and high pH, and it mainly degrades the hydrolysis product lactic acid oligomers. Some people speculate that lipase can degrade various low T_m, amorphous polymers with no chiral carbon atoms, and a large number of methylene groups between ester bonds, but cannot hydrolyze polyesters with typical chiral carbons, such as PHB and PLA (Akutsu-Shigeno et al. 2003). Williams D F from the University of Liverpool first pointed out in 1981 that protease K from *Tritirachium album* has a degrading effect on polylactic acid (Williams 1981). Since then, protease K has become a widely recognized PLA-degrading enzyme and is used in the degradation research of PLA and its mixtures. Inspired by the degradation effect of protease K, Oda and others measured the PLA degradation activity of 56 commercially available proteases. The research results show that some alkaline proteases, especially serine proteases, can degrade PLA and produce a large amount of lactic acid, while acidic and neutral proteases basically do not have the ability to degrade PLA (Oda et al. 2000). In addition to alkaline proteases, lipase (Martin et al. 2014) and keratinase (Kawai 2010; Kawai et al. 2011) have also been reported to have the ability to degrade PLA. The activity of depolymerase is related to the pH and temperature of the environment and the nature of PLA itself. The latest research results on PLA depolymerase show that protease preferentially degrades PLLA, while lipase and keratinase preferentially degrade PDLA. The performance of the enzyme can be improved by immobilization methods or by means of embedding in polymers, for example, the stability of the enzyme can be significantly improved, and the degradation behavior in situ can be achieved (Sun 2025; Cao et al. 2025).

Because the ester bond is prone to breakage under certain high-temperature conditions, PLA will also undergo thermal degradation at high temperatures. The degradation of PLA in the molten state is mainly due to the intramolecular ester exchange at the end of the PLA chain, which generates cyclic oligomers of lactic acid and propylene ester. This reaction is reversible, and the generated oligomers and propylene ester will reinsert into the linear long-chain polyester chain, causing the long-chain molecules to shorten, leading to a decrease in the molecular weight of the polymer and a wider distribution of molecular weight. Due to the different preparation methods and production processes of PLA, its thermal stability also varies greatly. The main influencing factors include the structure of the end group, the water content of the system, the molecular weight, the residual small molecules, and the processing technology, etc.

In addition to this, polylactic acid (PLA) also has good degradability in the human body. Its degradation process in the body, known as the dissolution process, is the process of transforming from a water-insoluble solid to a water-soluble substance. This mainly includes four stages: water absorption, ester bond breakage, diffusion of soluble oligomers, and dissolution of fragments, which is consistent with the degradation process in the natural environment. Macroscopically, the overall structure of PLA in the body is gradually destroyed, the volume gradually decreases, it becomes fragments, and finally completely dissolves and is absorbed by the human body or excreted out of the body. Microscopically, the molecular chain undergoes chemical decomposition, the molecular weight decreases, the degree of cross-linking decreases, the molecular chain breaks, the side chain breaks, etc., and finally becomes a water-soluble small molecule and enters the body fluid, is phagocytosed by cells, and is transformed and metabolized. The excellent mechanical properties, chemical stability, and good biocompatibility and absorbability of PLA make it stand out among many materials and play an irreplaceable role in biomedical fields such as tissue filling, cell culture, and drug carriers.

9.4.2 Polyhydroxyalkanoates

Polyhydroxyalkanoates (PHA) are aliphatic copolyesters synthesized by microorganisms using carbon sources for fermentation. They are a type of biopolyester and can be used as a carbon source or energy material for microorganisms under conditions of insufficient cell nutrition. Depending on the monomer structure, the types and properties of the polyester vary greatly. The most common ones are poly β-hydroxybutyrate (PHB), poly β-hydroxyvalerate (PHV), and their copolymer-poly β-hydroxybutyrate/poly β-hydroxyvalerate (PHBV). In various natural environments, such as soil, activated sludge, seawater, etc., PHA shows good biodegradability. Unlike PLA, in the natural environment, PHA can be directly hydrolyzed by extracellular enzymes secreted by microorganisms. At present, hundreds of microorganisms that can degrade PHA have been found, including bacteria, fungi, and actinomycetes, which can secrete extracellular depolymerase to decompose PHA, ultimately generating CO_2 and H_2O and releasing energy. The degradation process of PHA is affected by many factors, mainly including ① the side chain length of the 3-hydroxyalkanoates repeating unit; ② differences in material morphology due to changes in sequence structure and preparation methods; ③ the geometric shape of the PHA sample, such as film, particles, spheres; ④ degradation conditions (pH, temperature, enzymes) (Tarazona et al. 2020). Due to the different types of microorganisms and their substrates, the degradation pathways and degradation products of PHA also have certain differences. Taking the simplest structured poly β-hydroxybutyrate (PHB) in the PHA family as an example, the decomposition of PHB under the action of microorganisms can generally be divided into intracellular degradation and extracellular degradation. Intracellular degradation is the active degradation of the endogenous storage substances of the synthesizing bacteria themselves. It is a cyclic process based on changes in nutritional conditions. When the nutrition is unbalanced and there is a carbon source, the cells will accumulate a large amount of PHB, and when the nutrition is rebalanced, PHB will be decomposed.

Intracellular PHB exists in an amorphous state, and its molecules are mobile. Natural PHB granules have a special surface layer composed of proteins and phospholipids, which is sensitive to physical or chemical stress. Figure 9.7 represents the intracellular degradation process of PHB (Ci 2004). Amorphous intracellular PHB granules degrade under the action of depolymerase, forming a mixture of monomers and dimers, with the dimers subsequently forming monomers under the action of dimer hydrolase. In the metabolism of PHB, the most critical enzyme is 3-hydroxybutyrate dehydrogenase, which is a bidirectional regulatory enzyme involved in both synthesis and decomposition. When it catalyzes synthesis, it can be inhibited by high concentrations of CoA and activated by acetyl-CoA; conversely, when it catalyzes decomposition, it is activated by CoA and inhibited by acetoacetyl-CoA. It can be considered that the balance between the degradation and synthesis of PHB is the balance between CoA and acetyl-CoA (Ci 2004). The extracellular degradation of PHB is carried out by extracellular PHA depolymerase secreted by microorganisms to degrade PHB using exogenous polymers. After the depolymerase hydrolyzes the extracellular polymer, it produces water-soluble products. It can be observed that after hydrolysis, a clear, transparent circle appears around the colony on the plate, and this reaction can proceed under both anaerobic and aerobic conditions (Fig. 9.8). PHA-degrading strains have been isolated from various ecological environments, including soil, compost, aerobic and anaerobic sludge, fresh

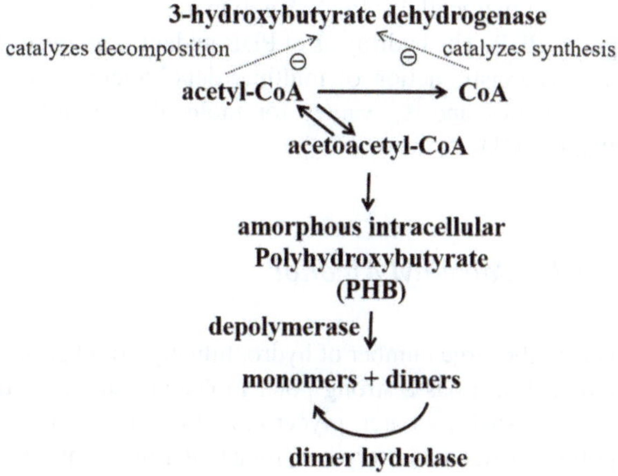

Fig. 9.7 Intracellular metabolic pathway of PHB. (Ci 2004)

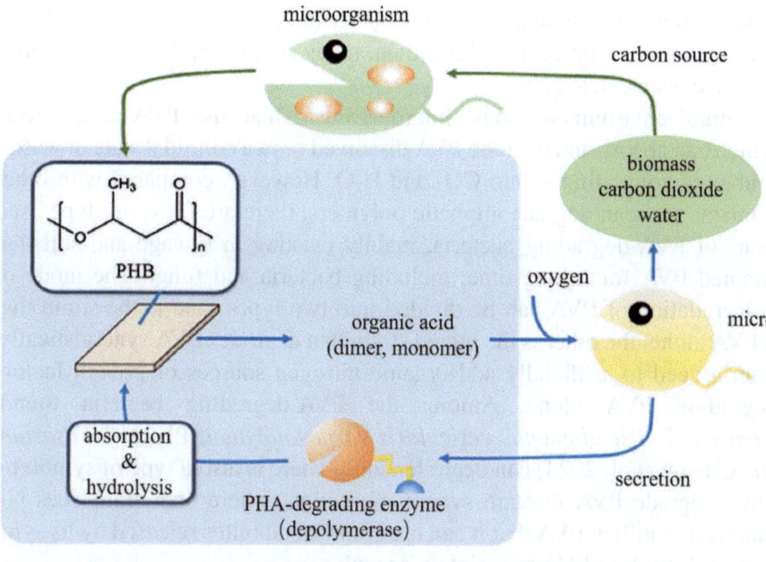

Fig. 9.8 Extracellular degradation pathway of PHA (see color insert)

water and seawater (including deep sea), estuary sediments, and air. Typical examples include *Alcaligenes faecalis, Pseudomonas lemoignei, Pseudomonas delafieldii,* and *Penicillium.*

PHA-degrading strains can be distinguished based on the types of polyesters they degrade. Most of the bacteria discovered so far can only specifically degrade short-chain and medium-chain polyhydroxyalkanoates, but some bacteria show a wide range of polyester degradation, and they can synthesize more than one PHA depolymerase. Taking the most extensively studied PHA-degrading bacterium, Paucimonas *lemoignei*, as an example, this bacterium can synthesize at least seven kinds of extracellular PHA depolymerases, one of which has specificity for amorphous PHB. The hydrolysis of PHB under the action of depolymerase often involves the synergistic action of multiple depolymerases, which have different substrate specificities and K_m values for molecules of different chain lengths (Ci 2004; Wang 2001).

9.4.3 Polyvinyl Alcohol

Due to the large number of hydrophilic hydroxyl groups in polyvinyl alcohol (PVA) molecules, it has a strong polar molecular structure and is easily soluble in polar solvents such as water, glycerin, and ethylene glycol. Because the BOD value of polyvinyl alcohol is very low, for a long time, people believed that polyvinyl alcohol could not be biodegraded until 1973 when Suziki et al. (Suzuki 1973) isolated a strain of Pseudomonas (*Pseudomonas broeoplis*) from the soil and found that it could produce a polyvinyl alcohol degrading bacterium that could oxidize and break down polyvinyl alcohol molecules. Since then, many studies on the biodegradability of PVA have been carried out.

In the natural environment, some microorganisms can use PVA as a carbon source, which can absorb and degrade PVA dissolved or in a colloidal state in wastewater, ultimately converting it into CO_2 and H_2O. However, compared with other microorganisms that can degrade aliphatic polymers, there are far fewer types and distributions of PVA-degrading bacteria, mainly existing in sewage and soil that have contained PVA for a long time, including bacteria and fungi. The mode of bacterial degradation of PVA can be divided into two types: one is the strain that degrades PVA alone; the other is the mixed strain that degrades PVA synergistically. Most bacteria need to artificially add organic nitrogen sources or growth factors when degrading PVA alone. Among the PVA-degrading bacteria found, *Pseudomonas* O-3, *Pseudomonas vesicularis varpovalolyticul* PH, and *Fusarium oxysporum* (Zhang et al. 2024) can degrade alone. There is also a type of symbiotic bacteria that degrade PVA through synergistic action, where one strain does not directly degrade or utilize PVA, but it can use some metabolites released by its symbiotic bacteria degrading PVA to maintain growth.

PVA degrading enzymes mainly include three types: polyvinyl alcohol oxidase (secondary alcohol oxidase), polyvinyl alcohol dehydrogenase, and β-diketone hydrolase (oxidative polyvinyl alcohol hydrolase). According to the Modern plastic yearbook report, when the polymer chain contains a $-CH_2CHOH-$ unit component, the following biochemical reactions can occur, as shown in Fig. 9.9.

The PVA dehydrogenases discovered so far are mainly a type of dehydrogenase that requires pyrroloquinoline quinone (PQQ) as a coenzyme. The strains that produce these enzymes need PQQ as a growth factor, so these bacteria need to live symbiotically with another strain that produces and secretes PQQ to degrade PVA. In terms of the substrates catalyzed, PVA dehydrogenase mainly catalyzes the dehydrogenation of secondary alcohols and PVA, and has no catalytic activity for

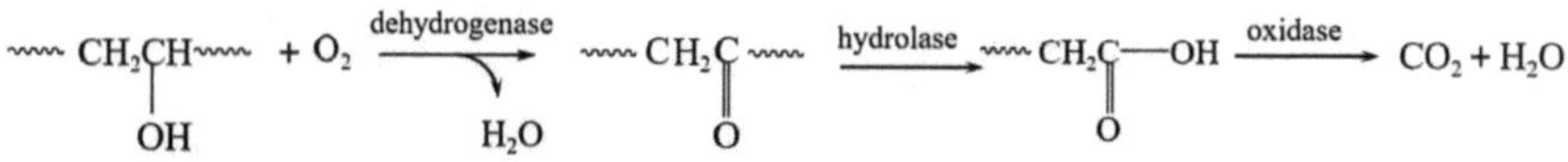

Fig. 9.9 PVA enzymatic hydrolysis mechanism

primary alcohols, so this enzyme is different from the quinone enzyme of ethanol dehydrogenase. β-Diketone hydrolase, also known as oxidative PVA hydrolase, was initially isolated from *Pseudomonas*. It was isolated from the fermentation broth of the bacterial strain. Judging from the substrates of this enzyme, it only has catalytic activity for oxidized PVA and has no catalytic activity for many low molecular weight ketone compounds. The catalytic requirements of PVA oxidase for PVA are generally long carbon chains (molecular weight greater than 300 Da) and exposed hydroxyl groups. Under the participation of O_2, the hydroxyl groups are oxidized into ketone groups, so PVA oxidase is also called secondary alcohol oxidase. At present, the catalytic mechanisms of these three enzymes are well understood, but different strains produce different combinations of PVA-degrading enzymes. One combination is polyvinyl alcohol oxidase and β-diketone hydrolase produced by cells, and another combination is polyvinyl alcohol dehydrogenase and β-diketone hydrolase produced by cells. At present, there are still disagreements among researchers about the mechanism of PVA degradation catalyzed by PVA-degrading enzymes. In recent years, there have been reports on the research of fungal degradation of PVA. The basidiomycete *Phanerochaete chrysoporium*, which degrades lignin, has a good effect on the degradation of polycyclic aromatic compounds and artificially synthesized macromolecular compounds, and other anti-heterotypic biomass. When it is used to degrade PVA, the lignin peroxidase (LiP) it produces can increase its unsaturation and degrade PVA by forming carbonyl double bonds. The first part of the fungal degradation process mainly occurs in the amorphous part of the PVA chain, and the remaining crystalline part is degraded very slowly. Inés et al. (Mejía et al. 1999) proposed a degradation route for fungi: first, free radicals are modified into epoxides, and then hydrolyzed to form double bonds. This hypothesis is obtained through infrared and ultraviolet spectroscopy analysis. The free radical cation is reorganized through 7 carbon atoms to form benzaldehyde (Fig. 9.10).

In solution, the degradation mechanism of PVA is mainly the random breakage of the polymer chain. The first step is the oxidation of the 1,3-hydroxyl group, catalyzed by oxidase and dehydrogenase enzymes, to form β-hydroxyketone or 1,3-diketone groups. The latter is easily catalyzed by β-diketone hydrolase, causing the carbon-carbon chain to break and form chains with carboxyl and methyl ketone endpoints. The internal random breakage of PVA is not affected by the structural factors of the polymer, but the hydrophobicity of the polymer, such as the amount of residual acetyl groups, does affect the activity of PVA dehydrogenase. PVA oxidase also shows a dependence on the microstructure of large molecules. Most of the enzymes separated are extracellular enzymes, but there is a special intracellular esterase that can hydrolyze the participating acetyl groups, which mainly occurs in

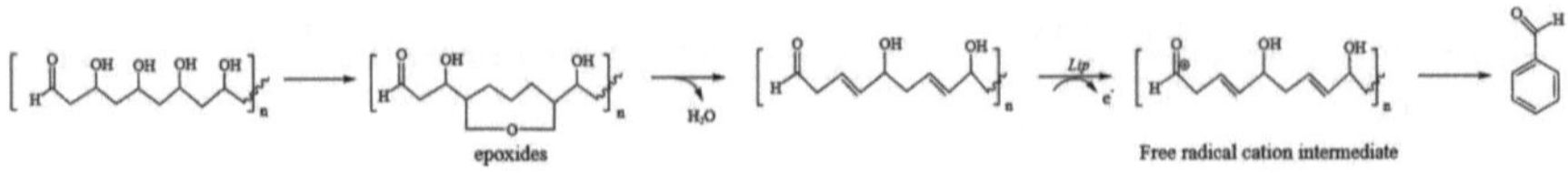

Fig. 9.10 Biodegradation pathway of PVA by *Phanerochaete chrysoporium*. (Mejía et al. 1999)

the final intracellular metabolism process of partially alcoholized PVA samples (Dong et al. 2005).

9.4.4 Polyamide 4

Ordinary polyamides are difficult to degrade in the natural environment. Their high resistance to degradation is mainly due to their highly symmetrical molecular structure and strong interchain hydrogen bonding forces, which lead to a highly ordered crystalline form that is difficult to degrade naturally. The emergence of polyamide 4 (PA4) has changed this situation. Hashimoto et al. (1994) first explored the biodegradability of PA4 in soil in 1994 and found that PA4 degraded in farm soil (composted for over 10 years) and completely disappeared within 4 months, while other polyamides still existed in the soil after nearly 15 months. By measuring the molecular weight, it was found that the molecular weight of PA4 remained almost unchanged during the degradation process, indicating that it was biodegraded rather than hydrolyzed. When ordinary campus soil (not composted for over 10 years) was used for degradation testing, PA4 could not be degraded, indicating that the microorganisms capable of degrading PA4 are selective and specific. Recently a highly efficient single bacterium named strain NR4 for PA4 degradation was isolated from the soil enriched and acclimated by PA4 films. A mechanism for PA4 biodegradation was explained by relevant enzyme-coding genes and metabolic pathways (Wang et al. 2023). Further research found that PA4 could be completely degraded in activated sludge within 28 days (Hashimoto et al. 2002), much faster than in soil, and the degradation process was accompanied by a large amount of CO_2 production, which is completely in line with the current requirements for biodegradable polymers.

Further research found that PA4 showed good biodegradability in natural environments such as soil, compost, activated sludge, and seawater. In seawater, the degradation rate of PA4 was the same as that of polyester PHB (Tachibana et al. 2013), and as the incubation time in seawater increased, a biofilm gradually formed on the surface of the PA4 membrane, indicating that microorganisms were growing on the PA4 membrane. Relative polyamide 4-degrading enzymes and structural properties have been reported in these years (Saito et al. 2024; Saito et al. 2023). To understand the degradation mechanism of PA4 more deeply, *Pseudomonas sp.* ND-10, ND-11 (Yamano et al. 2008) and *Pseudoxanthomonas* sp. TN-N1 (Yurika et al. 2022), which have obvious degradation effects on PA4, were screened and isolated from activated sludge. They were inoculated on a basic medium with PA4 as the only carbon source. By analyzing the degradation products, it was confirmed that GABA was produced as an intermediate degradation product under the action of this strain, and the final degradation products were CO_2 and NO_3^-, indicating that the strain degrades PA4 by using its extracellular enzyme to hydrolyze the amide bond. A strain of PA4-degrading bacteria MND-1 was isolated from seawater, which belongs to the family *Alteromonadaceae*. This strain can degrade PA4 alone without other microorganisms. The degradation products contain GABA and low molecular weight GABA, which are the same as the degradation products of PA4 in soil and activated sludge. This means that PA4 has the same degradation mechanism in seawater, soil, and activated sludge. Existing research shows that the excellent biodegradability of PA4 is related to its hydrophilicity. When PA4 is modified by block, copolymerization, grafting, and other structural modification methods, its biodegradability will decrease with the decrease of its hydrophilicity.

So far, the PA4-degrading bacterial strains found in the natural environment include *Fusarium* sp., *Stenotrophomonas* sp., and *Pseudomonas* sp. isolated from soil. *Pseudoxanthomonas* sp., *Pseudomonas* sp. isolated from activated sludge, and *Aestuariibacter* sp. and *Alteromonadaceae* sp. isolated from seawater. Sasanami et al. (Yurika et al. 2022; Yamano et al. 2017) successfully isolated and purified the PA4 degradation enzyme from PA4-rich Pseudomonas broth, which was confirmed to be a metalloproteinase with low activity against casein, showing optimal activity at 35 °C and pH 7, and can hydrolyze PA4 into GABA oligomers.

Nakayama et al. (Yamano et al. 2017) reported on the degradation of PA4 in organisms. They implanted PLA nonwoven fabric, PA4 nonwoven fabric, PA4 film, and PA4 injection-molded samples subcutaneously in rats, and after a month, they recovered the samples and measured the molecular weight, finding that PA4 degrades more easily in the body than PLA. Copolymers composed of PA4 and polycaprolactone also degraded in mice. In addition, PA4 implanted in mice did not affect the surrounding tissues, and its safety assessment did not show mutagenicity or cytotoxicity, proving that PA4 has good biocompatibility.

Although it is currently possible to attribute the biodegradation of PA4 to the hydrolysis of extracellular enzymes produced by certain bacterial groups in the natural environment, more information about the degradation enzymes is still unknown. At the same time, whether conditions such as light and heat in the natural environment have a certain catalytic effect on the degradation of PA4 is a question that needs to be answered in the future.

9.5 Life Cycle Assessment

9.5.1 Origin and Development of Life Cycle Assessment (LCA)

Life cycle assessment (LCA) is an important strategy currently used for analyzing the environmental performance of materials. The application of LCA in the field of materials plays a huge guiding and promoting role in the sustainable development of the material industry, improving the efficiency of resource and energy use, and reducing pollutant emissions. Because the resource utilization rate of biodegradable materials is high and the environmental pollution is small, it is conducive to environmental protection and sustainable utilization of resources. Therefore, LCA is of great significance to biodegradable materials.

The concept of LCA originated in the late 1960s (Bjørn et al. 2018), which was mainly from the perspective of protecting raw materials and energy, using various methods to calculate and analyze the supply and consumption of resources and energy. In 1969, the Midwest Research Institute (MRI) in the United States evaluated beverage packaging bottles for the Coca-Cola Company, and first applied LCA to the comprehensive evaluation of resources, energy, and environmental impacts (Cays and Cays 2021). Subsequently, other companies in the United States and Europe also carried out assessment of a number of products, such as packaging materials and containers (Hunt and Franklin 1996). In the early 1970s, most of life cycle assessments were carried out in the United Kingdom, Germany, and Sweden (Fink 1997). After the energy crisis in the mid-1970s, the governments of the United States, the United Kingdom, and other countries conducted a large number of studies on industrial production energy analysis (Jensen et al. 1997). In the late 1970s, research on LCA was not limited to energy flow analysis. By the mid-1980s, the development of LCA entered a climax, during which the assessment methods of environmental impacts made substantial progress, such as the threshold value method independently formed in Switzerland and the Netherlands, and the environmental priority method developed in Switzerland (Xiance 2010).

From the late 1980s to the early 1990s, the deepening understanding of resources and environmental benefits led to the development of the life cycle inventory analysis method (Guinée et al. 2011). In the 1990s, many issues in life cycle assessment were unified and recognized. In August 1990, at the International Environmental Toxicology and Chemistry Society (society of environment toxicology and chemistry, SETAC), participants discussed and studied the concept and theoretical framework of LCA in detail, and finally determined the concept of "life cycle assessment," thereby unifying international LCA research and making LCA research enter a stage of standardized and stable development (Activities of the LCA-Society of Japan (JLCA) and Development Towards the Future 1998; Nobuhiko Narita etc 2004). Due to the establishment of standardization, LCA has developed in multiple countries, including Japan, Spain, Italy, etc. (Xiance 2010).

From the two explanations of LCA, both defined in 1993 (Zuoren and Zhihong 2004) and 1997 (ISO International Standard 14040 1997), it can be seen that the definition of LCA has been continuously improved and enriched to easy to understand and apply. However, the basic ideas and methods are consistent (Yamano et al. 2017). The life cycle refers to a series of interconnected stages in the product system, from the generation of natural resources or raw materials to final disposal (ISO International Standard 14040 1997). The 1997 ISO 14040 standard defined the LCA evaluation process includes four components: determining the purpose and scope, inventory analysis, impact assessment, and result interpretation (Xiance 2010; ISO International Standard 14040 1997). In 2003, an ad-hoc group revised the standards

ISO 14040, 14041, 14042, and 14043 to improve readability. They maintained 14040 as a framework document, organized other requirements into new files. Thus, a new standard 14044 emerged and contained all requirements ("Environmental management—Life cycle assessment—Requirements and Guidelines") (Finkbeiner et al. 2006).

9.5.2 Significance of Life Cycle Assessment (LCA)

Because LCA involves the evaluation of the whole chain of products and contains many links, it has attracted the attention and application of various fields of human society. It brings new requirements and content to our economic and social operation, sustainable development strategy, and environmental management system. Its significance lies in the following three aspects (Zuoren and Zhihong 2004): Industrial enterprise sectors, management departments and consumers departments, management departments and consumers. In the industrial sector, LCA is beneficial for enterprises to identify and diagnose the ecology of product systems, evaluate and compare the impacts of product, evaluate the production process according to life cycle of the product, including environmental and energy resource factors, effectively reduce the impact on resources, energy and environment, and improve production efficiency, improve production process and reduce production costs. Government environmental management departments and international organizations apply LCA to focus on the interaction between related products, and uses LCA in various management agencies of society to supervise the production of products from different social levels and promote the sustainable development of the economy, society and environment. The use of LCA thinking help understand the relationship between industrial structure and social environmental problems, thus providing a basis for the formulation of regional and industry development policies; Finally, at the consumer level, it not only increases consumers' comprehensive understanding of the product, but also improves the quality of consumers, and comprehensively considers the source and future destination of the product in the process of selecting consumer goods. We can choose more rationally those products and companies that have less impact on resources, energy, and the environment (Xiance 2010).

Although the application of LCA is of great significance to human society, it is not perfect. At present, LCA has certain limitations and shortcomings in its research theory and practical application. It contains the following aspects: First, the objectivity problem, since the quantitative method of LCA evaluation will have an impact on the evaluation results; it is an inherent limitation whether the factors of each link of the product can be objectively presented. In addition, due to the limited information and data at each stage of the life cycle, many data, especially lack of the data of typical production, resulting in incomplete presentation of the whole research process; Second, the limitations of the calculation model, the method of converting inventory data into environmental damage is usually very complex and uncertain, and the assumptions of the life cycle inventory analysis or environmental impact evaluation model may not apply to some potential impacts or applications; Thirdly, the limitations of time and region, in different time and regional ranges, due to production technology, resources and energy structure and other factors, the same product gets different environmental inventory data, with the update of production technology and process, the evaluation results obtained also have reference value only for a certain period of time (Xiance 2010).

9.5.3 Case Analysis

1. *Life Cycle Assessment of Polyvinyl Alcohol* (Parascanu et al. 2021)

Organic aerogels, valued for their low density, thermal insulation, and porosity, are emerging as key materials in construction, energy, and related fields. While multiple synthesis methods exist, freeze-drying stands out as a cost-effective, eco-friendly, and scalable approach. This case conducted a gate-to-gate LCA comparing nanoclay-reinforced polyvinyl alcohol (PVA) aerogel production via freeze-drying at laboratory (LS) and pilot line (PL) scales. Key variables include functional units (1 kg, 1 m^3/m^2 adjusted for aerogel thickness), energy efficiency, and production capacity upgrades in the PL. The environmental impacts of primary drying electricity consumption, and process optimizations were systematically evaluated (Parascanu et al. 2021).

For the PL, upgrades in production (PLP) and energy efficiency (PLE) achieved 35% energy reduction and full-capacity operation. Ten scenarios were analyzed using two functional units (1 kg and 1 m^3 aerogel), with aerogels synthesized from PVA (2.5 wt.%)-bentonite (0.25 wt.%) hydrogel via a three-step freeze-drying cycle (freezing, primary/secondary drying). System boundaries, validated by primary data, focused solely on production stages at Spain's ITQUIMA, excluding raw material transport, equipment, and end-of-life phases (Parascanu et al. 2021).

For the PL, four kinds of upgrade scenarios were analyzed: (i) production upgrade (PLP), (ii) energy efficiency upgrade (PLE), (iii) combined upgrades, and (iv) no upgrades, resulting in ten total scenarios (Table 9.2) (Parascanu et al. 2021).

Background processes (BPs)—nonspecific, indirectly influenced activities supporting foreground operations encompass sub-processes like raw material extraction, facility dismantling, waste treatment, and supply-chain emissions. PVA simulated synthesis was applied to aerogel production, where lab/pilot-scale plants quantified yields, resource use, and emissions (Tables 9.3 and 9.4). Key outputs included steam (air emission) and wastewater from condenser cooling (tap water used during freezing), while distilled water for hydrogel preparation was experimentally quantified (Parascanu et al. 2021).

The Life cycle impact assessment (LCIA) converts inventory data into impact indicators (e.g., resource depletion, environmental damage) using methodology-aligned categories: abiotic depletion (ADP), ADP-fossil fuels (ADP-FF), global

Table 9.2 Description of the scenarios considered in the LCA for aerogel production (Parascanu et al. 2021)

Scenario	Description	Abbreviated name
1	1 kg of aerogel produced in a laboratory scale	LSkg
2	1 kg of aerogel produced in the pilot line, without upgrading	PLkg
3	1 kg of aerogel produced in the pilot line, upgrading only energy	PLEkg
4	1 kg of aerogel produced in the pilot line, upgrading only production	PLPkg
5	1 kg of aerogel produced in the pilot line, upgrading both production and energy	PLPEkg
6	1 m^3 of aerogel produced in the laboratory scale	LSm^3
7	1 m^3 of aerogel produced in the pilot line, without upgrading	PLm^3
8	1 m^3 of aerogel produced in the pilot line, upgrading only energy	PLEm^3
9	1 m^3 of aerogel produced in the pilot line, upgrading only production	PLPm^3
10	1 m^3 of aerogel produced in the pilot line, upgrading both production and energy	PLPEm^3

Table 9.3 Input and output data to produce 1 kg of aerogel panel at laboratory and pilot line scale for all the scenarios considered (experimental analysis) (Parascanu et al. 2021)

		LSkg	PLkg	PLEkg	PLPkg	PLPEkg
Input						
PVA	kg	1.31E+00	3.90E-01	3.90E-01	7.10E-01	7.10E-01
Distilled water	kg	5.23E+01	1.55E+01	1.55E+01	2.85E+01	2.85E+01
Bentonite	kg	3.30E-03	1.00E-03	1.00E-03	1.80E-03	1.80E-03
Tap water	kg	n.m.	4.50E+03	2.00E+03	8.50E+03	4.00E+03
Energy	kWh	1.37E+04	3.53E+02	2.43E+02	7.02E+02	4.81E+02
Emissions to the environment						
Steam	kg	5.27E+01	1.49E+01	1.49E+01	2.82E+01	2.82E+01
Waste water	kg	n.m.	4.50E+03	2.00E+03	8.50E+03	4.00E+03

n.m. = not measured

warming (GWP), ozone depletion (ODP), photochemical oxidation (POCP), acidification (AP), and eutrophication (EP). Energy demand is assessed via Cumulative Energy Demand (CED) V1.10, distinguishing renewable/nonrenewable sources based on fuel higher heating values. Results are presented as characterized values and normalized values, enabling quantification of product/system impacts across metrics (Tables 9.5 and 9.6) (Parascanu et al. 2021).

In nearly all impact categories, electricity production/consumption is the most environmentally detrimental factor, followed by tap/distilled water, and bentonite usage. The scenarios ranked by highest environmental impact are

Table 9.4 Input and output data to produce 1 m^3 of aerogel panel at laboratory and pilot line scale for all the scenarios considered (experimental analysis) (Parascanu et al. 2021)

		LSm^3	PLm^3	$PLEm^3$	$PLPm^3$	$PLPEm^3$
Input						
PVA	kg	2.50E+01	2.49E+01	2.49E+01	2.51E+01	2.51E+01
Distilled water	kg	1.00E+03	9.95E+02	9.95E+02	1.00E+03	1.00E+03
Bentonite	kg	6.20E-02	6.22E-02	6.22E-02	6.28E-02	6.28E-02
Tap water	kg	n.m.	1.57E+05	7.40E+04	1.83E+05	8.60E+04
Energy	kWh	2.61E+05	1.37E+04	9.61E+03	1.57E+04	1.09E+04
Emissions to the environment						
Steam	kg	1.01E+03	9.53E+02	9.53E+02	9.93E+02	9.93E+02
Wastewater	kg	n.m.	1.57E+05	7.40E+04	1.83E+05	8.60E+04

n.m. = not measured

Table 9.5 Comparison between different functional units (FU) (kg vs. m^3) to determine the environmental impacts of aerogel panel production at laboratory and pilot line scale (characterization method) (Parascanu et al. 2021)

Impact category	Unit	LSkg	LSm^3	PLkg	PLm^3
ADP	kg Sb eq	2.78E-04	5.30E-03	8.99E-06	3.55E-04
ADP (FF)	MJ	5.00E+04	9.52E+05	1.37E+03	5.54E+04
GWP	kg CO_2 eq	4.83E+03	9.21E+04	1.29E+02	5.09E+03
ODP	kg CFC-11 eq	6.10E-04	1.16E-02	1.61E-05	6.29E-04
POCP	kg C_2H_4 eq	1.39E+00	2.64E+01	3.75E-02	1.50E+00
AP	kg SO_2 eq	3.71E+01	7.07E+02	9.72E-01	3.79E+01
EP	kg PO_4^{3-} eq	2.56E+00	4.88E+01	6.79E-02	2.67E+00

Table 9.6 Comparison between different FU (kg vs. $m^{3)}$ to determine the environmental impacts of aerogel panel production at laboratory and pilot line scale (normalization method) (Parascanu et al. 2021)

Impact category	LSkg	LSm^3	PLkg	PLm^3
ADP	3.28E-12	6.25E-11	1.06E-13	4.19E-12
ADP (FF)	1.59E-09	3.03E-08	4.37E-11	1.76E-09
GWP	9.62E-10	1.83E-08	2.56E-11	1.01E-09
ODP	6.84E-12	1.30E-10	1.80E-13	7.04E-12
POCP	1.64E-10	3.12E-09	4.43E-12	1.77E-10
AP	1.32E-09	2.51E-08	3.45E-11	1.35E-09
EP	1.94E-10	3.70E-09	5.15E-12	2.02E-10

LSm^3 > PLm^3 > LSkg > PLkg, attributed to material density/thickness differences. The most affected categories—ADP (FF), AP, and GWP—are linked to both direct electricity use and background processes. These processes generate emissions exacerbating the aforementioned impacts. Scalability from lab to pilot scale reduces environmental impacts, primarily due to decreased heat loss during production.

2. *Life Cycle Assessment of Polylactic Acid* (Zhao et al. 2022)

This case aimed to evaluate the energy consumption and environmental effects by analyzing energy flows of PLA food plastic packaging over the total life cycle. By comparing with traditional PE materials, the most impact life cycle phase was identified, and the optimal strategy for reducing energy use and environmental pollution was determined. Total life cycle stages, from raw material procurement to final disposal, were considered. Based on Meng's 2010 PLA life cycle evaluation, this research set 1 metric ton of PLA food box plastic as the functional unit for evaluation. To estimate energy consumption and environmental impacts, this LCA was determined through a comprehensive analysis of data from raw material acquisition to final processing, considering the entire life cycle of PLA plastic packaging. The system boundary covered the whole process, from raw material acquisition to disposal. Figure 9.11 shows the five-stage life cycle flowchart: raw material acquisition, transportation, production, product use, and final disposal (Zhao et al. 2022).

This case utilized corn from the Liaoxi region in Northeast China, collecting computational data encompassing all information from planting to harvest stages. The production process of PLA plastic packaging followed the flowchart shown in Fig. 9.12. The PLA particles can be processed by extrusion. Subsequently, the final PLA plastic packaging will be processed by blow molding or injection molding (Zhao et al. 2022).

The production processes for all products are centralized within TEDA's fabrication plant. Inter-process product transfer is mechanized, with each step characterized by energy consumption and waste generation (Zhao et al. 2022).

Product use: This case considered only the impact of transportation, ignoring the impact of the product itself. In the use stage, the transportation is from the fabrication plant to a central urban distribution point (Zhao et al. 2022).

Final disposal: By the slow action of soil microorganisms, PLA plastic packaging products degrade into water and CO_2 in landfills. However, the natural degradation rate of PLA in soil is relatively low. Assuming complete combustion generates only carbon dioxide and water, the relevant data obtained from the LCA database of Sichuan University are integrated in Table 9.7 (Zhao et al. 2022).

The case employed the CML2001 method to analyze the environmental impacts of LCA. Seven impact categories were evaluated: abiotic depletion potential (ADP fossil fuels) in MJ equivalents, global warming potential (GWP) in kilograms of CO_2 equivalents, acidification potential (AP) in kilograms of SO_2 equivalents, eutrophication potential (EP) in kilograms of PO_4^{3-} equivalents, photochemical ozone formation potential (POCP) in kilograms of C_2H_4 equivalents, and human toxicity potential (HT) and terrestrial ecotoxicity potential (TETP) in kilograms of

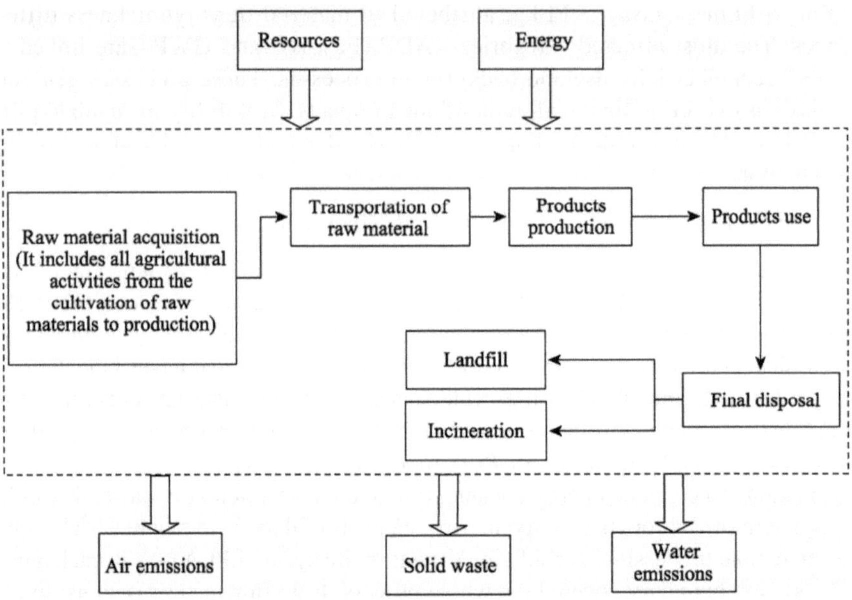

Fig. 9.11 LCA system boundary for PLA plastic packaging products (Zhao et al. 2022). (Reprinted with permission from JRE Editorial Office and the corresponding author of this paper. *Journal of Resources and Ecology* 13.3 (2022): 428–441)

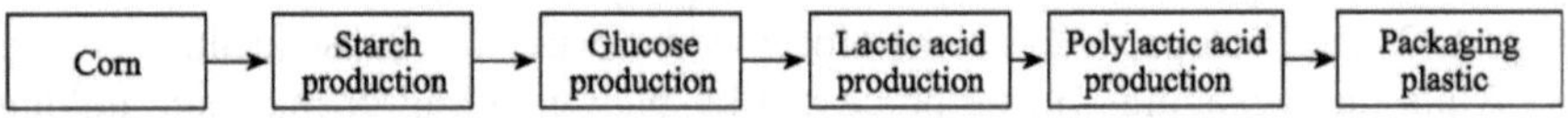

Fig. 9.12 Production flowchart of PLA plastic packaging (Zhao et al. 2022). (Reprinted with permission from JRE Editorial Office and the corresponding author of this paper .*Journal of Resources and Ecology* 13.3 (2022): 428–441)

(1,4)-DCB equivalents. Characterization transforms data from the life cycle inventory into environmental impact indicators based on predefined classifications. Emissions are multiplied by their respective characterization factors and then summed to determine the total impact. Relevant environmental and mineral impacts are detailed in Table 9.8 (Zhao et al. 2022).

To evaluate the relative significance of environmental impacts, the characterization results are normalized. Through standardized and weighting procedures, the environmental impact metrics are converted into globally comparable units across all LCA categories. This is achieved by dividing the corresponding reference values by the characterization results (Zhao et al. 2022).

The LCA results indicated that the production stage has the most significant environmental impact. The seven impact categories, ranked by environmental impact magnitude, are: GWP > ADP > HTP > AP > TETP > EP > POCP, as shown in Table 9.9 (Zhao et al. 2022).

Table 9.7 LCA inventory of one ton PLA plastic packing products (ISO14040:2006) (unit: kgt^{-1}) (Zhao et al. 2022)

Type	Substances	Life cycle stages						
		Raw material acquisition	Transportation of raw material	Product production	Product use	Final disposal		Total
						Landfill	Incineration	
Resources	Coal	2.05E+02	1.99E+00	2.88E+03	1.30E-01	7.75E-02	9.48E-03	3.08E+02
	Crude oil	1.61E+01	5.50E+01	4.84E+02	3.65E+00	2.14E+00	2.60E-01	5.61E+02
	Natural gas[a]	1.59E+00	3.31E+03	3.53E+01	2.20E-04	1.29E-04	1.58E-05	3.68E+01
Air emissions	CO_2[b]	-4.15E+03	0.00E+00	0.00E+00	0.00E+00	0.00E+00	0.00E+00	-4.15E+03
	CO_2	1.59E+03	1.40E+02	2.84E+03	7.42E+00	9.12E+00	5.91E+02	5.18E+03
	CO	9.10E-01	7.40E-01	1.28E+01	9.30E-01	5.40E-01	6.64E-02	1.60E+01
	CH_4	5.90E-01	1.43E+02	1.19E+01	1.95E-03	2.26E+00	1.40E-04	1.48E+01
	N_2O	2.17E+01	4.05E+03	0.00E+00	5.40E-04	3.17E-04	3.88E-05	2.17E+01
	SO_2	2.24E+00	1.50E-01	4.16E+01	4.97E-03	2.92E-03	3.57E-04	4.40E+01
	NOx	6.31E+00	3.15E+00	3.42E+01	6.40E-02	3.75E-02	4.59E-03	4.38E+01
	SOx	4.12E+00	0.00E+00	0.00E+00	0.00E+00	0.00E+00	0.00E+00	4.12E+00
	PO_4^{3-}	3.50E-01	0.00E+00	0.00E+00	0.00E+00	0.00E+00	0.00E+00	3.50E-01
	NO_3^-	4.76E+00	0.00E+00	0.00E+00	0.00E+00	0.00E+00	0.00E+00	4.76E+00
	NH_3	4.23E+00	0.00E+00	0.00E+00	0.00E+00	0.00E+00	0.00E+00	4.23E+00
	NMVOC	1.09E-02	5.60E-01	1.49E+00	1.80E-01	1.10E-01	1.29E-02	2.35E+00
	Dust	3.00E-01	1.47E+02	4.00E-01	2.53E+01	1.48E+01	1.81E+00	2.29E+02

(continued)

Table 9.7 (continued)

Type	Substances	Life cycle stages						
		Raw material acquisition	Transportation of raw material	Product production	Product use	Final disposal		Total
						Landfill	Incineration	
Water emissions	Cu	1.61E+00	0.00E+00	0.00E+00	0.00E+00	0.00E+00	0.00E+00	1.61E+00
	Zn	2.31E+00	0.00E+00	0.00E+00	0.00E+00	0.00E+00	0.00E+00	2.31E+00
	Cd	3.30E-01	0.00E+00	0.00E+00	0.00E+00	0.00E+00	0.00E+00	3.30E-01
	Pb	1.76E-02	0.00E+00	0.00E+00	0.00E+00	0.00E+00	0.00E+00	1.76E-02
	Waste liquid	5.47E+01	3.21E+01	6.08E+03	2.14E+00	1.25E+00	1.50E-01	6.17E+03
	Waste sludge	3.44E+00	3.20E-01	4.46E+02	2.15E-03	1.26E-02	1.54E-03	4.50E+02

Note: [a]Natural gas's unit is $m^3\ t^{-1}$; [b]CO_2 refers to the amount of carbon dioxide that corn absorbs from the atmosphere through photosynthesis. *NMVOC* non-methane volatile organic compounds

Reprinted with permission from JRE Editorial Office and the corresponding author of this paper .*Journal of Resources and Ecology* 13.3 (2022): 428–441

Table 9.8 The characterization factors of environmental impact (ISO14040:2006) (Zhao et al. 2022)

Impact category characterization factor	ADP	GWP	AP	EP	POCP	HTP	TETP
Coal	27.91	-	–	–	–	–	–
Crude oil	41.87	–	–	–	–	–	–
Natural gas	38.84	–	–	–	–	–	–
CO_2	–	1	–	–	–	–	–
CO	–	2	–	–	2.70E-02	–	–
CH_4	–	21	–	–	7.00E-03	–	–
N_2O	–	3.10E+02	–	0.27	–	–	–
SO_2	–	–	1	–	–	9.60E-02	–
NOx	–	–	0.7	0.13	2.80E-02	1.2	–
SOx	–	–	0.8	–	–	–	–
PO_4^{3-}	–	–	–	1	–	–	–
NO_3^-	–	–	–	0.42	–	–	–
NH_3	–	–	1.88	0.33	–	–	–
NMVOC	–	–	–	–	0.42	–	–
Dust	–	–	–	–	–	0.82	–
Cu	–	–	–	–	–	–	14
Zn	–	–	–	–	–	–	25
Cd	–	–	–	–	–	–	1.70E+02
Pb	–	–	–	–	–	–	33

Note: *ADP* abiotic depletion potential of fossil fuels, *GWP* global warming potential, *AP* acidification potential, *EP* eutrophication potential, *POCP* photochemical ozone formation potential, *HTP* human toxicity potential, *TETP* terrestrial ecotoxicity potential

Reprinted with permission from JRE Editorial Office and the corresponding author of this paper .*Journal of Resources and Ecology* 13.3 (2022): 428–441

Table 9.9 The detailed environmental impacts of each stage (Zhao et al. 2022)

Impact category		Raw material acquisition	Transportation of raw material	Product production	Product use	Final disposal	Total
ADP	Normalization result	1.70E-11	6.21E-12	2.68E-10	4.12E-13	2.71E-13	2.92E-10
	Environmental impact value	1.95E-12	7.14E-13	3.09E-11	4.74E-14	3.12E-14	3.36E-14
	Percent (%)	5.81E-00	2.13	91.82	0.14	0.09	100
GWP	Normalization result	9.92E-11	3.39E-12	7.38E-11	2.24E-13	1.54E-11	1.92E-10
	Environmental impact value	2.28E-11	7.79E-13	1.70E-11	5.16E-14	3.53E-12	4.42E-11
	Percent (%)	51.67	1.76	38.45	0.12	8	100

(continued)

Table 9.9 (continued)

Impact category		Raw material acquisition	Transportation of raw material	Product production	Product use	Final disposal	Total
AP	Normalization result	7.50E-11	9.86E-12	2.75E-10	2.08E-13	1.37E-13	3.60E-10
	Environmental impact value	3.00E-12	3.94E-13	1.10E-11	8.34E-15	5.49E-15	1.44E-11
	Percent (%)	20.84	2.74	76.33	0.06	0.04	100
EP	Normalization result	6.59E-11	2.59E-12	2.81E-11	5.35E-14	3.52E-14	9.67E-11
	Environmental impact value	4.61E-12	1.82E-13	1.97E-12	3.74E-15	2.46E-15	6.77E-12
	Percent (%)	68.15	2.68	29.07	0.06	0.04	100
POCP	Normalization result	5.70E-12	9.25E-12	5.44E-11	2.75E-12	2.24E-12	7.44E-11
	Environmental impact value	2.85E-13	4.63E-13	2.72E-12	1.38E-13	1.12E-13	3.72E-11
	Percent (%)	7.67	12.44	73.18	3.7	3.01	100
HTP	Normalization result	3.11E-12	4.81E-11	3.02E-11	8.06E-12	5.31E-12	9.48E-11
	Environmental impact value	6.23E-13	9.62	6.04E-12	1.61E-12	1.06E-12	1.90E-11
	Percent (%)	3.29	50.75	31.86	8.5	5.6	100
TETP	Normalization result	1.25E-10	0	0	0	0	1.25E-10
	Environmental impact value	1.37E-11	0	0	0	0	1.37E-11
	Percent (%)	100.00	0	0	0	0	100

Note: *ADP* abiotic depletion potential fossil fuels, *GWP* global warming potential, *AP* acidification potential, *EP* eutrophication potential, *POCP* photochemical ozone formation potential, *HTP* human toxicity potential, *TETP* terrestrial ecotoxicity potential.
Reprinted with permission from JRE Editorial Office and the corresponding author of this paper .*Journal of Resources and Ecology* 13.3 (2022): 428–441

References

Activities of the LCA-Society of Japan (JLCA) and Development Towards the Future (1998) Int J LCA 3(2):69–70

Adi S, Omer Y, Ayelet F (2024) Enzymatic degradation of polylactic acid (PLA). Appl Microbiol Biotechnol 108(1):413

Akutsu-Shigeno Y, Teeraphatpornchai T, Teamtisong K et al (2003) Cloning and sequencing of a poly(DL-lactic acid) depolymerase gene from *Paenibacillus amylolyticus* strain TB-13 and its functional expression in *Escherichia coli*. Appl Environ Microbiol 69(5):2498–2504

AS 5810-2010 (2010) Biodegradable plastics—biodegradable plastics suitable for home composting, Standards Australia

Bjørn A, Owsianiak M, Molin C et al (2018) LCA history. Life cycle assessment: theory and practice, pp 17–30

Bocchini S, Fukushima K, Blasio AD et al (2010) Polylactic acid and polylactic acid-based nanocomposite photooxidation. Biomacromolecules 11:2919–2926

Cao Y, Ma W, Chen S et al (2025) Immobilization of Proteinase K into PLA for self-biodegradable. Int J Biol Macromol 304:140873

Cays J, Cays J (2021) Do something: mid-twentieth century developments. In: An environmental life cycle approach to design: LCA for designers and the design market, pp 61–78

Ci S (2004) Study on the ultraviolet mutagenesis of DS9713a strain and the enzymatic properties of its PHB depolymerase. Northeast Normal University, Changchun

Dong L, Lei W, Xia M et al (2005) Biodegradation of polyvinyl alcohol. Chin J Biotechnol 25(7):28–33

Elsawy MA, Kim KH, Park JW et al (2017) Hydrolytic degradation of polylactic acid (PLA) and its composites. Renew Sust Energ Rev 79:1346–1352

Fink P (1997) The roots of LCA in Switzerland—continuous learning by doing. Int J LCA 2(3):131–134

Finkbeiner M, Inaba A, Tan RBH et al (2006) The New International Standards for Life Cycle Assessment: ISO 14040 and ISO 14044. Int J LCA 11(2):80–85

GB/T 1040-2006 (2006) Plastic—determination of tensile properties—Part 2: test conditions for moulding and extrusion plastic, General Administration of Quality Supervision, Inspection and Quarantine; Standardization Administration Committee

GB/T 19276.1-2003 (2003) Determination of the ultimate aerobic biodegradability of plastic materials in an aqueous medium—method by measuring the oxygen demand in a closed respirometer, Administration of Quality Supervision, Inspection and Quarantine of People's Republic of China; Standardization Administration of China

GB/T 19276.2-2003 (2003) Determination of the ultimate aerobic biodegradability of materials in aqueous medium—Method by analyzing released carbon dioxide. General Administration of Quality Supervision, Inspection and Quarantine of the People's Republic of China, China Standards Press

GB/T 19277.1-2011 (2011) Determination of the ultimate aerobic biodegradability of materials under controlled composting conditions using methods for the determination of released carbon dioxide Part 1: general methods, Administration of Quality Supervision, Inspection and Quarantine of People's Republic of China; Standardization Administration of China

GB/T 19277.2-2013 (2013) Determination of the ultimate aerobic biodegradability of plastic materials under controlled composting conditions .method by analysis of evolved carbon dioxide Part 2: gravimetric measurement of carbon dioxide evolved in a laboratory-scale test, Administration of Quality Supervision, Inspection and Quarantine of People's Republic of China; Standardization Administration of China

GB/T 19811-2005 (2005) Determination of the degree of disintegration of plastic materials under defined composting conditions in a pilot-scale test, General Administration of Quality Supervision, Inspection and Quarantine of the People Republic of China, China National Standardization Administration Committee

GB/T 22047 (2009) Plastics—determination of the ultimate aerobic biodegradability in soil by measuring the oxygen demand in a respirometer or the amount of carbon dioxide evolved, General Administration of Quality Supervision, Inspection and Quarantine of the People's Republic of China, China National Standardization Administration

GB/T 28206 (2011) Specification for compostable plastic, General Administration of Quality Supervision, Inspection and Quarantine of the People's Republic of China, China National Standardization Administration

GB/T 32106-2015 (2015) Plastics—determination of the ultimate anaerobic biodegradation of plastic materials in an aqueous system—method by measurement of biogas production, General Administration of Quality Supervision, Inspection and Quarantine of the People's Republic of China, AQSIQ;SAC

GB/T 33797-2017 (2017) Plastics—determination of the ultimate anaerobic biodegradation under high-solids anaerobic-digestion conditions—method by analysis of released biogas [S]. General Administration of Quality Supervision, Inspection and Quarantine of the People's Republic of China, National Standardization Administration of China. China Standards Press, Beijing

GB/T 38737-2020 (2020) Plastics—determination of the ultimate anaerobic biodegradation of materials in a controlled sludge digestion system by measuring the released biogas [S]. State Administration for Market Regulation, National Standardization Administration. China Standards Press, Beijing

Grause G, Chien M-F, Inoue C (2020) Changes during the weathering of polyolefins. Polym Degrad Stab 181:109364

Guinée JB, Heijungs R, Huppes G, Zamagni A, Masoni P, Buonamici R, Ekvall T, Rydberg T (2011) Life cycle assessment: past, present, and future

Hashimoto K, Hamano T, Okada M (1994) Degradation of several polyamides in soils. J Appl Polym Sci 54(10):1579–1583

Hashimoto K, Sudo M, Ohta K et al (2002) Biodegradation of nylon4 and its blend with nylon6. J Appl Polym Sci 86(9):2307–2311

Hunt RG, Franklin WE (1996) LCA-how it came about: personal reflection on the original and the development of LCA in the USA. Int J LCA 1(1):4–7

Iñiguez-Franco F, Auras R, Rubino M et al (2017) Effect of nanoparticles on the hydrolytic degradation of PLA-nanocomposites by water-ethanol solutions. Polym Degrad Stab 146:287–297

ISO 18830:2016 (2016) Plastics—determination of aerobic biodegradation of non-floating plastic materials in a seawater/sandy sediment interface—method by measuring the oxygen demand in closed respirometer. ISO, Geneva

ISO 19679 (under Development ISO/DIS 19679, 2019) (2016) Plastics—determination of aerobic biodegradation of non-floating plastic materials in a seawater/sediment interface—method by analysis of evolved carbon dioxide. International Organization for Standardization, Geneva

ISO 19679 (under Development ISO/DIS 19679, 2019) (2019) Plastics—determination of aerobic biodegradation of non-floating plastic materials in a seawater/sediment interface—method by analysis of evolved carbon dioxide

ISO 22404:2019 (2019) Plastics—determination of the aerobic biodegradation of non-floating materials exposed to marine sediment—method by analysis of evolved carbon dioxide

ISO International Standard 14040 (1997) Environmental management—life cycle assessment—principles and framework. International Organisation for Standardisational (ISO), Geneva

Jensen AA, Elkington J et al (1997) Life Cycle Assessment (LCA): a guide to approaches, experiences and information sources. In: Report to the European Environment Agency, Copenhagen

Kawai F (2010) Polylactic acid (PLA)-degrading microorganisms and PLA depolymerases, green polymer chemistry: biocatalysis and biomaterials. ACS Symp Ser 1043:405–414

Kawai F, Nakadai K, Nishioka E et al (2011) Different enantioselectivity of two types of poly(lactic acid) depolymerases toward poly(l-lactic acid) and poly(d-lactic acid). Polym Degrad Stab 96(7):1342–1348

Linyu J, Siyuan H, Dan L, etc. (2020) Current status of degradable materials and their research progress in the marine field. Packag Eng 41(19):108–115

Lors C, Pauline L, Chung HP (2025) State of the art on biodegradability of bio-based plastics containing polylactic acid. Front Mater 11:1476484

Luckachan GE, Pillai C (2011) Biodegradable polymers-a review on recent trends and emerging perspectives. J Polym Environ 19(3):637–676

Maoying W (2010) Plastic degradation and stabilizer (III): thermo-oxidative degradation and thermo-oxidative stabilization. Plastic Additives 3:48–54

Maraveas C, Kyrtopoulos IV, Arvanitis KG et al (2024) The aging of polymers under electromagnetic radiation [J]. Polymers 16(5)

Mario O-R, Joaquim S-R, Cristopher DS et al (2025) Thermophilic anaerobic biodegradation of commercial polylactic acid products. Bioresour Technol 132296

Martin RT, Camargo LP, Miller SA (2014) Marine-degradable polylactic acid. Green Chem 16(4):1768–1773
Mejía GAI, López OBL, Mulet PA (1999) Biodegradation of poly (vinylalcohol) with enzymatic extracts of phanerochaete chrysosporium. Macromol Symp 148(1):131–147
Nobuhiko Narita etc (2004) Current LCA Database Development in Japan—results of the LCA Project. Int J LCA 9(6):355–359
Oda Y, Yonetsu A, Urakami T et al (2000) Degradation of polylactide by commercial proteases. J Polym Environ 8(1):29–32
OECD (1994) OECD guidelines for the testing of chemicals. OECD
Parascanu MM, Esteban-Arranz A, de la Osa AR et al (2021) Comparison of nanoclay/polyvinyl alcohol aerogels scale production: Life Cycle Assessment. Chem Eng Res Des 176:243–253
Qi X, Ren Y, Wang X (2017) New advances in the biodegradation of Poly(lactic) acid. Int Biodeterior Biodegradation 117:215–223
Rongqiu L (2016) Study on the degradation behavior and rapid detection method of polylactic acid film. Southwest University of Science and Technology, Chengdu
Saito Y, Honda M, Yamashita T et al (2023) Marine bacterial enzyme degrades polyamide 4 into gamma-aminobutyric acid oligomers. Polym Degrad Stab 215:110446
Saito Y, Jin I, Yamada M (2024) Marine bacteria have multiple polyamide 4-degrading enzymes[J]. J Gen Appl Microbiol 70(3):136–140
Shi Ke S, Tingting WZ (2019) Progress in the biodegradability of degradable plastic polylactic acid (PLA). Plastics 48(3):42–47
Sun S (2025) Enzyme-embedded biodegradable plastic for sustainable applications: advances, challenges, and perspectives. ACS Appl Bio Mater 8:1785–1796
Suzuki T (1973) Some characteristics of Pseudomonas O-3 which utilizes polyvinyl alcohol. Agric Biol Chem 37:747–756
Tachibana K, Urano Y, Numata K (2013) Biodegradability of nylon 4 film in a marine environment. Polym Degrad Stab 98(9):1847–1851
Tarazona NA, Machatschek R, Lendlein A (2020) Unraveling the Interplay between abiotic hydrolytic degradation and crystallization of bacterial polyesters comprising short and medium side-chain-length polyhydroxyalkanoates. Biomacromolecules 21(2):761–771
Wang Z (2001) Study on the degradation mechanism of polyhydroxyalkanoates (PHAs). Northwest University, Xi'an
Wang L, Zhang Z, Zhang D et al (2023) Exploring microbial degradation of polyamide 4 in soils: unveiling degradation mechanisms, pathways, and the contribution of strain NR4. J Clean Prod 429:139535
Wei S, Jun Z (2006) The mechanism and influencing factors of UV degradation of PVC. Synthetic Resins Plastics 23(4):80–84
Williams DF (1981) Enzymic hydrolysis of polylactic acid. Eng Med 10(1):5–7
Xiance M (2010) Life cycle assessment of polycarbonate and polylactic acid. Beijing University of Technology, Beijing
Yamano N, Nakayama A, Kawasaki N et al (2008) Mechanism and characterization of polyamide 4 degradation by Pseudomonas sp. J Polym Environ 16(2):141–146
Yamano N, Kawasaki N, Ida S et al (2017) Biodegradation of polyamide 4 in vivo. Polym Degrad Stab 137:281–288
Yangyang H, Guangming Z, Ben L (2019) Research progress on photodegradable polymers. China Plastics 33(6):132–138
Yanxia L, Xianguo L, Xiaoyong D et al (2012) Study on the photodegradation mechanism of nonylphenol in water. Chin Soc Environ Sci 70(17):1819–1826
Yujun P, Dawei L (2018) Discussion on bio-based degradable materials. Today's Print 303(12):64–67
Yurika S, Masayoshi H, Haruka N et al (2022) Purification and characterization of an enzyme that degrades polyamide 4 into gamma-aminobutyric acid oligomers from *Pseudoxanthomonas* sp. TN-N1. Polym Degrad Stab 197:109868

Zaaba NF, Jaafar M (2020) A review on degradation mechanisms of polylactic acid: hydrolytic, photodegradative, microbial, and enzymatic degradation. Polym Eng Sci 60:2061–2075

Zhang X, Liu C, Chen H, Song J-y (2024) Identification of a PVA-degrading strain of Fusarium oxysporum and its degradation performance. Acta Agric Jiangxi 36(10):47–53

Zhao M, Yang Z, Zhao J et al (2022) Life cycle assessment of biodegradable polylactic acid (PLA) plastic packaging products—taking Tianjin, China as a case study. J Resour Ecol 13(3):428–441

Zuoren N, Zhihong W (2004) Ecological environment material. Mechanical Industry Press, Beijing, p 5

Chapter 10
Recycling and Comprehensive Utilization of Materials

Mingda Wang, Ziying Wang, and Liming Zhao

Polymer materials are cheap, lightweight, and durable materials that can be easily made into various products and have been widely used. Primarily derived from non-renewable resources like petroleum and natural gas, they are utilized in the production of disposable packaging and short-lived products, often discarded within a year. Obviously, our current use of polymer materials is unsustainable, causing a series of environmental problems and waste of resources during use and disposal. Therefore, the recycling and reuse of polymer materials is an important way to reduce these negative effects, and it has become one of the most active research and development areas in the polymer materials industry.

This chapter will briefly discuss several recycling methods for polymer materials, including but not limited to the recycling and reuse of biobased materials. Worldwide, there are significant differences in the amount and method of recycling depending on the type and use of polymer materials. In the future, advances in collection, sorting, and reuse technologies and systems will create great opportunities for the recycling of recyclable polymer materials.

M. Wang · Z. Wang
State Key Laboratory of Bioreactor Engineering, East China University of Science and Technology, Shanghai, China

L. Zhao (✉)
State Key Laboratory of Bioreactor Engineering, East China University of Science and Technology, Shanghai, China

Key Laboratory of Biobased Material Engineering of China Light Industry, East China University of Science and Technology, Shanghai, China
e-mail: zhaoliming@ecust.edu.cn

L. Zhao (ed.), *Research and Applications of Bio-based Degradable Materials*,
https://doi.org/10.1007/978-981-95-1188-4_10

10.1 Overview of Polymer Material Recycling

10.1.1 Life cycle of Polymer Materials

Despite the declining production costs and expanding application domains in the polymer materials industry, the treatment and recycling of polymer waste are poised to become pivotal challenges in the coming decades. However, in the next few decades, the treatment and recycling of polymer waste will become a key challenge for humanity. To recycle and reuse polymer materials, it is necessary to understand and design the life cycle of the materials. The life cycle of materials can be described by Fig. 10.1; that is, the raw materials are used for specific purposes and can be produced and processed. The source of raw materials can be fossil resources, biomass, or recycled materials. The previous chapter has detailed the biodegradation process of materials, and this chapter will focus on material recycling.

When the "usage" behavior occurs, the life cycle of polymer materials as consumer products begins. After use, the material realizes its use value, and when it is discarded as garbage, its life cycle ends. However, in this process, the physical properties of the polymer material do not change. Therefore, how to reasonably recycle the discarded materials, simplify the life cycle as much as possible, and realize the efficient recycling and circular use of materials is worth in-depth study.

The traditional fossil-based plastic recycling industry has been established and is developing. However, due to the introduction of the "green" plastic concept, traditional recycling technology faces many new problems that need to be solved urgently. In addition, there are issues such as separation costs, secondary pollution affecting the quality and processing technology of recycled materials. With the development of biomanufacturing technology, the application of various biobased materials is gradually popularized, and petroleum-based and biobased plastics will coexist for a long time, jointly producing sustainable, low-cost materials. Consequently, the widespread adoption of bioplastics and biocomposites will significantly influence the recycled plastics industry in the foreseeable future. Therefore, it is necessary to develop green and efficient recycling and comprehensive utilization technologies for polymer materials, including biobased materials,

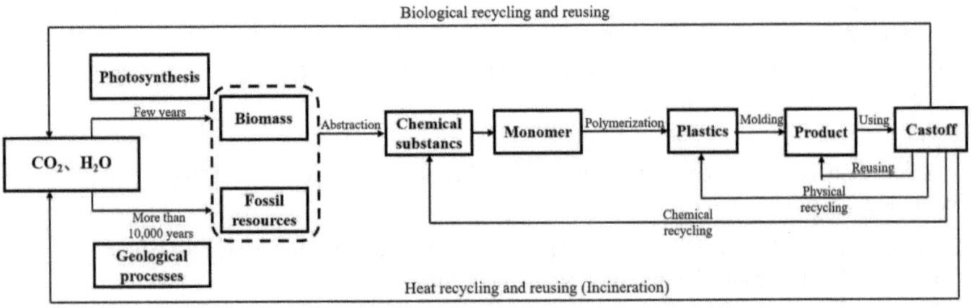

Fig. 10.1 Material life cycle loop. (Reprinted with permission from John Wiley and Sons, 2004 © 2004, Wiley Online Library (Kirkby and Azapagic 2005))

and it is necessary to develop technically feasible, effective, efficient, and economical recycling systems and final markets.

10.1.2 Polymer Waste Sources

Polymer waste can be roughly divided into two categories according to its source: One is the first type of solid plastic waste (SPW) generated during the manufacturing process, such as postindustrial (PI) waste, including injection runners, waste generated during the production process—fallen products, slices, and trimmings. These wastes will be collected and recycled in the industrial production link and will not enter the circulation link. Industrial waste typically offers advantages like cleanliness and a well-defined polymer composition. Moreover, this waste stream tends to flow unidirectionally, minimizing contamination from other polymers or non-polymers. In terms of recycling, these are usually high-quality polymer wastes. The other is postconsumer (PC) waste. The recycling of waste plastics also has different regulations depending on the country or region, and the strictness of the collection scheme also varies greatly. Generally, postconsumer plastic waste is a mixture; its composition is unknown, and it may be contaminated by organic parts (such as food residues) or non-polymer inorganic parts (such as paper). The composition of recycled waste products is more complex than that of industrial waste products. After recycling and initial processing, the subsequent treatment of postindustrial and postconsumer waste follows a similar process. Thus, the composition of the waste significantly impacts its recycling efficiency. Reusing is a better method of material recycling, which can obtain new raw materials in the recycling process through mechanical (usually leading to granulation) or chemical (usually leading to monomer structuring) methods. If polymer waste cannot be recycled, energy should be recovered as much as possible. The method of landfilling should be avoided as much as possible for urban solid waste, because although this method has a lower processing cost, it can cause serious pollution.

10.1.3 Recycling Methods of Polymer Materials

The most effective way to reduce material waste is to find alternatives. Many countries have issued “plastic bans,” prohibiting plastics from entering the packaging market to replace traditional fossil polymer materials, but this is neither realistic nor wise. With the development of the modern packaging industry, plastics have already replaced most traditional packaging materials, such as paper, glass, and metal, due to both the excellent performance of plastic materials and the low production cost. At the same time, the energy consumed in the production process of plastic packaging materials is much less than that of packaging materials such as paper, glass, and metal. However, plastic packaging materials cause resource loss in raw materials and processing, and the waste emissions in their production process can also harm the environment.

Recycling is the most effective and promising method to reduce solid waste. By recycling, the use value of materials is increased. Under the most ideal conditions, using reusable materials to replace disposable materials is an important and feasible idea to solve the pollution problem of polymer material waste. This method offers a shorter processing time and lower energy consumption in the recycling process. The treatment and recycling process of polymers can be divided into four major categories: re-extrusion (first category), mechanical (second category), chemical (third category), and energy recovery (fourth category). Mechanical recycling (i.e., secondary recycling or material recycling) involves physical processes, while chemical recycling and treatment (i.e., the third level, including the recovery of raw materials) are used to produce raw material chemicals for chemical products. Energy recovery includes the complete or partial oxidation of materials, and the production of heat, energy and/or gas fuel, oil, and carbon, as well as by-products that must be treated, such as ash. Figure 10.2 summarizes the recycling and reuse methods of polymer materials.

Among the many disposal and recycling technologies, landfill treatment, original use, and simple regeneration, these three physical methods have been widely used in practice and are currently relatively mature. On the contrary, the use of chemical means, modified regeneration, thermal decomposition, degradation, and thermal energy recovery technologies is still in the development stage.

As mentioned earlier, primary recycling refers to the recycling of industrial waste, which is usually the offcuts produced by polymer materials in the processing and molding process. The biggest feature of industrial waste is that it is pollution-free, and the composition and content are clear. After mixing the waste with unused materials for reuse, the recycling rate can generally reach 10–25%. Unless the product has special performance and quality requirements, most industrial waste will be

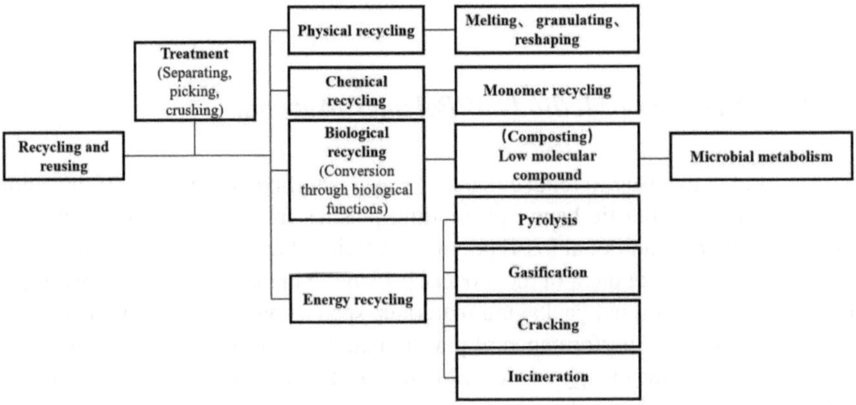

Fig. 10.2 Recycling and reuse methods

recycled once. From an economic point of view, the recycling rate of the first recycling and reuse of plastic by production enterprises is higher and easier to achieve.

Secondary recycling refers to the process of collecting, compressing, crushing, separating, purifying, and drying urban waste, and then reusing it in one or more forms. The objects of recycling are generally postconsumer waste. The difficulty in secondary recycling of different types of materials (metals, glass, paper, polymers, etc.) lies in separation. The most common types of urban waste polymers are PE, PS, PP (polypropylene), PVC (polyvinyl chloride), and PET (polyethylene terephthalate), as well as some biobased polymer materials. Most of these are incompatible with each other, and the mixture of two or more types of polymers can lead to a significant decline in material performance. The effective separation of polymer types is a prerequisite for the feasibility of plastic regeneration schemes. Collecting low-pollution single-type polymers or separating different types of plastics is a process with high efficiency and low cost. Therefore, secondary recycling will become the best solution.

The chemical method refers to the use of chemical methods to treat polymer waste, causing it to undergo depolymerization reactions. Ordinary products can be converted into monomers, oligomers, chemicals, and fuels through the chemical method. Depolymerizing waste polymers into monomers is the most economically valuable method, which surpasses all the recycling methods mentioned above. It can not only recycle but also re-obtain high-quality polymer products. The depolymerization process of polymethyl methacrylate (PMMA) polymers is relatively mature, and the recovery rate is very high.

Polyesters can readily decompose through processes such as alcoholysis and hydrolysis. Given the existence of depolymerization technology and its feasibility, chemical treatment of polymer waste holds significant allure. The problem of this chemical depolymerization is that some polymers (PS, PE, etc.) are difficult to control depolymerization, or the effect is not good, and impurities in the polymer will affect the depolymerization effect.

If the above recycling methods are not feasible, or the recycled products have no economic value, recycling should be carried out according to the energy value of the polymer. Polymers are typically derived from petroleum, boasting an energy content akin to hydrocarbons, double that of coal and paper, and more than four times that of regular waste. Nonetheless, they emit toxic gases, heavy metals, and harmful organic substances. Produced during combustion, they will cause harm to the environment. The feasibility of combustion depends on the energy utilization rate, whether emissions can be effectively controlled, and the harm to the environment.

With the continuous development of polymer recycling technology, the importance of infrastructure investment, the establishment of viable markets, and the participation of industry, government, and consumers is increasingly prominent. By using the life cycle assessment (LCA) method to evaluate domestic waste, it is possible to determine the environmental impact related to alternatives in the "from grave to grave" process of polymer materials, thereby finding the most sustainable solution. Currently, 90% of the plastics used are synthesized using nonrenewable fossil energy. Recycling is considered a sustainable practice, which means that waste management plans can use energy more sustainably (Fig. 10.3). Therefore, waste management plans must be incorporated into the production cycle of plastics and the treatment of plastic solid waste (PSW) (Kirkby and Azapagic 2005). This plan can help select and apply appropriate technologies, processes, and management procedures to achieve specific waste management goals. The goal of international waste management is to meet the needs of society, tap the potential of waste prevention, reuse, and recycling, control the waste generated in the treatment process, minimize environmental impact, and effectively use resources. The "International Waste Management Cycle" can be divided into six categories, namely waste production, waste treatment, classification and treatment from the source, collection separation and treatment, transfer station treatment and waste transportation, and disposal.

Every step of the recycling chain must consider the technical and economic feasibility of all material recycling processes, as well as the overall commercial

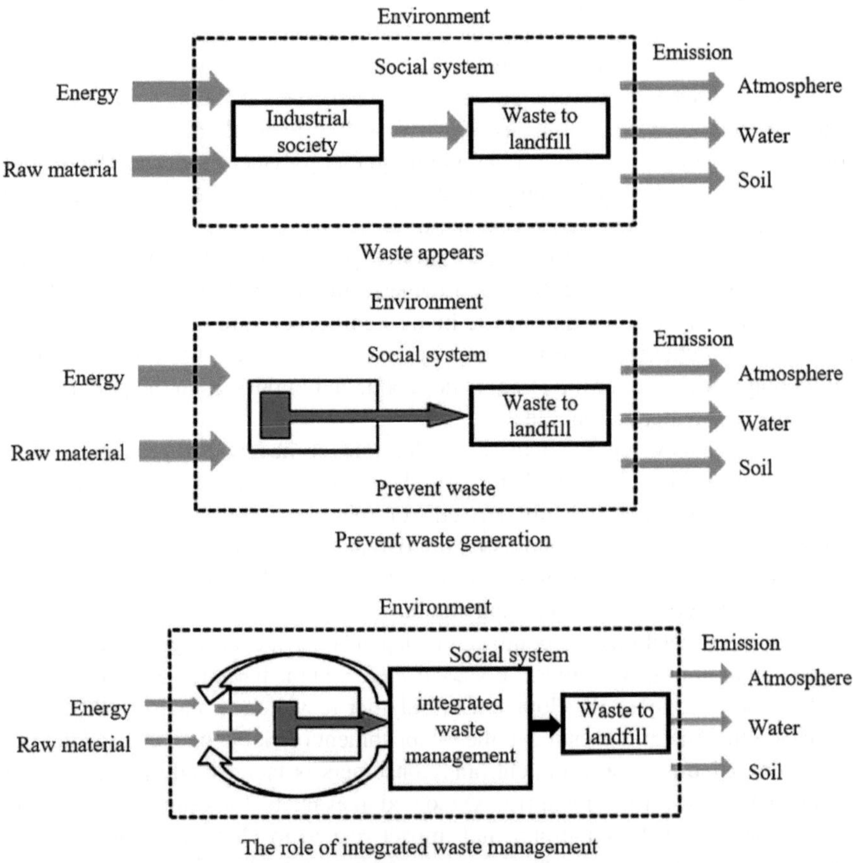

Fig. 10.3 Roles of waste prevention and integrated waste management. (Reprinted with permission from John Wiley and Sons, 2004 © 2004, Wiley Online Library (Kirkby and Azapagic 2005))

viability of advanced recycling methods. In chemical recycling and energy recovery, collection, processing, and sales are key to success. Currently, except for a few products, these technologies have not been widely applied. However, in the long run, they still have potential for development.

10.1.4 Open-Loop Recycling and Comprehensive Utilization

According to the concept of a circular economy, material recycling can be divided into open-loop and closed-loop modes. In the closed-loop cycle, recycled plastics are used to manufacture products identical to the original recycled products. These new products can be made entirely from recycled plastics or mixed with original plastics; this dilution form ensures the product can be sustainably recycled. Closed-loop recycling is one of the common recycling methods for most PET packaging products. In the open-loop cycle, recycled plastics can be used to produce plastics different from the original recycled ones.

Essentially, whether it is an open-loop cycle or a closed-loop cycle, they are objectively divided based on the newly produced products, and there is no additional value in the recycling process.

Based on the "open-loop" concept, it is of great significance to functionally regenerate from the perspective of comprehensive utilization of waste materials. Environmental functional materials developed using discarded polymer polymers as the main raw materials mainly include water-absorbing materials, porous materials, absorbent materials, catalytic materials, adhesive materials, reducing materials, and energy conversion materials, etc. These materials have good industrial prospects in environmental governance and engineering applications.

10.2 Mechanical Recycling

10.2.1 Reuse, Classification, and Primary Recycling

People use a myriad of plastic products in daily life, but many of them are disposable, serving only a single life cycle or circulation period, and may quickly become waste after purchase, such as food packaging. Recycling is often more ideal in most cases to mitigate energy consumption.

Several technologies have been developed to separate and classify plastic waste. For the recycling industry, this technology must achieve classification recognition in a short time, improve recycling efficiency, and reduce recycling costs, so as to have value for large-scale application. Most classification technologies are based on density sorting, but in fact, the densities of most plastics are similar, so the effects of processing and molding methods on material density should also be considered. The effectiveness of using density as a classification basis is controversial.

In response to some problems in density sorting, researchers have made some improvements. Because liquid separation is affected by factors such as wettability, density changes (related to porosity, fillers, pigments, etc.), shape factors of reduced particle size, and the freedom of the material from other materials, and because of poor wettability or surface contamination, bubbles may adhere to the plastic surface, causing individual flaky materials to float in the solution. Therefore, in the separation process, the use of a hydraulic cyclone can improve the efficiency of density separation, and the centrifugal force generated by the cyclone can improve the wettability of the material.

Another practical method for sorting plastic waste is through electrostatic separation by friction; that is, just by rubbing against each other, two types of resins can be distinguished. The frictional electrostatic separator distinguishes materials based on the phenomenon of surface charge transfer. Plastic mixtures rub against each other inside the drum to charge the materials; one material carries a positive charge, and the other material carries a negative or neutral charge (Al-Salem et al. 2009), thus achieving the purpose of separation. In the frictional electricity method, the purity of materials with a particle diameter of 2–4 mm is high, and the recovery rate is also high.

Classification is the most important step in the recycling process, and the removal of paint from waste plastics is one of the main problems faced by recycling plants. The presence of coatings can damage the performance of recycled plastics due to the generation of concentrated stress. Grinding can generally be used to remove coatings, such as removing chromium from coated plastics through simple grinding; sometimes, low-temperature assistance can be used to enhance the release process and prevent coated materials from embedding in plastic particles. Although this low-temperature method can promote release, it is still difficult to separate plastic particles from the coating. Abrasion is another method to remove paint and coatings, mainly used for larger-sized integral parts. Recyclers can also use the solvent evaporation method, immersing coated plastics in solvents to release the coated plastics; this method is suitable for removing disc coatings. High-temperature hydrolysis stripping refers to hydrolyzing the paint in hot water to release it from the plastic. However, these process technologies often require careful control of process conditions and cannot meet the requirements for large-scale promotion and application. Moreover, the degradation process of these processes (mainly photo-oxidation) will reduce the resale value of these recycled products.

At present, most of the plastic waste recycled is postindustrial waste, that is, waste generated in the industrial process. But the main recycling of postindustrial waste (primary recycling) is to reintroduce the waste, industrial plastics, or single-polymer plastics edges and parts back into the extrusion cycle to produce similar material products; the shape and performance of the used waste plastics are consistent with the original product. However, primary recycling is only suitable for semi-clean waste with clear components and content; this process is usually part of the plastic product processing production. For example, re-injecting molding of unqualified PE boxes, boxes that do not meet the specification requirements, will be stacked and reenter the cycle or the final stage of production.

The main recycling process also includes re-extrusion of plastics after consumption; households are often the main source of this waste flow. However, recycling household waste brings many challenges, one of which is selective and classified recycling. Collecting a relatively small amount of mixed plastics in resource-rich areas is economically valuable. However, in many countries, this has led to resource consumption and brought huge operating costs.

10.2.2 Mechanical Recycling Methods and Steps

Mechanical recycling, also known as secondary recycling, is a process of reusing plastic solid waste (PSW) by producing plastic products mechanically. Mechanical recycling of waste plastics is only suitable for single-polymer plastics such as PE, PP, PS, etc. Mechanical recycling provides a feasible way for the recycling of polymer materials, especially for foam and rigid plastics. In daily life, many products come from mechanical recycling processes, such as food bags, pipes, sinks, door and window profiles, and blinds and curtains. The main problem with processing mechanically recycled products is the quality of the waste. The more complex the waste, the more severe the pollution, the more difficult it is to recycle mechanically. Sorting, washing, and preparing plastic waste are crucial for producing high-quality, transparent, clean, and uniform products. The degradation and heterogeneity that exist in the mechanical recycling process are the main problems faced by mechanical recycling. During the polymerization process, many chemical reactions such as polymer addition, polymerization, and condensation are theoretically reversible due to the different structures of the polymer products, but the supply of energy or heat can cause photo-oxidation and mechanical stress, leading to photodegradation and mechanical degradation. To reprocess the recycled materials into new products, the waste must first be converted into new raw materials. This stage is usually referred to as "end of waste," which begins after the collection step is completed. Normally, this process includes the following steps: ① Sorting: by shape, density, size, color, or chemical composition; ② packaging: if the plastic is not processed at the sorting site, it is usually packaged for transportation; ③ cleaning: removing (usually organic) contaminants; ④ abrasion: reducing the size from product to thin slice; and ⑤ mixing and granulating: reprocessing the sliced particles into granules, depending on the composition, PI waste tends to separate in advance. Therefore, the classification rate of PC waste is higher than that of PI waste. Mechanical recycling includes many processing and preparation steps. Mechanical recycling is a costly process, so these steps and working time should be minimized as much as possible. The first step in mechanical recycling is usually to reduce the plastic to a more suitable shape (microparticles, powder, or thin slices). Generally, recycling schemes usually include the following steps (Zia et al. 2007), as shown in Fig. 10.4 (Aznar et al. 2006).

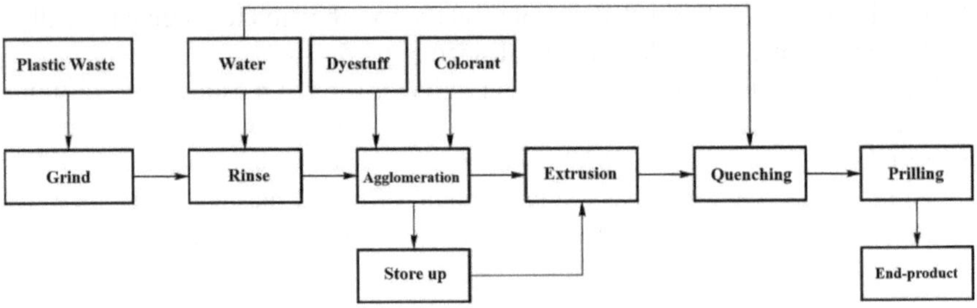

Fig. 10.4 Mechanical recycling steps. (Reprinted with permission from *Fuel Process Technol*, 2006, 87(5): 409–420.@ 2006, Elsevier Ltd (Aznar et al. 2006))

10.2.3 Separation Methods

Flotation, also known as sink-float separation, is a density-based separation technique (Ignatyev et al. 2014). This is the main method for separating shredded polymer films, usually using water as a flotation agent. High polymers with a density of less than 1 g/cm^3 (unfilled PP and PE) will float during the separation process, while other common high polymers (PS, PET, PVC, ABS, etc.) will sink. Similarly, flotation can be used to further separate tank polymers using a medium denser than water. However, many polymers in PC waste have a “density range” rather than a single density value, and these ranges often overlap, making them difficult to be effectively completely separated. Figure 10.5 shows the typical density range of packaging polymers (Callister and Rethwisch 2010).

Some non-melt contaminants inevitably affect the re-pelletizing or subsequent reprocessing of recycled materials, and melt filtration is an effective technique for removing non-melt contaminants from the melt. Components generally removed include wood, paper, aged rubber particles, and polymers with higher melting points (such as PET when processing PP at 220 °C). Melt filters have different mesh sizes; the smaller the mesh, the stronger the ability to remove contaminants.

In addition to the above, there are many other separation techniques used (or developed) for separating mixed polymers.

1. *Foam Plastics*

Also known as selective flotation separation, it is another method for separating polymers with similar densities (Censori et al. 2016). The basic principle of foam plastic flotation is to adhere (or not adhere) bubbles to the selected polymer surface, causing it to float (or sink). In the precursor process, the surface characteristics of the selected polymer will change from hydrophobic to hydrophilic, or increase its hydrophilicity (Fraunholcz 2004). Currently, this technology has not been widely applied in the industrial field and is undergoing large-scale research. This product is mainly used for separating mixed plastics with a density greater than water, such as PS, PVC, PET, PC, or POM binary mixtures (Wang et al. 2015).

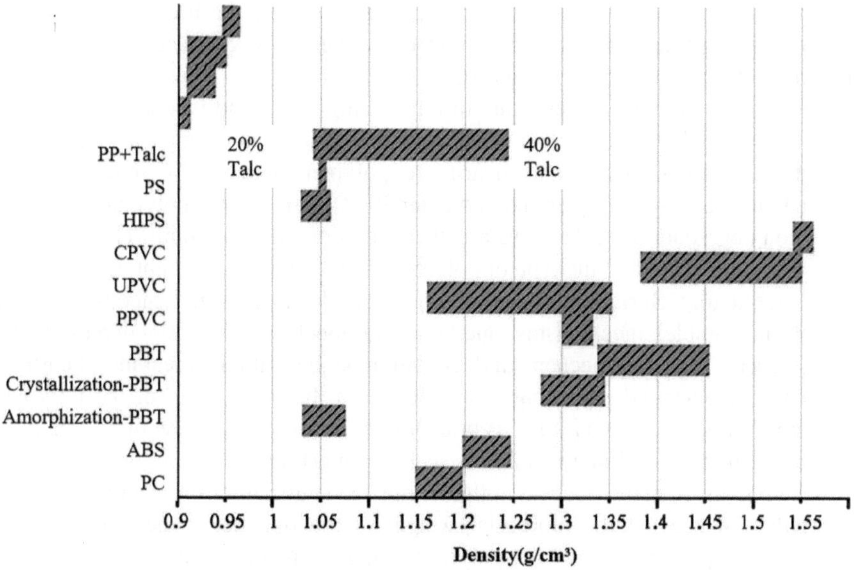

Fig. 10.5 Typical density range of packaging polymers. (Callister and Rethwisch 2010)

2. *Magnetic Density Separation (MDS)*

MDS is a technology improved on the basis of density, originating from the mineral processing industry. Using a magnetic liquid (containing iron oxide) as a separation medium, the density of the liquid can be changed with a special magnetic field (Rem et al. 2013). MDS can be used to separate multiple polymers in one step. This system has been successfully used to separate polypropylene and polyethylene from MPO (Serranti et al. 2015), as well as to separate PVC and rubber from construction and building waste (Luciani et al. 2015). However, this method is still based on the density method, and polymers with overlapping densities will still contaminate each other.

3. *X-Ray Detection*

It can be used to separate PVC containers, which have high chlorine content and are easy to identify (Arvanitoyannis and Bosnea 2001).

10.2.4 Challenges Faced by Mechanical Recycling

The main problem with mechanical recycling is that polymers will degrade under certain conditions, especially under heat, oxidation, light, ion radiation, hydrolysis, and mechanical shearing and action. There are mainly two types of degradation during the mechanical recycling process of polymer materials: posttreatment (thermo-mechanical degradation) and degradation during the lifespan (La Mantia 1996). Among them, the most important is the thermo-mechanical degradation of polymers during the reprocessing process.

During the melting process, due to the heating effect and mechanical shearing effect of the polymer, thermo-mechanical degradation can occur. Under the combined effect of temperature and shear force, polymers undergo different degrees of degradation reactions (Beyler and Hirschler 2002). In industrial polymers, the most common degradation mechanisms are chain scission and branching. The degree of degradation is related to the type of polymer, initial molecular weight, temperature, and other factors. During the degradation process, due to the simultaneous existence of various complex mechanisms, one or several mechanisms may play a dominant role. Generally, thermomechanical degradation begins with the breaking of carbon-carbon covalent bonds, forming free radicals on the main chain of the polymer. Certain chemical reactions, such as branching reactions, may cause these free radicals to break or cross-link (also known as branching) (Fig. 10.6).

Chain scission mainly occurs at the terminal side groups or on the main chain of the polymer, causing the molecular weight of the polymer to decrease, further reducing the performance of the polymer. Chain-branching reactions cause cross-linking between molecular chains, increasing molecular weight. Polypropylene is a typical chain scission reaction, while some types of PE are more prone to chain-branching reactions. Both reactions can release unsaturated low molecular weight volatile components to a certain extent. Within a certain degradation time, the molecular weight of polypropylene samples decreases due to thermomechanical degradation, as shown in Fig. 10.7 (Qian et al. 2011; Delva et al. 2014). In addition to the decrease in molecular weight, enhanced polydispersity and changes in molecular weight distribution were observed, indicating the presence of different chain lengths in the polymer. As shown in Fig. 10.7b, as the extrusion cycle increases, the melt flow index (MFI) of the material also increases linearly, indicating an increase in low molecular weight chain segments.

Changes in the molecular weight of polymers have a significant impact on the rheological and mechanical properties of polymers. Figure 10.8a is a trend chart of

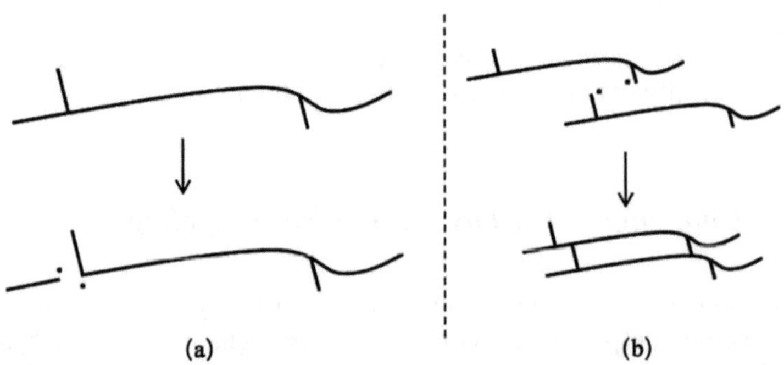

Fig. 10.6 Thermomechanical degradation of polymer chain free radical breakage and cross-linking. (**a**) Random chain scission and (**b**) cross-linking

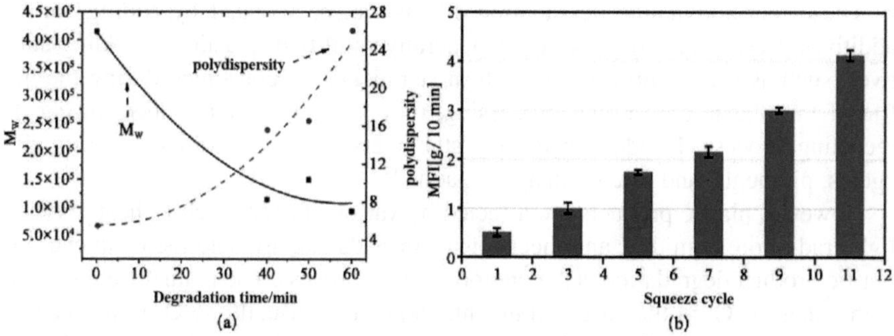

Fig. 10.7 Performance changes after thermomechanical degradation of polypropylene. (**a**) Evolution of molecular weight and polydispersity of PP samples; (**b**) evolution of MFI of PP samples after multiple reprocessing. (Qian et al. 2011; Delva et al. 2014). (Reprinted with permission from *Polym Bull*, 2011, 67: 1661–1670.@ 2011, Springer Nature)

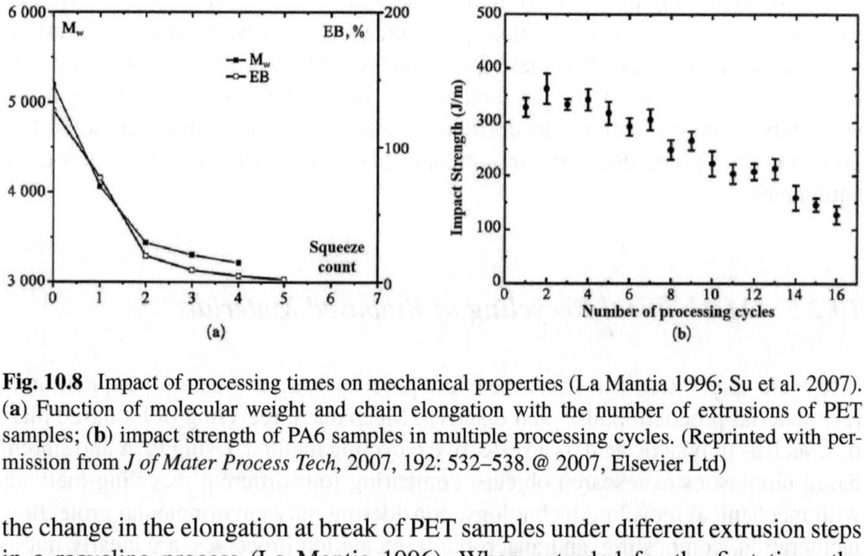

Fig. 10.8 Impact of processing times on mechanical properties (La Mantia 1996; Su et al. 2007). (**a**) Function of molecular weight and chain elongation with the number of extrusions of PET samples; (**b**) impact strength of PA6 samples in multiple processing cycles. (Reprinted with permission from *J of Mater Process Tech*, 2007, 192: 532–538.@ 2007, Elsevier Ltd)

the change in the elongation at break of PET samples under different extrusion steps in a recycling process (La Mantia 1996). When extruded for the first time, the molecular weight decreases significantly, and the elongation at break shows the same downward trend. Generally speaking, thermomechanical degradation significantly impacts mechanical properties, particularly elongation at break and impact strength. As shown in Fig. 10.8b, the recycling treatment cycle also affects the impact strength of PA6 (Su et al. 2007). The size and distribution of polymer molecular weight are closely related to its strength parameters. For some polymers, the mechanical strength may even increase after recycling. In addition to changes in mechanical properties and rheological properties, thermodynamic properties will also affect thermal properties (such as melting temperature, crystallization, etc.) and physical properties (surface characteristics, color, etc.).

These thermodynamic degradation effects can be reduced by adding various additives. Researchers improve the regenerative ability of plastics by adding additives, such as heat stabilizers. Since heat stabilizers are consumed during the first life of the plastic and during processing, they are generally re-added during the recycling process. In addition to heat stabilizers, impact modifiers, crosslinking agents, pigments, and fillers can also be added.

However, plastic products are affected by various factors such as heat, oxygen, light, radiation, humidity, and mechanical stress during use, and the main cause of plastic product degradation is the photooxidation process. The main difference here is that there is O_2 in the air environment, and there is usually no O_2 in mechanical recycling equipment. This will form oxygen groups on the polymer chain, thereby affecting the final performance of the material. Degradation during the life cycle and thermomechanical degradation can produce low molecular weight volatiles, and in the solid state, these volatiles are mostly wrapped in large polymer chains. However, during the reprocessing process, these volatile low molecules will become pollutants, hindering the effective reprocessing of the melt. These volatiles are tiny (oxidized) fragments of the original polymer, which can be identified by various analytical techniques (such as mass spectrometry) (Xiang et al. 2002). This not only harms the performance of the product, but also harms the processing process itself; especially some pollutants will corrode the processing equipment. To solve this problem, appropriate degassing treatment needs to be carried out on the recycling equipment.

10.2.5 Mechanical Recycling of Biobased Materials

Life cycle assessment will cover the entire process from raw material acquisition to raw material production, use, and disposal. Mechanical recycling of biobased plastics, such as polylactic acid, is an effective recycling method. Using PLA and starch-based bioplastics as research objects, comparing four different recycling methods with mechanical recycling technology, considering only environmental protection, incineration, composting, and anaerobic fermentation processes, are clearly not as good as mechanical recycling.

PLA is the most researched type of recyclable bioplastic. Under certain conditions, such as O_2 and H_2O, it is biodegradable. Although the biodegradability of polylactic acid (PLA) greatly reduces its waste's negative impact on the environment, the recycling of the material and research on PLA performance changes are still very important. This is firstly because industrial waste inevitably exists in various industrial processes, where waste is ground and mixed with pure polymers for recycling. PLA films can be processed into recyclable PLA particles directly by changing processing equipment and process parameters. These recyclable PLA particles can be mixed with the original PLA at a ratio of 20–50%, which not only reduces the cost of materials but also reduces the amount of waste in landfills, achieving environmental protection. Secondly, the PLA waste generated by repeated use can extend its service life. For example, PLA particles can be extruded with a twin-screw extruder and then pressed into samples with a laboratory injection molding machine. The tensile strength of PLA is not affected by the number of extrusion cycles, indicating that the total reduction in its relative molecular weight is very small, only decreasing by 5.2% after ten extrusions (Fig. 10.9). In addition, the decrease in the elongation at break is also very small (2.2–2.4%), and it is not related to the number of extrusions. As the number of extrusions increases, the impact strength significantly decreases (after ten extrusions, the impact strength is 20.2%), while the melt flow rate (MFR) and the permeation rate of water vapor and O_2 significantly increase. After ten extrusions, the MFR value of the sample is more than three times higher than that of the original sample. As the number of cycle increases, the thermal stability of PLA slightly decreases, the cold crystallization temperature and melting point slightly decrease, but there is no significant effect on the glass transition temperature of PLA (Enkiewicz et al. 2009).

Another study on PLA materials (containing 92% propylene glycol and 8% D-propylene glycol) showed that only the tensile modulus of elasticity remains constant with the thermal mechanical cycle of the injection molded product. Conversely, factors such as stress-strain, modulus, hardness, and rheology show a significant decrease at break. Figure 10.10 shows the change in PLA viscosity (η) with the number of injections (Pillin et al. 2008); a single injection can significantly reduce PLA viscosity (from 3960 to 713 Pa·s).

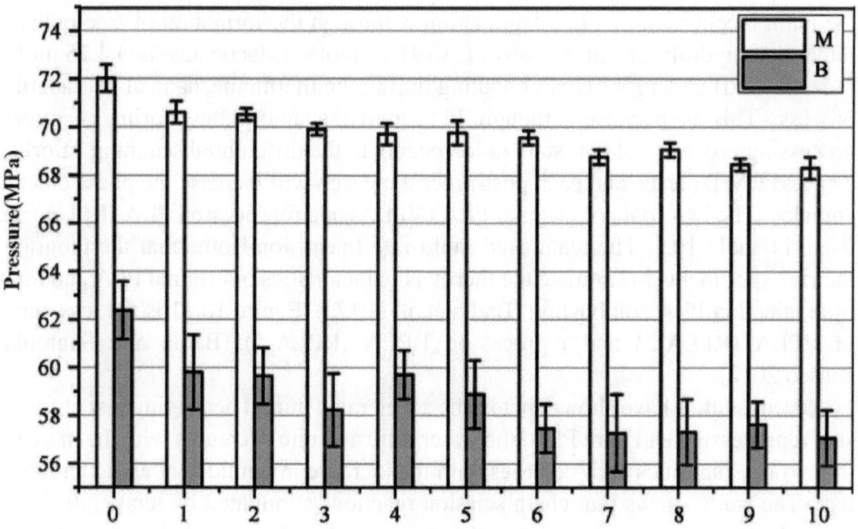

Fig. 10.9 Changes in the tensile strength (σ_M) and tensile stress at break (tensile stress σ_B) of polylactic acid with the number of extrusions (Enkiewicz et al. 2009). (Reprinted with permission from *Polym Test*, 2009, 28(4): 412–418.@ 2009, Elsevier Ltd)

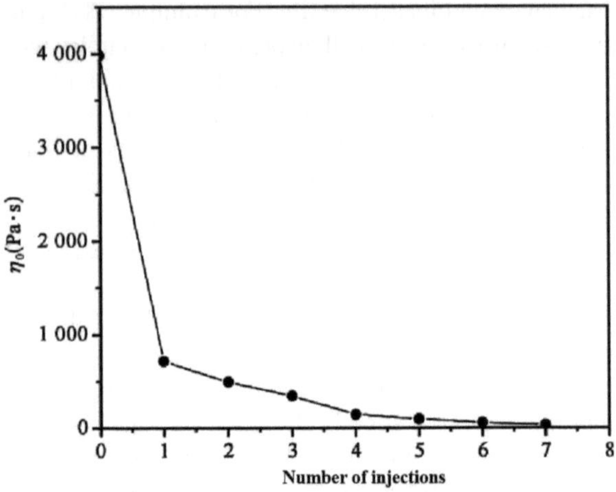

Fig. 10.10 Changes in the zero viscosity of PLA with the number of injections (Pillin et al. 2008). (Reprinted with permission from *Polym Degrad Stab*, 2008, 93(2): 321–328.@ 2008, Elsevier Ltd)

The viscosity of polylactic acid (PLA) changes with the number of injections. Repeated processing causes a sharp drop in viscosity, mainly because chain breaking during the processing causes PLA degradation, thereby significantly reducing the molecular weight. Quinone compounds have a good stabilizing effect, can capture free radicals, and maintain the length of the PLA chain at the processing temperature. To reduce the degradation of polylactic acid, quinone (PLA-Q) and cycloheptatrienone (PLA-T) are used together as stabilizers. The results show that the main mechanism of PLA degradation is through the formation of free radicals rather than hydrolysis. Amorphous PLA (95.75 mol% L-lactic acid and 4.25 mol% D-lactic acid) undergoes chain breaking during the thermomechanical degradation process. This is because although PLA remains amorphous during the post-processing process, cold crystallization occurs in the differential scanning calorimetry and DMTA tests, and each post-processing step will increase the phase change enthalpy. The DSC data of original PLA (VPLA) and reprocessed PLA (RPLA) are listed in Table 10.1. The team used multi-rate linear non-isothermal thermogravimetric experiments to simulate the thermal characteristics of original PLA and multiple injection PLA combustion (Badia et al. 2012a). Figure 10.11 is the loss curve of VPLA (RPLA-5) and reprocessed RPLA (RPLA-5) (Badia and Santonja-Blasco 2012).

Recent studies have shown that in the accelerated humid heat aging test of three-step reprocessing and pure PLA, the water diffusion rate decreases with the increase in reprocessing times and increases with the increase in humid heat aging temperature. The study shows that chain scission reactions stimulated by temperature can cause degradation. However, due to the presence of moisture in the polymer matrix, hydrolysis chain scission can also occur, which has a significant impact on the reprocessing and high-temperature treatment of PLA. After multiple posttreatments of polylactic acid, the effect of humid heat aging is particularly evident at high temperatures (Badia et al. 2012b).

Table 10.1 Evolution of DSC parameters in VPLA and RPLA (Badia et al. 2012a)

Material	T_g (°C)	T_{CCO} (°C)	T_{CC} (°C)	ΔH_{CC} (j/g)	ΔH_M (j/g)
VPLA	52.2 ± 0.1	106.2 ± 1.7	123.5 ± 0.2	2.21 ± 0.01	2.19 ± 0.04
RPLA-1	56.7 ± 0.1	105.6 ± 0.3	117.3 ± 0.3	23.37 ± 0.14	23.83 ± 0.67
RPLA-2	56.5 ± 0.2	102.0 ± 0.1	110.2 ± 0.4	28.52 ± 0.34	29.77 ± 1.70
RPLA-3	56.7 ± 0.3	101.2 ± 0.6	109.1 ± 1.0	28.32 ± 0.60	27.33 ± 0.97
RPLA-4	56.8 ± 0.1	100.1 ± 0.3	107.3 ± 0.5	27.41 ± 0.51	26.94 ± 0.52
RPLA-5	56.6 ± 0.1	99.6 ± 0.2	106.4 ± 0.1	28.53 ± 0.73	28.31 ± 1.04

T_g glass transition temperature, T_{CCO} and T_{CC} induction temperature and peak temperature in cold crystallization DSC, ΔH_{CC} cold crystallization enthalpy, ΔH_M specific melting enthalpy
Reprinted with permission from *Polym Degrad Stab*, 2012, 97(4): 670–678.@ 2012, Elsevier Ltd

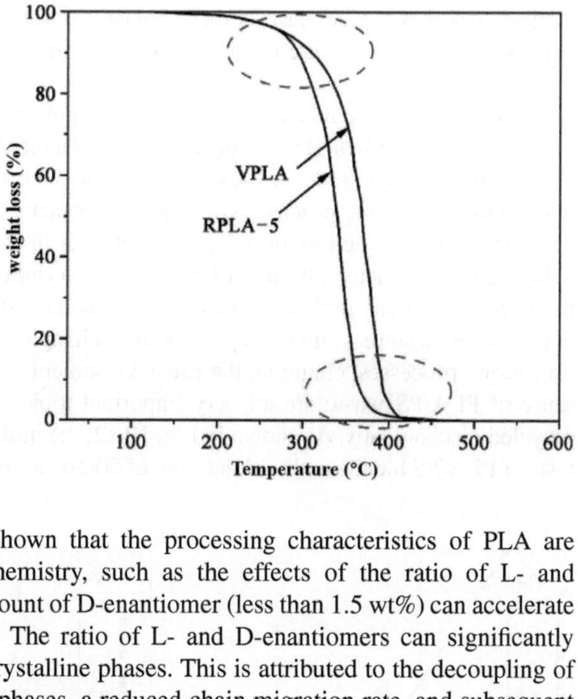

Fig. 10.11 Mass loss curves of original PLA and PLA (Badia and Santonja-Blasco 2012). (Reprinted with permission from *Bioresour Technol*, 2012, 114: 622–628.@ 2012, Elsevier Ltd)

Existing research has shown that the processing characteristics of PLA are closely related to stereochemistry, such as the effects of the ratio of L- and D-enantiomers. A small amount of D-enantiomer (less than 1.5 wt%) can accelerate the crystallization process. The ratio of L- and D-enantiomers can significantly influence the presence of crystalline phases. This is attributed to the decoupling of amorphous and crystalline phases, a reduced chain migration rate, and subsequent modifications to mechanical properties.

Compared to PLA, there is relatively less research on the mechanical recycling of other pure biobased polyesters. Polyhydroxyalkanoates (PHA) have good mechanical properties and biodegradability especially polyhydroxybutyrate (PHB) has extremely high mechanical properties, but it is expensive. Polyhydroxybutyrate-co-valerate (PHBV) is a copolymer of PHB, and in recent years, PHBV has been successfully recycled mechanically.

10.2.6 Mechanical Recycling of Biobased Material Mixtures

Research on the mechanical recycling of bioplastic mixtures, similar to pure bioplastics, is ongoing. As polymers from renewable resources replace some petroleum-based resins, these mixtures of biobased and petroleum-based polymers, known as mixed bioplastics, offer better sustainability than pure synthetic polymers. In addition to environmental protection, another advantage of mixtures is that they can overcome the limitations of pure plastics. For example, pure PLA has disadvantages such as poor toughness, impact resistance, and crystallization performance, and it tends to thermally degrade when in a molten state. This limitation restricts its current use in food packaging and impacts the recyclability of PLA. To overcome the shortcomings of pure PLA, mixing PLA with other polymers is a cost-effective method. Another typical case is thermoplastic starch (TPS), which has disadvantages such as low moisture content, low temperature resistance, and problems of brittleness and plasticizer migration after starch recrystallization. To solve these problems, TPS can be added to a hydrophobic polymer matrix, and then different mixtures of TPS with biodegradable and nonbiodegradable polymers can be studied.

In addition to synthetic plastics, there are also biobased polymers in the existing waste plastic recycling process, which is another factor of high research value in the recycling of bioplastic mixtures. The cost of PLA in the mixture of PLA and PS is balanced with the degradability of PS, and its performance is between pure PLA and pure PS, with broad application prospects in medical devices and packaging products. The material recycling and performance changes of PLA/PS waste in multiple processing processes, changing the use environment of the blend, and studying the reuse of PLA/PS waste are all very important topics. PLA mixtures can also be recycled mechanically. As shown in Fig. 10.12, the multiple extrusion and injection tests of PLA/PS blends with a blend ratio of 50/50 (weight ratio) show that the stress

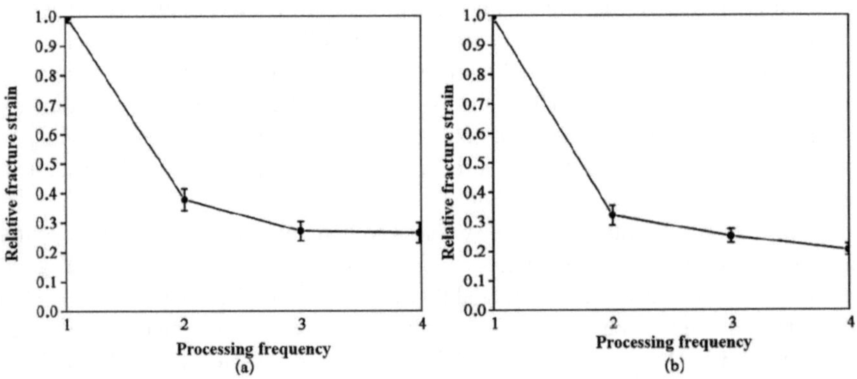

Fig. 10.12 Changes in the mechanical properties of PLA/PS blends with the number of processing times (Hamad et al. 2011). (**a**) Relative stress of chain breakage; (**b**) relative strain of chain breakage. (Reprinted with permission from *J Materials Sci*, 2011, 46(9): 3013–3019.@ 2011, Springer Nature)

and strain of the blend drop sharply within two treatment cycles, while the Young's modulus decreases by 26% after four processing cycles. After four processing cycles, a larger drop in fracture strain was observed, which was 73% and 79% respectively. The apparent viscosity of the PLA/PS blend can reach 3100 Pa·s, and it decreases with the increase of processing times. After each processing cycle, its apparent viscosity decreases by 15–30%, which is due to the reduction of cyclic molecular weight caused by each processing cycle (Hamad et al. 2011).

In addition, researchers have also conducted studies on the impact resistance modification of recycled PLA. According to the different proportions of the modifier (10–15 wt% or 4–8 wt%) in PLA, two types of commercial impact modifiers based on organic elastomers have been added to PLA. Adding a small amount of impact modifier can effectively improve the mechanical properties of the material. The results show that the regenerated material has high hardness, small deformation, and its water absorption and impact resistance are better than those of the raw material (Roberto et al. 2010). After multiple recycling cycles, some mechanical properties are still retained, but it will lead to a significant decrease in impact strength (see Table 10.2).

Mixing styrene polymers with TPS thermoplastic elastomers is an economical way to reduce greenhouse gas emissions (GHG). This type of blended bioplastic replaces some petroleum-based polymers with TPS, maintaining its performance, reducing costs, and improving the environment. In addition, this blend can be recycled with almost no loss of performance, and some blends can be completely biodegraded. However, the recyclability of many bioplastic blends has not yet been studied.

10.3 Chemical Recycling

Chemical recycling is a highly efficient recycling method rooted in the principle of "sustainable development." This method involves recycling waste as a precursor to create high-value-added products for various industrial and commercial applications, thereby opening up new application fields. However, it must be pointed out that due to current limitations in raw material costs, capital investment, and business scale, the polymers produced by chemical recycling are often more expensive than the raw materials and have not yet reached the conditions for industrial application.

Table 10.2 Cantilever beam impact strength (IS) and heat distortion temperature (HDT) of PLA/PS blend

Cycle	IS (J · m^{-1})	HDT (°C)
PLA, reference	141 ± 20	60.7 ± 1.2
First recycle	71 ± 7	68 ± 1
Second recycle	64 ± 3	69± ± 1
Third recycle	63 ± 10	66 ± 1

10.3.1 Chemical Substance Recycling Methods and Pathways

Chemical recycling refers to the process of converting plastic materials into smaller molecules (usually liquids or gases). These small molecules can serve as raw materials for producing petrochemical products and plastics. This process resembles a depolymerization reaction, inevitably altering the chemical structure before and after chemical recycling. The products obtained through chemical recycling can be used as fuel, thus forming a sustainable cycle with high economic benefits, high yield, and low energy consumption. Some chemical cycles are similar to the processes and methods used in the petrochemical industry, such as pyrolysis, gasification, liquid-gas hydrogenation, steam, or catalytic cracking.

The biggest advantage of chemical recycling is that it can recycle contaminated multi-component polymers through simple pretreatment. Given the high cost and complex process of chemical recycling, the target of chemical recycling should be relatively expensive materials. Based on the economic benefits generated from this, chemical recycling can become a feasible industrial solution. For most cheap raw materials, the production scale of petrochemical plants is much larger than that of plastic plants (six to ten times), so it is unrealistic to rely solely on processing waste materials for chemical recycling, and recycled materials can only be used as a supplement to regular materials.

The monomer recycling pathway is fundamentally different from other recycling pathways in chemical recycling products. The purpose of monomer recycling is to use the recycled plastic waste as raw material for polymer synthesis reactions, while other recycled products are usually used as fuel to fully utilize their potential calorific value. The latter method aims at the thermal utilization of the product as the recycling target, which is the same as the purpose of incinerating polymer waste for reuse.

The following takes PET cracking as an example to introduce some basic chemical recycling methods and steps. Chemical recycling of PET can completely depolymerize it into monomers such as terephthalic acid (PTA), dimethyl terephthalate (DMT), bis(2-hydroxyethyl) terephthalate (BHETA), and ethylene glycol (EG). Depending on the choice of degrading agent, PET depolymerization methods can be divided into alcoholysis, glycolysis, hydrolysis, and amine/ammonolysis. The main depolymerization product of alcoholysis is dimethyl terephthalate (DMT), the main depolymerization product of glycolysis is bis-hydroxyethyl terephthalate (BHET), the main depolymerization product of hydrolysis is terephthalic acid (PTA), and the main depolymerization product of amine/ammonolysis is the diamine compound of PTA. Figure 10.13 summarizes the choice of different pathways for PET chemical decomposition, as well as the products obtained from PET depolymerization.

Alcoholysis of PET is conducted under high temperature and high pressure conditions (180–280 °C, 20–40 atm), yielding dimethyl terephthalate (DMT) and ethylene glycol (EG). The products from sugar fermentation and amine/ammonia hydrolysis can serve as plasticizers, crosslinking agents, chain extenders, corrosion inhibitors, and other value-added products. These include the production of

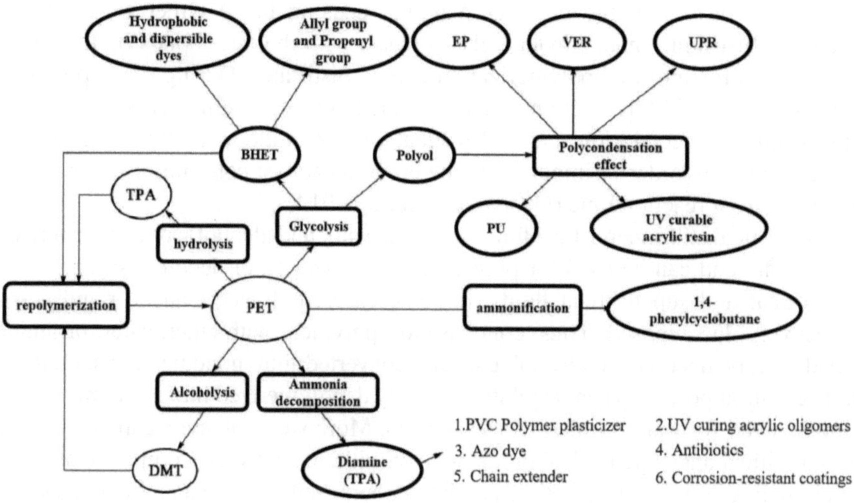

Fig. 10.13 Different PET decomposition methods and their products, derivatives, and value-added products

unsaturated polyester (UP), precursor resins for polyurethanes, textile dyes, epoxy resins, and vinyl esters.

The hydrolysis of polyester refers to the reaction of PET with water under high temperature, neutral, acidic, or alkaline conditions, causing the polyester chain to break and form TPA and EG. The main disadvantage of preparing TPA by hydrolysis is its low purity. Water is a weak nucleophilic reagent, and the reaction is generally slow and inefficient.

Sugar fermentation is the simplest and oldest method in PET depolymerization, and it has now become a commercial method for recycling PET, adopted by internationally renowned companies such as DuPont, Goodyear, Shell Polyester, Zimmer, and Eastman Kodak (Scheirs 1998). At a sugar fermentation temperature of 180–250 °C, PET undergoes an ester exchange reaction with an excess of ethylene glycol to produce bis(hydroxyethyl) terephthalate (BHET). The fermentation of PET can be carried out with different diols such as ethylene glycol (EG), diethylene glycol (DEG), propylene glycol (PG), polyethylene glycol (PEG), 1,4-butanediol, and hexanediol. The reaction is slow without a catalyst, so ester exchange catalysts are often used. In addition to forming monomers, specific low molecular weight oligomers, such as α,ω-dihydroxy substances (polyols), can also be formed, making sugar fermentation a common recycling method. The oligomers can also be used to synthesize polymers such as unsaturated polyester, polyurethane, vinyl ester, epoxy resin, etc. When the quality requirements for recycled PET are high, sugar fermentation is the best recycling choice. However, this method is not suitable for materials containing some low molecular weight copolymers, dyes, or pigments. Therefore, the most suitable material for recycling is PI waste (Scheirs 1998). The glycolysis of waste PET is divided into three stages: oligomers, dimers, and monomers.

Ethanol diffuses into the polymer material, causing the polymer to swell and increase diffusion. Then, ethylene glycol reacts with the ester bonds on the chain, degrading PET into low polymerization degree distillates. During the depolymerization process of PET, reaction parameters such as reaction time, reaction temperature, catalyst concentration, and PET reagent ratio are important influencing factors. When ranked in order of importance, the order is catalyst concentration > reaction temperature > reaction time (George and Kurian 2014).

Subcritical and supercritical fluids (such as water and alcohol) are excellent reaction media and can be used for plastic depolymerization or decomposition. Using subcritical and supercritical fluids, in some cases, polymers can be rapidly and selectively decomposed. Thus, condensation polymers with ether, ester, or amide bonds can be decomposed by solvents and converted into monomers. In fact, these condensation polymers can be relatively easily depolymerized into monomers without the need for water or alcohol as catalysts. Moreover, polymers can decompose in subcritical and supercritical fluids, with or without catalysts. Composite plastics, such as fiber-reinforced plastics, can be decomposed into smaller molecules and fiber materials.

10.3.2 Chemical Recycling of Pure Biobased Materials

In chemical recycling, biobased polymer molecules can be decomposed into small molecule monomers and then reenter the polymerization process. Taking polylactic acid as an example, the chemical reaction of polylactic acid undergoes two main processes: Firstly, PLA is hydrolyzed under high temperature to produce lactic acid; secondly, PLA is prepared into L-propyl lactate through thermal degradation, which is used for the polymerization of new PLA. The chemical recycling process of PLA involves hydrolysis into lactic acid, typically occurring under high temperature and high pressure conditions. The resulting lactic acid is of high purity and can be polymerized into pure PLA. This cycle of PLA production constitutes a closed-loop cycle, which, from a life cycle assessment perspective, can be referred to as a "cradle-to-cradle process." The yield mainly depends on operating conditions, such as temperature, reaction time, and water-material ratio, etc. The reaction time and temperature have a significant impact on the recovery efficiency (Table 10.3) (Faisal et al. 2006).

Chemical recycling has also been reported in other biopolymers. After thermal decomposition of PHA, it can be chemically recycled into vinyl monomers, and enzymatically treated to transform into oligomers. Cellulose ester is a biodegradable material synthesized from renewable biological raw materials, and the cellulose ester produced from cellulose esterification has become a material with important biological uses.

Biological recycling is a viable alternative to chemical recycling. However, biological recycling takes longer than mechanical recycling or incineration, and it requires the material to have a certain degree of biodegradability.

Table 10.3 The impact of different reaction times and temperatures on the conversion of PLA to lactic acid (Faisal et al. 2006)

Reaction time (min)	Recovery efficiency (%)				
	240 °C	250 °C	260 °C	300 °C	350 °C
5	11.9	21.6	53.9	67.4	55.5
8	35.2	80.1	70.8	71.9	46.7
10	61.7	90.7	84.8	83.8	42.0
15	85.5	91.1	89.2	83.0	30.1
20	–	92.5	91.5	82.5	22.0
30	–	88.4	80.1	79.9	12.7

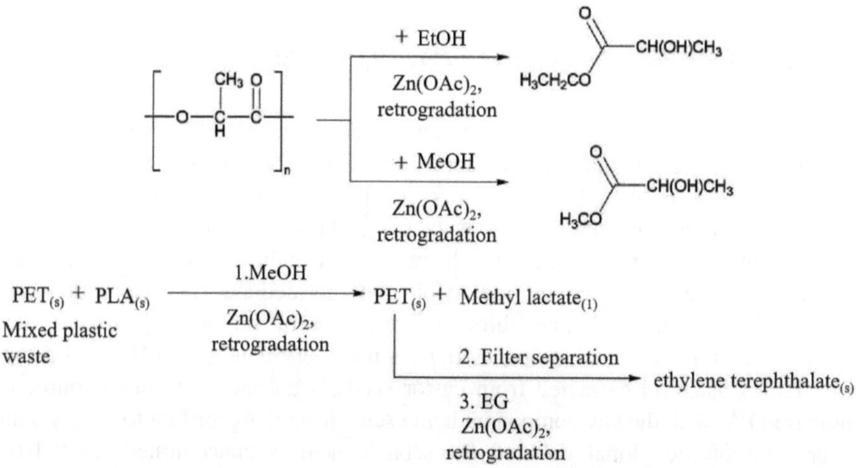

Fig. 10.14 Alcoholysis of polylactic acid (Sánchez and Collinson 2011). (Reprinted with permission from Eur Polym J, 2011, 47(10): 1970–1976.@ 2011, Elsevier Ltd)

Research on the chemical recycling of bioplastic mixtures mainly focuses on sorting methods, thermal–mechanical degradation of polymers, and sensitivity to material impurities. For example, due to similar densities, PLA and PET cannot be simply classified by manual sorting or density-based sorting methods, and the recycling efficiency is not high.

Existing research has achieved selective recycling of polylactic acid and PET, such as the method of filtering and recycling unreacted solid PET after depolymerization of polylactic acid. Among the three catalysts used for the glycolysis of PET waste, zinc acetate has the best solubility and the best effect. In addition, zinc acetate can effectively carry out alcoholysis on PLA (in methanol or ethanol), generating lactate esters. Since PLA is converted into a liquid monomer, and PET is still an unreacted solid under alcoholysis conditions, solid PET can be filtered out, making it a promising method for the chemical recycling of pure PLA and mixed PLA/PET. Figures 10.14 show the alcoholysis process of PLA (Sánchez and Collinson 2011).

Currently, methods for degrading PLA using PLA-degrading enzymes have also been developed. Using enzymes to recycle PLA from plastic waste containing PLA is more competitive economically than other methods of recycling PLA.

Due to structural differences, thermoplastic cellulose esters usually cannot be miscible with petroleum-based polymers. After cellulose acetate butyrate (CAB) is mixed with different petroleum-based resins, it can be immersed in acetone solvent, and the CAB matrix can quickly dissolve in CAB under the solid state of petroleum resin. The dissolved CAB is then added to water, precipitated into a solid form, filtered, and reused. However, the key issue in the chemical recycling of mixed components is still the separation between the components.

10.3.3 Recycling of Biobased PA, PUR, and PC

Most high-performance plastics are currently made from mineral raw materials, but biobased engineering plastics are rapidly developing, expanding the application fields and markets of biobased polymer materials. Biobased engineering plastics have two major advantages: The first is the low carbon footprint produced by renewable raw materials, and the second is the recycling capability and energy efficiency of the polymerization process. For example, the intermediates for the production of polyamide (PA) can be obtained through several possible biological pathways, such as those involving biobased adipic acid, hexamethylenediamine (HMDA), and caprolactam. Castor oil extracted from castor seeds is currently the main source of biobased PA, with the key component being sebacic acid. According to survey data, over 70% of the global demand for sebacic acid is concentrated in PA1010 and PA610.

Polyamide is a thermoplastic that can be ground into granules for further processing. The main problem with mechanically recycling PA is that it is mainly aimed at PA6, PA66, PA612, PA11, and PA12, among which there are almost no examples of recycling biobased PA. Biobased PA11 derived from castor seeds is industrialized by the French chemical company Arkema. This polymer has good compatibility with PA66 and can use the recycling method of PA66 to recycle carpet fibers and convert them into polymer monomers after the end of the carpet's life.

Thermoplastic polyurethane (TPU) is a linear polymer with different chain segments alternating. These chain segments consist of hard segments and soft segments. The hard segments are formed by the reaction of diisocyanate with short-chain diols (i.e., chain extenders), while the soft segments are formed by the reaction of diisocyanate with long-chain bifunctional diols (i.e., polyols). Due to the variation in the structure and/or molecular weight of the three reaction compounds, the number of possible combinations is almost infinite, so the structure of the polymer can be designed in many ways. Currently, biobased polyurethane (BPU) is attracting widespread attention because it can replace petroleum-based polyurethane. Since there is no precedent for preparing isocyanates using a biobased method, the current research focus is on replacing fossil polyols with biobased polyols. Castor oil can be used as the main raw material to produce biobased polyols, and soybean oil is also widely used. However, the mechanical properties of soybean oil-based BPU are lower, and it needs to be reinforced with glass fibers or hemp fibers. Epoxides can also be prepared using limonene oil extracted from citrus and CO_2 produced in industrial processes, such as copolymerizing oxo-limonene and CO_2 to prepare thermoplastic polycarbonate limonene ester with performance similar to polystyrene (PS). In addition, dioxo-limonene can be crosslinked with multifunctional amines (such as aminocaproic acid or citric acid) to produce carbonate dioxo-limonene, which allows the production of various crosslinked bio-PUs based on terpenes without using isocyanates. Like other plastics, BPU can be recycled through mechanical recycling or chemical recycling, but its recycling has always been a challenge.

In the recycling of BPU waste, the traditional method is to use mechanical regrinding and to use recycled material filling in new formulations. For example, BPU soft foam plastic can be recycled using scrap blocks and adhesives to make carpet pads, sports pads, cushion pads, and similar products. BPU parts from various sources (such as cars, refrigerators, and industrial decorations) can be made into granules, mixed with strong adhesives or polyurethane systems, and then made into sheets or molds under heating and pressure conditions. They are mainly used for sound insulation, such as in furniture and flooring. The chemical recycling of polyurethane includes sugar fermentation, hydrolysis, pyrolysis, hydrogenation, etc., but compared with the preparation from raw materials, this method is not economically competitive in most cases.

Polycarbonate (PC) has good properties such as impact resistance, transparency, high-temperature resistance, and dimensional stability. Unlike most other thermoplastics, PC can withstand large plastic deformations without breaking. Therefore, at room temperature, it can be processed into shape using sheet metal techniques such as bending brakes. Currently, biobased PC, replacing other polymer materials, has received widespread attention. Because traditional PC is synthesized from bisphenol A and phosgene, toxic gases such as phosgene and bisphenol A often cause pollution and operational hazards. In recent years, using L-tartaric acid as a raw material, PC monomers can be polymerized in three steps by four commercially available lipases from different sources at 80 °C. A new water-soluble PC with a side chain primary hydroxyl group was synthesized using trimethyl ester carbonate 2-(2-benzyloxyethoxy) as a raw material. PC can also be successfully synthesized by the melt polymerization of terpene diphenol and diphenyl carbonate without any catalyst. PC can be processed by various common thermoplastic processing methods, but PC has poor processability, low thermal conductivity, and high processing temperature (about 300 °C). In addition, PC shows strong adhesion to metals in the molten state, making it difficult to process and requiring high viscosity. Because the recycling and reprocessing steps will reduce the mechanical properties of PC, it may be necessary to use impact modifiers in recycled PC to achieve the desired performance level.

An alternative approach in recycling involves increasing the molecular weight to an adequate level. During the reprocessing process, condensation thermoplastics (such as PET, PBT, PA, BPA-PC, and TPU) and their mixtures will be hydrolyzed, alcoholized, and thermally decomposed. The main reason is that the processing of condensed thermoplastics undergoes rapid degradation reactions at high temperatures (about 300 °C), leading to a decrease in molecular weight and a severe decline in mechanical and thermal properties. This hinders the reprocessing of a large amount of industrial post-condensation waste for strict engineering applications. To overcome this obstacle, it is sometimes necessary to increase the molecular weight of the polymer. Using reactive chain extenders is one of the methods to increase the molecular weight of low molecular weight condensates. These chain extenders can react with the functional chain ends of the polymer in conventional mixtures, such as in a plastic melt extruder, to obtain regenerated raw materials with higher relative molecular mass.

10.3.4 Chemical Recycling of Biocomposite Materials

Biocomposite materials include petroleum-based matrices, biobased polymer matrices, glass fibers, natural fiber reinforcements, original polymers, recycled polymers, etc. For all biocomposite materials, it is important to consider their sensitivity to non-biodegradation, water absorption, and biodegradation processes to ensure the structural and functional stability of the material during its service life.

Research on thermoplastic rather than thermosetting biobased matrix composites is gradually being carried out, the reason being their recyclability. After mixing natural fibers with a biopolymer matrix, each component comes from renewable resources, and these biocomposites may be compostable, so they can be used as alternatives to glass fibers.

In addition to being biodegradable, the recyclability of biocomposites makes them more valuable. This recycling scheme can reduce raw material consumption, extend the use time of products, reduce carbon consumption, thereby extending their lifespan and reducing environmental impact. The mechanical properties of compostable flax/PLLA are comparable to those of glass fiber/PP, and its composite performance is superior to that of hemp/PP and sisal/PP composites. As shown in Fig. 10.15, after mechanical recycling, flax fibers with different fiber contents (weight percentages of 20% and 30%) can be retained. Researchers have studied the tensile properties of their biocomposites until the end of the third injection cycle. Post-processing leads to a decrease in molecular weight, a reduction in fiber length, and separation of fiber bundles (Antoine et al. 2008). However, after three cycles, the performance of the biocomposite (without the addition of the original polymer) is still an important indicator of the recyclability of the material. PLA/flax composites are mainly used in the automotive industry. Mechanical recycling of PLA/flax can lead to a reduction in tensile strength (23%), impact strength (8%), and E modulus (5%), and a corresponding reduction in the molar mass of PLA. Studies have shown that the reduction in fiber length is an important reason for the decline in fiber mechanical properties.

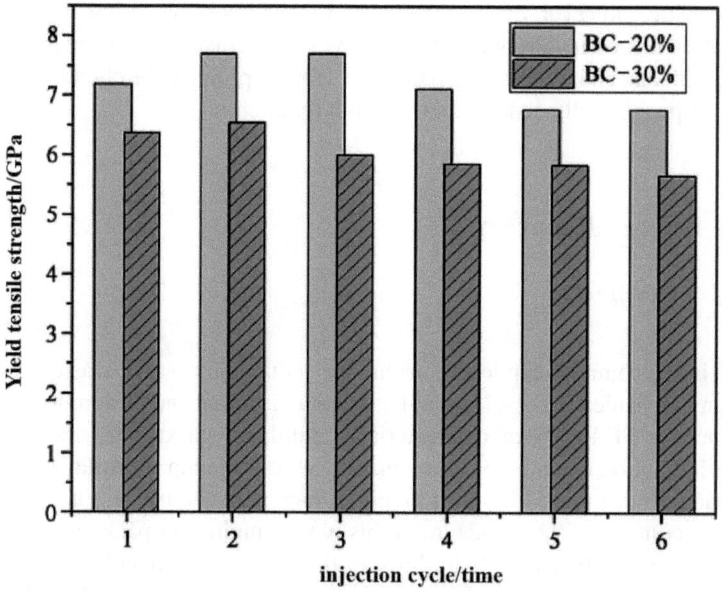

Fig. 10.15 Changes in the tensile yield strength of PLLA/flax composite fibers with 20% fiber (BC-20%) and 30% fiber (BC-30%) with the injection cycle. (Reprinted with permission from *Compos Part A: Appl Sci Manuf*, 2008, 39(9): 1471–1478. © 2008, Elsevier (Antoine et al. 2008))

In addition, some scholars have studied the recycling ability of biodegradable polymer matrices and their cellulose-reinforced composites. Chemical thermal mechanical pulp (CTMP) fibers are used to enhance three types of biodegradable substrates (PLLA, aliphatic polyester, and starch substrate). The research results of high-performance PLA composites show that PLA reinforced with jute fibers has higher impact strength, high-temperature resistance, good moldability, etc., and was applied to PC components and mobile phone shells in 2004 and 2006, respectively. Thermally reversible cross-linked PLA has excellent shape memory and recoverability, such as rewritable shape memory. Grinding the biocomposite and then using it as a reinforcing material is also a method of recycling. Now, there are renewable PLAs that have been enhanced by rice husks and jute fibers as a reusable reinforcing material.

It is generally believed that due to the higher sensitivity of composites to thermal mechanical degradation, biopolymer-reinforced composites have lower recyclability than pure thermoplastic substrates. For this reason, composting is considered the main waste treatment method for such materials. But as mentioned earlier, researchers have tried to recycle composites, such as compostable flax/poly-L-lysine (PLL). Typically, ensuring the recyclability of biocomposites reinforced with natural fillers (based on synthetic or biopolymers as thermoplastic matrices) while maintaining their mechanical properties across multiple post-processing cycles is crucial. The material will be repeatedly processed to enhance the interfacial adhesion between the filler and the matrix, thereby improving the thermal stability of the reprocessed biocomposite. However, due to the repeated recycling of biocomposites, the polymer materials have undergone varying degrees of degradation, a decrease in molecular weight, and an increase in crystallinity. How to promote the chemical recycling of biocomposites still needs further in-depth research.

10.4 Energy Recovery

10.4.1 Pyrolysis

Pyrolysis is a common application technique, suitable for plastic waste that has not yet been mechanically recycled, but has been incinerated and/or dumped into landfills, such as PE/PP/PS mixed materials, multilayer packaging, fiber-reinforced composites, polyurethane construction, and waste cleanup. Multilayer composite film materials are more difficult to recycle than their alternatives of metal, paper, and glass containers. The structure of this typical multilayer packaging material is shown in Fig. 10.16 (Tullo 2016). Most of these materials are polyethylene, which allows for larger packaging volume and a relatively intact structure. If higher toughness is required, packaging companies generally choose PET, the preferred resin for beverage containers. EVOH (EVOH) is often used in food packaging applications because it can more effectively block O_2 than PE, PET, or nylon.

Unlike mechanical recycling, pyrolysis recycling methods can handle high-polluting substances, such as residues from car shredders, as well as highly heterogeneous plastic mixtures, thereby increasing the flexibility of processing relative to raw material composition. Considering that feasible sorting is often difficult to achieve in actual recycling, or sorting all different types of plastics, the economic aspect becomes a major advantage of pyrolysis recycling.

In the absence of oxygen, pyrolysis is carried out under medium to high temperature (500 °C, 1–2 atm) conditions, and high temperature can destroy the

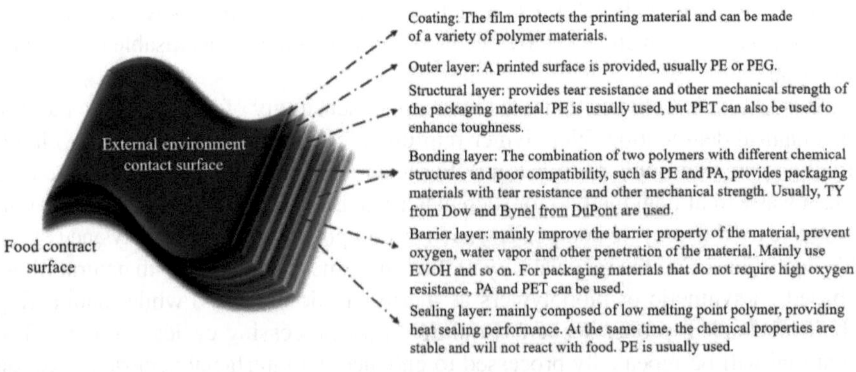

Fig. 10.16 Multilayer system structure of food packaging materials (Tullo 2016)

Table 10.4 Operation status of demonstration plants

Process	Location	Capacity	Status
Mao Shangji Zi	Japan	3 t/day	Operating
Beijing Royce Hotel	China	6 sections/day	Unknown status
Sapporo/Toshiba	Japan	14.8 carats/year	Operation
Altis	Japan	Unknown	Commercial application
Gaussler Evitz	Germany	1.85 km/year	Unknown status
Ever-changing World Technology	United States	10 milligauss/year	Demo version

macrostructure of the polymer and produce small molecule products (Angyal et al. 2007). Pyrolysis products can be divided into three parts: gas, liquid, and solid residues (Al-Salem et al. 2009). The process of the new type of vortex reactor technology is widely used, which can also be used in the design of traditional plastic waste pyrolysis plants with traditional sorting parts. Several representative plastic waste pyrolysis demonstration plants have been built and put into operation, as shown in Table 10.4 (Butler et al. 2011).

The difficulty of thermal decomposition lies in the complexity of the reaction. According to its main decomposition pathways, different polymers will produce completely different products. The presence of certain impurities will also affect the yield ratio of the products, e.g., the presence of certain oxygen-containing compounds will lead to the production of methanol or formaldehyde. PE, PP, PS, and PVC account for about 80% of polymer production (Fig. 10.17) (Brems et al. 2012). Although the cracking process is simple, it is only economically feasible for large-scale cracking processes.

Another major problem in the thermal decomposition process of polyvinyl chloride is the presence of PVC in the raw materials. Due to the high corrosion resistance requirements of acidic substances for equipment materials, even if the halogen content in the raw materials is very small, it cannot be used as fuel or petrochemical raw materials. The general specification requires that the chlorine content should not exceed 10 ppm (Bhaskar et al. 2003). In view of the characteristics of PVC pollutants in the SPW process, a method for cracking PVC pollutants at low temperature (300 °C) has been proposed (Sadat-Shojai and Bakhshandeh 2011). Under the conditions of pyrolysis reaction, plastic melting degradation, PVC degradation, and other types of plastics are almost unaffected, and the dechlorination effect can reach 98%. By adding $CaCO_3$, CaO, $NaHCO_3$, $NaCO_3$, $NaCO_3$ or NH_3, the residual chlorides in the wastewater can be neutralized.

As shown in Fig. 10.18, the primary pyrolysis reactor can be of various design types (Brems et al. 2013). Its main designs are bubble column fluidized bed, stirred tank reactor, and screw/ogden reactor, which have been widely verified. The fluidized bed reactor has become the best choice for plastic cracking due to its many advantages, such as uniform product quality and high conversion rate.

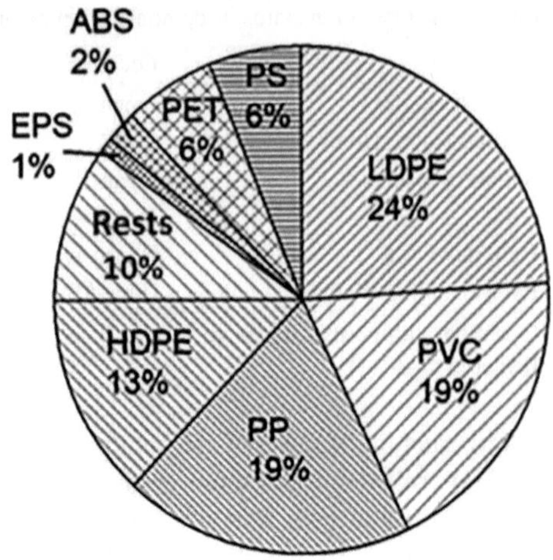

Fig. 10.17 European polymer classification average plastic consumption (Brems et al. 2012)

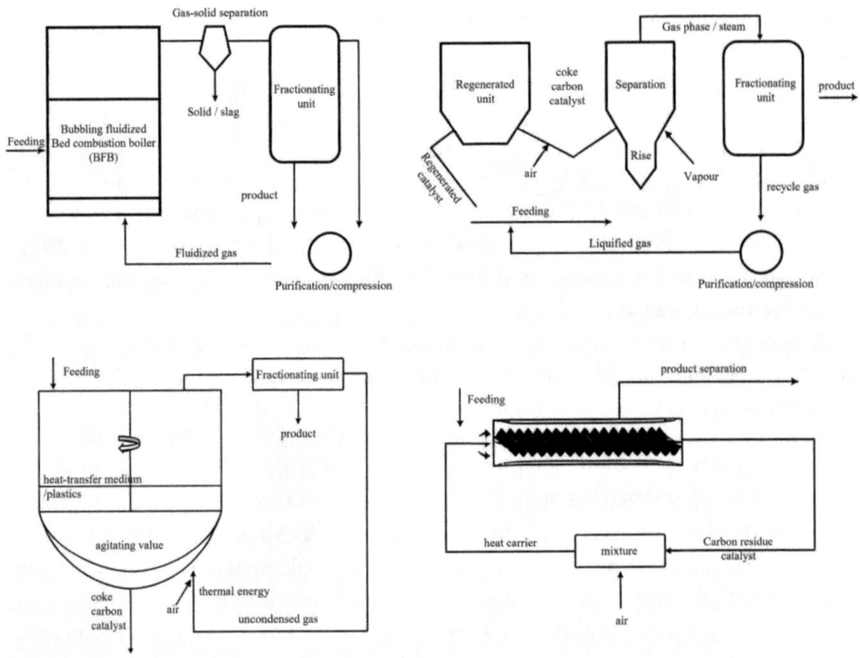

Fig. 10.18 Different designs of plastic waste pyrolysis (Butler et al. 2011). (Reprinted with permission from Waste Biomass Valori, 2011, 2(3): 227–255. © 2011, Springer)

The reaction products are mainly divided into three types: gas, liquid, and coke. The temperature of the reaction kettle greatly affects the mass distribution between the fractions. In addition to temperature, there are many factors affecting the composition of the products during the pyrolysis of plastics, including the macrostructure of the polymer, the degree of micro-mixing, the residence time of the gas, temperature, and fluidized gas, etc. To enhance production efficiency and maximize profits, detailed modeling of the production process is essential for achieving exceptional results. Although the degradation of polymers, the mechanism of free radical degradation, and kinetics are very complex, the importance of recycling waste plastics by pyrolysis is unquestionable. Compared with mechanical methods and chemical monomer recycling methods, pyrolysis is cheaper, and the process is simpler, so it is more practical to apply pyrolysis to industrial production.

10.4.2 Gasification and Hydrocracking

Gasification is a type of pyrolysis technology that can produce fuel or combustible gas from waste. During the gasification process, almost all raw materials composed of organic matter will be partially oxidized into a gaseous mixture. This process requires oxidants, which are usually a mixture of steam and pure oxygen or just air, but air can usually be used as a gasifying agent. During the gasification process, the raw materials undergo multiple reactions, which can both release and absorb heat, but the overall process is endothermic. The higher thermal efficiency produced can be used for power generation.

So far, TEXACO is the most common and famous gasification technology. It gently cracks (depolymerizes) plastic waste into synthetic heavy oil in the liquefaction step and some condensable and non-condensable gas fractions. The produced oil and condensate gas are injected into the entrained gasifier. Gasification is carried out under the action of O_2 and steam at 1200–1500 °C. Through various cleaning processes (including the removal of HCl and HF), clean and dry synthetic gas can be produced, mainly CO and H_2, as well as small amounts of CH_4, CO_2, H_2O, and some inert gases. Figure 10.19 lists the TEXACO gasification process (Brems et al. 2013).

Methanol synthesis gas is an important application direction of gasification recycling. As shown in Fig. 10.20, methanol is one of the largest chemical substances in the world, used to produce a variety of industrial chemicals such as formaldehyde, acetic acid, and methylamine. Methanol can be produced through various routes, including the CH_4 oxidation route and the synthetic gas catalytic route; the latter is most widely used in industry (Brems et al. 2013).

The conversion of synthesis gas to methanol needs to be carried out under high temperature and high pressure; the reaction itself is exothermic and is subject to equilibrium constraints. This is a key factor for reactor design, and care should be taken to prevent the conversion direction from getting out of control and the catalyst from severely deactivating. The resulting methanol can be further used in the production of olefins or petroleum-like products through methanol-to-olefins (MTO) and methanol-to-gasoline (MTG), respectively.

The first catalyst system used for methanol production was ZnO/CrO, developed by BASF in the mid-1920s. The reactor operates at high temperatures (350 °C) and pressures (250–300 atm). With the development of synthesis gas purification

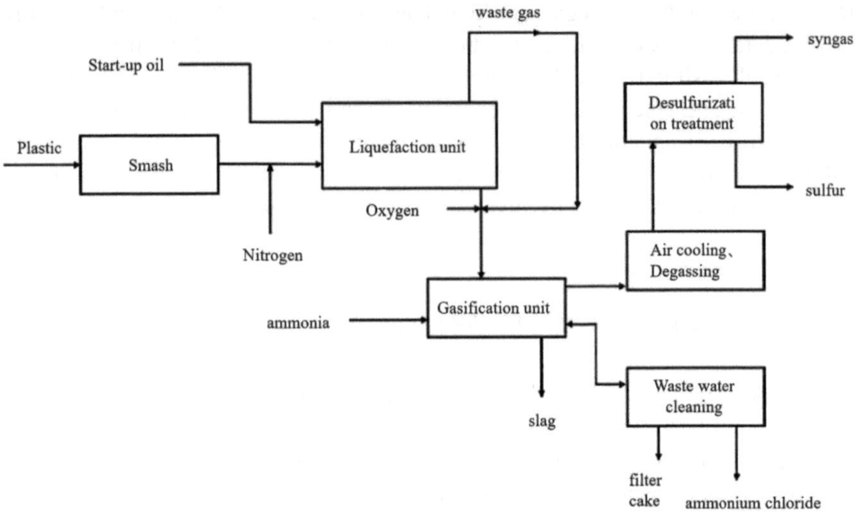

Fig. 10.19 TEXACO gasification process flow diagram (Brems et al. 2013)

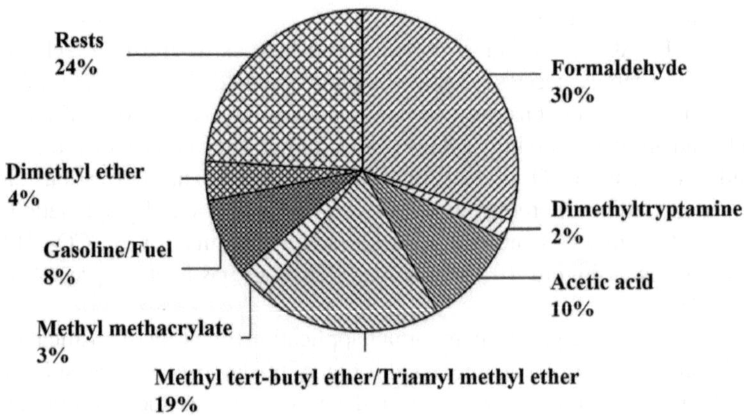

Fig. 10.20 Uses of methanol (Brems et al. 2013)

technology, the easily poisoned copper catalyst was redeveloped. In 1966, ICI (Imperial Chemical Industries) introduced a new catalyst with higher activity, $Cu/ZnO/Al_2O_3$. Thus, the reactor can operate under lower temperatures (220–275 °C) and pressures (50–100 atm) conditions. The general procedure of this process is shown in Fig. 10.21 (Abrol and Hilton 2012). The catalytic reaction is sensitive to sulfur. The preferred purity of synthesis gas should ensure that the sulfur content is below 0.1 ppm to maintain the activity of Cu and keep the activity of Cu for more than 5 years (Spath and Dayton 2003).

Although there have been reports on the industrial-scale application of methanol, further experimental work is needed for future breakthroughs in this method to

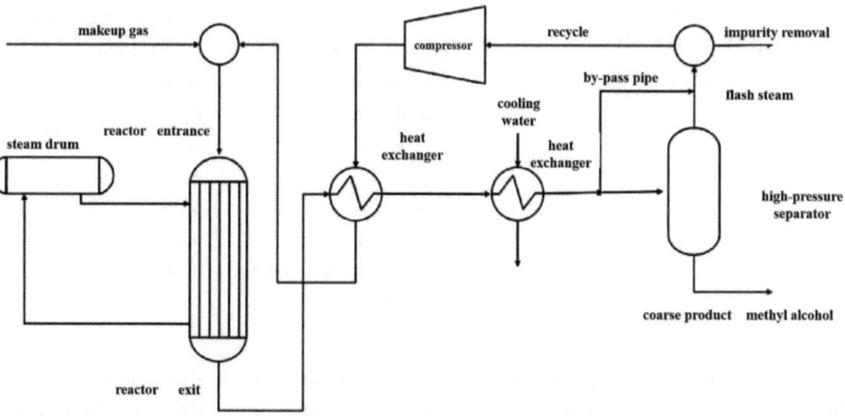

Fig. 10.21 General procedure for methanol production. (Reprinted with permission from Comput Chem Eng, 2012, 40(5): 117–131. © 2012, Elsevier (Abrol and Hilton 2012))

improve equipment design and optimize products. Progress in this field will help improve gasification reactors and expand their application range.

The biggest difference in the process of catalytic cracking of plastics is the addition of hydrogen. This process is carried out in a rising hydrogen pressure (about 70 atmospheres) and an environment of 375–400 °C, using Ni/S catalyst or NiMo/S catalyst. Because inorganic substances exist in the original SPW, plastics are first liquefied, and substances that cannot be distilled are filtered out. It is usually carried out by low-temperature pyrolysis, and then the liquid is transported to a fluidized bed with a catalyst. The presence of hydrogen can significantly improve product quality, increase the H/C ratio, and reduce the aromatic content. Through experiments with various types of catalysts, it has been proven that paraffin has a higher yield and can increase the yield of plastic pyrolysis liquid. Other major advantages are that it handles heteroatoms very well and does not produce toxic products, such as dioxins.

Although there are many reports on the hydrogenation cracking industry at the current stage, these technologies have either been discontinued (industrial scale has been discontinued) or are still in the research stage due to many economic and operational issues. At this stage, how to minimize costs and improve product quality is still in urgent need of in-depth research.

10.4.3 Toxicity of Pyrolysis Gasification Products

The definition of energy recovery is to generate heat, steam, and electricity by burning waste. In the raw material recovery process, due to the limitation of production costs, its utilization value is not high; therefore, it is more ideal to treat waste by incineration.

Table 10.5 Calorific values of single-polymer plastics

Item	Calorific value (MJ/kg)
Polyethylene	43.30–46.50
Polypropylene	46.50
Polystyrene	41.90
Kerosene	46.50
Gas oil	45.20
Heavy oil	42.50
Petroleum	42.30
Household PSW mixture	31.80

Since plastics come from crude oil, they generate a high amount of heat when burned, compared to petroleum and urban solid waste. Table 10.5 lists the calorific values of many single-polymer plastics. Because plastics have a high calorific value, they are a convenient source of energy. Energy recovery technology has received widespread attention and importance, but the key issue in practical application is the toxic and harmful substances produced by incineration, and the pollution these substances cause to the environment.

The toxicity of gaseous products and their production under various pyrolysis conditions is a major issue that needs to be addressed. Here, seven common plastics are listed, namely acrylonitrile–butadiene–styrene, nylon, polyester, polyethylene, polystyrene, PVC, and rigid polyurethane foam. For polyurethane foam, CO (CO) and hydrogen cyanide (HCN) are the main toxic substances found among over 100 gaseous products. For PVC, the main products of thermal decomposition are hydrogen chloride, benzene, and unsaturated hydrocarbons. If O_2 is present, common combustion products include CO, CO_2, and H_2.

The main toxic products in a PVC fire are hydrogen chloride and carbon monoxide; the former can cause strong irritation to human senses and lungs, and the latter can cause suffocation. If the combustion toxicity of PVC materials (based on their HCl content) is compared with pure HCl tests, it is found that the toxicity of many products after exposure seems to be explained by the HCl produced, and products containing chlorinated oil can cause serious component corrosion and environmental toxin release. Catalytic cracking can inhibit the formation of chlorinated hydrocarbons, reduce the reaction temperature, shorten the reaction residence time, and improve the selectivity of the product, so it is superior to traditional cracking processes.

10.5 Limitations, Challenges, and Opportunities

As mentioned earlier, plastic recycling is a complex and important issue. This problem is not only about the update of scientific technology and methods but also involves public environmental awareness, the construction of urban waste treatment facilities, and other social governance issues. Among them, "ecological design" of products and "cradle to grave" life cycle analysis are two powerful means to optimize plastic recycling.

In recent years, with the introduction of relevant national policies and regulations, consumers' understanding of climate change, environmental protection, and other issues has deepened, and the development of biobased materials will usher in new opportunities. Biobased does not mean biodegradable, and vice versa. Therefore, biobased polymers should not be the complete solution to plastic waste—they still need to establish a comprehensive recycling system. Because plants absorb CO_2 during photosynthesis, the carbon footprint of biobased polymers is lower than that of petrochemical polymers or alternatives. It is worth noting that recycling methods are not neutral or energy-free, as polymerization, processing, distribution, and recycling all require energy, and the current energy structure is mainly based on fossil fuels. The cycle of biodegradable plastics is another factor limiting their large-scale application. Even biobased polymers will not degrade or compost overnight; some biobased polymers also need several years to degrade, and many composting facilities will not use these materials because they leave some plastic residues in the compost.

Some people think that biobased materials are not recyclable and can only be biodegraded, which is obviously wrong. Biobased polymers are not necessarily biodegradable or compostable, such as PLA or PHA are only biodegradable or compostable under appropriate conditions, while other products, such as biobased PE, are not biodegradable or compostable, so they need to be recycled to extend their life cycle. Conversely, some biodegradable or compostable polymers, such as PCL, are based on petroleum production. Therefore, reasonable recycling methods are not necessarily related to the nature of the raw materials.

For biobased polymers that are not biodegradable or compostable, if they are chemically identical to petrochemical-based polymers at the molecular level, they can simply be included in the recycling methods for such petrochemical-based compounds. If these biobased polymers are discarded directly into the environment without sufficient waste treatment, they will have the same lifespan as the same petrochemical polymers because their chemical molecules are the same.

For recycling to become economically viable, polymers need to achieve a significant level of usage. However, currently, the application and promotion of biobased polymers remain limited, and achieving a substantial volume of recyclable plastic usage is still a distant goal. The recycling of mixed plastics containing biobased polymer components also presents numerous challenges.

Mechanical recycling provides an effective and simple method for the reuse of polymer materials, but the recycled polymers often suffer from contamination from various sources, resulting in poor mechanical properties of the produced raw materials. High water absorption rate and sensitivity to hydrothermal degradation are also prominent disadvantages of many bioplastics. Water absorption can reduce the interfacial strength between the reinforcing material and the resin, thereby affecting the production of biocomposites. At present, we still do not fully understand the interactions between various components in biocomposites and how they affect thermal stability and the formation of degradation products during processing.

When the limitations of mechanical recycling reduce economic value, other methods such as chemical recycling can be adopted, but chemical recycling usually requires high energy consumption, making the recycling cost very expensive. At the same time, chemical recycling often involves complex processes and is only suitable for certain specific biopolymers.

Mechanical, chemical, composting, and incineration recycling, these four different recycling methods each have their own characteristics, with both advantages and disadvantages. In some cases, both mechanical recycling and chemical recycling are not easy to achieve. For example, in biobased composites, due to the sensitivity of the biobased matrix and/or reinforcing fibers to heat treatment, if the matrix is biodegradable, composting is the best method. However, the composting process must be carried out at an industrial level. Although the quality of recycled materials and processing costs are key issues in the recycling process, reality shows that the process of recycling natural fiber-reinforced thermoplastics is very economical, but lacks repeatability. If no better method can be found, pyrolysis, combustion, and other value-added technologies can also be used as feasible solutions for managing biobased regenerated plastic waste. Incineration can partially recover the potential thermal energy of bioplastics and biocomposites, so it can be used as a final choice.

To maximize the effective use of "green" plastics, not only do scientists and engineering technicians need to solve various technical problems in the process, but more importantly, through various science popularization education, appropriate classification of plastic products, and system management, waste can be recycled, thereby improving the public's understanding of various plastics.

References

Abrol S, Hilton CM (2012) Modeling, simulation and advanced control of methanol production from variable synthesis gas feed. Comput Chem Eng 40(5):117–131

Al-Salem SM, Lettieri P, Baeyens J (2009) Recycling and recovery routes of plastic solid waste (PSW): a review. Waste Manag 29(10):2625–2643

Angyal A, Miskolczi N, Bartha L (2007) Petrochemical feedstock by thermal cracking of plastic waste. J Anal Appl Pyrolysis 79(1–2):409–414

Antoine LD, Isabelle P, Alain B et al (2008) Effect of recycling on mechanical behaviour of biocompostable flax/poly(l-lactide) composites. Compos Part A: Appl Sci Manuf 39(9):1471–1478

Arvanitoyannis IS, Bosnea LA (2001) Recycling of polymeric materials used for food packaging: current status and perspectives. Food Rev Int 17(3):291–346

Aznar MP, Ca Ballero MA, Sancho JA et al (2006) Plastic waste elimination by co-gasification with coal and biomass in fluidized bed with air in pilot plant. Fuel Process Technol 87(5):409–420

Badia JD, Santonja-Blasco L (2012) Reprocessed polylactide: studies of thermo-oxidative decomposition. Bioresour Technol 114:622–628

Badia JD, Stroemberg E, Karlsson S et al (2012a) Material valorisation of amorphous polylactide. Influence of thermo-mechanical degradation on the morphology, segmental dynamics, thermal and mechanical performance. Polym Degrad Stab 97(4):670–678

Badia JD, Santonja-Blasco L, Martinez-Felipe A et al (2012b) Hygrothermal aging of reprocessed polylactide. Polym Degrad Stab 97(10):1881–1890
Beyler CL, Hirschler MM (2002) Thermal decomposition of polymers. SFPE Handb Fire Protect Eng 2:110–131
Bhaskar T, Uddin MA, Kaneko J et al (2003) Liquefaction of mixed plastics containing PVC and dechlorination by calcium-based sorbent. Energy Fuel 17(1):75–80
Brems A, Baeyens J, Dewil R (2012) Recycling and recovery of post-consumer plastic solid waste in a European context. Therm Sci 16(3):669–685
Brems A, Dewil R, Baeyens J et al (2013) Gasification of plastic waste as waste-to-energy or waste-to-syngas recovery route. Nat Sci 5(6):695–704
Butler E, De Vlin G, Mcdonnell K (2011) Waste polyolefins to liquid fuels via pyrolysis: review of commercial state-of-the-art and recent laboratory research. Waste Biomass Valori 2(3):227–255
Callister W, Rethwisch D (2010) Materials science & engineering. Wiley
Censori M, La M, Carvalho F et al (2016) Separation of plastics: the importance of kinetics knowledge in the evaluation of froth flotation. Waste Manag 54:39–43
Delva L, Ragaert K, Degrieck J et al (2014) The effect of multiple extrusions on the properties of montmorillonite filled polypropylene. Polymers 6(12):2912–2927
Enkiewicz M, Richert J, Rytlewski P et al (2009) Characterisation of multi-extruded poly(lactic acid). Polym Test 28(4):412–418
Faisal M, Saeki T, Tsuji H et al (2006) Recycling of poly lactic acid into lactic acid with high temperature and high pressure water. Trans Ecol Environ 92:225–233
Fraunholcz N (2004) Separation of waste plastics by froth flotation—a review, part I. Miner Eng 17(2):261–268
George N, Kurian T (2014) Recent developments in the chemical recycling of postconsumer poly(ethylene terephthalate) waste. Ind Eng Chem Res 53(37):14185–14198
Hamad K, Kaseem M, Deri F (2011) Effect of recycling on rheological and mechanical properties of poly(lactic acid)/polystyrene polymer blend. J Mater Sci 46(9):3013–3019
Ignatyev IA, Thielemans W, Vanderbeke B (2014) Recycling of polymers: a review. Chem Sustain Chem 7(6):1579–1593
Kirkby N, Azapagic A (2005) Municipal solid waste management: can thermodynamics influence people's opinions about incineration. Wiley
La Mantia FP (1996) Recycling of PVC and mixed plastic waste. Chem Tec Publishing
Luciani V, Bonifazi G, Rem P et al (2015) Upgrading of PVC rich wastes by magnetic density separation and hyperspectral imaging quality control. Waste Manag 45(11):118–125
Pillin I, Montrelay N, Bourmaud A et al (2008) Effect of thermo-mechanical cycles on the physico-chemical properties of poly(lactic acid). Polym Degrad Stab 93(2):321–328
Qian S, Igarashi T, Nitta KH (2011) Thermal degradation behavior of polypropylene in the melt state: molecular weight distribution changes and chain scission mechanism. Polym Bull 67(8):1661–1670
Rem P, Di Maio F, Hu B et al (2013) Magnetic fluid equipment for sorting secondary polyolefins from waste. Environ Eng Manag J 12:951–958
Roberto, Scaffaro, Marco et al (2010) Preparation and recycling of plasticized PLA. Macromol Mater Eng 296(2):141–150
Sadat-Shojai M, Bakhshandeh GR (2011) Recycling of PVC wastes. Polym Degrad Stab 96(4):404–415
Sánchez AC, Collinson SR (2011) The selective recycling of mixed plastic waste of polylactic acid and polyethylene terephthalate by control of process conditions. Eur Polym J 47(10):1970–1976
Scheirs J (1998) Polymer recycling: science, technology, and applications. Focus Catal 2006(9):8
Serranti S, Luciani V, Bonifazi G et al (2015) An innovative recycling process to obtain pure polyethylene and polypropylene from household waste. Waste Manag 35(1):12–20
Spath PL, Dayton DC (2003) Preliminary screening-technical and economic assessment of synthesis gas to fuels and chemicals with emphasis on the potential for biomass-derived syngas

Su KH, Lin JH, Lin CC (2007) Influence of reprocessing on the mechanical properties and structure of polyamide 6. J Mater Process Technol 192:532–538
Tullo AH (2016) The cost of plastic packaging. Chemical & Engineering News Edition of the American Chemical Society
Wang CQ, Wang H, Fu JG et al (2015) Flotation separation of waste plastics for recycling—a review. Waste Manag 41(7):28–38
Xiang Q, Xanthos M, Mitra S et al (2002) Effects of melt reprocessing on volatile emissions and structural/rheological changes of unstabilized polypropylene. Polym Degrad Stab 77(1):93–102
Zia KM, Bhatti HN, Bhatti IA (2007) Methods for polyurethane and polyurethane composites, recycling and recovery: a review. React Funct Polym 67(8):675–692

Correction to: Research and Applications of Bio-based Degradable Materials

Liming Zhao

Correction to:
L. Zhao (ed.), *Research and Applications of Bio-based Degradable Materials*, https://doi.org/10.1007/978-981-95-1188-4

The original version of the book was inadvertently published with incorrect copyright holder for the chapters as "© The Author(s), under exclusive license to Springer Nature Singapore Pte Ltd. 2025" whereas it should be "© Shanghai Scientific & Technical Publishers 2025". The copyright holder for the chapters has now been updated in the book.

The updated version of this book can be found at
https://doi.org/10.1007/978-981-95-1188-4

L. Zhao (ed.), *Research and Applications of Bio-based Degradable Materials*,
https://doi.org/10.1007/978-981-95-1188-4_11

MIX
Papier aus verantwortungsvollen Quellen
Paper from responsible sources
FSC® C105338

If you have any concerns about our products,
you can contact us on
ProductSafety@springernature.com

In case Publisher is established outside the EU,
the EU authorized representative is:
Springer Nature Customer Service Center GmbH
Europaplatz 3, 69115 Heidelberg, Germany

Printed by Libri Plureos GmbH
in Hamburg, Germany